DYNAMIC WELL TESTING IN PETROLEUM EXPLORATION AND DEVELOPMENT

DYNAMIC WELL TESTING IN PETROLEUM EXPLORATION AND DEVELOPMENT

First Edition

HUINONG ZHUANG
Author and Editor

NENGQIANG LIU
Examiner and English Translator

ELSEVIER

Amsterdam • Boston • Heidelberg • London • New York • Oxford
Paris • San Diego • San Francisco • Singapore • Sydney • Tokyo

Elsevier

225, Wyman Street, Waltham, MA 02451, USA

The Boulevard, Langford Lane, Kidlington, Oxford OX5 1GB, UK

Radarweg 29, PO Box 211, 1000 AE Amsterdam, The Netherlands

Notice

No responsibility is assumed by the publisher for any injury and/or damage to persons or property as a matter of products liability, negligence or otherwise, or from any use or operation of any methods, products, instructions or ideas contained in the material herein. Because of rapid advances in the medical sciences, in particular, independent verification of diagnoses and drug dosages should be made

Library of Congress Cataloging-in-Publication Data

A catalog record for this book is available from the Library of Congress

British Library Cataloguing in Publication Data

A catalogue record for this book is available from the British Library

ISBN: 978-0-123-97161-6

For information on all Elsevier publications
visit our web site at store.elsevier.com

Working together to grow
libraries in developing countries

www.elsevier.com | www.bookaid.org | www.sabre.org

ELSEVIER   BOOK AID International   Sabre Foundation

CONTENTS

FOREWORD

The author of this book, Zhuang HuiNong, graduated from Peking University in 1962 and majored in aeronautical aerodynamics. It was just at the time of the discovery of the Daqing oil field and flourishing development of the petroleum industry in China. After graduation he was assigned to the Daqing oil field, where he immediately began to devote himself to the exploration and development of oil and gas fields. In response to the call of Academician Tong Xianzhang and other experts of older generations, he embarked on his research on the well test. With so many years of unremitting effort and perseverance, he has made considerable achievements. The difference between his major of aeronautical aerodynamics and his research object of well test and subsurface percolation mechanics is really earthshaking. Nevertheless, his solid academic knowledge and practical experience have finally resulted in rich fruits.

This book provides a new perspective for well test research: dynamic reservoir description—bringing dynamic performance analysis to a new and higher level. The author has long been working in oil fields and insisting combination of theory with practice. Soon after joining the Research Institute of Petroleum Exploration and Development, he devoted himself to the development of large and medium gas fields in China; during the recent decade, based on his analysis and research on well test data of various gas fields such as the JB, KL-2, QMQ, and SLG gas fields, he has identified many important gas field features not available from static geological data and has made great contributions to the appropriate development of those gas fields. These valuable information and experience have been merged into this book, which will serve as a good reference for the development of oil and gas fields in the future.

This book deals with all aspects of dynamic reservoir research and modern well test analysis and provides very good guidance and criteria of research on reservoir dynamic performance, especially on gas reservoir dynamic performance. Therefore, it is really a good book with great practicable value.

Qin Zhongjian
Engineering Academician of China,
Former Vice President of CNPC

Published in 2004, *Gas Reservoir Dynamic Description and Well* Test has been sold out. Many suggestions for improvement have been offered by experts and enthusiastic readers with regard to the content and misprints of the first edition. In addition, during the last four years after the first edition was completed, I have been lucky enough to be further involved in dynamic research on some large and medium gas fields in China, and therefore deepened understanding of the gas reservoir dynamic description. During this period I developed a new concept of "dynamic deliverability of gas wells" and deduced the "stable point LIT equation" with which to establish the initial deliverability and dynamic deliverability of the gas well; this method has been widely used in many gas fields. Therefore, it is necessary to make some supplements and amendments in some important areas of the first edition based on these new achievements.

In Chapter 1, "Introduction," an overview of gas reservoir dynamic description methods is added, and a "new concept of gas reservoir dynamic description" focuses on the establishment of gas well deliverability. Thus, the analysis of initial deliverability and dynamic deliverability, the establishment and tracing analysis of gas well dynamic models, the initial static pressure gradient analysis and the tracing analysis of subsequent dynamic formation pressure, and the interpretation of the influence of basic geological conditions on subsurface percolation are all integrated to form a complete system for gas reservoir dynamic descriptions applied for gas reservoir research in the process of gas field development.

In Chapter 3, Section 3.6 of "LIT deliverability equation from only one stable test data point" is added. In this section, a simple but practical method of the initial LIT deliverability equation from only one early stable test data point is introduced for the gas wells to resolve a number of puzzles met in the process of gas well research, including deliverability test data acquisition, deliverability analysis, and the application of deliverability indexes. Based on this initial LIT deliverability equation from only one stable test data point in the early stage of production, subsequent dynamic deliverability equations, dynamic IPR curves, dynamic absolute open flow potential, and dynamic reservoir pressure at the gas drainage boundary of each gas well are deduced and established. This approach has been used in vertical and horizontal wells in many gas fields.

Chapter 6 is partly adjusted. The latest achievements in the interference test between wells in the Su-6 infill well block of the SLG gas field are introduced. It is the first time that interwell pressure interference had been observed among the thousands of gas wells that have been put into production in this block. Based on this result, the critical connectivity spacing of this gas province with reserves of trillions of cubic meters was confirmed. Meanwhile, some previous examples of oil reservoirs are deleted.

Chapter 8 is comprehensively revised. Some examples of gas reservoir research by means of dynamic descriptions in recent years are highlighted, and these examples fully represent the new concept of gas reservoir dynamic descriptions. In this chapter, in addition to the JB gas field and KL-2 gas field described in the first edition, a gas reservoir dynamic description and follow-up research results in the SLG, YL, and DF gas field are also introduced. These contents also record the joint efforts of the leaders and experts of these gas fields, the Lanfang Branch of the Scientific Research Institute of Petroleum Exploration & Development, CNPC where I work and myself in resolving difficulties met in the process of gas field development, and embody our friendship and the fruits of our painstaking efforts and experiences, which I will never forget.

In addition to the supplements listed earlier, some improper equations and figures in the first edition have also been corrected. Once again, I express my heartfelt gratitude to those experts and readers who cherish and love this book for their encouragement and help.

HuiNong Zhuang
August 2008

Reservoir evaluation by means of well test or dynamic analysis was applied in foreign countries as early as 1940s. In China, Academician Tong Xianzhang, a well test expert of older generation in China, led a work team to Daqing Oilfield in 1960, and analyzed measured pressure data from early exploration wells by "Horner analysis method" which was just created not long before. As a result, he and his colleagues accurately calculated formation parameters such as initial pressure, permeability and skin factor etc., pioneered the research on well test in oil and gas fields in China.

After graduation in 1962 I went to Daqing Oilfield. Being deeply inspired by the older generation of experts and my cherished dream spouted in my childhood of dedicating myself to aviation, I threw myself into subsurface percolation mechanics research and then engaged in well test, the most practical area in subsurface percolation mechanics research. And unexpectedly I have been working in this field for more than four decades since then. I really can not remember, since I engaged in well test, how many wells I have tested, how many times I have climbed up to the top of the gin pole, how many pressure charts I have scanned, how many well test analyses I have done, and how many miles I have walked in various oil and gas fields in China. One day, eventually, I realized that what I had done was to provide a different description of hydrocarbon reservoirs other than geological research, so as to find something invisible to geologists merely through static approaches. This is why the title of this book is called "Dynamic Well Testing in Petroleum Exploration and Development".

For more than 40 years, dramatic changes have taken place in well test, from the simple pressure build-up curve analysis to the current "modern well test." Along with the manufacturing of high precision electronic pressure gauges, publication of a large number of theoretical research results and the development of perfect well test inter-pretation software, new concepts of dynamic reservoir descriptions have been developed and successfully used for reservoir interpretation.

Unlike oil reservoir, gas reservoir exploration and development focuses on integra-tion of upstream and downstream, especially early evaluation and research and dynamic descriptions. I am fully aware of this, being involved in the research during the last decade. Through dynamic descriptions, we can become more familiar with internal geological features of gas reservoirs and improve their development benefits. Otherwise, potential risks will be imposed on the subsequent development

This book is largely based on my understanding of well test methods and my practical experiences over the past 40 years. Those data and field examples used to verify theo-retical models, in particular, were acquired or analyzed by myself. In fact it is these practices such as the interference testing and pulse testing conducted in Shengli Oilfield many years ago, the dynamic data analysis and research for production testing wells in JB gas-field in the 1990s, the dynamic description and research in exploration and development preparation of KL-2 gas-field in recent years and the evaluation and research on dynamic models for production testing wells in the reservoir of Upper Paleozoic of the Ordos Basin etc. that have provided me with the best opportunity to study and understand well test.

In the preparation of this book, I have received enthusiastic support and help from all walks of life. Many experts of older generation in oil and gas exploration and development, including Wang Naiju, Shen Pingping, Meng Muyao, Pan Xingguo, Zhu Yadong, Liu Nengqiang, Chen Yuanqian, Yuan Qingfeng and some others have offered their valuable instructions and advice with great patience and helped me correct many mistakes in the first draft. I am grateful to Li Haiping, Assistant Chief Engineer of Exploration & Production Company of PetroChina, Jin Zhongchen, former Deputy General Manager and Chief Geologist Minqi of Changqing Oil Field Company of PetroChina, Zhang Fuxiang from Tarim Oilfield, Tan Zhongguo from Changqing Oilfield Company, and many other people for their assistance in improving relevant examples. My thanks also go to Langfang Branch Company of Scientific Research Institute of Petroleum Exploration & Development, CNPC, especially to Director Li Wenyang, whose painstaking efforts in creating a great scientific research atmosphere have inspired me to write this book while I contributed to the development of the industry as well as to the research on oil and gas field development for the rest of my life. In particular, I should owe my research on gas reservoir and well test to Dr Han Yongxin, who has offered great support for most of my research projects during the past decade, including those in JB gas-field, KL-2 gas-field, QMQ gas-field, SLG gas-field, HTH gas-field, SB gas-field and some other large and medium gas fields; it is in our jointed research works on interpretation of gas reservoir characteristics that we developed some new ideas and concepts.

It should be also noted that this book is simply the least I can do over the past 40 years. I just summarize and integrate my personal experience in the book. In retrospect I always recall my teachers. It was my teachers who taught me how to pursue my research in a down-to-earth and prudent manner, and how to behave with integrity. Although most of them have passed away, I still would like to dedicate my humble achievements to them.

Finally, I would like to thank my family, especially my wife Shi Caiyun, who reviewed the entire content of the first draft of this book and converted it into the electronic version. I am grateful to her for helping me fulfill this project in my old age.

HuiNong Zhuang
August 2003

Dynamic Well Testing in Petroleum Exploration and Development (English version) has been published by Elsevier. As the author, I am very glad to have such an opportunity to share my experience in well test research with overseas experts and readers.

The first edition (Chinese version) of this book was published in China in 2004, the title of which was *Gas Reservoir Dynamic Description and Well Test*. Readers can realize my thinking and explanation of the title from some chapters in this book. The gas reservoir dynamic description means that well test research not only interprets the reservoir parameters near the wellbore of an individual oil or gas wells, but also finds out more information about the reservoir feature and regularity of the deliverability decline by tracing analysis of the dynamic parameter variation in several wells of the reservoir at different times, making sufficient preparation of the development of the gas field. Gas reservoir dynamic description methods are introduced in Chapter 1. The concept of dynamic deliverability of gas wells and research methods of the regularity of deliverability decline are introduced in Chapter 3. Field examples of "dynamic description" done during preparation of the development of several gas fields are given in Chapter 8. This book (Chinese version) has been paid great attention by Chinese engineers engaged in gas reservoir dynamic analysis and/or well test research after its publication; during this period the natural gas industry of China was developing very quickly, and I myself was taking part in this great project. It was on the basis of summarization of dynamic performance analysis and well test research for newly discovered and newly developed gas fields that the book was further enriched and published as its second edition in December 2009.

Elsevier Press Company realized the value of this book and settled on its English version to be published after conferring with the author. I take this opportunity to express my deep and heartfelt thanks to Mr. John Fedor, Senior Acquisitions Editor with Elsevier, and Ms. Kathryn Morrissey, Senior Editorial Project Manager with Elsevier, for their helpful assistance in the whole publishing process.

I also thank my colleague and friend, Professor Liu Nengqiang, a famous well test expert in China. In order to make the quality of the monograph *Dynamic Well Testing in Petroleum Exploration and Development* (English version) to be good enough for future readers, the author invited him to be in charge of the examination, approval, and English translation of the monograph. His effort lays a solid foundation for the publishing of this monograph.

Finally, the author thanks SPT Energy Group Inc. for its assistance in translation of this book from Chinese into English.

HuiNong Zhuang
June 2012

HuiNong Zhuang, a professor and senior engineer, graduated from Peking University in 1962. After graduation he took part in research of the development program of the Daqing oil field in its early stage and then since 1965 served in the Shengli oil field where his interest was in the oil/gas well test. In the 1980s he took charge and operated interference tests and pulse tests in an carbonate reservoirs successfully; during this period he invented the interpretation type curves for the interference well test in dual porosity reservoirs and applied these type curves in field practice; took charge of the research of downhole differential pressure gauges and applied these gauges in data acquisition in fields; and consequently won the invention award from the China Nation Science and Technology Committee and was present at the First International Meeting on Petroleum Engineering in Beijing in 1982; his paper was published in the JPT. Since 1990 he served on the Research Institute of Petroleum Exploration and Development of CNPC, was concerned with the exploration and development of several large- or medium-scale gas fields in China, and has been doing dynamic performance research for the past 20 years. At present he is working at SPT Energy Group Inc. as its chief geologist. He has been devoting himself to dynamic performance analysis and well testing for more than 40 years.

Introduction

Contents

Dynamic Well Testing in Petroleum Exploration and Development   © 2013 Petroleum Industry Press. Published by Elsevier Inc.
ISBN 978-0-12-397161-6, http://dx.doi.org/10.1016/B978-0-12-397161-6.00001-2

1 THE PURPOSE OF THIS BOOK

The modern well test has been around since the beginning of the 1980s. In China, as implementation of the reform and opening-up policies, advanced well test methods, interpretation software, and advanced test instruments, tools, and equipment have been introduced almost simultaneously. Looking back at the advances made since the early 1980s, it is very exciting to see that the new developed knowledge and techniques have been applied successfully in the discovery, development preparation, and development operation of many major gas fields in China. However, it should also be noted that application of the modern well test sometimes in some places is still not good enough and still needs to be improved further.

The well test today is very different than it was two or three decades ago. Just as in all other fields, due to the application of computer and advance in science and technology, workers today seldom make calculations manually; well test analysts and reservoir engineers no longer frequently look up complicated formulae in well test books and perform tedious computations with calculators—the results can be obtained easily by simply tipping some menu items.

Because of this, does it mean that well test work has become much easier? The answer is "no"; on the contrary, as research activities go further and more in-depth, the well test does not become easier, but faces greater challenges.

First of all, well test analysis is required to provide not only simple parameters such as reservoir permeability, but also more detailed information about reservoirs, such as their types and boundary conditions, and ultimately to deliver a "dynamic model" of gas wells and gas reservoirs, that is, the dynamic model reflecting the conditions of the gas well and the gas reservoir truly and correctly, which can be used in gas field evaluation and performance prediction.

In China, there are very many reservoir types, and so well test analysis becomes much more difficult. As far as the reservoir type is concerned, there are sandstone porous reservoirs, fissured reservoirs in carbonate rocks, biothermal massive limestone reservoirs, and irregularly distributed block-shaped reservoirs in volcanic rocks; as far as the planar structure of a reservoir is concerned, there are well-extended, uniformly distributed large-area formations, fault-dissected reservoirs with complicated boundaries, and banded lithologic reservoirs formed by fluvial facies sedimentation; as far as the fluid type is concerned, there are common dry gas reservoirs, condensate gas reservoirs, and gas-cap gas reservoirs with oil rings and edge water or bottom water; and as far as reservoir pressure is concerned, there are gas reservoirs with normal pressure coefficients, extremely thick gas reservoirs with super high pressure and underpressured gas reservoirs. They are really of a rich variety and very colorful. This has undoubtedly brought about new challenges to Chinese well test analysts and reservoir engineers.

Moreover, the quality of pressure data nowadays is no longer like it was in the early 1980s. At that time, pressure data were acquired by mechanical pressure gauges and the number of pressure data read out from a pressure chart would be about one hundred or even less than that. The results interpreted from such pressure data contain not only quite a few contents but also quite a few debates about the interpretation. Today, however, the number of data acquired by electronic pressure gauges is usually as many as ten thousand, even a million; they consist not only of the pressure buildup interval but also the pressure "whole history," including all flow and shut-in intervals during the testing. Even very slight differences, if any, between the "well test interpretation model" obtained from analysis and the actual condition, that is, the tested reservoir and the tested well, will be shown at once in the verification process during interpretation so that no carelessness is allowed.

1.1 Well Test: A Kind of System Engineering

Therefore, we can say that the well test today no longer merely means several formulae and simple calculations, but rather a kind of system engineering. This system engineering includes several parts as follows:

1. The persons in charge of exploration and development must propose appropriate test projects in time
2. Making optimized well test design
3. Acquisition of accurate pressure and flow rate data onsite
4. Interpreting acquired pressure data by well test interpretation software and integrating geological data and test technique; perform reservoir parameters evaluation
5. Provide dynamic descriptions of gas wells and gas reservoirs by integrating the pressure and production history data acquired during production tests of gas wells
6. Create new well test models when necessary and add them into well test interpretation software for future application

1.2 Well Test: Multilateral Cooperation

The work listed previously should be done by different departments respectively, every one among them is associated with others, and each one affects the final results:

1. Only when leaders of the competent authorities have thoroughly recognized the important role of well test data in describing gas reservoir characterization and guiding development of the gas field can they arrange test projects timely and provide financial support for such projects to be executed.
2. Only by conducting optimized designs can we get better results with less effort and acquire pressure data that can explain and resolve our problems.
3. The acquisition of pressure data is usually done by service companies. The test crew of the service company, although working pursuant to the contract, should recognize what good data are and how to meet design requirements. The well test supervisor

must check data before acceptance according to the design requirements to ensure the success of data acquisition.

4. Data analysis will ultimately demonstrate the application value of test results. In this book, such analysis is summarized as a "dynamic reservoir description," which means using dynamic data acquired in gas wells, such as pressure and flow rate, as the main basis to evaluate the gas production potential of gas wells, while at the same time providing a description of geological conditions within the gas drainage area that affect gas deliverability and its stability, including reservoir structures, reservoir parameters, boundary distribution, and dynamic reserves controlled by this individual well, thereby guiding deliverability planning and development plan design for the gas field. This is usually accomplished through collaboration between dynamic performance analysts and reservoir engineers. Furthermore, only when such analysis results have been approved by the competent authorities can they play their due roles.

In order to study gas reservoirs from different perspectives or individual different positions, the purpose of this book is therefore to explain how to jointly comprehend well test data and understand gas reservoirs for its proper development.

1.3 Writing Approaches of this Book

The approaches adopted in writing this book are as follow.

1. The application of well test methods aims at not only gas wells but also gas reservoirs. Analyzing well test data should have the gas field or the gas reservoir in mind: it is in fact the goal that the author strives for.

2. Establish a graphical analysis method. The basis of the graphical analysis method is utilizing fundamental flow theories. Create a set of model graphs of pressure curve, establish organic connections between flow characteristics in reservoirs and well test curve characteristics so that interpreters can take a "quick look," that is, to understand reservoir conditions quickly and conveniently from measured well test curves.

3. Analyzing many field examples is another important feature of this book. This book introduces field examples of well test analysis application not only to gas well studies but also to gas field studies; not only some successful cases but also some failed ones from which some lessons were drawn, experiences were summarized, and ultimately successes achieved through such continuous experiences.

4. Although some basic formulae are introduced in one chapter, this book will neither explain how to apply them in calculation nor derive them. This book is written for those who understand these formulae and shows how to make interpretations with well test interpretation software. This book will help readers grasp the correct interpretation and analysis methods, especially the research methods for gas fields. Regarding the derivation and application of these formulae, some very good monographs are available for reference [1, 2].

Therefore, this book is a good reference for well test analysis applications. Readers are herein expected, with the help of this book, to comprehend the essence of well test analysis, to acquire and apply well test data properly, and then contribute a reliable description to the development of gas fields. It is the purpose of this book to help readers understand the well test comprehensively, make use of the well test properly, and establish and confirm dynamic models of gas fields correctly with the powerful means of the well test.

2 ROLE OF WELL TEST IN GAS FIELD EXPLORATION AND DEVELOPMENT

The well test is indispensable in the exploration and development of gas fields. During the entire process, starting from when the first discovery well in a new gas province is drilled, to verification of reserves of the gas field, and to the whole history of its development and production, the well test plays very important roles in many aspects, such as confirming the existence of gas zones, measuring the deliverability of gas wells, calculating the parameters of the reservoir, designing the development plan of the gas field, and providing performance analysis during development; in fact, none of those mentioned earlier can be done without a well test. Table 1.1 indicates in detail the roles of a well test during the different exploration and development stages.

2.1 Role of Well Test in Exploration

2.1.1 Drill Stem Test (DST) of Exploration Wells

After discovering a potential structure in a new prospect exploration area, the first exploration wells are drilled. During drilling, the show of gas and oil (SG&O) may be discovered by gas logging or logging while drilling. At this moment, it is not certain whether the SG&O really means that those hydrocarbon zones are the zones with commercial oil/gas flow. In order to be certain, a drill stem test needs to be run. If the zones have quite high productivity during the DST, a further test for measuring their pressures and flow rates and a transient test for estimating their permeability and skin factor should be done.

High gas productivity of an exploration well foretells the birth of a new gas field, and data of flow rate and pressure acquired in a DST are the direct evidence of the birth (Table 1.1).

2.1.2 Exploration Well Completing Test

Further verification of the scale and gas deliverability of the gas field is generally carried out by well completing tests. Well completing tests are usually run zone by zone when an exploration well has penetrated the target beds and well completion with casing or other

Table 1.1 Role of Well Test in Exploration and Development of Gas Fields[a]

Test analysis contents \ Implementation items	DST during drilling process of exploration wells	Completion gas well test of exploration wells	DST and completion gas well test of detailed prospecting wells	Reserves evaluation of gas-bearing area	Deliverability test and other transient well tests of development appraisal wells	Stimulation treatments such as acidizing or/and fracturing	Production test and extended test of development appraisal wells	Reserves verification of gas field	Numerical simulation of gas field and making development plan	Dynamic performance monitoring of gas field	Completion gas well test of adjustment wells
	Gas field exploration phase				**Developmental preparation phase**					**Gas field development phase**	
Verify dynamic reserves of gas reservoirs							★	■	■	★	★
Infer dynamic reserves controlled by individual gas well in gas reservoir					☆		★			□	
Identify connectivity between wells by interference test					☆		★	■	■	☆	
Identify distribution of impermeable boundaries in reservoir					☆	☆	★	■	■	☆	☆
Provide turbulence factor during producing of gas wells					★	★	★		■	☆	★
Determine related parameters of double porosity reservoirs			☆	□	★	☆	★	■	■	☆	★
Length and flow conductivity of induced fracture		☆	☆	□	★	★	★	□	■	☆	★
Evaluate drilling and completion quality by skin factor	★	★	★	■	★	★	★	■	■	★	★
Estimate reservoir permeability by transient well test	★	★	★	■	★	★	★	■	■	★	★
Confirm absolute open-flow potential by deliverability test	☆	★	★	■	★	★	★	■	■	★	★
Measure formation pressure of the reservoir	★	★	★	■	★	★	★	■	■	★	★
Find out gas bearing conditions in reservoir	★	★	★	■	★						★

[a] ★ — Items that must be implemented;   ☆ — Items that may be implemented;   ■ — Parameters that must be used;   □ — Parameters that may be used.

modes has been done. At this moment, the borehole wall of the well is solid, the test conditions are fairly mature, and there is enough time for testing so that various parameters of the reservoir can be estimated more accurately; different flow rates can be selected for the deliverability test so that the initial absolute open flow potential (AOFP) q_{AOF} of the reservoir can be calculated.

For some low permeability reservoirs, such as gas reservoirs in Carboniferous and Permian systems in the Ordos Basin, it often fails to obtain commercial a flow rate only by ordinal perforation completion, and it is necessary to recomplete the well by taking strong stimulation treatments such as acidizing and/or fracturing. In this situation, reestimating the skin factor and fracturing index is very important (Table 1.1).

Sometimes the expected gas production rate of a tested well or a tested reservoir cannot be obtained during the test after perforation; it may mean that the gas saturation is very low or there is no gas at all in the reservoir, but it is also possible that the permeability of the reservoir is so low and/or that the reservoir near wellbore was damaged so seriously that the gas cannot flow from the reservoir to the wellbore. To distinguish the real reasons of low production rates is extremely important for evaluation of the reservoir.

Skin factor is an important parameter indicating if a gas producing well has been damaged. Importance should be attached and much attention must be paid to every tested zone in which a transient pressure test can be run, especially to those zones with high permeability and low pressure, penetrated with dense drilling fluid and a long soak time, because those zones are probably damaged so seriously that their productivities are reduced too much. In this case, acidizing should be done to remove or reduce the damage; if the permeability of the tested zone is known from the well test to be very low, say less than 0.1 mD, fracturing may be necessary to improve its productivity.

Whether the tested zone needs to be stimulated and how the effect of the stimulation treatment is are both identified by the well test.

2.1.3 Reserves Evaluation
Once data of production rate, reservoir pressure, and permeability of an exploration well have confirmed the birth of a gas field, evaluation of reserves of the field should be commenced.

Several Issues Worth Noting in Reserves Evaluation
Volumetric methods are commonly used now for calculating reserves based on static data provided by geophysical prospecting, logging, and core analysis. Then the analogy method is applied to estimate the recoverable reserves by using a given recovery factor.

However, it has been discovered from practice in recent years that there is a serious risk in estimating reserves depending only on static data; and, at least, the following problems need to be noted.

1. **Reserves calculated by the volumetric method are erroneous for fissured reservoirs with group- and/or series-distributed fissures**

   Fissured reservoirs with group- and/or series-distributed fractures herein mean the ancient buried hill-typed fissured reservoirs with heterogeneously distributed fractures; their special character is that the oil or gas is stored in the fissure system with areal- and group- and/or series-distributed fractures, some local regions of the system have very high permeability, and the matrix rock is very tight, that is, in well test terms, a double media with very high storativity ratio ω, whose value can be as high as 0.3−0.5.

   This kind of reservoir can be best identified by the shape of well test curves:
   - The shape of pressure buildup curves, especially of a pressure derivative curve, is often very strange or unusual: it often has no obvious radial flow portion; it goes up and down steeply and so shows sharp fluctuations; and then approaches a trend of abrupt updip at a later time
   - The pressure drawdown curve declines rapidly, and the bottom-hole pressure cannot build up to its original value after shutting in
   - In most of this kind of gas wells, the water content ratio rises quickly after water breakthrough; the pressure buildup curve will become more complicated at early time if there is condensate oil in the reservoir

2. **Estimation of recoverable reserves in lithologic gas reservoirs formed by fluvial facies deposition**

   In the 1990s, many studies in the world showed that the recovery efficiency of some low permeability gas reservoirs formed by fluvial facies deposition is quite low. Further studies discovered that the existence of lithologic boundaries hinders the improvement of the recovery efficiency under the conditions of a normal well pattern [3]. It is possible to improve the recovery efficiency of this kind of reservoir by drilling infill wells.

3. **Integral reservoir characteristics shown by pressure distribution**

   If all gas wells in a gas field are located in an integral connected reservoir, when measuring their initial reservoir pressures and converting them from the measured depths into corresponding elevation depths, the relation of the initial pressures of these gas wells with the depths will be consistent with the pressure gradient measured in any single gas well in the reservoir. The overall characteristics of the gas field can be determined by such a simple principle.

   If a gas field is an integral reservoir, the calculation of reserves is undoubtedly fairly simple; if not, the causes therein must be found and analyzed, combining with its geologic characteristics carefully, and reflected in reserves estimation. Sometimes, the poor accuracy of tested pressure data would bring difficulty to analysis and identification or even make research insignificant. Therefore, paying more attention to this study and acquiring raw test data properly are undoubtedly the basis for all evaluation work. However, if the formations in the same horizon drilled by an exploration well are indeed not in the same pressure system, reserves estimation must be evaluated further.

Role of Well Test Method in Reserves Evaluation

During the exploration stage, well test data cannot be used directly to reserves calculation, but can supplement or correct it to a certain extent, including the following.

1. Provide deliverability as a basis of reserves calculation

The evaluated original gas in place of gas reservoirs means reserves under the condition that the flow rates of the gas wells meet the commercial flow standard. Whether the wells meet this standard or not must be evaluated by a well test. Sometimes the zones near the borehole have been damaged seriously during drilling and/or completion; therefore, the value of skin factor S of this well is very high and the flow rate is rather low even very low. However, it met the commercial gas flow standard after stimulation treatment for eliminating the damage. Also, both whether a gas well has been damaged and how much its absolute open-flow potential is after stimulation must be determined by the well test.

2. Provide characteristic coefficient of stabilized production for double porosity reservoirs

Geologic studies very often regard all carbonate reservoirs containing fractures as "double porosity" but do not distinguish such a double porosity reservoir from homogeneous sandstones in reserves analysis.

This special term of "double porosity" was suggested by former USSR scientist Baranbrat in 1960 when he was studying the mathematical model of well test for fissured reservoirs, and a flow model graph was also given by him [4]. Baranbrat proposed two parameters, storativity ratio ω and interporosity flow coefficient λ, to describe flow characteristics of this kind of reservoir. The storativity ratio ω means the ratio of hydrocarbon stored in fissures to that stored in the whole reservoir, that is, in both fissures and the matrix of it. The greater the ω, the more hydrocarbon stored in fissures. Because the oil and gas in fissures can flow very easily into the well and be produced, it is therefore the fissures that bring a high flow rate at the beginning of the production. However, if the ω value is high, as time elapses a little further, due to little hydrocarbon being supplemented from the matrix, the deliverability will drop sharply; however, if the ω value is very low, for example, $\omega = 0.01$ or even lower, which means more hydrocarbon is stored in the matrix, the deliverability of the reservoirs will be very stabilized.

Another parameter, interporosity flow coefficient λ, is also very important. It means the flow conductivity of hydrocarbon from the matrix to the fissures. If the

λ value is fairly high, when the pressure in fissures decreases due to the fluid there flows into the well, the fluid in the matrix will be supplemented into the fissures promptly so that the well will maintain stable production. However, if the λ value is very low, even if quite a lot of hydrocarbon does exist in the matrix, the matrix still cannot feed the fissures sufficiently for a very long time, even as long as several years after an extremely sharp drop of fissure pressure. For this reason, such reserves have no commercial value at all.

Therefore, it can be seen that for reservoirs with double porosity characteristics, the parameters ω and λ calculated from the well test are really very important indices for diagnosis of the stabilized production characteristics of the reserves. ω and λ can be determined only by well test. Moreover, determination of these two parameters imposes very stringent requirements on well testing conditions. This is discussed further in Chapter 5.

There are a large number of fissured carbonate reservoirs in China. Some oil/gas wells have very high deliverability at the very beginning. Encouraged by this phenomenon, field managements may think that they have found a gold mine. However, they may fail to analyze the roles of parameters ω and λ properly. For example, some wells start flowing at a rate of 100 thousand cubic meters of natural gas per day, but just last a few days only and then become depleted. This is indeed an important lesson to be learned.

3. Provide information about planar distribution of the reservoir for reserves evaluation

If reservoirs of a gas field extend continuously on a horizontal plane, only the outer boundary must be demarcated in reserves estimation, resulting in more room for maneuvering in placing development wells. When the well spacing is quite large in the early exploration stage, an effective thickness distribution map can only be drawn by the method of interpolation with a few thickness values of drilled wells, but this map cannot reflect the true distribution characteristics of the reservoir. However, well test data, especially long-term well test data, can authentically reflect the change of extension of the reservoir. For example, the area and shape of a block oil/gas field where the tested well locates in can be confirmed by well test analysis; the distance of the gas—water contact to the tested well located in a gas field with edge water can also be estimated by well test analysis. Take the JB gas field as an example: the conclusion that its Ordovician reservoirs are spread continuously and extremely heterogeneous was obtained from the analysis of pressure buildups and interference tests between wells run during short-term production tests; these results provided powerful evidence of the planar distribution characteristics of the reservoirs, thus freeing the managements' mind of apprehensions for the reserves to ultimately pass examination and approval by the National Reserve Committee of China. This example is discussed in detail later on in this book.

4. Provide original reservoir pressure data for reserves evaluation

In addition to being related to static parameters of reservoirs such as area, thickness, porosity and gas saturation, the reserves of a gas reservoir are also proportional to its original reservoir pressure; especially for overpressured gas reservoirs, the influence of the original reservoir pressure is even more prominent. Therefore, the original reservoir pressure must be determined accurately before reserves evaluation of a gas reservoir starts.

It had been required somewhere sometime that the controlled reserves of an individual well must be calculated using data of every well test. Such a requirement is improper for it has too much oversimplified reserves calculation from well test analysis or too much overestimated well test method, and so no useful conclusions can usually be met.

When entering into a pseudo-steady flow period at a medium—late stage of development of a gas field, many methods can be used to check the reserves. This is discussed further in more detail later on in the following chapters.

2.2 Role of Well Test in Predevelopment

The dependence on well test data in this stage is undoubtedly more serious.

A foreign company, for example, decided to develop a gas field in cooperation with a Chinese partner. The reserves of this gas field had been examined and verified. The company insisted spending a year of time and much manpower and money to conduct dynamic tests and analysis on more than 10 wells. Initially, the necessity of doing so was suspected, but later, it was proved to be effective. It is just this dynamic performance research that results in what had become the decisive basis for making development plans.

Many uncompartmentalized gas fields have been discovered in China over recent years—the number of them is more than that of the ones discovered ever before. Dynamic performance research during the predevelopment stage is also gradually being put on the agenda. It is therefore especially important to focus on dynamic performance research based on previous experiences and lessons.

2.2.1 Deliverability Test of Development Appraisal Wells

Deliverability values of individual wells are taken as the primary basis for making development plans. The AOFP is usually used to indicate the deliverability level. The inflow performance relationship (IPR) curve is further required to be plotted from the initial deliverability analysis.

Just like what is discussed in Chapter 3 of this book, several deliverability test methods are used onsite to determine the absolute open flow potential. The deliverability test methods applied in exploration and predevelopment stages are different: in the exploration stage, some simple methods, for example, the single point test method, can be used only for identifying if the deliverability of the gas well has met the industrial gas flow standard, and setting up the lower limit of it for reserves evaluation; while in the predevelopment stage, the deliverability test is not only for accurate calculation of deliverability indices and the planar distribution of the reservoirs in the gas field, but also for finding out the long-term stability characteristics of the deliverability.

It will be introduced in Chapter 3 of this book that, for some low-permeability lithologic gas reservoirs formed by fluvial facies sedimentation, because the effective drainage area controlled by an individual well is limited and the flowability of reservoirs is poor, the transient open-flow potential evaluated during the early stage of exploration would be very different from the commonly referred deliverability under stable production conditions. Sometimes, such a difference could be 10 times or even larger. Some Chinese and overseas research results suggest that if the reservoirs are confirmed to be like this, a new development strategy should be adopted. In addition, those fissured reservoirs with group- and/or series-distributed fractures in buried hill gas fields obviously cannot be put into production with a conventionally designed stable production rate.

Therefore, during the predevelopment stage, a systematic and rigorous deliverability test of development appraisal wells is extremely essential for a gas field, especially for a large uncompartmentalized gas field. Test analysis and calculations must not only give conventional initial absolute open flow potential, but also evaluate and provide dynamic deliverability indices during the production process; even provide a proper "production rate arrangement in its whole life" by well test analysis and deliverability prediction conducted by software when necessary [5].

2.2.2 Transient Well Test of Development Appraisal Wells

Development appraisal wells in large uncompartmentalized gas fields in China are usually studied by short-term production tests today. During the short-term production tests, high precision electronic pressure gauges are used to measure or monitor the bottom–hole pressure (flowing pressure and shut-in pressure) throughout the entire process. Such tests can not only determine the deliverability of gas wells, but also provide shut-in pressure buildup curves and the entire pressure history. Just like what is shown in Table 1.1, much important information about a gas reservoir can be obtained from the tests.

1. Information about distribution of gas-bearing areas and gas-bearing formations in the gas fields.
2. Initial reservoir pressure p_R.
3. Initial absolute open flow potential and dynamic absolute open flow potential of main gas zones, as well as planar and vertical distribution of the deliverability.

4. Effective permeability of gas zones and the relationship between effective permeability (from well test analysis) and permeability from logging analysis.

5. Information about damage of gas wells; if acidizing and/or fracturing stimulation treatment is needed; and the skin factor after stimulation.

6. For fractured wells: estimate the effect of fracturing treatment and calculate the length, flow conductivity and skin factor of the generated fracture.

7. For double porosity reservoirs: when significant double porosity characteristic curves appear, storativity ratio ω and interporosity flow coefficient λ values are analyzed, and special properties of the reserves and stabilized production characteristics are evaluated.

8. Non-Darcy flow coefficient during the production of gas wells is provided. In the design of gas field development plan, non-Darcy flow coefficient D must be used whenever selecting parameters related to the relationship between flow rate and producing pressure differential. Non-Darcy flow is formed due to turbulent flow near the bottom hole, and the skin due to non-Darcy flow is a major part of the pseudo-skin. The reasons resulting in turbulent flow are very complicated. Non-Darcy flow coefficient D can only be determined by the well test as errors are always generated when it is estimated by theoretical methods.

9. Information about reservoir boundaries can be obtained if the pressure buildup test lasts long enough. Also, if information about boundaries is obtained soon after beginning the test, that the boundaries are not far from the tested well can be predicted.

   Well test interpretation software today usually contains well test models comprising different types of boundary combinations. Furthermore, numerical well test software is able, by considering the specific geological characteristics of the gas zones, to assemble the reservoir model with proper shaped boundaries and formation parameter distributions and to provide related theoretical well test curves. Vivid descriptions of a specific tested object can be obtained by matching the theoretical well test curves with measured ones. It is especially worthy to note that such a description comes from the vivid exhibition of gas zones in the process of production and so reflects the features much closer to the reality.

10. If conditions allow, the planar and vertical connectivity of layers in the reservoir can be studied through an interference test between wells or vertical interference test.

    An interference test between wells is very difficult to run in gas fields. This is simply because the compressibility of natural gas is much greater than that of oil or water. Moreover, the permeability of gas zones is usually very low, and the well spacing is large, so that a successful test often takes a long time. In the JB gas field, for instance, the interference test between well L-5 and other wells lasted 10 months. This test delivered extremely valuable achievements: it verified the interwell communication within the gas bearing area and also revealed obvious heterogeneity characteristics.

11. In principle, the dynamic reserves of gas wells and the gas bearing area can be predicted on the basis of these successful well tests.

    The "in principle" here is that the dynamic reserves predicted by the results of dynamic tests are only the reserves in the area that have been influenced by the dynamic tests, but do not contain the reserves outside this area.

    If a gas well is located within a closed or nearly closed lithologic block, the reserves within the block affected by this well can be estimated by analysis of dynamic characteristics. However, data of this well mean nothing for judgment of another very closely adjacent region partitioned by the boundary.

    If the well is located within part of a continuously distributed reservoir, dynamic data cannot cut the boundary of the region controlled by any adjacent wells, and therefore dynamic data can only provide information about the mutual connection of these wells.

2.2.3 Well Test of Production Test Wells

If the production test wells have already connected to the pipeline network and so can produce continuously for several months, they can provide much richer information that can be used for the design of development plans. In particular, the dynamic models of the gas wells can be improved through pressure history verification.

1. During a long-term production test, the influence of boundaries around the gas well will be gradually reflected in the decrease of bottom-hole flowing pressure. The dynamic model of gas wells can be improved by verifying pressure history, adding and/or modifying boundary influences, adjusting the location and distance of the boundary to the well, and so on.

2. A perfect modified dynamic model not only verifies and confirms the formation parameters near the well but also determines the area and dynamic reserves of the area controlled by the well and so can be used for dynamic performance prediction.

3. A perfect modified dynamic model of a gas well in a constant-volume block can be used to calculate the average reservoir pressure during production and calculate the variation of dynamic deliverability indices.

2.2.4 Selection and Evaluation of Stimulation Treatment

Selecting the stimulation treatment measure is a very critical element in the development plan. However, evaluating if a gas well needs stimulation treatment and the effectiveness of such stimulation treatment can only be done by well test analysis. In some foreign countries, the field owner must, when engaging a service company to implement gas well stimulation treatment, first provide the parameters about its geology and completion and those from well test evaluation of the well so that the stimulation measure can be designed; after stimulation treatment, in order to evaluate the effectiveness of the treatment, the owner must also request third parties, well test service company or relevant consulting company, to appraise the result of the treatment by well test analysis.

2.2.5 Verifying Reserves and Making the Development Plan

Only after completing dynamic performance analysis and research mentioned earlier does the time for making the formal development plan really come.

1. The reserves have been verified by dynamic performance research, in which parameters provided by a transient well test were used.
2. Reservoir parameters have been corrected. Permeability k, for example, is not the permeability from logging interpretation but the effective permeability; skin factor S, non-Darcy flow coefficient D, double porosity parameters ω and λ, and so on are also the parameters actually acquired from the formation. In addition, the description about reservoir boundaries is a particularly very critical condition for numerical simulation.
3. The production test history can be used to match and correct the parameters used for numerical modeling.

When the requirements mentioned here are all met, numerical simulation analysis can be carried out and a practical and feasible development plan can be made.

2.3 Role of Well Test in Development

Conventional well test methods can be done almost throughout the entire development process of a gas field to provide dynamic monitoring, without any difficulties brought from swabbing and so on such as the case in oil fields.

For a normally producing gas well, however, unless permanent bottom-hole pressure gauges are used, it is obviously inappropriate to perform a well test by operations of running pressure gauges in hole and putting them out of the hole while opening and shutting in the well frequently. In fact, because the formation conditions have already been known thoroughly through early research, retesting the well is required only in case of anomalous events happening during production of the gas well.

However, the following tests are absolutely necessary.

1. Regular monitoring of downhole flowing pressure and static pressure for inferring dynamic deliverability indices of the gas well.
2. For newly drilled adjustment wells, the basic formation parameters must be obtained from well test analysis and the initial deliverability equation of them must be established before putting into production (see Table 1.1).

It is noted through the aforementioned analysis that items listed at the top right corner in Table 1.1 are blank. It means these items are not feasible. More deepening or intensive understanding of the reservoir can only be obtained through well test only as the gas field research is being deepened continuously. It is not practical to expect that all these parameters can be determined simply through well tests during the early stage of exploration. For example, it is impossible to determine the exact initial absolute open flow potential of an exploration well simply through short-term DST; it is also impossible to do an overall analysis of boundaries or to determine the double porosity parameters of reservoirs simply through very short-term well tests in the exploration wells. Even if well test analysts do give those parameters mentioned previously, such parameters are merely speculative and cannot suffice to be the basis for further analysis. However, as more gas wells are put into production tests or production, and as the flowing of these wells goes on, the radius of influence increases, and pressure buildup testing lasting quite a long time is carried out, research work will continue to intensify. Some parameters, which could not be obtained previously, can and should be determined through well test analysis at this time; such parameters include initial and dynamic deliverability indices, boundary distance L_b and shape, block sizes A, double porosity parameters ω, λ, and β, double permeability formation parameters κ, composite formation parameters M_C and ω_c, non-Darcy flow coefficient D, reservoir connectivity parameters ε and η, and the dynamic reserves of block. With this knowledge, the gas reservoir dynamic model can be established and used effectively for the dynamic performance analysis of gas zones and gas reservoirs. These are just the stage character and overall characteristics of a well test.

3 KEYS OF WELL TEST ANALYSIS

Well test research started in the 1930s. By the time of the 1970−1980s, it devolved into the "modern well test." As advances in theoretical research on flow mechanics and continuous improvement of well test software, the role of the well test on gas field exploration and development expands and deepens continuously.

What are the key elements in well test research? What has been driving advances and development of well test research? How does the well test serve gas field studies? All these questions can be roughly answered by Figure 1.1.

3.1 Direct Problem and Inverse Problem

Well test research roughly resolves two types of problems: direct and inverse.

Direct and inverse problems are defined from the viewpoint of information theory. A direct problem means describing the performance of a known formation in terms of its gas production rate and reservoir pressure on the basis of flow mechanics theory, whereas resolving an inverse problem means that if the variations of gas production rates and bottom-hole pressures of one or several wells in a gas reservoir during their flowing and shutting in process have been measured, finding out inversely the static conditions of the gas zones, including the values of formation parameters, the structure of permeable areas in the reservoir, the planar distribution of gas zones, and so on.

This book herein explains the procedure of resolving these problems by well test research with the hope that readers of this book, especially those interested in

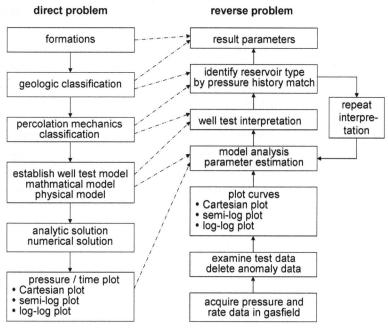

Figure 1.1 Illustration of well test research.

participating in well test research, can correctly "locate" the jobs they are participating or interested in and straighten out the relationship between well test research and geologic research.

Previous well test research failed to distinguish different types of formations or believed that all formations are "homogeneous media" identically. The semilog straight line analysis methods [Miller—Dyes—Hutchison (MDH) method and Horner method] invented in the 1950s found that flow will enter the radial flow stage that reflects reservoir conditions when wellbore storage disappears; in this stage, pressure variation shows a straight line in semilog paper, that is, the coordinates of pressure vs logarithm of time, and an inverse proportion relationship exists between the slope of the straight line m and the formation permeability k:

$$k = \frac{0.00121q\mu B}{mh}.$$

This is simply the basis of the "conventional well test interpretation method" that uses the well test method to determine formation parameters inversely [6, 7].

However, measured curves are far more complicated, especially in the case of carbonate formations, multilayer formations, or formations with complex boundaries; in these cases it is often very difficult to find out proper straight line portions. Furthermore, straight line sections alone can hardly describe other characteristic parameters of the reservoir. Therefore, in the 1970s, the type curve match method was created [8–11].

In the early 1980s, Bourdet invented pressure derivative-type curves [33]. On pressure derivative-type curves, each kind of flow in the formation corresponds to a special characteristic pattern, while each kind of flow is determined by the special geological conditions of the specific formation. Therefore, an organic connection is established between the geologic characteristics and the graphical characteristics.

So far, the combination of log–log analysis (i.e., pressure and its derivative-type curve match analysis) and semilog analysis (i.e., conventional analysis method) has formed the dominant theoretical foundation of "modern well test interpretation" and becomes the dominant analysis method of well test interpretation software. There were a wide variety of calculation formulae and analysis plots used before—provided they can be integrated into the modern well test interpretation model, they can be added into the interpretation software and widely used. However, as analysts become more dependent on well test analysis software, some other methods, such as the Y-function method for judging the presence of faults, the Masket method for calculating formation parameters, and various unique point methods for conducting interference test analysis, are increasingly losing their chance of being used.

3.2 How to Understand Direct Problems

The process of establishing the relationship between characteristics of the formation and those of well test plots starts from solving the direct problem. The research tasks of resolving the direct problem can be summarized in several parts as follow.

3.2.1 Analyzing the Formation Where the Oil/Gas Well Locates and Classifying it Geologically

The geologic bodies across China where the gas fields locate are very complicated, and whose rough classification is given in Table 1.2. For easy comparison, Table 1.2 also lists typical examples of gas fields in China. In fact, the types of gas reservoirs are far more than these, and even many different types may exist simultaneously in one gas field [12].

3.2.2 Classifying, Simulating, and Reproducing Formation from the Viewpoint of Flow Mechanics

It is seen that the generating conditions of various reservoirs are very different; if described by flow mechanics equations, they must be simplified and classified into some major categories, and the description must be used only within a certain scope. Sandstone reservoirs, for example, are usually simplified into a model of an infinitely homogeneous porous medium. Strictly speaking, the existence of this kind of reservoirs

Table 1.2 Types and Examples of Gas-Producing Zones

Type of gas reservoir	Typical examples of gas fields in China
Large homogeneous sandstones	TN gas field, SB gas field, YA13-1 gas field
Fault-dissected local homogeneous sandstones	Fault block gas fields in SL, LH, and ZY oil fields, HTB gas field
Rocks; some parts of which are homogeneous sandstones	P5 gas field
Carbonate rocks; some parts of which are like homogeneous	Some areas of southern region of JB gas field
Carbonate rocks showing significant heterogeneity	Most areas of JB gas field
Fissured carbonate rocks showing significant double porosity characteristics	L5 area in center region of JB gas field
Extremely thick fluvial facies sedimentary sandstones	KL2 gas field, DN2 gas field, DB gas field
Thin-layer sandstones having lithologic boundaries formed by fluvial facies sedimentation	Carboniferous/Permian gas fields in Ordos Basin
Condensate gas fields in fault-dissected sandstone	YH and YTK gas fields etc. in Tarim Basin
Gas caps in sandstone with oil rings and edge or bottom water	QL and XLT gas fields
Carbonate rocks with group- and/or series-distributed fractures	QMQ, SQ, and CN YAC gas fields, etc.
Biothermal limestone bodies	LJZ, PFW and SL gas fields, etc.
Volcanic massifs of eruptive facies	XS gas field in DQ oil field
Offshore shoal, sandbar	DF gas field

in nature is impossible. However, the well testing duration is limited and so the range of pressure influence is also limited; therefore, within such limited scopes of time and space, the target being studied can be considered being roughly consistent with an infinitely homogeneous formation.

Based on the knowledge mentioned previously, reservoirs can be further simplified and classified from the viewpoint of flow mechanics, including the following.

Basic Medium Types
- Homogeneous medium, including sandstones, fissured carbonate rocks showing homogeneous behavior, etc.
- Double porosity medium, including sandstones and carbonate rocks comprising natural fissures
- Double permeability medium, mainly means layered sandstones

These media are usually assumed as laminar two-dimensional distribution.

Bottom-hole Boundary Conditions (i.e., Inner Boundary Conditions)
- General completion condition of wellbore storage and skin
- Completion condition of having induced fracture connecting the well hole
- Partially perforated completion conditions
- Completion conditions of horizontal wells or deviating wells

Outer Boundary Conditions
- Infinitely outer boundary
- Impermeable external boundaries of single straight line or of some patterns formed by several impermeable boundaries
- Closed external boundary: closed small faulted blocks or lithologic traps
- Heterogeneous boundaries formed by variation of lithology or fluid properties
- Semipermeable boundaries, congruent boundaries of river channels formed by fluvial facies sedimentation in different periods
- Constant pressure boundaries (in oil reservoirs only)

Assumption of Fluid Properties
- Oil, gas, water, or condensate gas
- Any combination of oil, gas, and water

Any assemblage of any four elements, each one of them is selected from one of the four aforementioned conditions, constructs a physical simulation for a certain gas reservoir and reproduces the behavior of specific gas field during the research process.

3.2.3 Constructing the Well Test Interpretation Model and Resolving the Related Problem

The so-called well test interpretation model should contain both a physical model and a mathematical model.

The contents listed in Section 3.2.2 are just the descriptions of physical models. At the same time, these physical models can also be expressed in mathematical forms. For example, the flow in different types of medium can be expressed by different differential equations; different boundary conditions can also be expressed by different mathematical expressions. These are so-called mathematical models.

In the 1960s, the physical models mentioned earlier were materialized during the study of well test problems. Man-made sandstone bodies were built and used in the laboratory to be a reduced physical microminiature formation or a model. The model was saturated with oil or water, and the flow rate change was implemented by drilling holes in the model. The pressure change at individual points on the model was measured. Such a practice, however, not only was very difficult in constructing the model and very costly, but also could hardly simulate the elastic transient process. Therefore, it was abandoned long ago.

Establishing mathematical equations correctly is only the beginning of the process of resolving direct problems, while solving these equations is really more important.

In the past, the analytic method was applied basically for resolving the equations. Because these equations, under the hypothesis of Darcy's law, are mostly partial differential equations, mathematical manipulation methods such as the Laplace transform must therefore be used to convert them into ordinary differential equations in the Laplace space for resolving them and then invert the solution back into the real space. Moreover, the solution procedure usually aimed only at some relatively simple boundary conditions, such as those shaped circular and square. Today these equations can also be resolved directly by numerical methods, but it takes a fairly long time, and sometimes model adjustment cannot complete immediately.

Theoretical research of the well test has mostly targeted construction of the well test model and resolved the related mathematical problem, as many famous researchers such as van Everdingen, Ramey, Gringarten, Agarwal, Earlougher, and Bourdet did. It can be summarized as

Identify typical geologic model → *Construct well test model (formulate its mathematical equations)* → *Resolve the equations by analytic or numerical method* → *Draw pressure variation curve, that is, make type curves*

This is the whole process of resolving direct problems.

3.2.4 Expression Forms of Research Results of Resolving Direct Problems in Well Test

The ultimate expression form of the research results of resolving direct problems in the well test is type curves, that is, the plot of relation between pressure and pressure derivative and time, used for well test analysis. For instance, the type curve for homogeneous reservoirs is as shown in Figure 1.2.

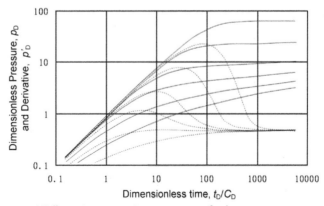

Figure 1.2 Well test interpretation type-curve for homogeneous reservoirs.

Different type curves for different formations have been obtained. Both the theoretical model corresponding to actual formation and the parameters of the formation can be obtained through the type-curve match. This is the theoretical basis of understanding the formations by the well test method.

3.3 Describing Gas Reservoirs: Resolving Inverse Problem

The great majority of engineers involved in well testing are trying to resolve the inverse problems rather than resolving the direct problems discussed earlier. What resolving inverse problems means is:

Make well testing arrangement and well test design based on the requirements of gas field exploration and development → Acquire pressure and flow rate data onsite → Analyze or interpret acquired data → Analyze or interpret gas reservoir characteristics and calculate reservoir parameters → Construct the dynamic gas reservoir model to be used for gas reservoir description.

This process is explained here step by step.

3.3.1 Well Test Design

Just as a complete set of design drawings of a building must be available before constructing, careful design must be done before performing well test research of a gas field. Well test design includes the following.

1. Define the problem to be resolved by the well test according to the requirements of gas field research. For example, make clear the deliverability of the gas well, calculate gas reservoir parameters, understand boundary situations or connectivity between wells, and so on; see Table 1.1.
2. Properly arrange jobs of testing and rough time schedules for existing gas wells and those gas wells expected to be completed.
3. Make test designs for each specific tested well.
   - Collect geologic and logging data
   - Collect drilling and completion data
   - Simulate the test to get well test curves with well test interpretation software
   - Make well test operation plan, including time schedule, types of instrument to be used, method and depth of running gauges in hole, requirements of data acquisition, and so on
   - Well flowing and shutting-in duration, flow rates arrangement, measuring requirements, and so on.

3.3.2 Acquiring Pressure and Flow Rate Data Onsite

Acquisition of pressure and flow rate data onsite is one of the most important components and is the foundation of the modern well test. Acquisition of downhole pressure, from the requirement of the modern well test, must be carried out with high-precision electronic pressure gauges, which generally offer a precision up to 0.02% FS and a resolution of 0.01psi (0.00007 MPa), and must be able to record one million data points each time with a high sampling rate of one point per second. In this way, the downhole pressure changes during several months can be recorded accurately and completely in only one run and can meet the requirement of pressure derivative-type curve analysis.

According to its original meaning, the pressure referred here means pressure at the middle point of the pay zones. Theoretically, therefore, the pressure gauge must be put there, right at the midpoint of the pay zones, to measure the pressure. Moreover, the pressure thus measured must reflect the "whole process of pressure change" rather than merely a certain flowing or shutting-in period to be analyzed only.

The measured pressure data should be drawn into three plots, namely:

Cartesian plot of pressure history vs time

Log—log plot of pressure and its derivative vs time; the pressure here is either flowing pressure for drawdown or shut-in pressure for buildup

Semilog plot of pressure vs time; the pressure here is the same as described earlier

The three kinds of plots are the results of resolving the direct problem as introduced earlier and also the starting point of resolving the inverse problem. However, these three kinds of plots at this stage are not theoretical outcomes, but from data acquired onsite in the field.

It must be noted here that all these plots are drawn by computer with well test interpretation software. Nobody wants to plot any figures manually today. Sometimes, just for the purpose of learning, every engineer should try to do the interpretation, including plotting these figures by hand, once in a while.

3.3.3 Graphical Analysis in Well Test Interpretation

Well test research or well test interpretation is, after all, a kind of "graphical analysis":

- Identify the type of well test model by comparison of graphical characteristics
- Confirm the best suitable model through type curve match
- Confirm reservoir type and calculate parameters by selection of the best suitable model

Just as what was reiterated earlier, the so-called "well test interpretation model" actually means an idealized replica of an actual reservoir. If the performance of the well test interpretation model, that is, the relation of the pressure changes and flow rate vs time, is completely consistent with that of the tested well and the actual reservoir, the well test model can be used to represent the tested well and the actual reservoir, and the parameters of the model can be regarded as those of the tested well and the actual reservoir.

3.3.4 Well Test Interpretation Combining Actual Formation Conditions

It can be said that well test interpretation performed by modern well test interpretation software has already been simplified to easy screen operations. However, the well test interpretation process is by no means merely mouse clicks. This is because existing well test interpretation software has no function of artificial intelligence yet and even cannot resolve the "ambiguity" problem in well test interpretation. In other words, it is possible that two or more kinds of completely different reservoirs may show similar patterns on the pressure and its derivative log—log plot so that much attention must be paid during interpretation. Judgment would become even especially difficult when the quality of measured pressure data is poor, data points are jumping up and down, and/or the test duration is not long enough.

The role of well test engineers and reservoir engineers involved in well test analysis is to combine the operation of each well test analysis with the geologic situations and actual conditions of the gas field, including conditions of production, operation, and technique of data acquisition. If any of the conclusions obtained from analysis by well test interpretation software are different from the reality of the gas field, the well test interpreter must abandon the obtained results and redo the analysis from the very beginning until satisfactory conclusions are obtained.

The several arrows with dotted line in Figure 1.1 illustrate the operation process of well test analysis combined with the actual situation of the gas field.

The process of resolving an inverse problem by running well test interpretation software must also pass pressure history match verification. This verification compares the pressure history calculated from the obtained theoretical model with the actual measured pressure history. If both are consistent, the selected model is considered to be correct; if not, parameter adjustment is necessary and must be made.

If a fair long-term production test has been carried out in the well, pressure history match verification will be more effective. Here, it is especially important to note that the actual pressure history must be acquired in order to perform this verification. Pressure data in the critical period to be analyzed, say one pressure buildup, must be acquired continuously and with high precision; the acquisition of flowing pressure in the flowing period is equally essential. A sufficient number of pressure points should be measured for other long-term pressure monitoring.

3.3.5 Recommend Knowledge Obtained from Well Test Interpretation to be Applied in Gas Field Development

The goal of well test interpretation is not only to calculate the parameters of individual wells but also ultimately to resolve the problems in gas field development. Shell Oil Company, for example, through analyzing gas field conditions by well test methods during joint development of the Changbei gas field in China, has recognized from transient well tests that

- The permeability of gas zones is fairly low ($k = 0.6$ mD), causing gas wells to offer a low initial gas deliverability

- The reservoir was rectangular (600×2000 meters) formed by a channel deposit and with lithologic boundaries; some of the boundaries are a flow barrier
- The gas-bearing area and reserves controlled by the tested well are 2.2 km^2 and $100-500$ million cubic meters, respectively
- A fracture has been generated by hydraulic fracturing stimulation, the half length of the fractures is $x_f = 70$ m, and the fracture skin factor is $S_f = 0.5$

As soon as confirmed, these results were immediately accepted by the engineers in charge of the development plan design. The parameters and other information obtained from well test analysis listed earlier were employed in their numerical simulations, and rectangular grid-shaped fluid flow barriers were assumed in their model. In order to overcome the influence of flow barrier boundaries on the recovery percentage of reserves, the horizontal well design scheme was optimized in the drilling plans designed.

This is a typical example of using results of dynamic research directly in development design.

3.4 Computer-Aided Well Test Analysis

Just as all subjects in science and technology, well test research is done completely under the support of computers and related software. Nowadays popular professional books on well testing often attach some calculation examples when explaining the application of some formulae. This is very necessary for understanding the role of formulae and for providing parameters. Presently, however, these formulae have become the background and reference for running computer software. Generally, well test engineers will not use these formulae to calculate directly and manually. Therefore, they should pay more attention to the following.

- Assist the departments in charge of exploration and development with making a well test research plan properly during gas field exploration and development, and making clear what problems should be targeted during different phases and what data should be acquired without delay when having opportunities.
- Make well test designs properly with the support of well test interpretation software before data acquisition.
- Monitor the acquiring of pressure data properly. Acquire pressure data continuously and timely without delay. It is especially important to warn that there will almost be no opportunities of acquiring such data. Missing any acquisition or making any mistakes in acquisition will bring the serious result of losing the opportunity of understanding the reservoir through the tested well and the tested zone.
- Review and analyze acquired data, and eliminate abnormal variations due to problems caused by technique conditions and wellbore factors.
- Perform well test interpretation based on pressure data with well test interpretation software, combined with geologic characteristics of the reservoir, and evaluate the

deliverability of the tested well, planar structure parameters and completion effect of the reservoir. For key wells, especially when pressure data are acquired by surface readout electronic pressure gauges, the evaluation process is usually not accomplished just in one operation. Establish a perfect dynamic model of the gas well and gas reservoir during well test interpretation, combined with geological and reservoir engineering research results, thereby forming an organic component of the gas reservoir engineering study.

- Provide this dynamic model for numerical modeling and development plan design. If necessary, numerical well test software being improved gradually can also be used to predict future production variations of gas wells and gas field and even adjacent regions where no well has been drilled. Such a prediction is called spatial expanding and time-elapsing well test research by some researchers.
- Monitor the dynamic performance of the entire gas field during the gas field development process. Verify the reserves of the gas field when its production has reached a certain stage.

In fact, research of the entire development process of a gas field is just research of its dynamic performance characteristics, and this research is conducted primarily by well tests with well test interpretation software as its major means.

4 CHARACTERISTICS OF MODERN WELL TEST TECHNOLOGY

4.1 One of the Three Key Technologies of Reservoir Characterizations

The three key technologies are geophysical prospecting, well logging and well test.

Well test means getting the formation parameters, describing the formation, and confirming and predicting the variation of well deliverability indices by testing and analyzing the pressure and flow rate, that is, dynamic data of a gas or an oil well. Generalized well testing includes something more, such as collecting and analyzing fluid samples, temperature measurement, and measuring and analyzing dynamic data such as the content of sand, water, and gas condensate. Figure 1.3 summarizes the major links of application of the three key technologies during the process of making a development plan of a gas field.

It is true that a well test has less extensive coverage on both horizontal and vertical directions than geophysical exploration and cannot distinguish subzones as detailed as well logging; furthermore, a well test can only pay its roles to wells that produce a certain amount of fluid. However, for reservoirs from which a well or some wells produce oil or/and gas, the well test can actually reflect their distinct information, even information deep inside the formation. Furthermore, geophysical exploration and well logging can only acquire some necessary information during exploration and the period before the gas field has been put into production, whereas acquisition and research of a well test can be done in the whole process of exploration and development of the gas field.

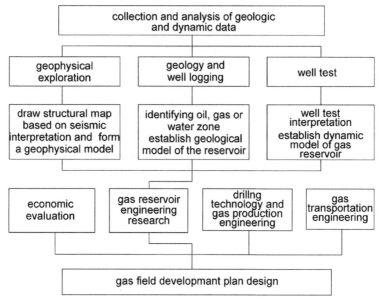

Figure 1.3 Gas reservoir research process.

4.1.1 The Distinct Information Here Includes the Following

1. Geophysical exploration and well logging deduce the existence of the gas/oil reservoir and measure permeability and other parameters indirectly by acoustic reflection, variation of electric, magnetic, or some other physical quantity indicators. Although these parameters are corrected by core analysis data, they are still different from their real values, especially for fractured or fissured reservoirs, and the difference may be great.

   Different from geophysical exploration and well logging, a well test can measure flow rate and pressure directly and obtain the real flow conductivity of the reservoir (such as permeability) from the relation of flow rate and pressure. If repeat formation tester and production logging data have been obtained, the differences between subzones can be distinguished, and therefore logging analysis results can thus be corrected by well test data.

2. Information about formation reflected by parameters from well logging analysis is only a very small area around the wellbore. While the area detected by the well test will enlarge gradually and become broader and broader, due to the pressure influence radius increases continuously as test time goes on, and so the involved area of well test research is much larger, the parameters obtained from well test research represent their actual values in a broad area, which makes it incomparable and very useful.

3. Well test technology can detect the impermeable boundaries of dozens or even hundreds of meters away from the tested well while well logging technology cannot. Geophysical exploration can roughly determine the existence of a fault by its down throw, but cannot detect whether the fault is sealing or isolating the fluid flow. Moreover, it is hard for geophysical exploration to determine the existence of lithological boundaries, especially in cases of thin zones.

4. Well test technology can determine the deliverability and absolute open flow potential of gas wells, but any other technology can do nothing about it.

5. Through shut-in pressure variation, can well test analysis evaluate the well completion effect, identify whether the well has been damaged by drilling fluid or/and well completion fluid or not. Additionally, after fracturing well test can give the half length of the fracture and the damage condition of the fracture face. All of these cannot be accomplished by geophysical exploration or well logging.

6. After confirming from well test interpretation, the theoretical reservoir model can be used to predict directly the future dynamic performance of the reservoir. Application of a numerical well test makes it possible to predict the production process of the whole gas field. However, geophysical exploration or well logging can only provide some parameters for the preparation of numerical modeling.

In summary, geophysical exploration, well logging and the well test cannot be replaced by one another. The well test can describe vividly and more clearly the reservoir characteristics under flow conditions, which make it more distinguished and useful. Especially during the middle and late stages of gas field development, well test appears more and more important.

4.1.2 Deficiencies of Well Test Technology

Well test technology has its deficiencies, for example:

1. Only wells able to produce oil and/or gas can be tested; it is not sure that typical test data can be acquired in every test or satisfied interpretation can be done for every test.

2. During a well test, especially in the early phase of exploration, a special testing pipe network is needed. Otherwise, produced natural gas must be burnt or vented to the atmosphere, which will cause energy waste.

3. In order to know the internal conditions of the reservoir, the test time must be prolonged. This can only be done in the production test when the gas well has been connected with the surface pipe network.

4. The well test process needs the cooperation of several departments, which increases the difficulty of data acquisition.

To sum up, well test research in a gas field has its own unique characters, and it is indispensable and irreplaceable. However, it is seen from some domestic gas field development plans that the basic research for well testing is still quite poor.

4.2 Methods of Gas Reservoir Dynamic Description

4.2.1 Dynamic Reservoir Description with Deliverability of Gas Wells at the Core

For many years, the contemporary well test, including the modern well test, concentrated on the analysis and study of pressure buildup curves:

1. In the aspect of theoretical study on flow mechanics, much effort has been made to establish various types of transient well test models.
2. In the aspect of well test analysis methods, a variety of type curve analysis methods have been developed by studying various type curves representing geologic features of different reservoirs.
3. Complete well test interpretation software has been developed, and so on.

However, the main achievements in pressure buildup analysis concentrate in further confirmation of reservoir parameters, which is only supplement of the understanding of static geologic conditions and can only provide some reliable information for the design of the development plan. But this is far from enough.

What is the most important issue in the development of gas fields? It is surely the deliverability of gas wells and gas reservoirs: What is the initial flow rate of a gas well? Will it stabilize? What is the stable flow rate and how long can this flow rate be maintained? Can the flow rate be enhanced by stimulation treatment, and how high is the potential of enhancement? How much is the predicted cumulative production of a gas well under the industrial development conditions? A series of questions related to deliverability are waiting to be answered.

This point is just neglected in the past well test study, which led to the continual adoption of empirical deliverability analysis methods developed in the middle of the last century. These methods are still being used even though they have shown various flaws for a long time; sometimes the predicted deliverability indices by them are far from the real ones and even lead to an incorrect direction in decision making of the gas field development.

As a result, the goal and the position of deliverability test analysis in a gas reservoir study should be reset and new ideas, new concepts and new thoughts should be put forward in a dynamic study so that the deliverability test analysis can be linked more closely with production demand.

4.2.2 New Thoughts in Gas Reservoir Dynamic Description

Figure 1.4 shows new thoughts and a flow sheet for a gas well and gas reservoir dynamic description centered on gas well deliverability.

The gas reservoir dynamic description should contain the following four main parts:

- To understand the influence of the gas reservoir geologic structure on gas flow
- To establish gas well and gas reservoir dynamic models and dynamic performance prediction

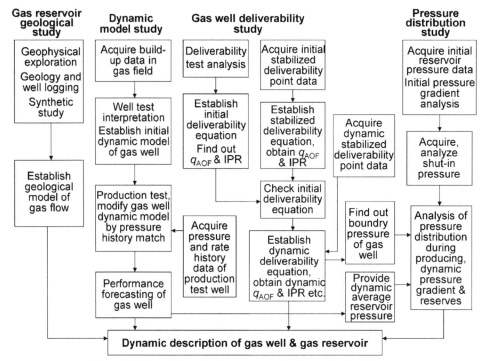

Figure 1.4 New thoughts in gas reservoir dynamic description.

- To study the initial deliverability of gas wells and its decline in regularity
- To study pressure distribution variation and dynamic reserves of gas reservoirs

Evaluation and Prediction of Gas Well Deliverability: The Core of Gas Reservoir Dynamic Description

After a gas field is put into production, a close supply-and-demand relationship exists between the gas field owners and the customers so what the problem operators care about most is undoubtedly the deliverability of the gas wells and the gas field:

1. How much are the initial maximum deliverability and rational stable production rate.
2. How to monitor and evaluate the dynamic deliverability of gas wells, that is, the regularity of deliverability decline.
3. In order to make a practical production plan, predict the dynamic reserves and deliverability of the gas wells and gas reservoirs under the production scale required by the market.
4. Predict when the gas wells must be shifted to pressure-boosting production and the production indices and decline process under the pressure-boosting conditions.

   Since the middle of the last century, many studies on gas well deliverability evaluation methods have been conducted and resulted in many classical well test methods, such as the back-pressure test, isochronal test and modified isochronal test, which have been incorporated into the petroleum industrial standards and are still widely used in fields today. However, these classical methods were created over six decades ago; due to limitations of the technical conditions and the cognition level to gas reservoirs at that time, they have some problems and cannot meet the onsite demand at present in many aspects.

1. The success rate of data acquisition was low due to the influence of water loaded in the wellbore and so on; the credibility of the AOFP even interpreted from data that seemed normal is doubtful.

2. Due to the limitation of operation conditions and cost, only a few gas wells can be tested with normative deliverability test methods, the tested wells account for a minority, in some gas fields, the number of tested wells accounts for 10% or even lower of the total wells, and so there is no essential basis in making a production plan for the rest of the majority of the gas wells.

3. The standardized deliverability test methods can only measure the initial absolute open flow potential in the initial production period of a well, but the potential of gas wells decreases continuously; for some gas wells in special lithologic reservoirs, the initial deliverability is very likely short-lived. To resolve this problem, the author put forward a new concept of "dynamic deliverability of a gas well," that is, actual deliverability indices during production; while those existent methods can do nothing in determining the dynamic deliverability indices.

Therefore, after many years of study and practice, the author came up with a method called the "stable point LIT deliverability equation." In this method, stable production data point during the initial production period of a well is acquired and substituted into a deliverability equation derived strictly from theory to determine the related parameters in it, and consequently obtain the initial deliverability equation, draw the initial IPR curve, and calculate the initial absolute open flow potential. The method can also take the gas production rate and flowing pressure during a selected suitable stable production period to establish a dynamic deliverability equation, calculate the dynamic open flow potential and the dynamic reservoir pressure at the gas drainage boundary, draw the dynamic IPR curve, and thus evaluate the deliverability indices after reduction of deliverability.

It is easy to see that this method can evaluate all gas wells in a gas field, resulting in an overall evaluation of the gas reservoir. Chapter 3 discusses this subject in detail.

Study of Gas Reservoir Dynamic Model

The aforementioned study on the deliverability of gas well and gas reservoir only gives evaluation of the gas deliverability of the well at present, but other questions regarding what kind of geologic factors causes this kind of deliverability performance, what are the reasons of deliverability decline and so on cannot be answered only by a deliverability test itself. In particular, no deliverability analysis method can predict the upcoming deliverability decline process.

To answer these questions, the study of a gas well dynamic model must be performed. Chapters 5, 6 and 7 of this book discuss this subject in detail, and the subject can be summed up into the following aspects.

- A preliminary dynamic model for a near wellbore reservoir is established through analysis of a pressure buildup curve.

  The preliminary dynamic model includes the following: permeability-thickness product kh; flow parameters, such as permeability k etc.; skin factor S, fracture half-length x_f, fracture conductivity F_{CD}, and completion parameters such as skin factor of fracture S_f; and impermeable or anisotropic boundary near gas wells. This kind of reservoir model established only by short-term pressure buildup analysis is not complete; the boundary far away from the tested well very often cannot be identified.

- Establish a complete dynamic model for gas wells by a production test.

  After a production test or relatively long time production, the effect of a boundary would appear gradually during the decrease of bottom-hole producing pressure. At this time, the pressure history match verification method, combined with the geologic structure of the reservoir, is used to correct boundary parameters of the model gradually so as to modify the dynamic model. The complete dynamic model for gas wells can also be established by using the latest method of deconvolution [74, 75].

Now the model established by well test interpretation not only describes the related geologic and completion parameters near the wellbore and boundary condition parameters, but also has the function of dynamic prediction, that is, it can predict, under given production indices, the variation of bottom-hole flowing pressure of the gas well during a period of time in the future; this, in fact, can be regarded as the prediction of the deliverability at that time.

Study of Pressure Distribution

Reservoir pressure is one of the three important factors that control the deliverability of gas wells and the only one that decreases continuously with production going on; in fact, the decline of reservoir pressure dominates the reduction of deliverability.

The traditional dynamic performance study during gas field development always pays much attention to the test and analysis of reservoir pressure. For example, in the gas fields in the Sichuan province of China, during the overhaul of the equipment in the gas-processing plant, all gas wells in the gas block are usually shut in at the same time, the static shut-in pressure of each well is measured, and with these data the decrease and distribution of reservoir pressure are analyzed, and the connectivity between wells and dynamic reserves controlled by individual wells is studied.

The initial reservoir pressures measured in a new exploration area can be used to analyze the initial pressure gradient to identify the overall features of the gas reservoir. Chapter 4 of this book discusses this issue in detail.

However, for a medium- or large-scale gas field supplying gas to customers, overall well shut-in in order to measure reservoir pressure is not practical and is even impossible.

Two practical methods can estimate the reservoir pressure of gas wells effectively without shutting in.

1. Estimation from dynamic model. This method can estimate the average reservoir pressure of a constant volume block at any time; details are introduced in Chapter 8 of this book.

2. Estimation from deliverability equation. This method can estimate the reservoir pressure at the gas drainage boundary of the tested well; the estimation method is elaborated in Chapter 3 of this book.

After the reservoir pressure of every well in the gas block has been acquired, the following study can be conducted:

- Distribution and variation of reservoir pressure in the gas field, which is helpful in understanding their effect on the decline of deliverability
- Estimation of dynamic reserves controlled by an individual well
- Analysis of the connectivity between gas wells

Understand the Effect of Geologic Conditions on Gas Flow

All dynamic performance of a gas reservoir originates from the geologic structure of the reservoir, for example:

- The flowing pressure of gas wells located in a continuously distributed homogeneous sandstone reservoir declines very slowly, and the gas production rate of them is stabilized or lowers very slightly and slowly. Conventional deliverability test analysis methods and pressure buildup analysis equations for gas wells in a homogeneous reservoir, such as back-pressure test, isochronal test, modified isochronal test, and straight line analysis for radial flow, that is, Horner or MDH methods, are all suitable for this kind of well. In development plan design and numerical simulation, the development indices calculated for various well patterns basically match the actual reservoir conditions too.

- In banded sandstone controlled by meandering river subfacies in fluvial sedimentation formation, the effective gas-bearing reservoir is separated into thin and low permeable bands. When gas wells are drilled in this kind of formation, the extremely low permeability not only hinders the deliverability of the wells, more seriously, the lithologic boundary restricts the dynamic reserves controlled by the individual wells, which leads to the rapid decrease of reservoir pressure and bottom-hole flowing pressure during production, accompanied by the depletion of deliverability.

- In some reservoirs of special lithology, such as the buried hill carbonate formation in the QMQ gas field or the volcanic strata in YC formation of the XS gas field, various kinds of fractures are the main storage space and flow channels, due to their complex structure, strong heterogeneity and various shaped impermeable boundaries exist in the circumference, each separated flow areas are very limited. More severe challenges will be met during establishing dynamic model for this kind of reservoir.

It is obviously impossible to do a dynamic description perfectly for all kinds of gas reservoirs mentioned previously without deep understanding of the static geologic conditions of the reservoirs.

In a word, only on the basis of the aforementioned four aspects of study and integrating understanding from all other aspects organically can we get a thorough understanding of gas wells and gas reservoirs from their dynamic performance. This goes beyond both the simple interpretation and analysis of a transient well test and the supply of a small amount of basic parameters for development plan design; it puts forward and completes an overall dynamic image of the gas reservoir focusing on deliverability after the gas field has been put into production, and this dynamic image can predict the future trend of the gas reservoir.

CHAPTER 2

Basic Concepts and Gas Flow Equations

Contents

Just as what was introduced in Chapter 1, this book tries to help reservoir engineers involved in gas field development to utilize knowledge of the well test and relevant computer software to resolve problems in gas reservoir descriptions and to establish

Dynamic Well Testing in Petroleum Exploration and Development   © 2013 Petroleum Industry Press. Published by Elsevier Inc.
ISBN 978-0-12-397161-6, http://dx.doi.org/10.1016/B978-0-12-397161-6.00002-4

a bridge linking well test interpretation results and the solution to gas reservoir development problems, thereby allowing well test results to play their due roles and make up its shortcomings and deficiencies that have long been existing in this field.

The interpretation and application of well test data fall into the category of "solving inverse problems" in the well test discipline. Solving an inverse problem does not require starting from deriving and solving the partial differential equations describing subsurface flow. However, readers are expected to have an overall understanding of well test theory, to know that it is strictly established on the basis of mechanics of fluids flow in porous media, that the methods and formulae applied in the well test were established through careful research and development done by several generations of scientists, and that these methods and formulae are therefore reliable and trustworthy.

In addition, the experiences gained by engaging in well testing for so many years also remind us that many basic concepts frequently encountered in our daily work seem sometimes specious. These concepts must be further elaborated clearly, their original definitions and basis must be found out, and they must be explained accordingly under gas field conditions and discussed when necessary.

1 BASIC CONCEPTS

1.1 Steady Well Test and Transient well Test

There are two kinds of well test, that is, steady well test and transient well test, if classified by the stability conditions of the flowing or working system during testing.

1.1.1 Steady Well Test

During steady well testing, the flow rates (including gas rate, oil rate, and/or water rate) and pressures (including bottom-hole flowing pressure and wellhead pressure) of the tested well (oil, gas, or water well) must keep steady under each working system (usually the choke size) or, from the requirement of engineering, their fluctuation must be less than a certain limit, which is regarded as basic steady state. The steady well test process is shown in Figure 2.1.

Figure 2.1 often serves as the diagram of basic results of the steady well test. It shows the following:
1. The flow duration of individual selected flow rate.
2. Whether the flow rate has reached steady state within the selected time interval, and how much the steady flow rates are.
3. Under the selected flow rate conditions how much the producing pressure differential (the difference between formation pressure and flowing pressure) roughly is, and how much the ratio of the producing pressure to the formation pressure is.

The steady well test of gas wells is also called the back-pressure test, which is an important method for determining the deliverability of gas wells. Data obtained from

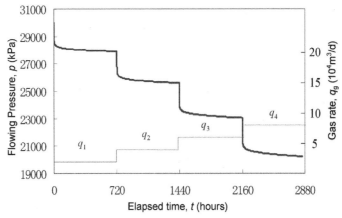

Figure 2.1 Variation of gas rate and pressure during steady well test.

a steady well test can be used to draw the relationship plot of flowing pressure (or producing pressure differential) and the flow rate, as shown in Figure 2.2a and 2.2b.

The result obtained from the steady well test in oil wells is the productivity index J_o or the specific productivity index J_{oR}, while it is generally the absolute open flow potential q_{AOF} and the inflow performance relationship curve (IPR) that are determined from the back-pressure test of gas wells.

1.1.2 Transient Well Test

The transient well test is a well test method widely used during the exploration and development of oil and gas fields. Its procedure involves changing the working system of oil, gas, or water wells, such as opening the well under the shut-in condition or instantaneously shutting in the well that is originally producing so as to cause redistribution of pressure in the reservoir and then measuring the variation of bottom-hole pressure during the whole process. Based on pressure variation data, combined with the flow rate and the properties of the oil and gas and the reservoir, the characteristic parameters of the tested well and the area influenced by the testing, that is, in the tested zone, are studied. These parameters include formation permeability k, flow coefficient kh/μ, formation pressure p_R, skin factor S, and characteristics of inner and outer boundaries.

Commonly used transient well test methods include pressure drawdown test, pressure buildup test, pressure fall-off test, injection well test, and multirate test.

1.2 Well Test Interpretation Models and Well Test Interpretation Type Curves

Just as described in Chapter 1, a well test analysis model simply means using physical or mathematical methods to reproduce the hydrocarbon flow process in the actual reservoir.

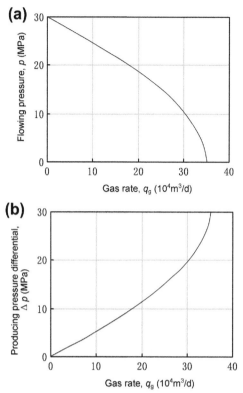

Figure 2.2 (a) The index curve (p_{wf} vs q_g) or IPR obtained from the back-pressure test of a gas well. (b) The index curve (Δp vs q_g) obtained from the back-pressure test of a gas well.

A physical model of a well test is the physical reproduction of flow in oil or gas layers. Such reproduction first offers a kind of physical description about the physical conditions of the oil and/or gas layers and fluid (oil, gas, and water) conditions in the layers, and it can also be achieved by material objects in a laboratory. For example, man-made or natural cores can be used to construct a formation model, which are saturated with oil and/or gas and/or water and pumped to simulate injection and withdrawal, and the pressure variation at individual points of which are simultaneously measured. There are some other physical models, such as the electric model and vertical pipe model.

A mathematical model of the well test, however, reproduces the flow process in a gas field using differential equations with proper internal and external boundary conditions (i.e., bottom-hole and formation conditions) and the initial condition of the well test. Also, it is the resolution of these differential equations that reflects the variation of pressure with time that reproduces the flow process. For well test analysis, the variation of pressure with time is usually expressed in graphical forms, that is, by a Cartesian plot of pressure vs time, semilog plot of pressure vs time, and log—log plot of pressure and its derivative vs time. Also, because the log—log plot can best reflect the flow characteristics, log—log plots are often used to represent well test analysis models in modern well test analysis.

A well test analysis model is used to interpret actual well test data, that is, to resolve the inverse problems; the log—log plot for this typical analysis model is called a well test interpretation type curve. There are different type curves for different formations. For example:

1. Different type curves classified by formation conditions: Homogeneous reservoir type curves, dual-porosity reservoir type curves, and dual-permeability reservoir type curves.
2. Different type curves classified by downhole conditions: Type curves for wells with wellbore storage and skin effects, type curves for wells with induced fractures, type curves for wells with partially perforated formations, and type curves for horizontal wells.
3. Type curves classified by external boundary conditions of reservoirs: Type curves for reservoirs with infinitely large external boundaries, type curves for reservoirs with various shapes of impermeable external boundaries or supply boundaries, and type curves for reservoirs with heterogeneous formation changes.

Many kinds of type curves can be obtained by mutually combining the formation conditions and internal/external boundary conditions listed earlier. These type curves were, as early as the end of the 1970s to the early 1980s, printed into colored plots as shown in Figure 2.3, but are seldom produced and applied any more now. Today, these type curves are all stored in well test interpretation software packages or, most of the time, are generated directly in real time by the computer programs in well test interpretation software when needed.

The meanings of ordinates p_D, p_D', and abscissa t_D/C_D in the type curve are explained in the following section.

1.3 Dimensionless Quantities and Pressure Derivative Curve in Well Test Interpretation Type Curves

Both ordinate and abscissa of the aforementioned type curves are dimensionless quantities or dimensionless variables. For example, dimensionless time is expressed as t_D or t_D/C_D, dimensionless pressure as p_D, and dimensionless pressure derivative as p_D'.

All physical quantities have their dimensions. For example, the dimension of length indicated by m is (L), the dimension of area indicated by m^2 is (L^2), the dimension of gas flow rate indicated by m^3/day is (L^3/t), and so on. However, there

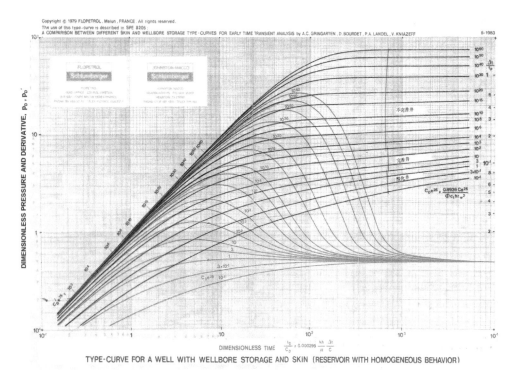

TYPE-CURVE FOR A WELL WITH WELLBORE STORAGE AND SKIN (RESERVOIR WITH HOMOGENEOUS BEHAVIOR)

Figure 2.3 An example of printed well test interpretation type curves.

are also some quantities that have no dimensions, such as gas saturation, S_g, porosity ϕ, and skin factor S.

For operational convenience, some quantities that have dimensions are often converted into dimensionless. For instance, time t is changed to t_D after being converted into dimensionless, and the expression is

$$t_D = \frac{3.6 \times 10^{-3}k}{\phi \mu C_t r_w} \cdot t \tag{2.1}$$

The right-hand side of Equation (2.1) contains not only time t but also permeability k, porosity ϕ, viscosity μ, total compressibility C_t, and effective wellbore radius r_w. If the unit of time is hr (hour), then the unit of k, ϕ, μ, C_t, and r_w in Equation (2.1) is exactly hr^{-1} after operation. Thus, t_D simply becomes a dimensionless quantity.

There is not only one method for defining dimensionless quantities of physical variables. Different formulae are often used, based on different needs, to define dimensionless quantities of the same quantity. For example, different dimensionless times are used in different formulae for different situations. In Equation (2.1), effective wellbore radius r_w can be replaced by the supply radius r_e, the area of oil or gas reservoir A, the half-length of induced fracture x_f, and so on for different purposes. Their expressions are as follow, respectively:

$$t_{De} = \frac{3.6 \times 10^{-3}k}{\phi \mu C_t r_e^2} \cdot t \tag{2.2a}$$

$$t_{DA} = \frac{3.6 \times 10^{-3}k}{\phi \mu C_t A} \cdot t \tag{2.2b}$$

$$t_{Dxf} = \frac{3.6 \times 10^{-3}k}{\phi \mu C_t x_f^2} \cdot t \tag{2.2c}$$

$$\frac{t_D}{C_D} = 2.261 \times 10^{-2}\frac{kh}{\mu C} \cdot t \tag{2.2d}$$

Other examples are pressure p and pressure derivative p', after being converted into dimensionless, they are expressed as p_D and p'_D:

$$p_D = \frac{0.5428kh\Delta p}{q\mu B} \tag{2.3a}$$

$$p'_D = \frac{0.5428kh\Delta p'}{q\mu B} \tag{2.3b}$$

It is especially important to note that the pressure derivative in well test analysis is defined as pressure differentiated by the logarithm of time. The characteristics of the pressure derivative are very critical for well test analysis, and the expression of pressure derivative is

$$\Delta p' = \frac{d\Delta p}{d\ln t} = \frac{d\Delta p}{dt} \cdot t \tag{2.4a}$$

Therefore,

$$p'_D = \frac{dp_D}{d\ln t_D} = \frac{dp_D}{dt_D} \cdot t_D \tag{2.4b}$$

or

$$p'_D = \frac{dp_D}{d\ln(t_D/C_D)} = \frac{dp_D}{d(t_D/C_D)} \cdot \frac{t_D}{C_D} \cdot \tag{2.4c}$$

Throughout this book hereinafter, the meaning of the pressure derivative written as either p_D' or $\Delta p'$ will always be defined as in Equation (2.4) and will not be explained repeatedly.

The introduction of dimensionless quantities brings many benefits. The greatest advantage is that the expression form of type curves can be simplified significantly. For the same type of formations, regardless of either the magnitude of their parameter values or the system of units being used in their expressions, they can share the same type curves so that such type curve offers universal suitability.

The method of representation of dimensionless quantities is described specially in Appendix 4.

1.4 Wellbore Storage Effect and its Characteristics on Type Curves

1.4.1 Implications of Wellbore Storage Effect

At the very beginning of starting up or shutting in an oil or a gas well, the wellhead flow rate is not equal to the sandface flow rate. Take a gas well as an example: the wellbore is filled with compressed natural gas before opening the well. At the beginning of starting up the well, it is the expansion effect of natural gas in the wellbore that drives the gas to flow out of the wellhead and so the gas flow rate is q_g (m^3/day), but meanwhile the sandface flow rate is still 0. As the gas amount flown out increases, the bottom–hole pressure decreases and a difference is created between bottom–hole pressure and formation pressure so that the sandface flow rate increases from 0 gradually to the level of wellhead flow rate q_g (m^3/day) ultimately, as shown in Figure 2.4.

The situation is just the opposite during the shut-down process. If surface (i.e., wellhead) shut down is adopted, the surface flow rate will drop to 0 immediately when the valve in the Christmas tree is closed. However, because more natural gas is needed for compression of the gas in the wellbore due to the existing pressure difference, there is still

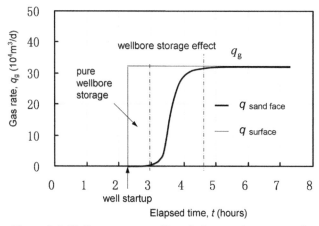

Figure 2.4 Wellbore storage effect during starting up a well.

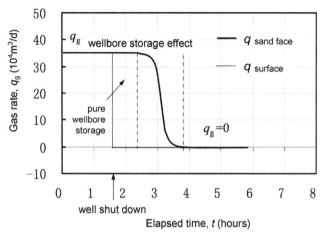

Figure 2.5 Wellbore storage effect during shut-down process.

natural gas that continues to flow from the formation to the wellbore until the balance between bottom-hole pressure and formation pressure is reached. Only at this time will the sand surface flow rate become 0. This process is as shown in Figure 2.5.

The seriousness of the wellbore storage effect is indicated by wellbore storage coefficient C, which is defined as

$$C = \frac{dV_w}{dp} \approx \frac{\Delta V_w}{\Delta p}, \tag{2.5}$$

where C is wellbore storage coefficient, m^3/MPa; ΔV_w is change of volume of natural gas stored in the wellbore, m^3 (the volume of natural gas under standard conditions); and Δp is change of wellbore pressure, MPa.

The physical significance of the wellbore storage coefficient is the capability of storage fluid of the wellbore through compression of the fluid in it, or the capability of discharging fluid by expansion of the fluid in the wellbore due to pressure decreasing under the condition that the wellbore is filled with natural gas or other fluids. Or, more specifically speaking, it refers to the increased fluid volume (the surface volume) in the wellbore to increase one unit of bottom-hole pressure during the fluid compression process after shutting in.

Obviously, the influence of the wellbore storage effect has retarded and disturbed the process of understanding the formation by observing the change of bottom-hole pressure. In order to reduce or eliminate the influence of wellbore storage, tools and methods for downhole shut in have been proposed and developed. When such tools are used, however, special requirements are imposed on the wellbore structure and operation procedure.

For pure gas wells, the formula for calculating the C value is

$$C = \frac{q_g B_g}{24 \Delta p} \cdot \Delta t \qquad (2.5a)$$

$$B_g = \frac{p_{SC} Z \overline{T}}{\overline{p} T_{SC}}$$

where C is the wellbore storage coefficient, m^3/MPa; q_g is natural gas flow rate, $10^4\,m^3/$ day; B_g is volume coefficient of natural gas; $\overline{p}$ and $\overline{T}$ are the average pressure (MPa) and the temperature (K) in the wellbore; p_{SC}, p_{SC} are the standard pressure (MPa) and standard temperature (K); Z is the deviation factor of real gas; Δp is the pressure decrease from the moment of starting up the well to the moment of t during the pure wellbore storage period, (MPa); and Δt is flow time during the pure wellbore storage period, (hr).

1.4.2 Order of Magnitude of Wellbore Storage Coefficient

The wellbore storage effect does not come from the influence of formation, but it has much to do with well test interpretation. Correct analysis of the influence of the wellbore storage effect can ensure obtaining correct formation parameters. Conversely, if the obtained wellbore storage coefficient cannot match well with actual conditions, such as the wellbore structure and test procedures, the obtained formation parameters are also very likely wrong. Table 2.1 shows the range of order of magnitude of a wellbore storage coefficient under different testing conditions.

If during the shut-in process of a gas well the load water generated at the bottom hole or the retrograde condensation phenomenon appears in condensate gas wells, the wellbore storage coefficient may not be a constant, and the so-called variable wellbore storage effect will take place. The C value may decrease, for example, in the event of retrograde condensation, and may also increase, for example, when the water production heightens the level of the liquid column in the wellbore.

Table 2.1 List of Estimated Order of Magnitude of Wellbore Storage Coefficient

Category	Order of Magnitude of C Value (m³/MPa)	Wellbore Structure and Test Procedures
Very high	>10	Extremely deep gas wells, normal or fairly low formation pressures, surface shut in
High	1−10	Deep gas wells, normal formation pressure, surface shut in
Medium	0.1−1	Shallow gas wells, deep gas wells with super-high formation pressure, surface shut in
Low	<0.1	Gas wells with downhole shut in

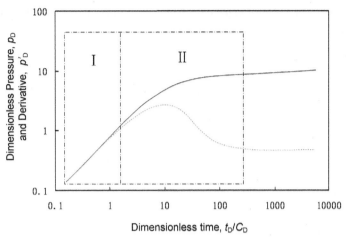

Figure 2.6 Characteristics of wellbore storage effect sections on type curve.

In these cases, the variable wellbore storage options in well test software should be used for type curve match analysis.

1.4.3 Characteristics of Wellbore Storage Effect on Well Test Interpretation Type Curves

The wellbore storage effect appears during the early stage on the type curve, as shown in Figure 2.6.

The wellbore storage effect section can be divided into two sections. Section I is the pure wellbore storage effect. This section indicates that during either the starting-up or the shutting-in process, the wellhead flow rate is nearly totally dominated by the compressibility of fluid in the wellbore; in other words, the sandface flow rate is still close to 0 after the well is just starting up; while the sandface flow rate after shutting in has not changed yet, that is, the sandface flow still keeps its original rate.

In this period, the slope of the type curve of pressure difference vs time equals 1, that is, $\dfrac{\Delta \log p_D}{\Delta \log(t_D/C_D)} \approx 1$; and the slope of the type curve of pressure derivative vs time also equals 1, that is, $\dfrac{\Delta \log p'_D}{\Delta \log(t_D/C_D)} \approx 1$.

Section II is the transition one. In this section, the sandface flow rate gradually approaches the wellhead flow rate; in other words, it increases gradually and reaches the wellhead flow rate eventually after the well is started up (Figure 2.4) or decreases gradually to 0 after shutting in (Figure 2.5).

When the sandface flow rate becomes equal to the wellhead flow rate, the flow that reflects formation conditions is completely established and radial flow starts.

1.5 Several Typical Flow Patterns of Natural Gas and their Characteristics on Interpretation Type Curves

1.5.1 Radial Flow

Radial flow is the most frequently treated transient flow state.

Assuming that the gas zone is a homogeneous, uniform thickness and that the oil/gas well has drilled the entire oil/gas zone, then after opening the well and starting up production, fluid in the reservoir will flow from the perimeter to the bottom hole along radial directions in the horizontal plane. On any horizontal plane perpendicular to the wellbore in the reservoir, the flow lines are always ray beams that converge to the wellbore from all directions; the equal pressure curves on the horizontal plane in the reservoir are all concentric circles and their center is the wellbore axis, as shown in Figure 2.7. Such flow is called radial flow. The corresponding position of the radial flow section on the type curves is as shown in Figure 2.8.

The major characteristics of radial flow on type curves include that the pressure derivative curve segment is a horizontal straight line and that the value of ordinate of this horizontal line on dimensionless coordinates is exactly 0.5.

Radial flow is a kind of typical transient flow state. As time elapses, the equal pressure curves are changing continuously, even though the flow lines are always ray beams that flow toward the bottom hole directly. In fact, once the transition flow starts, even during the early pure wellbore storage effect or transition periods, the flow curves are also always ray beams directing toward the bottom hole, but the equal pressure curves (i.e. isobaric lines) at those periods are relatively dense concentric circles in the area near the well. When the wellbore storage effect is finished, the flow regime changes to "radial flow."

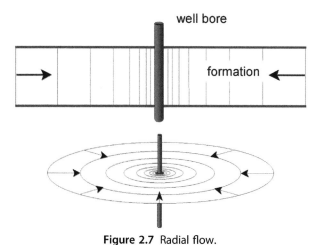

Figure 2.7 Radial flow.

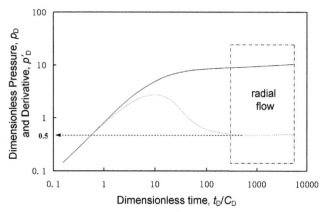

Figure 2.8 Characteristic line of radial flow on type curve.

1.5.2 Steady Flow

For an oil well producing at a constant flow rate, if the pressure distribution in the area around the well remains constant during its late stage, the flow condition is called steady flow. The start time of steady flow is denoted as t_{ss}; when $t > t_{ss}$, $\dfrac{\partial p}{\partial t} = 0$ at any point in the oil reservoir.

For natural water drive oil reservoirs with very large water bodies around them or for oil reservoirs being exploited via water flooding, the oil/water boundary may likely become a "constant pressure boundary" that allows the well producing in or approximately reaching the steady flow condition. The pressure distribution in the reservoir is as shown in Figure 2.9.

It should be pointed out here that such a steady flow state will generally not appear in gas wells, nor will the so-called "constant pressure boundary" exist.

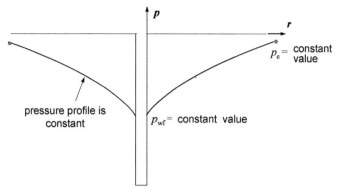

Figure 2.9 Pressure distribution during steady flow.

1.5.3 Pseudo-Steady Flow

The so-called pseudo-steady flow is, in fact, a kind of transient flow. When an oil or gas well located within a closed block is producing at a constant flow rate, the pressure drop will spread to all surrounding boundaries eventually in the late stage. Since the pressure at every point in the closed block will decrease at the same speed, the flow is pseudo-steady state flow, as shown in Figure 2.10.

If the start time of pseudo-steady state flow is t_{ps}, when $t > t_{ps}$, $\dfrac{\partial p}{\partial t} = $ constant for every point in the reservoir. It can be seen from Figure 2.10 that the shape of pressure distribution curves at this period remains unchanged and that the pressure distribution curves at different times are parallel with each other, but their heights are different.

Figure 2.11 shows the characteristic curve of pressure change during the pseudo-steady flow period. Within this period, the pressure derivative curve shows a unit-slope straight line for pressure drawdown, whereas on the pressure buildup curve, the pressure derivative decreases rapidly and sharply.

1.5.4 Spherical Flow and Hemispherical Flow

When a well was partially perforated at the top or in the middle of a thick layer, the flow will be as shown in Figures 2.12 and 2.13.

Figures 2.12 and 2.13 show the profiles that cut cross the well longitudinally, while transverse profiles still show radial flow patterns similar to that in Figure 2.7.

Both spherical flow and hemispherical flow are short-lived flow stages after well opening and producing startup of "partially perforated" wells. As the production time elapses, the flow far from the well will gradually become parallel with the top and the bottom of the reservoir, and the influence of spherical flow or hemispherical flow on the bottom-hole pressure will gradually disappear.

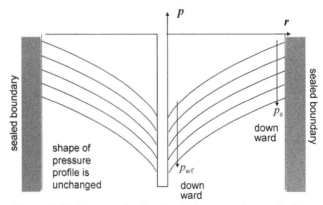

Figure 2.10 Pressure distribution during pseudo-steady flow.

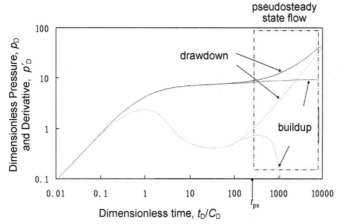

Figure 2.11 Characteristic curve of pseudo-steady flow on type curves.

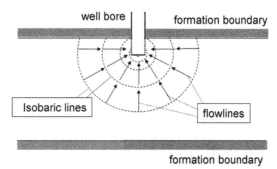

Figure 2.12 Illustration of hemispherical flow.

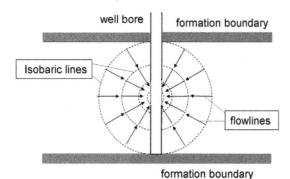

Figure 2.13 Spherical flow and hemispherical flow.

The characteristic line of spherical flow or hemispherical flow on the well test analysis type curves is as shown in Figure 2.14. It is a straight line with a slope of $-1/2$. The section before it is the partial-layer radial flow, and the section after it is the whole-layer radial flow, as shown in Figure 2.15.

1.5.5 Linear Flow

Linear flow often takes place in formation. Linear flow means that the directions of flow lines basically parallel each other on the formation plane in a certain local area so that the equal pressure curves are straight lines and form a plane. The major formations and completion conditions causing linear flows include the following.

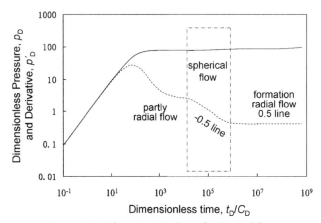

Figure 2.14 Characteristic line of spherical flow.

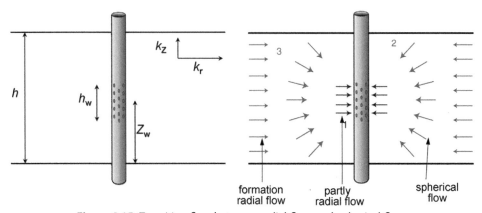

Figure 2.15 Transition flow between radial flow and spherical flow.

Linear Flow Caused by Parallel Impermeable Boundaries in the Formation such as Graben Formed by Faults and Narrow Channel Sandstone Formed by Sedimentation of Fluvial Facies and so on as shown in Figures 2.16 and 2.17

The linear flow in the channel reservoir mentioned earlier is shown in Figure 2.18.

The characteristic line of linear flow on well test interpretation type curves is as shown in Figure 2.19: the pressure derivative in this section is a half-unit slope straight line. However, it should be noted that such linear flow, because it is caused by the boundary influence, appears in the late time region of the transient well test curve.

Linear Flow Caused by Hydraulically Created Fracture

For a well drilled in deep reservoir being fractured hydraulically, one single vertical fracture intersecting the well will usually be created at the bottom hole, the fracture half-length is x_f, and its height is nearly the same as the formation thickness. Because the

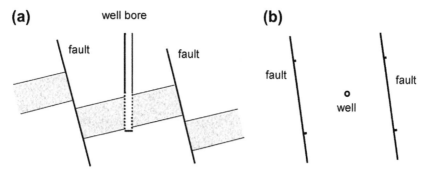

Figure 2.16 Grabens formed by faults. (a) Longitudinal diagram. (b) Structural plane diagram.

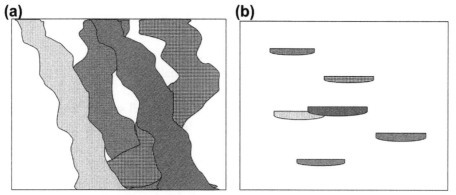

Figure 2.17 Illustration of banked sandstone formed by sedimentation of fluvial facies. (a) Plane view of channel sandstone. (b) Longitudinal profile of channel sandstone.

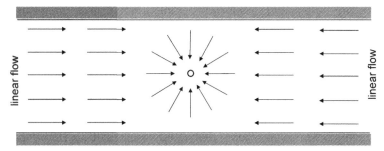

Figure 2.18 Linear flow in channel reservoir.

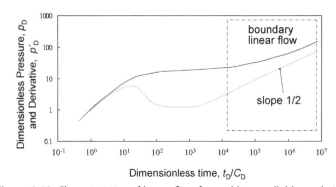

Figure 2.19 Characteristics of linear flow formed by parallel boundaries.

permeability inside the fracture is generally very high, it is called an "infinite conductivity fracture." The flow is as shown in Figure 2.20.

As shown in Figure 2.20, if the induced fracture is sufficiently long, linear flow perpendicular to the fracture face will appear near the fracture; as shown in Figures 2.20a and 2.20b, the flow lines are parallel, and therefore the equal pressure surface should be a plane. A characteristic diagram of linear flow on type curves is shown in Figure 2.21.

The characteristics of linear flow are as follow:

1. Pressure derivative curve is a straight line with a half-unit slope.
2. Pressure curve is also a straight line with a half-unit slope.
3. The ratio of the ordinate difference between the parallel straight lines, that is, the pressure and pressure derivative curves, to the length of one log cycle of ordinate is lg2 = 0.301.
4. Because the induced fracture is closely connected with the wellbore, such linear flow appears immediately following the wellbore storage flow section. This is different from the linear flow formed by parallel boundaries shown in Figure 2.19 in occurrence time.

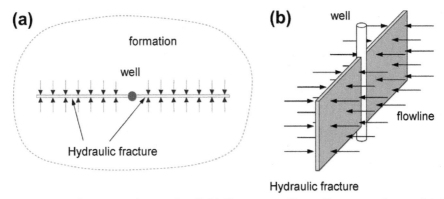

Figure 2.20 Linear flow near a fractured well. (a) Plane view of linear flow near a fractured well. (b) Longitudinal profile of linear flow near a fractured well.

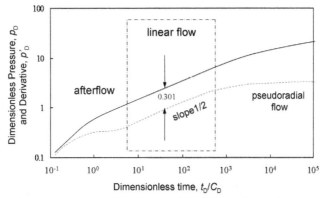

Figure 2.21 Linear flow characteristics of a hydraulic fracture.

Linear Flow Caused by Horizontal Well

After a horizontal well that drilled through the middle position of the reservoir has been started up, its flow can be divided into three major stages:

1. Radial flow perpendicular to the formation plane.
2. Linear flow perpendicular to the wellbore and parallel with the top and the bottom boundaries of the reservoir.
3. Pseudo-radial flow.

Illustration of its flow regime is as shown in Figure 2.22, and the corresponding sections of its characteristic curve on the well test type curves are as shown in Figure 2.23.

Conditions for the presence of a linear flow section in a horizontal well include:

1. The horizontal well section is sufficiently long.
2. The entire horizontal well section is in the effective reservoir.

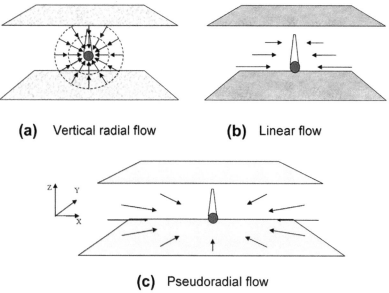

Figure 2.22 Flow regimes of a horizontal well.

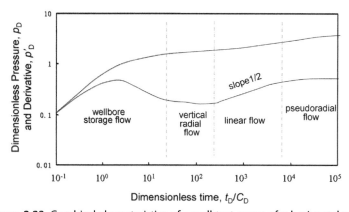

Figure 2.23 Graphical characteristics of a well test curve of a horizontal well.

3. The reservoir, which was drilled through by the horizontal well section, is generally homogeneous.
4. The test duration is sufficiently long.

The characteristics of the linear flow in horizontal wells are generally consistent with the situation of induced fracture as discussed earlier. Because linear flow exists after vertical radial flow in horizontal wells, the difference is that its characteristics are likely something between the two situations described earler (i.e., linear flow caused by parallel impermeable boundaries and linear flow caused by hydraulic fracture); in cases of a very thick layer, in particular, it is more like the characteristics of linear flow caused by parallel channel boundaries.

1.5.6 Pseudo-Radial Flow

Pseudo-radial flow is a kind of radial flow appearing at later time. For fractured wells, if the fracture half-length is not very long, pseudo-radial flow will appear during the extended test, as shown in Figure 2.24. For those horizontal wells whose horizontal sections are not very long, pseudo-radial flow may also appear during late time, as shown in Figure 2.25.

As seen from Figures 2.24 and 2.25, when looking from a position far from the well, either the horizontal section or an induced fracture can always be regarded as a vertical well with an expanded influence range. Therefore, approximate radial flow is observed

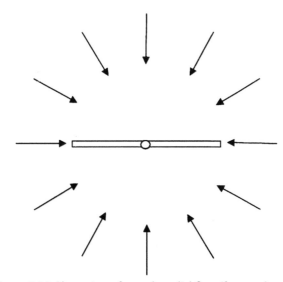

Figure 2.24 Plane view of pseudo-radial flow (fractured wells).

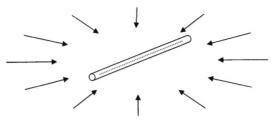

Figure 2.25 Plane view of pseudo-radial flow (horizontal wells).

from a position far from the well. For such radial flow, its pressure derivative should approach as a horizontal straight line at later time as shown in Figures 2.21 and 2.23.

1.5.7 Flow Condition in Formation Having been Improved or Damaged

Because homogeneity characteristics only exist in a certain local range in any formation, whereas heterogeneity characteristics are absolute and universal, it happens very often that the flow condition of hydrocarbon in formation becomes better or worse to a certain extent, for example:

1. One or more impermeable boundaries are present near the well, which will obstruct the flow (Figure 2.26).
2. Deteriorated zones with lower permeability are present near the well, which will retard the flow of the fluid (Figure 2.27).
3. Better zones with higher permeability are present in the periphery of the reservoir, which forms a kind of composite reservoir and will reduce the pressure gradient for the flow (Figure 2.28).

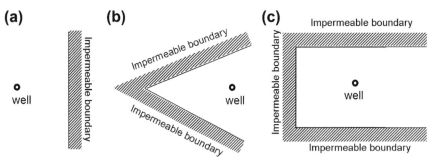

Figure 2.26 Impermeable boundaries obstructing gas flow. (a) Single impermeable boundary. (b) Sharp-angled impermeable boundary. (c) Combination of multiple impermeable boundaries.

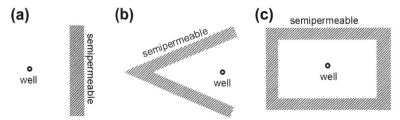

Figure 2.27 Flow barrier (semipermeable) boundaries retarding gas flow. (a) Single flow barrier boundary. (b) Sharp-angled flow barrier boundary. (c) Combination of multiple flow barrier boundaries.

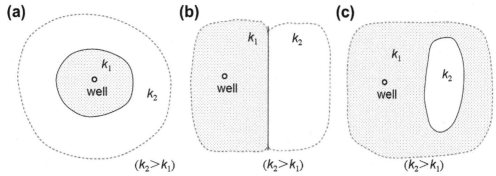

Figure 2.28 Areas with higher permeability are present in the periphery of the reservoir that betters gas flow. (a) Radial composite reservoir, permeability of outside area is higher than that of inside area. (b) Linear composite reservoir, permeability of other side area is higher than that of the area the well locates. (c) Permeability of a local part of reservoir is higher than that of other area of reservoir.

4. Deteriorated zones with lower permeability are present in the periphery of the reservoir, which also forms a composite reservoir and will increase the pressure gradient for the flow (Figure 2.29).
5. Oil ring or edge water is present around the well located at the gas cap position, which will restrict the gas flow (Figure 2.30) and so on.

Figures 2.26, 2.27, 2.29, and 2.30 can be generally classified to the situation that flow becomes deteriorated due to a worse perimeter or due to the flow barrier of impermeable boundaries; only Figure 2.28 shows the situation of a better flow condition in the periphery improving the gas flow.

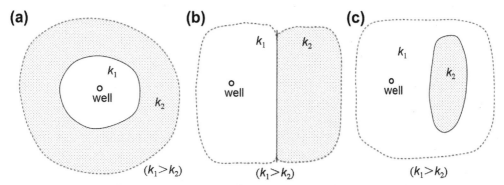

Figure 2.29 Areas with deteriorated permeability are present in the periphery of the reservoir that hinders gas flow. (a) Radial composite reservoir; permeability of outside area is lower than that of inside area. (b) Linear composite reservoir; permeability of other side area is lower than that of the area the well locates. (c) Permeability of a local part of reservoir is lower than that of other area of reservoir.

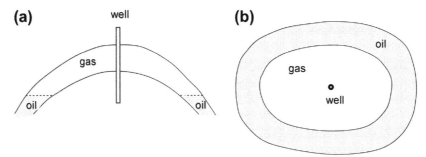

Figure 2.30 Oil ring or edge water is present around the gas well that restricts gas flow. (a) Longitudinal plan. (b) Plane view.

When flow is restricted due to deteriorated peripheries, the pressure gradient at the position in formation where flow is restricted will increase so that the derivative curve in the type curves will deviate from the horizontal straight line of radial flow and go upward.

When flow is bettered due to improved flow condition of peripheries, the pressure gradient at these positions will decrease so that the pressure derivative curve will decline.

Figure 2.31 shows the characteristics on type curves when flow is restricted or improved.

There are many causes leading to restricted flow, but all of them show similar characteristics on type curves. Analysis combined with geological characteristics must be conducted in order to identify the causes of its occurrence.

1.6 Skin Effect, Skin Factor and Equivalent Borehole Radius

For an oil or a gas well that has drilled and penetrated the formation and has been cemented and completed normally, its pressure distribution is as shown by the solid line

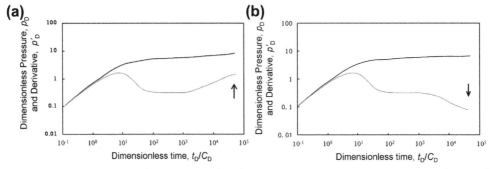

Figure 2.31 Characteristics of type curves when flow condition in periphery is restricted or bettered. (a) When flow is restricted, the pressure derivative curve goes upward. (b) When flow is bettered, the pressure derivative curve declines.

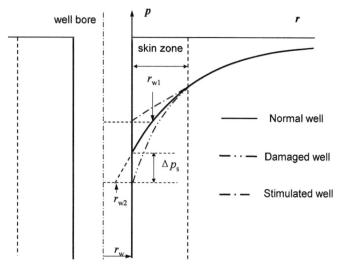

Figure 2.32 Pressure distribution in skin zone of undamaged and damaged wells.

in Figure 2.32. However, due to some causes, such as mud invading into the formation during drilling, cement invading into the formation during cementing, and partial penetration, the formation near the wellbore would have been damaged to a certain degree so that after starting up of the oil or gas well, the pressure gradient will increase, as shown by the dash-dot-dot lines in Figure 2.32.

This damaged zone is called the skin zone. An additional pressure drop Δp_s exists that is caused by the damage when oil or gas flows through the skin zone.

Δp_s has different meanings for formations with different permeability. For example, when the value of permeability k is very low, the producing pressure differential itself would be very large; if there is a fairly small Δp_s, the production of the gas well would not be affected dramatically. Conversely, if the k value is very high, for the same Δp_s, the gas flow rate may change by several times. Therefore, the true degree of damage cannot be shown until the Δp_s has been made dimensionless. This dimensionless coefficient is called the skin factor, symbolized by S and defined as

$$S = \frac{542.8kh}{qB\mu} \cdot \Delta p_S. \tag{2.6}$$

It is seen that
- $S = 0$ when $\Delta p_s = 0$, the well is not damaged
- $S > 0$ when $\Delta p_s > 0$, the well has been damaged
- $S < 0$ when $\Delta p_s < 0$, the bottom hole has been improved

For a damaged well, it is just like the wellbore has been reduced, therefore, an equivalent wellbore radius is introduced:

$$r_{we} = r_w \times e^{-S} \qquad (2.7)$$

The meaning of r_{we} can be seen from Figure 2.32, and

- when $r_{we} = r_w$, that is, $S = 0$ and $\Delta p_s = 0$, the well is not damaged
- when $r_{we} < r_w$, that is, $S > 0$ and $\Delta p_s > 0$, the well has been damaged
- when $r_{we} > r_w$, that is, $S < 0$ and $\Delta p_s < 0$, the well condition has been improved

1.7 Radius of Influence

Once an oil or a gas well is started up and starts flowing, the bottom-hole pressure begins to drop. At the same time, the pressure drop will extend gradually toward the deeper parts of the formation to generate a pressure drop funnel; as time elapses, this pressure drop funnel will expand continuously, as shown in Figure 2.33.

As shown in Figure 2.33, according to the definition of radius of influence, at the moment t_i, the boundary of the pressure drop funnel has extended to position r_i from the well. In other words, everywhere positioned r from the well in the formation with $r < r_i$ has already been disturbed by the producing well; while for those portions where $r > r_i$, the formation has not been disturbed, and pressure has not changed at all.

According to Reference [11], the formula for calculating the radius of influence is

$$r_i = 0.12\sqrt{\frac{kt}{\mu \phi C_t}}. \qquad (2.8)$$

For example, when $k = 10$ mD, $\mu = 0.02$ mPa·s, $\phi = 0.1$ and $C_t = 1.674 \times 10^{-2}$ MPa^{-1}, the values of the radius of influence r_i at different time t determined by Equation (2.8) are listed in Table 2.2.

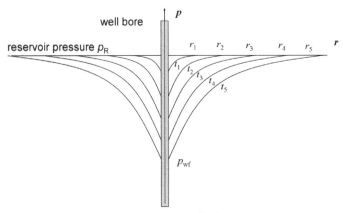

Figure 2.33 Radius of influence.

Table 2.2 Example of Relationship between Radius of Influence and Flowing Time

Opening time t, hr	50	100	150	200	250	300
Radius of investigation r_i, m	463.7	655.8	803.2	927.4	1036.9	1135.9

However, this concept still contains some meanings to be quantified:

1. The so-called radii of influence should refer to those positions where the pressure has been affected and disturbed. Wherein, the criteria for discrimination whether or not the pressure has been disturbed should be based on whether or not this pressure disturbance can be detected. However, the aforementioned definition has not quantified how much exactly this "detectable pressure value" is.

2. It is known from pressure analysis that, for identical formations, because the producing well serves as the only source of disturbance, its flow rate q is an important factor affecting the amount of disturbance. As seen from Equation (2.8), however, the radius of influence has nothing to do with the flow rate q at all. It means that the factor of flow rate q has failed to be taken into account.

3. The "detectable quantity" itself is also changing constantly as the testing instrumentations are improved and modernized constantly and their resolution and accuracy are enhanced continuously.

By simulation analysis using well test interpretation software, it is very easy, under the aforementioned parametric conditions and regardless of the duration of well producing, to determine the value of pressure disturbance on the front edge of its radius of influence:

- when the gas flow rate $q_g = 10 \times 10^4$ m^3/day, the disturbing pressure on the front edge is $\Delta p = 0.1128$ MPa
- when the gas flow rate $q_g = 1 \times 10^4$ m^3/day, the disturbing pressure on the front edge is $\Delta p = 0.01128$ MPa

The pressure drop of 0.1128 MPa is already a considerable quantity; the pressure change of 0.01128 MPa, however, is merely the minimal recordable pressure change by early mechanical pressure gauges. Of course, compared to the electronic pressure gauges widely used today that offer a resolution up to 0.00007 MPa (0.01 psi), 0.01128 MPa is already a very large quantity.

For this reason, the "radius of influence" is only a relatively vague concept that qualitatively describes the expansion of the range of influence and cannot be exactly used to conduct quantitative analysis.

Moreover, the "radius of influence" should never be confused with the "radius of investigation." Here is an obvious example: if information about deep positions in a reservoir, for example, the presence of fault boundary with distance L_b from the well, is to be obtained through a transient well test, the required time should not be

$$t_1 = \frac{\mu \phi C_t}{0.0144k} \cdot L_b^2.$$

However, it should be time t_2 that takes the influence of the mirror well used to substitute the fault presence arriving at the tested well into account:

$$t_2 = \frac{\mu \phi C_t}{0.0144k}(2L_b)^2 = 4t_1$$

In other words, if the time when the radius of influence r_i expends to fault L_b is t_1, four times of t_1 must be spent to make the arrival of the signal of the fault in order to detect the presence of the fault, that is, to investigate the information about the fault of L_b from the well. Furthermore, even at the moment $4 \times t_1$, such information just barely begins to appear.

There is a simulated example that clearly shows the relationship between pressure behavior and type curve characteristics. The geological condition of the example is: a producing well is drilled at one end of a rectangular closed block, as shown in Figure 2.34. The bottom-hole pressure behavior is tested after the well starts production from the stationary state.

The plot of behavior characteristics of the example is divided into two parts:
- One part indicates the spread of pressure drop funnel and how it spreads.
- The other part uses an arrow to point out the corresponding position and characteristics of test data on the log–log plot.

a. The wellbore storage flow section

At the instant of starting up the gas well, the wellhead flow rate reaches q_g, but the production of gas is caused primarily by the expansion of high-pressured gas in the wellbore. Thereafter, as the wellbore pressure decreases, the gas in the formation begins to flow into the wellbore, thus creating a pressure drop funnel around the well.

The corresponding characteristic line on the type curve is a unit-slope straight line at the beginning and then becomes a transition curve whose derivative shows a peak value and then declines (Figure 2.35).

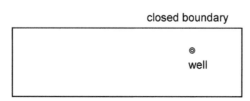

Figure 2.34 Position of a gas well within a rectangular closed reservoir.

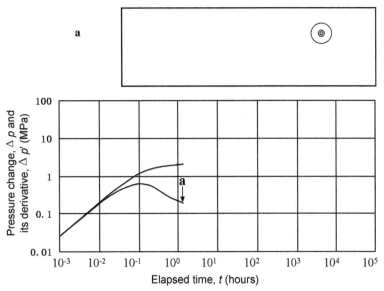

Figure 2.35 Pressure drop funnel and the type curve characteristics in wellbore storage flow section.

b. Radial flow section

As the test time elapses, the flow enters the radial flow section. In this period, the pressure drop funnel has already expanded to the first impermeable boundary, but the boundary influence has not got enough time yet to return to the well. Therefore, the flow constantly maintains the radial flow state.

On the characteristic curve, the pressure derivative is a horizontal straight line and shows typical radial flow characteristics (Figure 2.36).

c. First boundary reflection section

After the range of influence of the pressure drop funnel reaches the first boundary, the pressure influence will reflect toward the opposite direction and finally return back to the well. In this period, the bottom-hole pressure will appear as the boundary reflection.

On the characteristic curve, the pressure derivative will deviate from the horizontal straight line of radial flow and rise upward. On the ordinate direction, the range of a such rise is about 0.5 cycles of the ordinate scale and approaches another horizontal straight line (Figure 2.37).

d. First boundary reflection keeping section

In this section the pressure drop funnel continues to expand and arrives at the second and third boundaries. However, except for the influence of the first boundary, subsequent boundary influences have not yet reached at the well.

On the characteristic curve, the pressure derivative line stays on another fairly high horizontal straight line (higher than the radial flow line by 0.5 cycles) and still shows the influence of a single impermeable boundary (Figure 2.38).

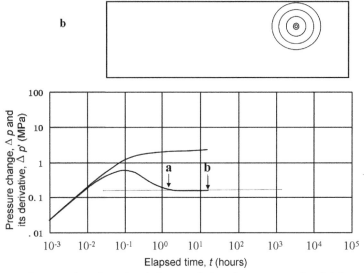

Figure 2.36 Pressure drop funnel and characteristics of type curve of radial flow section.

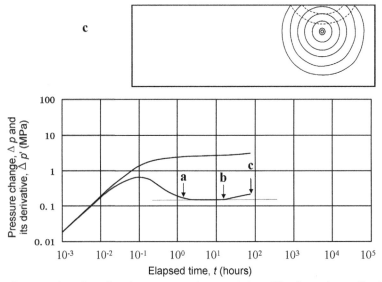

Figure 2.37 Pressure drop funnel and type curve characteristics of first boundary reflection section.

e. Three boundary reflection section

As seen from the well location, because the second and third boundaries are equidistant to the tested well, the pressure disturbance reaches the second and third boundaries simultaneously, then reflects backward simultaneously, and finally reaches the bottom hole of the tested well. In this period, the bottom–hole pressure will be

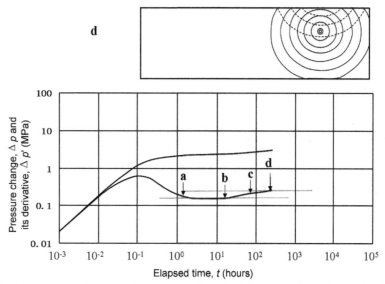

Figure 2.38 Pressure drop funnel and type curve characteristics of first boundary reflection keeping section.

affected by disturbance once again. On the characteristic curve, the pressure derivative deviates from the second horizontal straight line created by the first boundary influence and goes upward once again to represent the further influence of the boundaries (Figure 2.39).

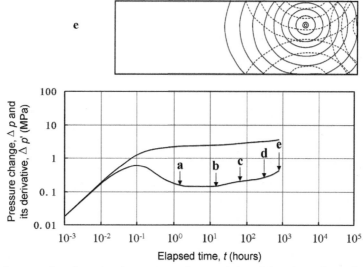

Figure 2.39 Pressure drop funnel and type curve characteristics of three boundary reflection section.

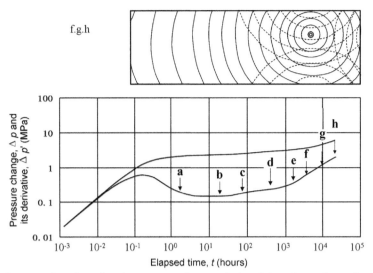

Figure 2.40 Pressure drop funnel and type curve characteristics of three boundary reflection keeping sections.

f, g, h. Three-boundary sustained reflection keeping section

As seen from the flow regimes, the influences of the first, second, and third boundaries will be kept continuously. Although the pressure drop funnel will gradually arrive at the fourth boundary, it will not generate disturbing influences on the bottom-hole pressure of the tested well during a short period of time. As seen from Figure 2.40, if data are acquired continuously, the test time when the influence of three boundaries appears has already approached 8000 hours, or 333 days. For most situations, a test with this long duration is already very difficult to run.

The characteristic of the pressure derivative curve is that at the late time it is approaching the half-unit slope straight line gradually.

If the boundary configuration and corresponding dimensions of the rectangular reservoir can be basically confirmed by data of geological and production testing, the change of pressure can be further predicted on this basis. Strictly speaking, however, the reservoir type and parameters cannot be confirmed ultimately and comprehensively only based on acquired pressure data alone (Figure 2.40).

i. Section when the reflection of three boundaries is kept but the reflection of the fourth boundary has not yet returned to the well

Within a fairly long period of time after the pressure drop funnel arrived at the furthermost fourth boundary, the boundary reflection has not yet returned to the well. During this period of time, the pressure derivative curve basically keeps the half-unit slope straight line as seen from Figure 2.41, this period lasts very long, ranging approximately between 10^4 and 10^5 hours, that is, about 10 years. Of course, it is no longer the duration that transient well test can bear, and it can only be estimated by

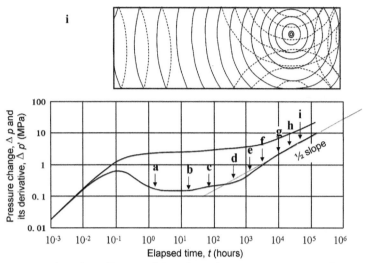

Figure 2.41 Situation when the reflection of three boundaries is kept but the reflection of the fourth boundary has not yet reached the well.

a theoretical model. The appearance of a pressure disturbance of the fourth boundary will lead the bottom–hole pressure to enter a pseudo-steady state in a closed block.

On the characteristic curve, the pressure derivative curve will eventually be affected by the closed boundary and goes upward further, gradually becoming a unit-slope straight line (Figure 2.41).

j. Closed boundary reflection section

In this flow section, the pressure drawdown of the entire block has entered the pseudo-steady state. The pressure everywhere inside the block declines uniformly.

On the pressure characteristic curve, the pressure derivative curve becomes a unit-slope straight line. At this moment, the time coordinate has extended to 10^5-10^6 hours, that is, it has extended from 11.4 to 114.2 years. This has already exceeded the development life of the oil or gas field. Offering no practical significance, it can only be discussed theoretically (Figure 2.42).

It can be seen clearly from this example what the pressure behavior characteristics of a well located in a closed block looks like, if this well can keep producing at a constant flow rate and is not disturbed by any other factors. This may be helpful for us to understand the radius of investigation in oil and gas fields—the detection and analysis of impermeable boundaries.

1.8 Laminar Flow and Turbulent Flow

Two different states, laminar flow state and turbulent flow state, often happen during the flow of viscous fluids, including natural gas, oil, and water. A laminar flow state means

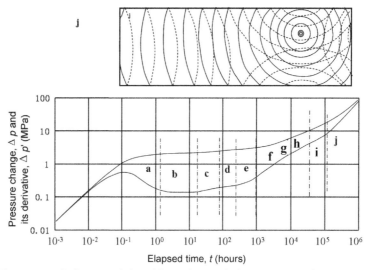

Figure 2.42 Reflection of closed boundary and characteristics of type curve.

that the flowing velocity at every individual point within the flowing space has a stable distribution profile, whereas the velocity distribution at individual points in the turbulent flow state is usually disordered and chaotic, without a stable velocity distribution profile. Figures 2.43 and 2.44 show schematically the velocity distribution situations of laminar flow and turbulent flow in a one-dimensional conduit flow.

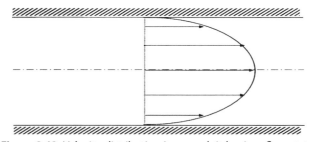

Figure 2.43 Velocity distribution in a conduit laminar flow state.

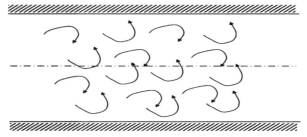

Figure 2.44 Velocity distribution in a conduit turbulent flow state.

The forming of laminar flow state and turbulent flow state is related to various parameters, such as the viscosity μ and velocity v of the fluid, the cross-sectional area S that the fluid passes through, and the roughness of the internal surface of the conduit. The characteristic parameter group comprised parameters μ, v, and S is called the "Reynold's number." When the Reynold's number is less than a certain critical value, the flow will show a laminar state, while when such a critical value is exceeded, the flow will change into turbulent flow. There is greater frictional resistance in turbulent flow than in laminar flow because it causes more energy loss.

The flow in porous media features something similar to the transformation from laminar flow to turbulent flow in conduit flow; moreover, the flowing space in this case is much more complicated than conduit flow. This is because the pore structure itself is very complicated; not only is any unique characteristic size unavailable, but also the diameter of the pores that the fluid passes through during flowing is changing. Therefore, it is generally believed that flow in the formation shows a laminar state at low velocities and conforms to Darcy's law; as the flowing velocity increases, the influence of inertia flow is first added into the flow, forcing the flow to deviate from Darcy's law. As the flowing velocity increases further, it will gradually transform into turbulent flow.

The frictional resistance coefficient and Reynold's number during flow defined by Cornell and Katz [14] are

Frictional resistance coefficient:

$$f_{CK} = \frac{64\Delta p}{\beta \rho v^2 \Delta x} \tag{2.9}$$

Reynolds number:

$$R_{eCK} = \frac{\beta k \rho v}{1.0 \times 10^9 \mu} \tag{2.10}$$

The relationship between them both obtained from experiments is as shown in Figure 2.45.

As can be seen from Figure 2.45,

1. When the Reynold's number $R_{eCK} < 1$, that is, the flowing velocity is low, an approximate linear relationship exists between f_{CK} and R_{eCK}. In other words, a direct proportion relationship exists between pressure gradient $\frac{\Delta p}{\Delta x}$ and flowing velocity v. This means that the flow here conforms to Darcy's law and the flow features a laminar flow state.

2. When the Reynold's number $R_{eCK} > 1$, experimental data points deviate from the straight line. The influence of inertia flow presents in these cases. The flow at that time has been affected by inertia flow; some researchers (e.g., Hubbert) believe that

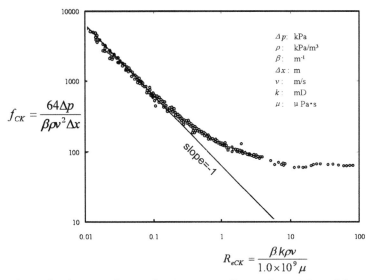

Figure 2.45 Relationship between frictional resistance coefficient f_{CK} and Reynold's number R_{eCK} in porous media.

the transformation from laminar flow to turbulent flow involves a very wide range of flowing velocity and that the turbulent flow phenomenon cannot be observed until Reynold's number R_{eCK} exceeds 600.

What plays a decisive role in determining the influence of turbulent flow is the turbulence coefficient β. On the basis of experiments and a study of 355 sandstone samples and 29 limestone samples, Jones has derived an empirical formula for calculating β [15]:

$$\beta = 1.88 \times 10^{10} k^{-1.47} \phi^{-0.53}. \tag{2.11}$$

Moreover, Equation (2.10) also shows that factors affecting the value of the Reynold's number include the viscosity μ of the fluid and its flowing velocity v. For natural gas zones, a small viscosity and a high flowing velocity are often present. Also, such high-speed flow becomes important only near the bottom hole of gas wells. Therefore, in his research work, Lee has combined such an influence of turbulent flow into the skin factor and proposed a pseudo-skin factor S_a comprising the influence of turbulent flow, which is expressed as [16]:

$$S_a = S + Dq_g, \tag{2.12}$$

where S_a is the pseudo-skin comprising the influence of turbulent flow; S is the true skin of the gas well; q_g is flow rate of the gas well, m^3/day; and D is the non-Darcy flow coefficient, $(m^3/day)^{-1}$.

D can be expressed as

$$D = \frac{7.18 \times 10^{-16} \beta k M p_{sc}}{h r_w T_{sc} \mu_{g,wf}},$$ (2.13)

where β is turbulence coefficient, m^{-1}; k is reservoir permeability, mD; M is molar mass of gas, kg/kmol, $M = \gamma_g \times 28.96$; γ_g is relative density of gas; p_{sc} is standard pressure, $p_{sc} = 0.1013$ MPa; T_{sc} is standard temperature, $T_{sc} = 293.2$ K; r_w is wellbore radius, m; h is effective thickness of reservoir, m; and $\mu_{g,wf}$ is viscosity of gas near the bottom hole, mPa·s.

When related parameters and the expression of β [Equation (2.11)] are substituted in, Equation (2.13) is reduced to

$$D = \frac{1.35 \times 10^{-7} \gamma_g}{k^{0.47} \phi^{0.53} h r_w \mu_g}.$$ (2.14)

The non-Darcy flow coefficient D is a very important characteristic parameter for the production of gas wells and it can, in principle, be calculated by Equation (2.14). For example, for a gas well, assuming that the formation and well parameters are $K = 57$ mD, $\gamma_g = 0.85$, $r_w = 0.09$ m, $\mu_g = 0.0244$ mPa·s, $h = 12.2$ m, and $\phi = 0.1$, its non-Darcy flow coefficient can then be calculated as

$$D = \frac{\left(1.35 \times 10^{-7}\right)\left(0.85\right)}{\left(57^{0.47}\right)\left(0.1^{0.53}\right)\left(12.2\right)\left(0.09\right)\left(0.0244\right)} = 2.17 \times 10^{-6} \left(m^3/day\right)^{-1}$$

or, with a Chinese statutory unit, is expressed as $D = 2.17 \times 10^{-2}$ $(10^4\ m^3/day)^{-1}$.

During calculation of the non-Darcy flow coefficient D by Equation (2.14) due to some uncertain factors, for example, the inability to accurately determine the permeability and effective thickness of the formation, a certain error is usually present. The ultimate solution would be actual measuring the D value by the transient well test method.

The principle based on which the D value is actually measured is that the skin factor obtained during each transient well test analysis is always a pseudo-skin as expressed in Equation (2.12). Such pseudo-skin has something to do with the gas flow rate. As long as different pseudo-skins S_a are measured under different flow rate conditions and the related straight line between S_a and q_g is obtained, the true skin factor S and non-Darcy flow coefficient D can simply be derived, as shown in Figure 2.46.

Some more special discussions on measurement and analysis of non-Darcy flow coefficient D are presented later.

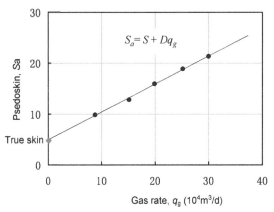

Figure 2.46 Determining non-Darcy flow coefficient D and the true skin with pseudo-skins under different gas flow rates.

2 GAS FLOW EQUATIONS

2.1 Definition of Reservoir as a Continuous Medium

The object we are studying is the formation in which natural gas is stored, and it is very complicated when seen from the view of a "close look."

For sandstones, natural gas exists in a large number of pores of different sizes and flows through interconnected pores. These pores not only have different sizes but also, as passages, connect from large-diameter pores to micropore throats or even to blind ends.

For carbonate rocks, natural gas exists in not only matrix pores but also microfissures, dissolved pores, solution cavities, or even connected groups of fracture systems.

Studying the flow of natural gas from the "close look" perspective is generally neither necessary nor possible. What we want to know is the macroscopic properties of natural gas flow in the formation; therefore, the "continuous medium" hypothesis used commonly in general fluid mechanics has been introduced in mechanics of fluids flow in porous media.

In the formation (e.g., ordinary sandstone formation), take arbitrarily a small unit cell volume ΔV that consists of both pore volume ΔV_p and volume of rock particles ΔV_R, and its porosity is defined as

$$\phi = \frac{\Delta V_p}{\Delta V}.$$

When the ΔV is sufficiently large and is greater than the critical value ΔV_0, such as ΔV_1 and ΔV_2 in Figure 2.47, the value of ϕ is roughly a constant, that is, $\phi = C$. However, if ΔV is too small, it is taken either from pores, where ϕ would approach 1, or from rock particles, where ϕ would approach 0, as shown in Figure 2.47.

When studying the flow process in the formation, the volume ΔV_0 is taken as the volume of the typical unit cell, that is, a point in the formation. All descriptions of flow

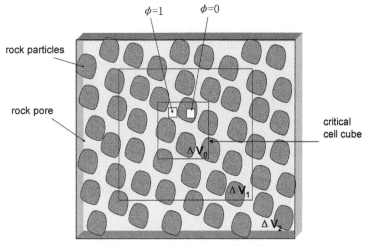

$\phi=1$ $\phi=0$

rock particles

rock pore

critical
cell cube

ΔV_0

ΔV_1

ΔV_2

Figure 2.47 Formation unit cell.

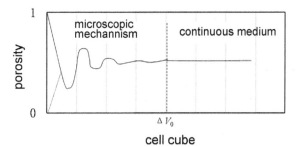

microscopic
mechannism

continuous medium

porosity

1

0

ΔV_0

cell cube

Figure 2.48 Determination of the value for a formation unit cell.

conditions are done using the cell as the unit. Factors such as the porosity of unit cell ϕ, permeability of unit cell k, fluid pressure within the unit cell p, and the flowing velocity of the fluid within the unit cell v are all physical parameters describing flow conditions. The physical quantities of its adjacent cell bodies, p, v, ϕ, and k, are continuous, thus forming a continuous medium. Figure 2.48 shows how to determine the value of a critical volume of the unit cell.

In general, for any widely spread sandstone reservoirs, the properties of an individual unit cell within a certain range are roughly consistent; such reservoirs are called homogeneous media.

2.2 Flow Equations

2.2.1 Deriving Flow Equations Based on Three Basic Equations

It is generally believed that three basic equations determine the flow state of fluid: momentum equation, continuity equation, and state equation. Sometimes, when considering the rheological properties of the fluid, the structural equation is added; for nonisothermal systems, the energy equation is added.

Momentum equation: in general fluid mechanics, this equation is expressed as Newton's law. It is the expression describing the relationship of the force applied onto the fluid unit and the mass of the fluid in the unit and velocity of fluid movement. In mechanics of fluids flow in a porous medium, the momentum equation is expressed as Darcy's law.

Continuity equation: it is one of the expressions of the mass conservation law. This equation describes the relationship between the flow parameters of the flow unit that we study and those of the adjacent units. In other words, it describes how the pressure and flowing velocity in the flow unit change continuously in the formation as the "continuous medium."

State equation: this equation describes the relationship among the parameters, that is, pressure, temperature, and density within the flow unit. It is one of the expressions of the energy conservation law.

2.2.2 Average Flowing Velocity and Flow Velocity of Unit Cell

Because the shapes of pore structures in porous medium are extremely irregular, talking about the actual flowing velocity of fluid inside a specific pore structure is meaningless. For this reason, the concept of average velocity of a unit cell is introduced. The average velocity at point K simply means the average value of the entire pore volume velocity field within the unit cell centered at point K.

Considering a cross section ΔA_0 perpendicular to the direction of fluid average flow within the unit cell at point K, assuming that the flow rate that flows through cross section ΔA_0 in a unit of time is Δq, then the average flowing velocity at point K can be defined as

$$v' = \frac{\Delta q}{\phi \Delta A_0},$$
(2.15)

where ϕ is the porosity.

However, v' is inconvenient for us in studying the flow process. The flowing velocity of fluid passing through porous medium in either the fluid flow in or the flow out process observed from outside of the porous medium is v, and this v is expressed as

$$v = \frac{\Delta q}{\Delta A_0}.$$
(2.16)

This velocity v is called flow velocity. If $\phi = 10\%$, the flow velocity v is only one-tenth of the average flowing velocity v'. Flow velocity will be used throughout all subsequent discussions about the flow process in this book.

2.2.3 Darcy's Law Applied for Flow of Viscous Fluid

All fluids being discussed in petroleum production, that is, natural gas, crude oil, and water, are viscous fluids. All viscous fluids offer a certain viscosity, and the viscosity is denoted by μ. When any viscous fluid flows over a solid surface, such as the inner wall of a pipeline or the pore surface of a porous medium, the layer of the fluid clinging on the solid surface will adhere on to the solid surface and the velocity of which will be zero. Therefore, a velocity difference of the fluid flowing is created between the surface and the center of the pore, and a tangential tensile force is thereby generated that will ultimately produce resistance against the flow.

If the flowing velocity is low, the flow shows a laminar state, as shown in Figure 2.43; and if the flowing velocity is high, the flow shows a turbulent state, as shown in Figure 2.44.

Darcy found the law of flow under the laminar state by experiments in 1856, which is called Darcy's law and is expressed as

$$v = -\frac{k}{\mu}\cdot\frac{\mathrm{d}p}{\mathrm{d}x}, \tag{2.17}$$

where v is flow velocity of fluid, k is permeability of porous medium, μ is viscosity of fluid, and $\dfrac{\mathrm{d}p}{\mathrm{d}x}$ is the pressure gradient along the x direction.

If the flow is not along the x direction and its velocity component in each direction x, y, and z can be expressed as in Equation (2.18), this is just another kind of Darcy's law:

$$v_x = -\frac{k_x}{\mu}\left(\frac{\partial p}{\partial x}\right)$$

$$v_y = -\frac{k_y}{\mu}\left(\frac{\partial p}{\partial y}\right) \tag{2.18}$$

$$v_z = -\frac{k_z}{\mu}\left(\frac{\partial p}{\partial z}\right)$$

The one-dimensional Darcy's law expressed as Equation (2.17) can also be rewritten as

$$-\frac{\mathrm{d}p}{\mathrm{d}x} = \frac{\mu}{k}\cdot v. \tag{2.19}$$

However, when the flow velocity of gas increases, the flow will deviate from Darcy's equation. It is generally believed that such a deviation is first caused by the inertia effect—at that time the Reynolds number is about 1–600; thereafter, as the flow velocity increases further, the influence of turbulent flow will become dominant [17] and Darcy's Equation (2.19) will change to

$$-\frac{\mathrm{d}p}{\mathrm{d}x} = \frac{\mu}{k}\cdot v + \beta\rho v^2. \tag{2.20}$$

Equation (2.20) can also be rewritten as

$$-\frac{\mathrm{d}p}{\mathrm{d}x} = \frac{\mu v}{k}\left(1 + \frac{k\beta\rho v}{\mu}\right) = \frac{\mu v}{\delta k} \tag{2.21}$$

or

$$v = -\delta\cdot\frac{k}{\mu}\cdot\frac{\mathrm{d}p}{\mathrm{d}x}, \tag{2.22}$$

where $\delta = 1/\left(1 + \dfrac{k\beta\rho v}{\mu}\right)$ is called the laminar flow−inertia flow−turbulent flow coefficient [18].

For one of any flow directions in anisotropic media, Darcy's law is expressed as

$$v = -\frac{k}{\mu}\cdot\delta\nabla p, \tag{2.23}$$

where

$$\delta = \begin{bmatrix} \delta_x & 0 & 0 \\ 0 & \delta_y & 0 \\ 0 & 0 & \delta_z \end{bmatrix}. \tag{2.24}$$

Equation (2.23) is the generalized momentum balance equation, or generalized Darcy's law.

2.2.4 Continuity Equation

The fluid in one unit cell in the formation (as shown in Figure 2.47) is flowing with a flow velocity of v; the v is a "vector" in a vector field. The components of velocity v are expressed as v_x, v_y, and v_z in the rectangular coordinate system; as v_r, v_θ, and v_z in the cylindrical coordinate system; or as v_r, v_θ and v_σ in the spherical coordinate system.

The continuity equation represents the mass conservation law. It means that for any unit cell storing some fluid in the formation, during any interval Δt, subtracting the fluid quantity flew out of the cell from that flew into the cell results the increment of quantity inside the cell body. In the form of generalized coordinates, it is written as

$$-\frac{\partial}{\partial t}(\phi\rho) = \nabla(\rho v), \tag{2.25}$$

where ϕ is porosity of porous media, ρ is density of fluid, and ∇ is divergence operator.
In the rectangular coordinates system, the expression of divergence is

$$\nabla(\rho v) = \frac{\partial(\rho v_x)}{\partial x} + \frac{\partial(\rho v_y)}{\partial y} + \frac{\partial(\rho v_z)}{\partial z}.$$

Therefore, the continuity equation is expressed as

$$-\frac{\partial}{\partial t}(\rho \phi) = \frac{\partial(\rho v_x)}{\partial x} + \frac{\partial(\rho v_y)}{\partial y} + \frac{\partial(\rho v_z)}{\partial z}. \tag{2.26}$$

Figure 2.49 shows the inflow and outflow in a cell body.
Equation (2.25) is expressed as

$$-\frac{\partial}{\partial t}(\phi \rho) = \nabla(\rho v) = \frac{1}{r} \cdot \frac{\partial}{\partial r}(r \rho v_r) + \frac{1}{r} \cdot \frac{\partial(\rho v_\theta)}{\partial \theta} + \frac{\partial(\rho v_z)}{\partial z} \tag{2.27}$$

in the cylindrical coordinate system and as

$$-\frac{\partial}{\partial t}(\phi \rho) = \nabla(\rho v) = \frac{1}{r^2} \cdot \frac{\partial}{\partial r}(r^2 \rho v_r) + \frac{1}{r \cdot Sin\theta} \cdot \frac{\partial}{\partial \theta}(\rho v_\theta \cdot Sin\theta) + \frac{1}{r \cdot Sin\theta} \cdot \frac{\partial(\rho v_\sigma)}{\partial \sigma} \tag{2.28}$$

in the spherical coordinate system.

Equation (2.25) can be simplified under one-dimensional flow conditions; the
expression becomes

$$-\frac{\partial}{\partial t}(\phi \rho) = \frac{\partial}{\partial x}(\rho v) \tag{2.29}$$

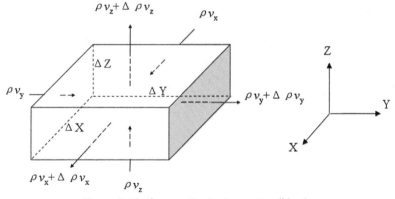

Figure 2.49 Flow continuity in a unit cell body.

in the rectangular coordinate system or

$$-\frac{\partial}{\partial t}(\phi\rho) = \frac{1}{r}\cdot\frac{\partial}{\partial r}(r\rho v) \tag{2.30}$$

in the cylindrical coordinate system for radial flow.

2.2.5 State Equation of Gas

At each point in the gas zone, that is, within each unit cell mentioned earlier, the density of gas changes with its pressure. Such change is described by the state equation of real gas. The state equation combines the continuity equations [Equations (2.26)−(2.30)] expressed by density and Darcy's law [Equation (2.23)] expressed by pressure together to form the flow equation describing the gas flow in formation.

The state equation describing the relationship among pressure p, temperature T, and volume V (or density ρ) of real gas has been studied by many scientists a long time ago, and the most popular forms are Equations (2.31) and (2.32); every parameter in these formulae should refer to a certain cell body being studied or a certain bulk containing the cell body being studied, and the state inside the bulk is homogeneous:

$$pV = nZRT = \frac{m}{M}ZRT \tag{2.31}$$

or

$$\rho = \frac{m}{V} = \frac{Mp}{ZRT} \tag{2.32}$$

where p is pressure inside the bulk being studied in the gas zone, MPa; V is volume of the bulk being studied in the gas zone, m^3; n is quantity of the matter (natural gas) in the bulk being studied, mol; m is mass of the matter (natural gas) in the bulk being studied, g or kg; M is molecular weight of the matter (natural gas) in the bulk being studied, for example, the molecular weight of air is $M = 28.97$ kg/kmol; Z is the deviation factor, or the Z factor of the gas, the chart provided by Standing and Katz in 1942 (Figure 2.50)[19] is herein attached for reference; R is universal gas constant; its value is different under different unit systems: R is 8.3143×10^{-3} [MPa$\cdot$m^3(kmol$\cdot$K)$^{-1}$] in CSU units; R is 8.3143×10^3 [Pa$\cdot$m^3 (kmol$\cdot$K)$^{-1}$] in SI units; and R is 10.732 [psi$\cdot$ft^3 (lbmol$\cdot$°R)$^{-1}$] in English oilfield units; and $\rho = D$ is density of gas, kg/m^3.

Gas compressibility C is often used in flow equations of gas and is defined as

$$C = -\frac{1}{V}\cdot\frac{\partial V}{\partial p} \tag{2.33}$$

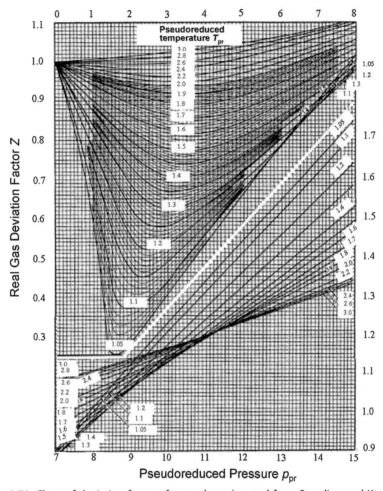

Figure 2.50 Chart of deviation factor of natural gas (quoted from Standing and Katz [19]).

From the definition of density expression 2.32, Equation (2.33) can be rewritten as

$$C = \frac{1}{\rho} \cdot \frac{\partial \rho}{\partial p} \tag{2.34}$$

or

$$C = \frac{1}{p} + Z \frac{\partial}{\partial p}\left(\frac{1}{Z}\right) = \frac{1}{p} - \frac{1}{Z} \cdot \frac{\partial Z}{\partial p} \tag{2.35}$$

Deviation factor Z is a correction coefficient and describes the deviation between real gas and ideal gas. The values of Z can be obtained by looking up at Figure 2.50.

In Figure 2.50, the abscissa variable is pseudo-reduced pressure p_{pr}, and the parametric variable is pseudo-reduced temperature T_{pr}. For a specific gas zone, once p_{pr} and T_{pr} are determined, the Z value can be looked up from the chart. p_{pr} and T_{pr} are expressed as

$$p_{pr} = \frac{p}{p_{pc}} \tag{2.36}$$

$$T_{pr} = \frac{T}{T_{pc}}. \tag{2.37}$$

p and T are the formation pressure and the formation temperature of the gas zone, respectively; p_{pc} and T_{pc} are called the pseudo-critical pressure and pseudo-critical temperature of the gas zone and can be obtained by calculation based on the components of natural gas and the molar contents of each component of it. For their calculation methods, please refer to Appendix A of Reference [20].

In fact, just like explained in the section of "The purpose of this book" in Chapter 1 "Introduction" of this book, due to the widespread application of computer and software today, the aforementioned calculations are very seldom performed manually. By using any ordinary well test interpretation software, the desired results can be obtained easily as long as the components of natural gas and the corresponding formation pressure and temperature are inputted.

2.2.6 Subsurface Flow Equations of Natural Gas

Subsurface flow Equation (2.38) can simply be obtained by combining the momentum equation of the subsurface flow of natural gas, that is, Darcy's law [Equation (2.23)] and the continuity equation [Equation (2.25)] simultaneously:

$$\frac{\partial}{\partial t}(\phi\rho) = \nabla(\rho v) = \nabla\left[\rho \cdot \frac{k}{\mu} \cdot \delta \nabla p\right]. \tag{2.38}$$

Equation (2.38) contains density ρ, porosity ϕ, viscosity μ, permeability k, and the laminar flow–inertia flow–turbulent flow coefficient δ; in this equation, the independent variable is time t and the dependent variable is pressure p; however, density ρ and some other parametric variables also change with pressure. Therefore, a useful partial differential equation cannot be obtained unless putting into the state equation describing the relationship between ρ and p.

In general, this partial differential equation is nonlinear for dependent variable p. The analytic method can be used to solve this partial differential equation only when it has been linearized under some particular conditions.

Flow Equation of Liquid and that of Gas Under High-Pressure Conditions

The compressibility C of gas under high-pressure conditions, just like that of liquid under normal temperature conditions, can be regarded as a constant so

$$\rho = \rho_0 e^{C(p-p_0)} \tag{2.39}$$

Thus, Equation (2.38) can be written as

$$\frac{\partial}{\partial t}\left[\phi\rho_0 e^{C(p-p_0)}\right] = \rho_0 e^{C(p-p_0)}\nabla\left[\frac{k}{\mu}\cdot\delta\cdot\nabla p\right] + \left[\frac{k}{\mu}\cdot\delta\cdot\nabla P\right]\cdot\nabla\left[\rho_0 e^{C(p-p_0)}\right] \tag{2.40}$$

or, after arranging,

$$C\phi\cdot\frac{\partial p}{\partial t} + \frac{\partial\phi}{\partial t} = \nabla\left[\frac{k}{\mu}\cdot\delta\cdot\nabla p\right] + C\left[\frac{k}{\mu}\cdot\delta\cdot\nabla p\right]\cdot\nabla p. \tag{2.41}$$

Gas Flow Equations Under Normal Conditions

The gas flow equation under normal conditions, Equation (2.42), can be obtained by substituting the state Equation (2.32) of real gas under normal conditions into Equation (2.38):

$$\frac{\partial}{\partial t}\left(\phi\cdot\frac{M}{RT}\cdot\frac{p}{Z}\right) = \nabla\left[\frac{M}{RT}\cdot\frac{p}{\mu Z}\cdot k\delta\cdot\nabla p\right] \tag{2.42}$$

and $\dfrac{M}{RT}$ can be regarded as a constant in isothermal conditions; therefore, Equation (2.42) can be simplified further as

$$\frac{\partial}{\partial t}\left(\phi\cdot\frac{p}{Z}\right) = \nabla\left(\frac{p}{\mu Z}\cdot k\delta\cdot\nabla p\right). \tag{2.43}$$

Fundamental Assumptions for Gas Flow Equations

Several fundamental assumptions have already been made during derivation of the aforementioned equations:

1. The gas zone is isothermal. This assumption has already been applied when deriving state Equation (2.33) and Equations (2.34), (2.35), and (2.39), and when deriving generalized flow Equations (2.41) and (2.43) as well. During general gas field development, this assumption of "isothermal gas zone" is consistent with actual formation conditions.
2. The gas zone is horizontal or nearly horizontal so that the influence of gravity is ignored during the flow process. This assumption has already been applied when deriving continuity Equations (2.26)–(2.30). It is actually acceptable for most gas zones.

3. The gas in the zones is single phase. This assumption has already been applied when deriving both Darcy's law and state equation and demonstrated in flow Equations (2.41) and (2.43). It is entirely suitable during the exploration phase and early development phase of gas fields; for condensate gas fields and those gas fields with oil ring or bottom water, however, the suitability conditions must be considered during application.

4. The gas-bearing porous medium is homogeneous, isotropic, and slightly compressible, and porosity ϕ is a constant. For homogeneous sandstone reservoirs, when the test time is not very long, this assumption is suitable for a certain limited range within the formation.

5. Permeability k is a constant and does not vary with pressure, p. This assumption can be satisfied in most formations, unless in some formations with extremely low permeability and the presence of so-called "start-up pressure" and "non-Darcy's flow effect" are proved there by sufficient laboratory or field data.

6. The gas flow is laminar, that is, $\delta = 1$. This assumption can be satisfied in most areas of the formation except the small area very near the wellbore. Also, the additional pseudo skin effect is often used to simulate the influence of turbulent flow happening in the small area near the wellbore.

The aforementioned fundamental assumptions are often not yet sufficient to linearize the flow equation and make the equation solvable successfully. For this reason, the equation must be simplified further by making more assumptions about the parameters, such as fluid viscosity μ, fluid compressibility C, and pressure gradient in the flow field, for different situations respectively.

More Assumptions Regarding Gas Flow Conditions and Further Simplification of Flow Equation

1. Simplification of flow equation for slightly compressible fluids

All items containing $\dfrac{\partial \phi}{\partial t}$ and $C(\nabla p)^2$ can be ignored if the following are further assumed:

- The compressibility of fluid C is very small and is a constant.
- The pressure gradient ∇p in the flow field is very small.
- The viscosity of fluid μ does not vary with pressure, that is, is a constant.

Then Equation (2.41) is simplified to

$$\nabla^2 p = \frac{\phi \mu C}{k} \cdot \frac{\partial p}{\partial t}. \tag{2.44}$$

Although this is a partial differential equation for gas reservoirs, it is just the same with the equation for oil reservoirs because its variable is also pressure. Also, results obtained from well test analysis are also just the same. This is the reason why a well test for oil wells and gas wells is usually called a joint "well test" collectively and are not distinguished from each other.

2. Equation simplification when parameters group $\dfrac{p}{\mu Z} = $ constant

When the gas is under a high-pressure condition, $\dfrac{p}{\mu Z}$ can be regarded as a constant, as shown in Figure 2.51. Figure 2.51 shows a particular analysis example of a gas sample, its molar mass $M = 17.5315$ kg/kmol, specific gravity $\gamma_g = 0.605$, pseudo-critical pressure $p_{pc} = 4.581$ MPa, and pseudo-critical temperature $T_{pc} = 198.33$K. It is seen that $\dfrac{p}{\mu Z} \approx$ constant when pressure p is greater than 25 MPa.

Under the fundamental assumptions 1−4 mentioned earlier, Equation (2.43) can be rewritten for high-compressibility gas as

$$\frac{\partial}{\partial t}\left(\frac{p}{Z}\right) = \frac{k}{\phi}\cdot\nabla\left(\frac{p}{\mu Z}\cdot\nabla p\right). \tag{2.45}$$

The left side of it can be expanded as

$$\frac{\partial}{\partial t}\left(\frac{p}{Z}\right) = \frac{1}{Z}\cdot\frac{\partial p}{\partial t} + p\cdot\frac{\partial}{\partial t}\left(\frac{1}{Z}\right)$$

$$= \frac{1}{Z}\cdot\frac{\partial p}{\partial t} + p\cdot\frac{d}{dp}\left(\frac{1}{Z}\right)\cdot\frac{\partial P}{\partial t}$$

$$= \frac{p}{Z}\cdot\left(\frac{1}{p}\cdot\frac{\partial p}{\partial t}\right) + p\left(-\frac{1}{Z^2}\cdot\frac{dZ}{dp}\right)\cdot\frac{\partial p}{\partial t}$$

$$= \frac{p}{Z}\cdot\frac{\partial p}{\partial t}\left[\frac{1}{p} - \frac{1}{Z}\cdot\frac{dZ}{dp}\right]$$

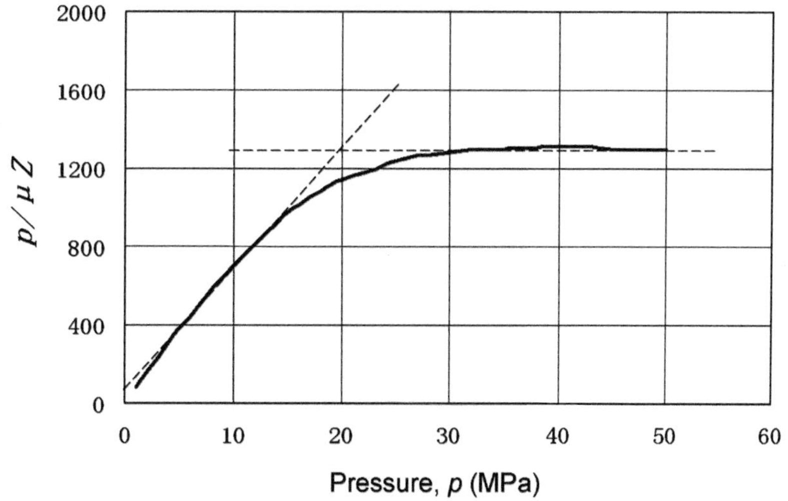

Figure 2.51 Relationship between $p/\mu Z$ and pressure p.

As can be seen from Equation (2.35), the parameter group $\left(\dfrac{1}{p} - \dfrac{1}{Z}\cdot\dfrac{\mathrm{d}Z}{\mathrm{d}p}\right)$ is simply compressibility C; therefore,

$$\frac{\partial}{\partial t}\left(\frac{p}{Z}\right) = \frac{pC}{Z}\cdot\frac{\partial p}{\partial t}. \tag{2.46}$$

Combining Equations (2.46) and (2.45) will give

$$\frac{k}{\phi}\cdot\nabla\left(\frac{p}{\mu Z}\cdot\nabla p\right) = C\cdot\frac{p}{Z}\cdot\frac{\partial p}{\partial t} \tag{2.47}$$

By differentiation, items at the left side of Equation (2.47) are resolved to give a result as follows:

$$\frac{k}{\phi}\cdot\nabla\left(\frac{p}{\mu Z}\cdot\nabla p\right) = \frac{p}{\mu Z}\cdot\nabla^2 p + \nabla\left(\frac{p}{\mu Z}\right)\cdot\nabla p$$

$$= \frac{p}{\mu Z}\cdot\nabla^2 p + \frac{\mathrm{d}}{\mathrm{d}p}\left(\frac{p}{\mu Z}\right)\cdot(\nabla p)^2$$

$$= \frac{p}{\mu Z}\cdot\nabla^2 p + \frac{\mathrm{d}}{\mathrm{d}p}\left(1/(\mu Z/p)\right)\cdot(\nabla p)^2$$

$$= \frac{p}{\mu Z}\cdot\nabla^2 p + \left[\frac{-1}{(\mu Z/p)^2}\cdot\frac{\mathrm{d}}{\mathrm{d}p}\left(\frac{\mu Z}{p}\right)\right]\cdot(\nabla p)^2$$

$$= \frac{p}{\mu Z}\cdot\nabla^2 p - \left[\left(\frac{p}{\mu Z}\right)^2\cdot\frac{\mathrm{d}}{\mathrm{d}p}\left(\ln\frac{\mu Z}{p}\right)\cdot\frac{\mu Z}{p}\right]\cdot(\nabla p)^2$$

$$= \frac{p}{\mu Z}\cdot\left[\nabla^2 p - (\nabla p)^2\cdot\frac{\mathrm{d}}{\mathrm{d}p}\left(\ln\frac{\mu Z}{p}\right)\right]$$

Therefore, Equation (2.47) is transformed as

$$\nabla^2 p - \frac{\mathrm{d}}{\mathrm{d}p}\left[\ln\left(\frac{\mu Z}{p}\right)\right]\cdot(\nabla p)^2 = \frac{\mu\phi C}{k}\cdot\frac{\partial p}{\partial t} \tag{2.48}$$

Because $\mu Z/p$ has been assumed to be a constant, the second item of the left side of Equation (2.48) is 0 so the equation is simplified into the form just the same as Equation (2.44), that is,

$$\nabla^2 p = \frac{\mu\phi C}{k}\cdot\frac{\partial p}{\partial t}. \tag{2.49}$$

Another Assumption for Gas Flow Conditions and Expression of Flow Equation with p^2

Al-Hussainy has made another assumption for gas flow equations [21].

1. The (μZ) value is a constant when formation pressure is relatively low (refer to Figure 2.52). The parameters used in Figure 2.52 are the same as those in Figure 2.51. It is seen that when pressure is less than 20 MPa, the (μZ) value is nearly constant; thereafter the value tends to increase. This fact is also shown clearly in Figure 2.51 in that it shows nearly a straight line in the early section.
2. The pressure gradient is very small, and therefore $(\nabla p^2)^2 \to 0$.

Note that $p \cdot \nabla p = \frac{1}{2} \nabla p^2$ and $p \cdot \partial p = \frac{1}{2} \partial p^2$, arranging Equation (2.47) will give

$$\nabla^2 p^2 - \frac{d}{dp^2}\left[\ln(\mu Z)\right] \cdot (\nabla p^2)^2 = \frac{\mu \phi C}{k} \cdot \frac{\partial p^2}{\partial t} \tag{2.50}$$

Considering the aforementioned assumptions simultaneously, Equation (2.50) can be simplified as [21]:

$$\nabla^2 p^2 = \frac{\mu \phi C}{k} \cdot \frac{\partial p^2}{\partial t}. \tag{2.51}$$

It is seen that Equation (2.51) has the same form as Equations (2.44) or (2.49), except that p^2 is used to replace p, and so a pressure square expression of the equation is obtained. Based on these assumed conditions, this equation is suitable when pressure is low.

Expression of Flow Equation with Pseudo Pressure

In order to linearize and solve the gas flow equation discussed earlier, some additional assumptions are made in addition to the fundamental assumptions 1–4. However, these assumptions restrict the application of the equations. For example, pressure expressions

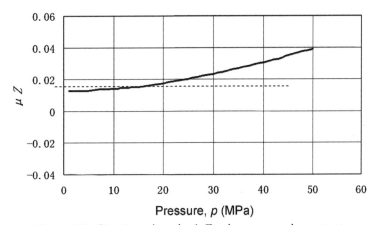

Figure 2.52 Situation when the (μZ) values are nearly constant.

(2.44) and (2.49) are suitable for high formation pressure conditions only, whereas the pressure square expression (2.51) is suitable for low formation pressure situations only. However, the span of pressure change during the tests in fields, especially deliverability tests, is usually very large. Thus the pressure or pressure square equations would become unsuitable in the whole process of testing. For this reason, Al-Hussainy specifically defined the pseudo pressure ψ of gas in 1965. Also, the gas flow equation with pseudo pressure ψ is suitable for the entire range of pressure change in whatever situation [22].

Pseudo pressure is defined as

$$\psi = \int_{p_0}^{p} \frac{2p}{\mu Z} dp. \tag{2.52}$$

p_0 in this expression is a reference pressure selected arbitrarily, and $p_0 = 0$ is usually selected. The flow equation becomes

$$\nabla \psi = \frac{\partial \psi}{\partial p} \cdot \nabla p = 2 \frac{p}{\mu Z} \cdot \nabla p \tag{2.53}$$

or

$$\frac{\partial \psi}{\partial t} = 2 \frac{p}{\mu Z} \cdot \frac{\partial p}{\partial t} \tag{2.54}$$

By substituting Equation (2.54) into Equation (2.46), it is rewritten as

$$\frac{\partial}{\partial t} \left(\frac{p}{Z} \right) = \mu C \cdot \frac{p}{\mu Z} \cdot \frac{\partial p}{\partial t}$$

$$= \frac{\mu C}{2} \cdot \frac{\partial \psi}{\partial t} \tag{2.55}$$

By further substituting Equation (2.53) into the right side of Equation (2.45), it can be further rewritten as

$$\frac{k}{\phi} \cdot \nabla \left(\frac{p}{\mu Z} \cdot \nabla p \right) = \frac{k}{\phi} \cdot \nabla \left(\frac{1}{2} \nabla \psi \right) = \frac{1}{2} \cdot \frac{k}{\phi} \cdot \nabla^2 \psi. \tag{2.56}$$

Thus Equation (2.45) is finally rewritten as

$$\nabla^2 \psi = \frac{\mu \phi C}{k} \cdot \frac{\partial \psi}{\partial t}. \tag{2.57}$$

It can be seen from comparison of Equation (2.57) with Equations (2.44), (2.49), and (2.51), except pseudo pressure ψ is used in Equation (2.57) to replace pressure p in

Equation (2.44) and Equation (2.49), or pressure square p^2 in Equation (2.51), their forms of expression are completely the same and their solving processes are the same as well. For the pseudo pressure Equation (2.57), however, except fundamental assumptions 1–4, all other additional assumptions have been ridded, therefore it can be used in any situations of gas flow and so has broad suitability.

There are several different names of pseudo pressure, such as "potential function of real gas" and "modified pressure square" (which comes from its dimension of $[p^2/\mu]$).

If the components and the pressure and temperature of a certain kind of gas are all known, the relationship of Z and p and that of μ and p can be established, and then the relationship of ψ and p can be obtained by integrating Equation (2.52). Figure 2.53 shows a typical curve of this relationship.

Nowadays all well test interpretation software contains programs for converting pressure p into pseudo pressure ψ. Therefore, field well test engineers are not required to do the conversion manually. Well test analysis can be done completely using pseudo pressure when running the software; what a analyst needs to do is simply enter the gas components or related characteristic parameters, such as the relative density of gas, and then click the pseudo pressure option. The units of pseudo pressure are listed in Table 2.3.

Readers interested in trying to calculate pseudo pressure from pressure manually can read the calculation example 2-1 in Reference [20].

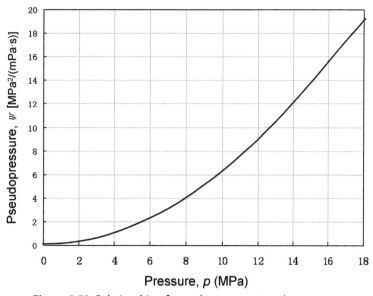

Figure 2.53 Relationship of pseudo pressure ψ and pressure p.

Table 2.3 List of Units of Pressure and Pseudo Pressure under Different Unit Systems

Unit System	China Statutory Unit System	SI Unit System	Oil Field Unit System
Pressure, p	MPa	kPa	psi
Pseudo pressure, ψ	$MPa^2/mPa \cdot s$	$kPa^2/\mu Pa \cdot s$	psi^2/cP

Pressure and pressure square expressions can also be obtained by analyzing and simplifying pseudo pressure expression (2.52):

1. Assuming that the (μZ) value is a constant, that is, $\mu Z = \mu_0 Z_0$. μ_0 and Z_0 are the viscosity and deviation factor under a certain initial condition. Thus, Equation (2.52) can be rewritten as

$$\psi = 2 \int_{p_0}^{p} \frac{p}{\mu Z} \, dp = \frac{2}{\mu_0 Z_0} \int_{p_0}^{p} p \, dp = \frac{1}{\mu_0 Z_0}(p^2 - p_0^2) \tag{2.58}$$

Therefore, Equation (2.57) is converted directly into Equation (2.51).

2. If $\frac{p}{\mu Z}$ = constant, say $\frac{p}{\mu Z} = \frac{p_0}{\mu_0 Z_0}$, where p_0, μ_0, and Z_0 are all values under a certain initial condition. Thus Equation (2.52) can be rewritten as

$$\psi = \frac{2p_0}{\mu_0 Z_0} \int_{p_0}^{p} \mathrm{d}p = \frac{2p_0}{\mu_0 Z_0} \cdot (p - p_0) \tag{2.59}$$

Substituting it into Equation (2.57), this equation can be converted directly to Equation (2.44) or Equation (2.49), that is, the expression of pressure.

This can also be seen clearly from Figure 2.54. The heavy solid line in Figure 2.54 is the relationship curve of pseudo pressure and pressure of a certain kind of natural gas. When pressure is relatively low, a quadratic relationship in the form of $\psi = a_1 p^2$ is obviously present, whereas when pressure is relatively high, a linear relationship of $\psi = a_2 p + b_2$ is obviously present. This has more vividly proved the feasibility of simplifying pseudo pressure to pressure p or pressure square p^2 as described earlier.

It can be seen from Equation (2.58) that the dimension of pseudo pressure is no longer that of pressure, but is the dimension of pressure square divided by viscosity. In China statutory unit system, its unit is $MPa^2/(mPa \cdot s)$. Such a concept seems vague sometimes during plotting and analysis. Therefore, some commercial well test interpretation software use the normalized pseudo pressure; for example, the software Workbench uses the normalized pseudo pressure expressed by

$$\psi_N = \frac{\mu_0 Z_0}{p_0} \int_{p_0}^{p} \frac{2p}{\mu Z} \, \mathrm{d}p.$$

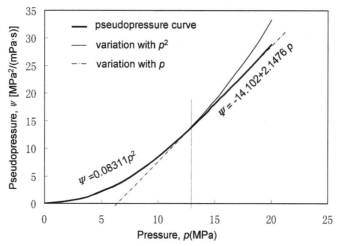

Figure 2.54 Pseudo pressure and its approximation: Pressure and pressure square.

In this way, the unit of pseudo pressure is converted back to the same of pressure, MPa; and the software EPS uses the rate-normalized pseudo pressure instead.

2.2.7 Dimensionless Expressions of Gas Flow Equations

Flow equations expressed by pressure, pressure square, or pseudo pressure as described earlier can be presented in general forms:

$$\nabla^2\phi = \frac{1}{\eta}\cdot\frac{\partial\phi}{\partial t},\tag{2.60}$$

where the meaning of ϕ is $\phi = p$ in pressure expression; $\phi = p^2$ in pressure square expression; $\phi = \psi$ in pseudo pressure expression; and $\eta = \dfrac{k}{\mu\phi C}$. Many calculation methods of parameter η in the process of solving equations have been provided by many researchers and are not introduced in detail here. For further information, please refer to Reference [20].

Equation (2.60) would, for a particular flow model in different coordinate systems, offer more concrete expression forms. Following are several examples.

For one-dimensional linear flow in a rectangular coordinate system, its expression is

$$\frac{\partial^2\phi}{\partial x^2} = \frac{1}{\eta}\cdot\frac{\partial\phi}{\partial t}.\tag{2.61}$$

Examples of this kind of flow are the early flow in fractured wells and the late flow state in a channel-shaped lithologic formation created by the sedimentation of fluvial facies (refer to Figures 2.18 and 2.20).

For plane radial flow in a polar coordinate system, its expression is

$$\frac{1}{r} \cdot \frac{\partial}{\partial r}\left(r \cdot \frac{\partial \phi}{\partial r}\right) = \frac{1}{\eta} \cdot \frac{\partial \phi}{\partial t}. \tag{2.62}$$

This kind of flow is met most frequently in normal completion gas wells in formations with uniform thickness (refer to Figure 2.7).

For spherical flow in a polar coordinate system, its expression is

$$\frac{1}{r^2} \cdot \frac{\partial}{\partial r}\left(r \cdot \frac{\partial \phi}{\partial r}\right) = \frac{1}{\eta} \cdot \frac{\partial \phi}{\partial t} \tag{2.63}$$

This kind of flow state appears during early flow in partially penetrated wells (refer to Figures 2.12 and 2.13).

During the process of solving Equation (2.60), different coefficients often result due to adopting different unit systems. This would bring about unnecessary troubles to researchers and users. To solve this problem, the variables are made dimensionless.

The definition of a dimensionless variable has already been explained at the beginning of this chapter, and several examples of dimensionless expressions have also been presented [refer to Equations (2.1)−(2.4)]. The forms of expression and coefficients of various types of dimensionless variables are described in Table 5.3 in Chapter 5 of this book and so are not repeated here.

When dimensionless quantities are applied, the gas flow equation is expressed in a very simple form:

$$\nabla^2 p_D = \frac{\partial p_D}{\partial t_D}. \tag{2.64}$$

2.2.8 Boundary Conditions and Initial Conditions for Solving Gas Flow Equations

It is well known that the initial conditions and boundary conditions must be given when solving differential equations. Take plane radial flow as an example, in the polar coordinate system, its flow equation is Equation (2.62), and the initial and boundary conditions for solving it are as follow.

Initial Condition

It is generally believed that for the formation being studied, the pressure at every point in the formation is identical initially and is a constant, that is,

$$p = p_i \text{ for any } r \text{ when } t = 0.$$

Internal Boundary Condition

An internal boundary condition means the bottom-hole boundary condition. At the bottom hole, the flow rate (gas flow rate or gas injection rate) passing through the wellbore is taken as the boundary condition.

According to Darcy's law,

$$\frac{k}{\mu} \cdot \frac{\partial p}{\partial r}\bigg|_{r=r_w} = \frac{q}{2\pi r_w h},$$

that is,

$$r \cdot \frac{\partial p}{\partial r}\bigg|_{r=r_w} = \frac{q\mu}{2\pi kh}. \tag{2.65}$$

When the gas flow rate is converted into that under standard state, it becomes

$$r \cdot \frac{\partial p}{\partial r}\bigg|_{r=r_w} = \frac{q_{sc}\mu}{2\pi kh} \cdot \frac{p_{sc}}{\bar{p}} \cdot \frac{T\bar{Z}}{T_{sc}}. \tag{2.66}$$

External Boundary Conditions

External boundary conditions mean the conditions at formation edges far away from the well that restrict the flow. Such a condition can be:

- Pressure maintains its original value at the infinity, which is expressed as

$$p = p_i\big|_{r \to \infty}$$

- There is a no-flow boundary at $r = r_1$, expressed as

$$r \cdot \frac{\partial p}{\partial r}\bigg|_{r=r1} = 0$$

- There is a constant pressure boundary at $r = r_1$, expressed as

$$p\big|_{r=r1} = C$$

For complicated external boundaries, the settings of boundary conditions must also be adjusted accordingly.

3 SUMMARY

The basic concepts and flow equations of the well test have been introduced in this chapter. "Basic concepts," such as the meanings of various technical terms encountered during well test interpretation, geological conditions under which various flow states exist, and the characteristics of various flow states on the interpretation type curves were introduced and explained one by one. The author believes that all these contents are very important; this idea is formed from many years of experience from the author's well test practice and is also the primary basis for ultimately establishing the "well test interpretation graphical analysis method."

As for "flow equations," the subsurface flow equations of natural gas have been derived based on the fundamental equations and following the usual procedures in commonly used books. This is not the keystone of this book, as it is merely the theoretical preparation made for building mathematical models for solving direct problem (refer to Figure 1.1). It is just from these fundamental equations that most graduate students pursuing a well test specialty degree start their research projects. Most readers of this book are assumed to be engineers engaged in well test data analysis and gas reservoir engineering research. Their primary task is to solve inverse problems, that is, analyzing formation parameters based on measured well test data so as to study gas fields. Whatever technical field the reader is involved in, however, mastering such knowledge is always very necessary for understanding the connotation of the well test discipline.

CHAPTER 3

Gas Well Deliverability Test and Field Examples

Contents

Dynamic Well Testing in Petroleum Exploration and Development   © 2013 Petroleum Industry Press. Published by Elsevier Inc.
ISBN 978-0-12-397161-6, http://dx.doi.org/10.1016/B978-0-12-397161-6.00003-6

A new idea concerning gas reservoir dynamic description was proposed in Chapter 1. It pointed out explicitly that the gas reservoir dynamic description concentrates on gas well deliverability evaluation, that is, for a reservoir in production or to be put into production, the most important issue for the operators is nothing like flow rate, absolute open flow potential (AOFP), and reasonable flow rate of each individual gas well, whether or not the rate can be kept stable and how the rate declines.

Previous gas well test studies lay particular stress on transient test analysis, resolving partial differential equations aiming at various formations and establishing various transient test analysis models, whereas deliverability test methods and analysis urgently needed in production onsite are still semiempirical methods developed before the 1950s. Field practices have shown that such methods are far from meeting practical needs in the production, following are some examples:

1. Deliverability equations obtained from field deliverability tests sometimes have abnormal values of $B < 0$ or $n > 1$; in such cases, it is impossible to establish a valid deliverability equation with data obtained by great efforts.

2. Much work must be done in performing those classical deliverability tests, which brings a problem of that only a few gas wells can be tested, while there is no way to give the deliverability of those untested gas wells for production planning of a gas field.

3. All classical deliverability test methods can only result in initial deliverability indices of a gas well, while there is no way to understand the deliverability decline process, even there is no this concept.

In this chapter, these classical methods, which have been placed into Chinese national regulations, are discussed in detail, and a "stable point laminar—inertial—turbulent (LIT) deliverability equation" is introduced, aiming at deficiencies of these classical methods.

Proceeding from three important factors influencing gas well deliverability, related formulae are deduced, the established process of the related equation is described in detail, and the feasibility of the method in vertical and horizontal wells is verified by its applications in some large and medium gas fields, such as the KL 2 gas field, SLG gas field, YL gas field, and DF gas field.

Based on the initial stable point LIT deliverability equation of gas wells, two new concepts, "dynamic deliverability equation" and "dynamic deliverability indices", that is, dynamic AOFP and dynamic inflow performance relationship (IPR) curve, are put forward in this chapter, combined with field examples, and data acquisition, analysis, and application are explained.

1 GAS WELL DELIVERABILITY AND ABSOLUTE OPEN FLOW POTENTIAL (AOFP)

As introduced in Chapter 1, gas well deliverability is the crucial issue in the gas reservoir dynamic description. For a gas field in production or planned to be put into production, the major concerns of management personnel are the initiated flow rate and the stable flow rate of each individual gas well, the stable production over the whole field and how many wells should be drilled to meet the production plan and so to ensure normal and stable gas supply to the downstream industries.

1.1 Meanings of Gas well Deliverability

Gas Well deliverability, as the name suggests, means the gas flow capability of a gas well. In the early stage, this gas flow capability is determined by the open flow test method and is regarded as "measured AOFP." However, this method has actually various disadvantages—not only is a large amount of natural gas wasted during testing, but also the well may be damaged by sand and/or water production. In addition, the measured flow rate is not really the maximum one (when bottom-hole pressure is 1 atm, or 0.1013 MPa) because friction resistance of the pipe string always exists during a gas well test.

At the end of the 1920s, the back-pressure test method was developed by Pierce and Rawlines [23] of the United States Bureau of Mines, was further improved at the end of the 1930s [24], and has been widely applied in gas fields since then. The method adopts different gas chokes to flow the tested well in a certain subsequence and simultaneously monitors and records the gas flow rate and bottom-hole flowing pressure, from which a "stabilized deliverability curve" is established and AOFP of the gas well calculated.

During the back-pressure test, because the gas flow rate and the bottom-hole flowing pressure of each choke are both required to be stabilized simultaneously, the test duration is relatively long, leading to a large amount of gas being blown out and wasted, especially for exploration gas wells in low permeability reservoirs. However, a very long test duration is often not acceptable to operators. Therefore, isochronal test and modified isochronal test methods were developed.

The isochronal test method was developed by Cullender in 1955 [25]. With this method, stabilized flowing pressure is unnecessary under each choke during testing so as to save testing time and reduce wasted gas. However, several times of flow and shut in are required, and pressure must be built up to the original formation pressure in each shut in. Compared with the back-pressure test, operation procedures are more complex and the testing duration is still quite long; especially for gas wells with accumulated fluids in the bottom hole, many technological problems would appear during testing.

The isochronal test method was improved further by Katz and colleagues in 1959 and became the modified isochronal test method later [26]. The modified isochronal test method has its advantage: it is unnecessary that pressure builds up to the original value during each shut-in period, the testing duration is thus relatively shorter than the isochronal test, and is especially applicable to low permeability gas zones.

As far as the analysis method is concerned, pressure itself was used for analysis in the early stage before the 1960s, then in consideration of compressibility of real gas, the pressure-squared expression method in resolving the partial differential equations was proposed by Russell and co-workers and real gas pseudo pressure expression by Al-Hussainy and colleagues was used (refer to Chapter 2), and on the basis of which LIT deliverability equation was generated. This LIT deliverability equation can better describe turbulent flow effects during gas flow in the formations and so can calculate the AOFP of gas wells more accurately.

However, from the development of the deliverability test method, we can see that all the test methods were generated before the mid-20th century; for instance, the back-pressure test method has been applied for nearly 80 years, that all the methods were generated from the necessity of production planning; as far as the methods themselves are concerned, due to the limitation of technical conditions and the understanding of gas flow at that time, they are inevitably characterized by trial and estimation, without strict theory.

It should be pointed out that large and medium scale gas fields discovered in China in the past years showed many very special characteristics, for example, some sandstone reservoirs with low or ultralow permeability cannot contribute commercial flow until treated by fracturing stimulation; permeable parts in some fissured limestone reservoirs are developed directionally and locally; there are some very thick and abnormal high-pressure sandstone gas reservoirs with developed fissures and some eruptive volcanic rock reservoirs—they are very complicated in the view of structural characteristics, reservoir lithology, or fracture development characteristics; and, furthermore, judged from their dynamic characteristics, there are plenty of reserves stored in fluvial sandstones with obvious banded impermeable or ultralow permeability boundaries. Take

Permo—Carboniferous formations in the Ordose Basin as an example; the major reservoir specialties are as follow.

1. Low permeability sandstone reservoir. The effective permeability of most gas zones is lower than 1 mD and is 2—3 mD only in some high-yielding wells; the lowest one is even less than 0.1 mD.
2. There are induced fractures at the bottom hole. Because fracturing treatment must be performed before a normal gas test due to low permeability, dynamic evaluation of the formation is basically for fractured wells.
3. Fluvial depositional formation with banded impermeable boundaries.

How to make a good development plan for these special lithologic reservoirs on the basis of deliverability tests is a very serious problem that needs to be studied further.

It is very difficult to describe comprehensively a gas reservoir or gas field with the classical deliverability well test methods for these special lithologic reservoirs. This chapter discusses in detail the development, existent problems, and related research results of the deliverability test methods on the basis of practice and experience during the last years in China.

1.2 Gas Well Deliverability Indices

Indexes of deliverability, such as wellhead flow rate and AOFP, have been widely used, but a thorough understanding of them is not enough yet.

1.2.1 Deliverability of a Gas Well

Generally, the deliverability of a gas well refers to the productivity of the gas well. It can either mean wellhead flow rate of the gas well with a given choke or be expressed by AOFP or IPR curve of the gas well; in terms of more generalized expression, wellhead flow rate and its change under a given bottom-hole or wellhead pressures, for example, wellhead flow rate when the wellhead pressure is not lower than export pressure.

1.2.2 Absolute Open Flow Potential of Gas Wells

Absolute open flow potential, or q_{AOF} as its name suggests, is the upper limit flow rate of a gas well. It is usually defined as the flow rate assuming the bottom-hole flowing pressure, that is, the sandface pressure measured by gauge is zero, or the absolute pressure is one atmospheric pressure (1 atm). Obviously, this condition cannot be realized due to the existence of wellbore friction. In other words, AOFP is an index impossible to verify directly with current technical conditions.

1.2.3 Validity of AOFP

There are no feasible and reliable methods for evaluating the validity of AOFP calculated by various deliverability test methods because direct verification of AOFP is impossible. In case of the same gas well tested with different methods, a test analyzed by different methods, or a test analyzed by the same analysis methods with different pressure variables, the calculated AOFP values may be different and even vary different sometimes. Inevitable unreal results lead to the wrong direction when some management personnel take the indices tendentiously according to their own mind as the basis of their decision making.

1.2.4 Initial and Dynamic AOFP

The AOFP of a gas well commonly determined in accordance with related industrial standards means calculated AOFP value from the test data obtained when the well is just put into production, which should be regarded as "initial AOFP." According to the commonly used design methods of the development plan, take one-fourth to one-fifth of the AOFP value to be the design "rational" flow rate of that gas well when put into production. However, no reasonable proofs are discovered by looking up related literatures. According to the author's analysis of real field data and theoretical research for many years, the understanding of stabilized deliverability of a gas well is not explicit or is even wrong sometimes.

There are three important factors influencing gas well deliverability: formation flow capacity of the gas zone kh, gas well completion state, that is, skin factor S, and formation pressure p_R. Due to continuous depletion of formation pressure, along with gas well production, gas well deliverability is changing and is reflected by continuously decreasing "dynamic deliverability," that is, dynamic IPR curve and dynamic AOFP.

A declining process of dynamic deliverability is determined directly by the effective drainage area and dynamic reserves of block where the well locates; the smaller the dynamic reserves, the quicker the deliverability declines.

1.3 Initial Deliverability, Extended Deliverability, and Allocated Production of Gas Well

There are several indices for describing the productivity of a gas well, and the meanings of these deliverability indices are changing with production time.

1.3.1 Initial Deliverability Index

The initial deliverability index indicates the gas productivity within a short interval (e.g., 1–3 days) after a gas well is primarily started up, when formation pressure remains its original level, and the flowing pressure may be still in the decline process and is not stable; hence, they are normally too high.

The test in exploration wells, especially when using the single point test method, often gives this transient AOFP.

The main factors influencing initial deliverability are formation flow capacity of the gas zone kh and well completion quality, that is, skin factor S; in addition, the initial deliverability is closely related to the time of the test data points, which will be illustrated in detail later.

1.3.2 Extended Deliverability Index

When an isochronal or modified isochronal test is conducted, an extended test point (or pseudo stable point) shall be arranged after the transient test, and extended AOFP (or pseudo AOFP) is obtained from test data. The obtained deliverability index is the extended deliverability index. As the stabilized bottom-hole flowing pressure in extended flow (usually lasts over 10 days to several dozen days) is always lower than transient pressure in the transient test, even though the flow rate of extended points is steady, similar to that of the initial transient test, the calculated AOFP is far lower than the initial AOFP.

The extended deliverability index depends not only on formation parameters, but also, and more important, on near-wellhole boundaries configuration and recoverable reserves controlled by the tested well.

If a back-pressure test is performed and the flow duration of each rate is long enough, an index similar to extended deliverability also can be gained.

1.3.3 Allocating Flow Rate Index

Being an economic index, allocating the flow rate depends on gas productivity of the gas well; it can be adjusted within the allowable deliverability range according to the related economic requirement.

In a gas field, each gas well can be arranged to maintain stable production at a low flow rate for many years or some wells can be planned to produce at a high flow rate; when a natural decline of the flow rate takes place in these wells, some new wells can be arranged to be put into production. The most important things are economic profit and ultimate recovery. In addition, actual production planning of a gas well depends on market demand, effects of seasonal changes upon commercial gas consumption, the regulating ability of gas storage, and so on.

Two gas wells may have absolutely different naturally declining rates, despite the same initial AOFP. The natural decline rate depends on the reservoir boundary. If the drainage area of a well is large, the rate will naturally decline slowly, otherwise the rate will decline fast. Only when the dynamic reserves controlled by each single well are confirmed on the basis of geological and dynamic performance researches can a prediction be made for a stable flow rate and a production declining rate so that a rational allocating flow rate is possible consequently. For low-yield gas wells, such stimulation treatments as acidizing and fracturing can be considered to enhance single well productivity and flow rate, and even a noncommercial gas well can be transformed into a commercial one. However, it is impossible now, in view of current technique conditions, to enhance the rate mainte-nance capacity of a single gas well and realize stable production by this stimulation treatment unless adopting special drilling techniques (e.g., horizontal well or multilateral well) and combining reliable geological studies so that more producing reserves are connected.

2 THREE CLASSICAL DELIVERABILITY TEST METHODS

Three classical methods refer to common deliverability test methods developed in the middle of the last century: back-pressure test method, isochronal test method, and modified isochronal test method.

2.1 Back-Pressure Test Method

The back-pressure test method was developed in 1929 and was improved by Rawlines and Schellhardt in 1936 [23, 24]. The well is produced to stabilized pressure at more than three increasing rate, and record bottom-hole flowing pressure during gas flow rate simultaneously. The corresponding relation between gas flow rate and flowing pressure is show in Figure 3.1. Related data are listed in Table 3.1.

Draw data listed in Table 3.1 onto a plot as Figure 3.2 and named deliverability plot, then from which AOFP can be calculated.

In the deliverability plot, the ordinate indicates producing a pressure differential in terms of pressure squared, $\Delta p_i^2 = p_R^2 - p_{wfi}^2$, in which p_R, p_{wfi}, and q_{gi} are formation pressure, bottom-hole flowing pressures, and gas flow rates with related chokes in succession. Normally the four test data points can be regressed into a straight line; if p_{wf} (absolute pressure) is 0.1 MPa, suggesting that the bottom-hole blowout pressure is equivalent to atmospheric pressure (1 atm). When the gas flow rate reaches its upper limit value, the gas well flow rate is AOFP, expressed by q_{AOF}. Generally speaking, it is

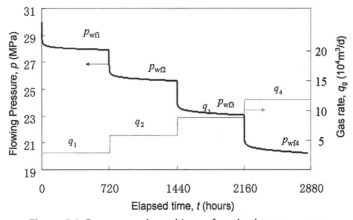

Figure 3.1 Pressure and rate history for a back-pressure test.

Table 3.1 Example of Pressure and Flow Rate for Back-Pressure Test

Flow and shut-in sequence	Stabilization start time (after well opening) (hour)	Formation pressure, p_R (MPa)	Flowing pressure, p_{wf} (MPa)	Gas flow rate, q_g (10^4 m^3/day)
Initial shut in		30		
First flow	720		27·9196	2
Second flow	720		25·6073	4
Third flow	720		23·0564	6
Fourth flow	720		20·2287	8

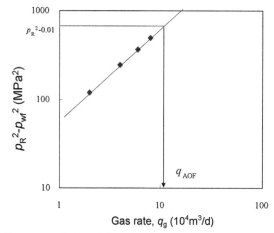

Figure 3.2 Deliverability plot from a back-pressure test.

impossible to measure AOFP directly, for it is impossible that bottom-hole pressure decreases to the atmospheric pressure, so AOFP can only be calculated by a related equation or graphic.

The back-pressure test requires a stable gas flow rate, stable bottom-hole flowing pressure, and constant formation pressure during each flow period with a choke, but stabilized flowing pressure is difficult to achieve in practice. The well may be kept in long-term production for pressure stabilization, sometimes resulting in a decline of formation pressure. This problem hinders application of the back-pressure test method.

2.2 Isochronal Test Method

Due to the disadvantages of the back-pressure test mentioned previously, the "isochronal deliverability test method" was suggested by Cullender in 1955 [25]. With the isochronal test, the well is again produced at more than 3 increasing rates but a shut-in period is introduced between each flow and the flowing pressure is measured at the same time.

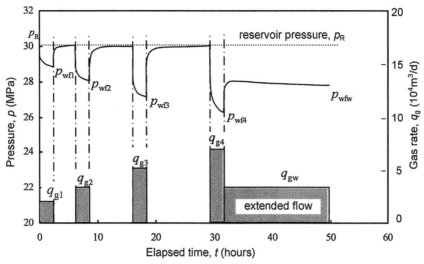

Figure 3.3 Pressure and rate history for an isochronal test.

There is no requirement of stabilized flowing pressure, but the producing duration at each rate must be equal and well shut in must last until the pressure reaches its original value. After the tests for gas flow rate and pressure transient points are completed, continue to produce at a relative lower flow rate until the flowing pressure stabilizes. The corresponding relation of gas flow rate and pressure using the computer software simulation method is shown in Figure 3.3.

Measured pressure and gas flow rate data are given in Table 3.2.

Application of the isochronal test method saves flowing time and decreases relief gas being wasted significantly. However, the test duration cannot be shortened effectively because pressure must build up to the original value during each shut in after each well flowing.

Table 3.2 Example of Pressure and Flow Rate for Isochronal Test

Flow and shut-in sequence	Duration of flows and shut ins (hour)	Formation pressure, p_R (MPa)	Bottom-hole flowing pressure, p_{wf} (MPa)	Gas flow rate, q_g (10^4 m^3/day)
Initial shut in		30	/	
First flow	2.5	/	28.1873	2
First shut in	4	30	/	
Second flow	2.5	/	26.1575	4
Second shut in	7	30	/	
Third flow	2.5	/	23.9153	6
Third shut in	10	30	/	
Fourth flow	2.5	/	21.4440	8
Extended flow	18	/	25.5044	4

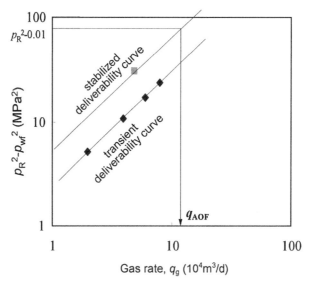

Figure 3.4 Deliverability plot for an isochronal test.

Each gas flow rate q_{gi} is corresponding to flowing pressure differential $\Delta p_i^2 = p_R^2 - p_{wfi}^2$, from which the relationship between gas flow rate and flowing pressure differential is obtained. For the final stabilized deliverability data point, the gas flow rate is q_{gw}, and the flowing pressure differential is $\Delta p_w^2 = p_R^2 - p_{wfw}^2$.

Figure 3.4 is a deliverability plot from the isochronal method. A transient deliverability curve is obtained from the regression of four transient deliverability data points in Figure 3.4. In order to obtain the stabilized deliverability curve, draw a parallel straight line of transient deliverability curve at the stabilized deliverability data point—this is the stabilized deliverability curve. Similarly, AOFP can be calculated by this graphic method.

2.3 Modified Isochronal Test Method

The modified isochronal test method was developed by Katz and colleagues in 1959 [26]. This method overcomes the shortcomings of the isochronal test method. It has been proved theoretically that for each shutting in during the isochronal test, pressure building up to its original value is unnecessary and thus transient test duration is shortened significantly. The corresponding relation of gas flow rate and pressure is shown in Figure 3.5. Related data are given in Table 3.3.

It can be seen from Figure 3.5 that the modified isochronal test not only reduces the duration of producing and wasted gas, but also shortens the total test duration. When plotting with test data, the pressure differential corresponding to gas flow rate q_{gi} is

$$\Delta p_i^2 = p_{wsi}^2 - p_{wfi}^2, \tag{3.1}$$

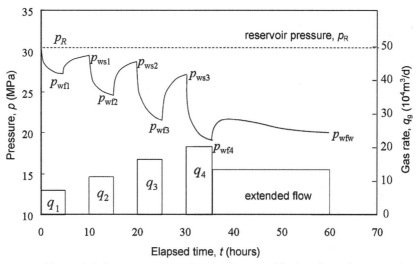

Figure 3.5 Pressure and rate history for a modified isochronal test.

that is,

$$\Delta p_1^2 = p_R^2 - p_{wf1}^2, \text{ corresponding to } q_{g1}$$
$$\Delta p_2^2 = p_{ws1}^2 - p_{wf2}^2, \text{ corresponding to } q_{g2}$$
$$\Delta p_3^2 = p_{ws2}^2 - p_{wf3}^2, \text{ corresponding to } q_{g3}$$
$$\Delta p_4^2 = p_{ws3}^2 - p_{wf4}^2, \text{ corresponding to } q_{g4}$$
$$\text{and } \Delta p_w^2 = p_R^2 - p_{wfw}^2, \text{ corresponding to } q_{gw}$$

A deliverability plot for the modified isochronal test can be made with these data, and it is the same as that of the isochronal test (Figure 3.4). Also, AOFP (q_{AOF}) can be calculated similarly.

Table 3.3 Example of Pressure and Flow Rate for Modified Isochronal Test

Flow and shut-in sequence	Duration of flows and shut ins (hour)	Shut-in bottom-hole pressure, P_{ws} (MPa)	Flowing pressure, p_{wf} (MPa)	Gas flow rate, q_g (10^4 m³/day)
Initial shut in		30(p_R)	/	
First flow	5	/	27.9145	2
First shut in	5	29.9139	/	
Second flow	5	/	24.7785	4
Second shut in	5	29.7887	/	
Third flow	5	/	20.3950	6
Third shut in	5	29.6372	/	
Fourth flow	5	/	14.0560	8
Extended flow	25	/	19.3545	6

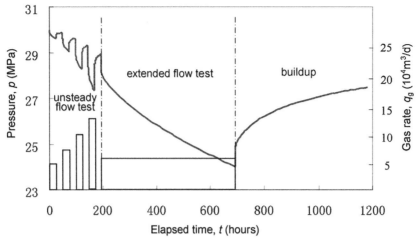

Figure 3.6 Pressure and rate history for an improved modified isochronal test.

Combined with practical conditions in China, some improvements of modified isochronal test procedures and the AOFP calculation method have been made; differences between the improved method and the classical ones are (1) adding a shut in after the fourth flow to acquire more data of buildup and (2) adding a final buildup after extended flow to get the final buildup pressure, by which we cannot only calculate the reservoir parameters and outer boundary distances, but also identify if the reservoir pressure has decreased, so that the producing pressure differential of the extended test can be adjusted, and the shut-in stabilized pressure p_{SS} can be obtained.

An improved modified isochronal test is shown in Figure 3.6.

2.4 Simplified Single Point Test

There are two different analysis methods of a simplified single point test.

2.4.1 Stable Point LIT Deliverability Equation

The method of establishing a stable point LIT deliverability equation after the gas well is put into production testing is discussed in detail in Section 6. It is a new method first proposed by the author on the basis of many years of studies and field practices. The method is applicable to establishing initial deliverability equation of gas wells, plotting initial IPR curves, estimating initial AOFP, and tracing investigation of dynamic deliverability indices and deliverability decline of gas wells after put into production.

2.4.2 AOFP Calculation With Single Point Test Method

The other one is AOFP calculation with the single point test method. It is to figure out the variation regular pattern of deliverability equation coefficients by statistics on the basis of a great deal of deliverability test results in a gas field and to establish a formula of calculating AOFP that is adaptive to this gas field or to adopt the existing AOFP calculating formula in literatures and select one single data point during the production test of a gas well so as to calculate AOFP. Refer to Section 8 of this chapter for related formulae and application.

2.5 Schematic Diagram of Calculating Pressure Differential for Various Test Methods

The relationship between pressure differentials and test point pressures is marked schematically in Figure 3.7 to avoid any errors that may occur in calculations during analysis.

3 TREATMENT OF DELIVERABILITY TEST DATA

3.1 Two Deliverability Equations

There are two commonly used deliverability equations at present:

1. Exponential equation (also called "simple analysis")
2. Binomial equation (also called "laminar−inertial−turbulent flow analysis" or "LIT analysis").

The deliverability equations are illustrated using the pressure-squared expression described later.

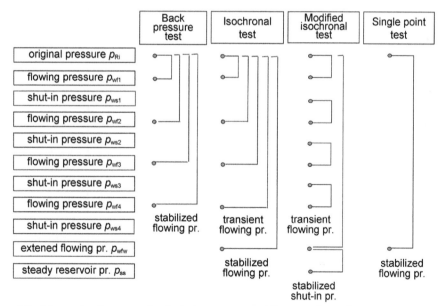

Figure 3.7 Schematic diagram of calculating pressure difference for different deliverability test analyses.

3.1.1 Exponential Deliverability Equation

A formula of flow rate and pressure differential is put forward according to practical experiences based on many observations by Rawlins and Schellhardt in 1936 [24]:

$$q_g = C\Delta p^2 = \left(p_R^2 - p_{wf}^2\right)^n \tag{3.2}$$

where q_g is gas flow rate, $10^4\,m^3/day$; p_R is formation pressure, MPa; p_{wf} is bottom-hole flowing pressure, MPa; C is coefficient of deliverability equation $(10^4\,m^3/day)/(MPa^2)^n$; and n is exponent of deliverability equation, a decimal between 0.5 and 1.0.

By taking a logarithm on both sides, it is shown that

$$\log q_g = n \log \left(\Delta p^2\right) + \log C. \tag{3.3}$$

Seen from Equation (3.3), if gas flow rate q_g and pressure-squared differential Δp^2 are plotted into logarithmic coordinates, a straight line is gained, the slope of the straight line is n, and the intercept is $\log C$.

According to common practices, the equation is often plotted with gas rate q_g as an abscissa against pressure difference Δp^2 as ordinate, and Equation (3.3) can be changed into

$$\log\left(\Delta p^2\right) = \frac{1}{n} \log q_g - \frac{\log C}{n}. \tag{3.4}$$

The deliverability equation is still expressed by a straight line in this case, but the slope becomes $1/n$. Obviously Figures 3.8 and 3.2 are essentially the same in plotting.

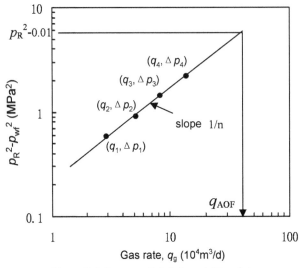

Figure 3.8 Exponential deliverability plot.

Figure 3.8 presents a common plotting method in exponential deliverability analysis. A deliverability equation like Equation (3.2) can be attained by analysis of this plot. Let $p_{wf} = 0.101$ MPa, then $\Delta p^2 = p_R^2 - 0.101^2$, and AOFP ($q_{AOF}$) is calculated when putting it into the equation:

$$q_{AOF} = C\left(p_R^2 - 0.101^2\right)^n. \tag{3.5}$$

Most modern well test interpretation software has deliverability analysis function; if correspondent gas flow rate q_i and flowing pressure p_{wfi} are input into a computer, it will automatically generate a deliverability plot and display the calculated q_{AOF} value. The index n in the equation is called the "turbulent flow intensity index," for that $n = 1$ indicates gas flow in the reservoir is totally laminar flow, whereas that $n = 0.5$ suggests totally turbulent flow. Generally, $0.5 < n < 1.0$, which means that gas flow in the reservoir is partially laminar flow and partially turbulent flow.

Figure 3.9 shows a parameter input interface of FAST for calculating deliverability, and Figure 3.10 gives a real example of a deliverability plot drawn with measured data.

3.1.2 LIT Equation

Laminar–inertial–turbulent flow analysis [20] is widely used for deliverability analysis in most areas of the world except North America. LIT analysis is also called

Sandface AOF

	Time (hr)	Sandface Pressure (kPa)	Pressure Squared (10^6 kPa^2)	ΔP_2/Rate (10^6kPa^2)/(10^3m^3/d)	Gas Rate (10^3m^3/d)
Shut-in		27137.01	736.42		
Flow #1		26510.13	702.79	1.012	33.059
Shut-in		26968.22	727.28		
Flow #2		25749.87	663.06	1.241	51.477
Shut-in		26666.51	711.10		
Flow #3		24715.64	610.86	1.446	69.037
Shut-in		26140.51	683.32		
Flow #4		23703.69	561.66	1.421	85.051
Shut-in		25577.41	654.20		
Flow #5					
Shut-in					
Flow #6					
Ext. Flow:		21577.99	465.61	8.831	40.634
Stab. Shut-in:		27137.01	736.42		
Stab. Flow:					

Figure 3.9 Parameter input interface of well test interpretation software FAST for deliverability analysis (pressure-squared form).

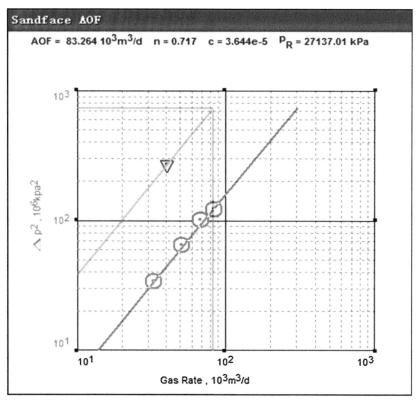

Figure 3.10 Exponential deliverability plot drawn by well test interpretation software FAST (in pressure-squared form).

binominal deliverability analysis. It is proposed by Forchheimer and Houpeurt on the basis of strict theoretical deduction of the solution of flow equations and is expressed by

$$p_R^2 - p_{wf}^2 = Aq_g + Bq_g^2, \tag{3.6}$$

where the coefficients A and B indicate the parts of laminar and turbulent within the gas flow in the reservoirs, respectively.

Equation (3.6) is often expressed as follows for linear regression:

$$\frac{p_R^2 - p_{wf}^2}{q_g} = A + Bq_g. \tag{3.7}$$

The left side of the equation, $\dfrac{p_R^2 - p_{wf}^2}{q_g}$, is also called "normalized pressure squared" and is expressed as Δp_N^2. A linear equation can be constructed with the plot of Δp_N^2 vs q_g, as shown in Figure 3.11.

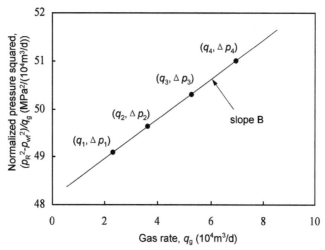

Figure 3.11 Schematic diagram of LIT analysis.

Let $p_{wf} = 0.101$ MPa in the LIT equation, or $\Delta p^2 = \Delta p^2_{max}$, and put it into Equation (3.6) to calculate AOFP:

$$q_{AOF} = \frac{-A + \sqrt{A^2 + 4B\Delta p^2_{max}}}{2B}. \qquad (3.8)$$

3.2 Difference between Two Deliverability Equations

Both exponential and LIT deliverability equations are mathematical expressions obtained from the match of measured data of pressures and corresponding flow rates and then used to predict flow rate values under any certain production condition. When the flowing pressure is atmospheric pressure, the gas flow rate is just AOFP.

Being deduced from the percolation mechanics equation, the LIT deliverability equation is highly adaptive to different formation and more accurate, which is specifically illustrated in Section 3.4. However, being an empirical formula, the exponential deliverability equation affords relatively less accurate results. It is discussed in detail as follows.

3.2.1 If Gas Flow Rate of Tested Well During Testing is Higher Than 50% of AOFP, Calculation Results of Two Deliverability Equations are Similar

A set of deliverability test data is given in Table 3.4. Data were acquired during a back-pressure test with four flow rates of a gas well in a homogeneous sandstone reservoir. Flow rates of the selected test points were 2, 4, 6, and 8×10^4 m³/day, respectively.

Table 3.4 Back-Pressure Test Result (Example No. 1)

Well opening and shut-in procedures	Formation pressure, p_R (MPa)	Bottom-hole flowing pressure, p_{wf} (MPa)	Gas flow rate, q_g (10^4 m³/day)
Initial shut in	30		
Initial flow		28.046	2
Second flow		25.868	4
Third flow		23.465	6
Fourth flow		20.799	8

Exponential and LIT deliverability equations established with the aforementioned test data are as follow:

$$q_g = 0.0194 \left(p_R^2 - p_{wf}^2 \right)^{0.978} \tag{3.9}$$

$$p_R^2 - p_{wf}^2 = 56.07 \, q_g + 0.282 q_g^2 \tag{3.10}$$

Equation (3.9) can be rewritten as

$$\left(p_R^2 - p_{wf}^2 \right) = 56.327 q_g^{1.022}$$

Based on equation analyses,

$$q_{AOF} \left(\text{binomial} \right) = 14.9 \times 10^4 \text{m}^3/\text{day}$$

$$q_{AOF} \left(\text{exponential} \right) = 15.03 \times 10^4 \text{m}^3/\text{day}.$$

It can be seen that there is little difference of AOFP values calculated by the two equations. This example is in a homogeneous reservoir; the highest measured gas flow rate is more than a half of AOFP and the pressure-squared difference is also nearly a half of reservoir pressure-squared difference.

Figure 3.12 shows two relation curves of flow rate vs flowing pressure obtained from the two different deliverability equations. This relation curve of flow rate vs flowing pressure is usually named an "inflow performance relation curve," or IPR curve for short, as it indicates different gas flow rates under different bottom-hole flowing pressures.

As can be seen in Figure 3.12:

- Perfect superposition of two deliverability curves within the range of test data points;
- A small deviation of exponential deliverability curve from binomial deliverability curve beyond the range of test data points;
- Where Δp is equal to Δp_{max}, that is, bottom-hole flowing pressure decreases down to 1 atm (= 0.1013 MPa), the exponential deliverability curve indicates the AOFP (q_{AOF}) is 15.03×10^4 m³/day, which is only about 2% higher than that of the binomial deliverability curve, 14.9×10^4 m³/day; the difference between them is very small.

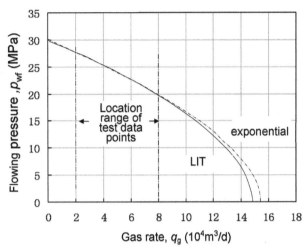

Figure 3.12 IPR curves generated by different deliverability equations (example No. 1).

3.2.2 Greater Error Generates from Exponential Deliverability Equation if Pressure Differences are Small of all Test Points

In an example listed in Table 3.5, the maximum pressure pressure-squared difference of test points (128.5 MPa^2) is smaller than 15% of that of the formation pressure pressure-squared difference (900 MPa^2),); the exponential deliverability curve deviates from the LIT deliverability curve seriously when the producing pressure difference is big.

Exponential and LIT equations constructed with the aforementioned test data are as follow:

$$q_g = 0.2122\left(p_R^2 - p_{wf}^2\right)^{0.748} \tag{3.11}$$

and

$$\left(p_R^2 - p_{wf}^2\right) = 8.1599\, q_g + 0.9961 q_g^2. \tag{3.12}$$

Table 3.5 Back-Pressure Test Result (Example No. 2)

Well opening and shut-in procedures	Formation pressure, p_R (MPa)	Bottom-hole flowing pressure, p_{wf} (MPa)	Gas flow rate, q_g (10^4 m^3/day)
Initial shut in	30		
Initial flow		29.663	2
Second flow		29.164	4
Third flow		28.529	6
Fourth flow		27.776	8

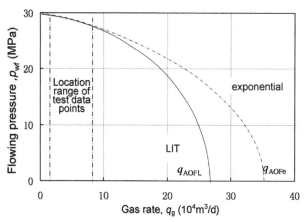

Figure 3.13 IPR curves generated by different deliverability equations (example No. 2).

Equation (3.11) can be rewritten as

$$\left(p_R{}^2 - p_{wf}{}^2\right) = 7.945 \, q_g{}^{1.3369}$$

and from the related deliverability equations we obtain

$$q_{AOF} \left(binomial\right) = 26.24 \times 10^4 \, m^3/day$$

$$q_{AOF} \left(exponential\right) = 34.40 \times 10^4 \, m^3/day$$

There is a dramatic difference between them. Figure 3.13 shows the difference of their IPR curves.

It can be seen from Figure 3.13 that perfect superposition of two curves occurs within the range of test data points, which indicates that the deliverability equation is derived correctly. However, there is an obvious deviation of the exponential deliverability equation from the binomial deliverability equation when bottom-hole flowing pressure declines, causing an error approximately 30% higher for calculated AOFP.

3.3 Three Different Pressure Expressions of Deliverability Equation

As introduced in Chapter 2, pseudo pressure (ψ) is regarded as the most proper pressure expression in describing the gas flow process. As shown in Equation (2.52), pseudo pressure is

$$\psi = \int_{p_0}^{p} \frac{2p}{\mu Z} dp.$$

The exponential deliverability equation is expressed as

$$q_g = C_\psi (\psi_R - \psi_{wf})^n \qquad (3.13)$$

and the pseudo pressure expression of the LIT deliverability equation is

$$\psi_R - \psi_{wf} = A_\psi q_g + B_\psi q_g^2. \qquad (3.14)$$

It should be pointed out that the pseudo pressure expression of the deliverability equation is applicable to various gas with different components or in different formation pressure and temperature environments, but the pressure should be transformed into pseudo pressure, that is, integral operation according to Equation (2.52) must be done before deliverability analysis with pseudo pressure starts. Also, before this integral operation, gas components or related parameters must be collected so as to obtain the relation between the values μ, Z and pressure p.

The operation process seems slightly troublesome, but it does not need to be done by analysts manually today, as well test interpretation software is commonly used. Only by clicking the pseudo pressure option key can the digital input interface (pressure square) shown in Figure 3.9 be transformed into a pseudo pressure operation interface, and related pseudo pressure analysis results are obtained; please refer to Figures 3.14, 3.15, and 3.16. It should be noted that data in a light color indicated during the pseudo pressure operation process are not necessary to be applied directly. Reservoir engineers only need to know the final results, such as AOFP (q_{AOF}), IPR curve, and deliverability equation and related deliverability plots.

As illustrated in Chapter 2, if the formation pressure is relatively low, pseudo pressure, in the deliverability analysis, can be approximately replaced by pressure square: $\psi = a_1 p^2$ (refer to Figure 2.54), and analysis results are also precise enough; if all pressure data are relatively high, pseudo pressure is nearly a linear function of pressure: $\psi = a_2 p + b_2$ (refer to Figure 2.54), pseudo pressure, in a deliverability analysis, can be approximately replaced by pressure itself, and analysis accuracy is also precise enough.

For many deliverability tests, the pressure variation range of test points may be very broad, including quite low and quite high pressure data. In these cases, precision requirements for analysis cannot be met no matter what pressure square p^2 or pressure p is used; and only with pseudo pressure in the analysis can the analysis precision requirement be surely met. A field example is shown in Figure 3.17a; When LIT deliverability analysis is performed using the pressure square method, factor B in Formula Equation (3.6) is smaller than zero, and calculation of AOFP (q_{AOF}) is impossible. In plotting of the deliverability plot, software treats B as zero and a horizontal straight line is obtained. This plot is useless for correct deliverability analysis.

If the analysis is performed with pseudo pressure instead, a normal deliverability plot for the same data is obtained and is shown in Figure 3.17b.

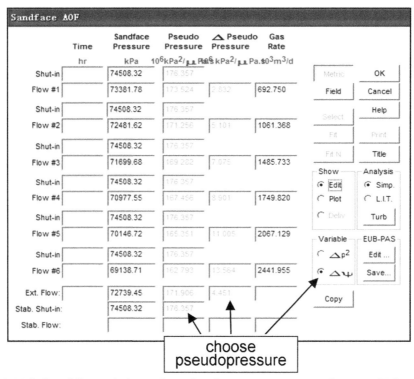

Figure 3.14 Deliverability analysis interface of well test interpretation software FAST (in terms of pseudo pressure).

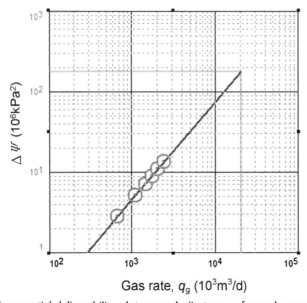

Figure 3.15 Exponential deliverability plot example (in terms of pseudo pressure; by FAST).

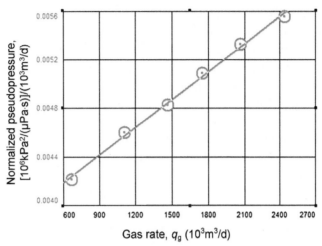

Figure 3.16 Binomial deliverability plot example (in terms of pseudo pressure; by FAST).

Seen from Figure 3.17b, a normal LIT deliverability curve appears in terms of pseudo pressure, which can be used for deliverability analysis, and the deliverability equation is

$$\psi_R - \psi_{wf} = 0.2019q_g + 1.061 \times 10^{-4}q_g^2.$$

With this equation, AOFP is calculated to be 131.98×10^3 m^3/day. SI units are used in the software, including

p_F is the Pressure, kPa

q_F is the Gas flow rate, 10^3 m^3/day

ψ_F is the Pseudo pressure, kPa^2/(μPa·s)

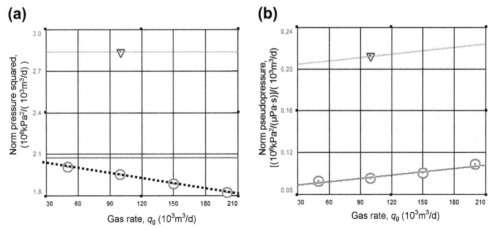

Figure 3.17 Abnormality of deliverability curve drawn with pressure square (a) and normal deliverability curve with pseudo pressure (b).

C_F is the Exponential deliverability equation coefficient, $(10^3 \text{ m}^3/\text{day})/[\text{kPa}^2/(\mu\text{Pa}\cdot\text{s})]^n$

A_F is the LIT equation coefficient, $[10^6 \text{ kPa}^2/(\mu\text{Pa}\cdot\text{s})]/(10^3 \text{ m}^3/\text{day})$

B_F is the LIT equation coefficient, $(10^6 \text{ kPa}^2/\mu\text{Pa}\cdot\text{s})/(10^3 \text{ m}^3/\text{day})^2$

These SI units are different from Chinese CSU units. In the Chinese CSU unit system, the units of the aforementioned parameters are

p_{CSU} is the Pressure, MPa;

q_{CSU} is the Gas flow rate, $10^4 \text{ m}^3/\text{day}$

Ψ_{CSU} is the Pseudo pressure, $\text{MPa}^2/(\text{mPa}\cdot\text{s})$

C_{CSU} is the Exponential deliverability equation coefficient, $(10^4 \text{ m}^3/\text{day})/[\text{MPa}^2/ (\text{mPa}\cdot\text{s})]^n$

A_{CSU} is the LIT deliverability equation coefficient, $[\text{MPa}^2/(\text{mPa}\cdot\text{s})]/(10^4 \text{ m}^3/\text{day})$

B_{CSU} is the LIT deliverability equation coefficient, $[\text{MPa}^2/(\text{mPa}\cdot\text{s})]/(10^4 \text{ m}^3/\text{day})^2$

Among the units of the parameters in the aforementioned deliverability equation in the Chinese CSU unit system and the SI unit system, the unit of n is just the same, and the units of coefficients C, A, and B have the following transformation relations:

$$C_{CSU} = 10^{3n-1} C_F \tag{3.15}$$

$$A_{CSU} = 10^4 A_F \tag{3.16}$$

$$B_{CSU} = 10^5 B_F \tag{3.17}$$

After a conversion of units, the deliverability equation deduced for the previous example in terms of CSU units is translated into

$$\psi_R - \psi_{wf} = 2.019 \times 10^3 q_g + 10.610 q_g{}^2,$$

where Ψ is pseudo pressure, $\text{MPa}^2/(\text{mPa}\cdot\text{s})$; and q_g is gas flow rate, $10^4 \text{ m}^3/\text{day}$.

It should be pointed out that $B < 0$ is not totally due to aforementioned reason, that is, it did not use pseudo pressure. An abnormality of the deliverability equation is sometimes caused by loading water or relieving of the damage in the bottom hole. This is discussed in detail in Section 3.6.

Please refer to the related appendix content for the conversion of equation coefficients in different unit systems.

In a word, expressions of the deliverability equation under different pressure forms can be concluded as follows:

1. Exponential deliverability equation:
   - Pseudo pressure

$$q_g = C_\psi (\psi_R - \psi_{wf})^n \tag{3.13}$$

- Pressure square

$$q_g = C_2\left(p_R^2 - p_{wf}^2\right)^n \tag{3.2}$$

- Pressure

$$q_g = C_1\left(p_R - p_{wf}\right)^n \tag{3.18}$$

2. LIT equation:
   - Pseudo pressure

$$\psi_R - \psi_{wf} = A_\psi q_g + B_\psi q_g^2 \tag{3.14}$$

- Pressure square

$$p_R^2 - p_{wf}^2 = A_2 q_g + B_2 q_g^2 \tag{3.6}$$

- Pressure

$$p_R - p_{wf} = A_1 q_g + B_1 q_g^2 \tag{3.19}$$

Analysts can select and apply these formulas according to practical testing conditions onsite. If well test interpretation software is used, pseudo pressure Equations (3.13) and (3.14) are preferred for analysis to enhance analysis precision and avoid any abnormalities.

3. Analysis of a field example: Well kL205

This field example given here is a back-pressure test of Well KL205, which gives the IPR curve, that is, the relationship between flow rate and flowing pressure. The deliverability equations are obtained from the pseudo pressure method and pressure-squared method, respectively. It significantly demonstrates the difference in calculation results using the two different methods. The deliverability equation with the pseudo pressure method is

$$\psi_R - \psi_{wf} = 37.290 q_g + 0.07634 q_g^2$$

and the deliverability equation with the pressure-squared method is

$$p_R^2 - p_{wf}^2 = 2.229 q_g + 3.859 \times 10^{-3} q_g^2$$

IPR curves from both deliverability equations are plotted simultaneously in Figure 3.18.

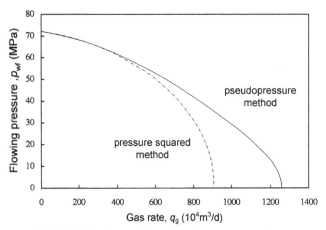

Figure 3.18 IPR curves of Well KL205 obtained from pseudo pressure and pressure-squared methods.

In Figure 3.18, the solid line indicates an IPR curve plotted by the pseudo pressure method, and the dotted line indicates the other one plotted by the pressure-squared method. It can be seen that there is a big difference between them. The pseudo pressure method has relatively higher precision because it sufficiently considers influence factors of gas properties in different pressure conditions. If permitted, it is better to adopt results obtained from the pseudo pressure method; those calculated by the pressure-squared method can be taken for reference. The difference of results obtained from both methods will be smaller if the flow rate range is wide enough, that is, the highest gas flow rate during testing is higher than half of AOFP. However, as discussed previously, for the pressure-squared method, errors will be inevitable when pressure is high.

4 PARAMETER FACTORS INFLUENCING GAS WELL DELIVERABILITY

To establish a deliverability equation is to make sure the relationship between gas flow rate and producing pressure drops. If actual corresponding values of gas flow rate q_g and flowing pressure are obtained during testing, a numerical relation between them is certainly found out, and this numerical relation can also be derived based on percolation mechanics theory, and how formation and fluid parameters influence the deliverability can be analyzed from the theoretical equation.

Summing up, three main factors influence the deliverability of a gas well:

1. Near wellbore formation coefficient kh
2. Formation pressure p_R and producing pressure difference Δp
3. Completion quality indicated by skin factor S

These are discussed one by one here.

4.1 Expressions of Coefficients A And B in Deliverability Equation of a Well In Infinite Homogeneous Reservoir

The relationship of flow rate and pressure in terms of pressure square in a transient state in an infinite homogeneous reservoir can be obtained by resolving percolation mechanics Equation (2.51) as

$$p_{Ri}^2 - p_{wf}^2 = \frac{42.42 \times 10^3 \overline{\mu}_g \overline{ZT} p_{sc}}{kh T_{sc}} \cdot q_g \left[\lg \frac{8.091 \times 10^{-3} kt}{\phi \overline{\mu}_g C_t r_w^2} + 0.8686 S_a \right] \quad (3.20)$$

where p_{Ri} is original formation static pressure, MPa; p_{wf} is bottom–hole flowing pressure, MPa; q_g is surface flow rate of gas well, 10^4 m^3/day; k is effective permeability, mD; h is effective thickness, m; $\overline{\mu}_g$ is average viscosity of gas, mPa·s; $\overline{Z}, \overline{T}$ is average Z factor and average formation temperature; p_{sc} T_{sc} is standard gas pressure and temperature, $p_{sc} = 0.1013$ MPa, $T_{sc} = 293.16$K; ϕ is porosity of gas zone, fraction; C_t is total compressibility, MPa^{-1}; t is time, hr; S_a is apparent skin factor, $S_a = S + Dq_g$; S is true skin factor; D is non-Darcy flow coefficient (m^3/day)$^{-1}$; and r_w is effective well radius, m.

If expressed by the existing LIT equation, Equation (3.20) can be changed into

$$p_{Ri}^2 - p_{wf}^2 = Aq_g + Bq_g^2. \quad (3.21)$$

By comparing Equations (3.20) and (3.21), expressions of A and B are

$$A = \frac{42.42 \times 10^3 \overline{\mu}_g \overline{ZT} p_{sc}}{kh T_{sc}} \left(\lg \frac{8.091 \times 10^{-3} kt}{\phi \overline{\mu}_g C_t r_w^2} + 0.8686 S \right) \quad (3.22)$$

$$B = \frac{36.85 \times 10^3 \overline{\mu}_g \overline{ZT} p_{SC}}{kh T_{SC}} \cdot D \quad (3.23)$$

In Equation (3.21), the coefficient A and gas flow rate are both multipliers, as are coefficient B and gas flow rate, which means that the smaller A or B values, the larger the corresponding gas flow rate value under the condition of the same producing pressure drop. Undoubtedly, values of A and B become very small for a high deliverability well, thus if the value of A or B becomes small due to the parameters decreasing in Equations (3.22) and (3.23), gas deliverability will be increased accordingly.

4.1.1 Analysis of Expression of A [Equation (3.22)]

1. The bigger the flow coefficient $\frac{kh}{\overline{\mu}_g}$, the smaller the value of A.
2. The smaller the diffusivity coefficient $\frac{k}{\overline{\mu}_g \phi C_t}$, the smaller the value of A, which seems in conflict with 1., but $\frac{k}{\overline{\mu}_g \phi C_t}$ is under logarithm operation and so has a relatively

minor influence. The effects of $\dfrac{k}{\mu}$ is mainly reflected in 1.; that is, the bigger the $\dfrac{k}{\mu}$, the higher the deliverability of the well.

3. The smaller S value is, that is, the less serious formation damage is, the smaller A value is, and gas deliverability is increased consequently.

4. The value of A varies with time under transient state after the well is opened. When time t goes on, the value of A grows bigger accordingly, resulting in a continuous decrease of gas deliverability. It indicates that gas deliverability declines continuously even if the bottom-hole pressure is basically unchanged. Figure 3.19 presents a variation curve of coefficient A of the deliverability equation in a gas well vs time. This well is drilled in strip-like fluvial lithologic formations; its deliverability declines as the gas production proceeds. The values of A in the plot are calculated by measuring flowing pressure data. It can be seen that a gradual increase of the value of A with time causes a continuous decrease of deliverability, which is considered a main reason bringing errors in calculating deliverability with transient pressure data acquired during short-term flowing of the well.

4.1.2 Analysis of Expression of B [Equation (3.23)]

1. The value of B is inversely proportional to $\dfrac{kh}{\mu_g}$, that is, the bigger the flow coefficient $\dfrac{kh}{\mu_g}$, the smaller the value of B; and it is consistent with the influence of $\dfrac{kh}{\mu_g}$ on the value of A.

2. The value of B is directly proportional to non-Darcy flow coefficient D, that is, the bigger the value of D, the smaller the deliverability.

Being a parameter related to the gas flow state in the formation, the D value is influenced by such factors as reservoir pore structure and penetration ratio. Also, D is

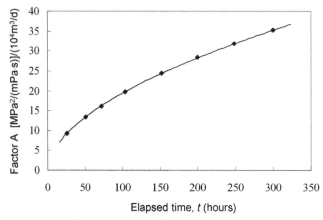

Figure 3.19 Variation of the value of coefficient A in deliverability equation vs time.

a parameter that is not estimated easily by common methods but can only be estimated effectively by deliverability tests onsite.

4.2 Deliverability Equation When Gas Flow Entering into Pseudo-Steady State

For a closed gas reservoir, the relationship between pressure difference and gas flow rate can be expressed as the following formula when flow enters into a pseudo-steady state:

$$p_R^2 - p_{wf}^2 = \frac{36.846 \times 10^3 \overline{\mu}_g \overline{ZT} p_{SC} q_g}{kh T_{SC}} \left(\ln \frac{0.472 r_e}{r_w} + S_a \right), \tag{3.24}$$

where

$$S_a = S + D q_g. \tag{3.25}$$

Change Equation (3.24) into the LIT equation:

$$p_R^2 - p_{wf}^2 = A q_g + B q_g^2,$$

where

$$A = \frac{29.22 \overline{\mu}_g \overline{ZT}}{kh} \left(\lg \frac{0.472 r_e}{r_w} + \frac{S}{2.302} \right), \tag{3.26}$$

$$B = \frac{12.69 \overline{\mu}_g \overline{ZT}}{kh} \cdot D, \tag{3.27}$$

where p_{sc} and T_{sc} are 0.101 MPa and 293.16K, respectively.

It can be seen from these equations that

1. The values of A and B are both influenced by flow coefficient $\dfrac{kh}{\mu_g}$; the bigger $\dfrac{kh}{\mu_g}$ is, the smaller A and B values are, indicating that a higher gas flow rate and AOFP are possible under the same pressure difference.
2. The value of A is influenced by skin factor S. Well deliverability is decreasing when S is increasing.
3. The value of B is also influenced by non–Darcy flow coefficient D: Well deliverability decreases due to considerable turbulent flow. Non-Darcy flow coefficient D is multiplied with q_g^2 in the equation so its influence on deliverability is smaller than that of A.
4. Producing a pressure difference will not change with time because the flow has entered into a pseudo-steady state, thus deliverability calculation is irrelevant with time in the equation. However, field practice is often quite different; estimated deliverability will still decline slightly, for the flow in the tested well usually cannot

immediately enter into a pseudo-steady state, even though the near boundaries have influenced the pressure, but the influence of some farther boundaries has not yet reached the tested well so the calculated deliverability may be lowered.

These analyses are expressed with pressure square, but they will also give absolutely similar results with pseudo pressure, and the related equations are listed as follow.

For a pseudo-steady flow in a round closed reservoir:

$$\psi(p_\mathrm{R}) - \psi(p_\mathrm{wf}) = A_\psi q_\mathrm{g} + B_\psi q_\mathrm{g}^2, \tag{3.28}$$

where

$$A_\psi = \frac{29.22\overline{T}}{kh}\left(\lg\frac{0.472r_\mathrm{e}}{r_\mathrm{w}} + \frac{S}{2.302}\right)\left[\mathrm{MPa}^2/(\mathrm{mPa\cdot s})\cdot(10^4\ \mathrm{m}^3/\mathrm{day})\right] \tag{3.29}$$

$$B_\psi = \frac{12.69\overline{T}}{kh}\cdot D\left[\mathrm{MPa}^2/(\mathrm{mPa\cdot s})\cdot\left(10^4\ \mathrm{m}^3/\mathrm{day}\right)^2\right] \tag{3.30}$$

and $p_\mathrm{sc} = 0.101$ MPa, $T_\mathrm{sc} = 293.16$K.

For unsteady-state flow in infinite homogeneous formation:

$$\psi(p_\mathrm{R}) - \psi(p_\mathrm{wf}) = A'_\psi q_\mathrm{g} + B'_\psi q_\mathrm{g}^2 \tag{3.31}$$

where

$$A'_\psi = \frac{14.61\overline{T}}{kh}\left(\lg\frac{8.09\times10^{-3}kt}{\phi\overline{\mu}_\mathrm{g}C_\mathrm{t}r_\mathrm{w}{}^2} + 0.8686S\right) \tag{3.32}$$

$$B'_\psi = \frac{12.69\overline{T}}{kh}\cdot D \tag{3.33}$$

and $p_\mathrm{sc} = 0.101$ MPa, $T_\mathrm{sc} = 293.16$K.

These equations aim for only simple homogeneous formations. Reduced skin factor is negative for fractured wells; if the testing duration is long enough, the flow has entered a pseudo radial flow regime and this analysis method is basically applicable.

5 SHORT-TERM PRODUCTION TEST COMBINED WITH MODIFIED ISOCHRONAL TEST IN GAS WELLS

Currently, the deliverability test is often properly extended for the purpose of short-term production tests in the oil fields of middle and western China. It can be used not only for calculation of deliverability, but also for verifying rate maintenance of the tested gas well.

5.1 Pressure Simulation of Tested Wells

To make this clear, a set of parameters similar to those onsite are selected to simulate the pressure history and perform an analysis base on the simulation. Simulation conditions include:

$k = 3$ mD, $h = 5$ meters, $\mu = 0.02$ mPa·s, $S = 0$, $x_f = 60$ meters (for fractured well), $S_f = 0.1$(for fractured well), and $L_{b1} = L_{b2} = 70$ meters (for channel reservoir).

Figures 3.20, 3.21, and 3.22 give pressure history graphs of a gas well under three different conditions: the well is in an infinite homogeneous reservoir, in an infinite homogeneous reservoir and has been fractured, and in a homogenous channel reservoir with lithological boundaries and has been fractured.

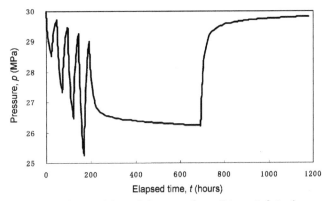

Figure 3.20 Pressure history during deliverability test of a well in an infinite homogeneous reservoir.

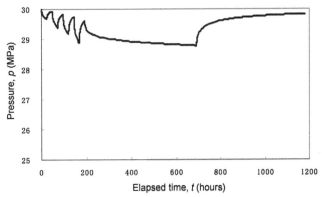

Figure 3.21 Pressure history during deliverability test of a fractured well in an infinite homogeneous reservoir.

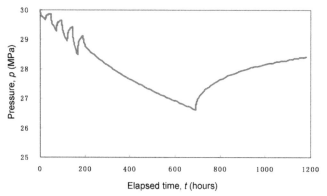

Figure 3.22 Pressure history during deliverability test of a fractured well in a channel reservoir with lithological boundaries.

Figure 3.23 shows pressure history under three different conditions plotted in the same graph in view of comparative analysis. Figure 3.23 shows:

1. In an infinite homogeneous formation, due to its low permeability, the producing pressure difference is quite large during well production. During extended flow at the rate of 8×10^4 m^3/day, the producing pressure difference reaches 15.5 MPa; but in 500 hours following the final shut in, bottom-hole pressure builds up basically to the original formation pressure. So when calculating deliverability, the original pressure or extrapolated pressure from the final buildup can be used as the formation pressure, which brings only a very little difference in results.

2. For the fractured well in an infinite homogeneous formation, although formation permeability is low, bottom-hole conditions have been improved by fracturing and so producing pressure difference decreases a lot; when the well is producing at the rate of

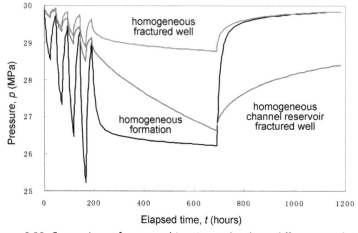

Figure 3.23 Comparison of pressure histories under three different conditions.

$8 \times 10^4 \, \mathrm{m^3/day}$, the producing pressure difference is only about 2 MPa. The pressure builds up basically to the original formation pressure after 500 hours following the final shut in, indicating that the original pressure can be used as the formation pressure in calculating deliverability.

3. For channel formation with lithological boundaries, the pressure history is quite different. On the one hand, the bottom-hole fracture decreases, producing a pressure difference, similar to that of common fractured wells in short-term production; on the other hand, the boundary effect causes bottom-hole flowing pressure to decline continuously and pressure cannot build up to the original formation pressure during shut in after the extended test.

Two problems are posed in deliverability calculations for such a situation of pressure declining continuously in reservoirs with boundaries:

1. No matter how much time is taken to conduct extended flow, no stable deliverability point can be detected at all. On the contrary, the longer extended flow lasts, the lower flowing pressure will be and the smaller the calculated AOFP value will be.

2. It can be seen from a shut-in pressure buildup test that formation pressure also declines after a long time of production. The original pressure is not considered applicable to deliverability calculations, and average formation pressure within the area of influencing radius of the tested well should be used instead.

For such special lithologic formations, special deliverability analysis methods should be adopted.

5.2 Improvement of AOFP Calculation Methods In Modified Isochronal Test

5.2.1 Classical Method

The modified isochronal test for obtaining AOFP, as introduced in many references, was shown in Figure 3.5, and repetitively shown again here in Figure 3.24.

The test consists of two sections, that is, transient deliverability test section AB and extended test section BC.

Pressure differences of transient points are calculated by following formulas:

$$\Delta p_1^2 = p_R^2 - p_{wf1}^2$$

$$\Delta p_i^2 = p_{ws(i-1)}^2 - p_{wfi}^2 \ (i = 2, 3, \ldots) \tag{3.34}$$

where p_{ws} is shut-in pressure at the end of the ith shut in, MPa; p_{wfi} is flowing pressure at the end of the ith flow, MPa; and Δp_i^2 is the ith difference of pressure square, $\mathrm{MPa^2}$.

For the stabilized deliverability point:

$$\Delta p_w^2 = p_R^2 - p_{wfw}^2, \tag{3.35}$$

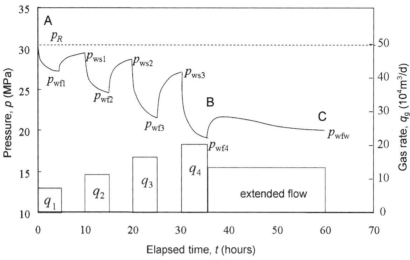

Figure 3.24 Pressure and rate history for a modified isochronal test.

where p_R is the original formation pressure, MPa; and p_{wfw} is flowing pressure of extended deliverability point C.

Only the formation pressure before the beginning of the test is available in calculating the pressure difference of extended deliverability point C, as the classical modified isochronal well test has no shut-in pressure buildup section after extended flow. This analysis method is applicable to wells in infinite homogeneous formations and fractured wells in infinite homogeneous formations. It can be seen from Figures 3.20 and 3.21 that even in a formation with very low permeability (only 3 mD), the pressure builds up basically to the original level after 500 hours of shutting in.

However, it is quite different in channel formation with lithological boundaries. It can be seen from Figure 3.22 that for fractured wells in a homogeneous formation with boundaries that the pressure builds up only to 27.48 MPa after shutting in 500 hours, 2.5 MPa less than the original pressure. This indicates that application of the original formation pressure is obviously inadequate in calculating the deliverability of the extended test point.

5.2.2 Improved Calculation Method

An "improved calculation method" is brought forward here, that is, formation static pressure measured after an extended test is taken as the formation pressure in calculating the pressure difference for extended deliverability determination.

In consideration of influencing range, an influence radius r_i after 500 hours shutting in is approximated as about 630 meters, and the pressure at this time represents the pressure on supply boundaries basically. This pressure is chosen as the formation pressure for calculation of the pressure difference of the extended test, making calculation results meet the actual situation.

5.2.3 Comparison of Two Calculation Methods

Calculation results by classic method and improved method are shown in Table 3.6. Note: data listed in Table 3.6 are results calculated with the LIT equation.

Analysis of Table 3.6 Contents

As seen from Table 3.6:

1. For a normal well or fractured well in infinite homogeneous formation, AOFP obtained by the classical method or improved method are very close; the difference is less than 5%.
2. For a normal well in channel formation with boundaries, calculated AOFP by the classical method is commonly 30% even much lower than that by the improved method.
3. It should be especially pointed out that AOFP of formations with boundaries calculated by the classical method is always variable, it decreases with time because during an extended test the flowing pressure keeps declining, and the original pressure is still taken as the static formation pressure.
4. With the improved calculation method, when flow almost enters into a pseudo-steady state during the test, the flow pressure keeps declining continuously, but the formation pressure declines accordingly as well, and the error in calculating AOFP will be reduced significantly.

Table 3.6 Comparison of AOFP by Different Calculation Methods

Reservoir and well type	Calculated AOFP (10^4 m^3/day)		Relative difference (%)
	Improved method (using measured shut-in pressure)	Classical method (using original formation pressure)	
Normal well in infinite homogeneous reservoir	8.9849	8.9279	0.6
Fractured well in infinite homogeneous reservoir	28.2991	27.0643	4.6
Fractured well in homogeneous reservoir with boundaries	20.6863	15.4587	33.8

Special Attention to Improved Deliverability Calculation Method

This method is a supplement to the classical method in some special conditions. It is a correction of test results, aiming at a rapid decline of formation pressure during the test due to serious boundary effects or the characteristics of a constant-volume reservoir where the tested well locates. The analysis method of this kind of problems has been discussed [20]. In addition, the calculation method is placed into the modified isochronal test method in Figure 3.7 of this chapter. Certainly, such a modification can also be considered being made for other deliverability test methods.

Some gas wells in China were once observed in back-pressure tests with different chokes (from big to small chokes sequentially) that the flowing pressure of a small choke in the late stage is lower than that of a big choke in the early stage, as shown in Figure 3.25.

It can be seen from Figure 3.25 that if deliverability is calculated by the classical method, abnormal phenomena is inevitable: a high flow rate in the early stage corresponds to a small producing pressure difference, whereas a low flow rate in the late stage corresponds to a big producing pressure difference, and so AOFP is unable to be calculated. The main reason for this phenomenon is formation pressure depletion during the test. It can be obviously seen in Figure 3.25 that formation pressure is about 41 MPa at the beginning of the test, but has declined down to 36.335 MPa when the shut-in pressure is measured for buildup at the end of the test; the pressure loss is about 4.6 MPa. The pressure decline is approximately in the trend of the double dots line in Figure 3.25. If a decreased formation pressure value is used to correct the deliverability analysis result, a roughly estimated deliverability value can be obtained.

Determination of formation pressure after the deliverability test. The determination of formation pressure after the deliverability test should be intended for gaining the pressure at the gas supply boundary. If the boundaries controlled by the tested well have

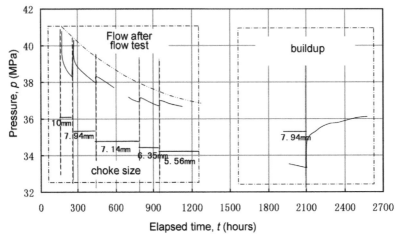

Figure 3.25 Measured pressure history during back-pressure test of a well in a constant-volume gas reservoir.

been clearly defined geologically, we can calculate an equivalent influence radius r_i according to the defined area and then derive formula $t_s = \dfrac{70\phi\overline{\mu}C_t r_i^2}{k}$ inversely from the formula $r_i = 0.12\sqrt{\dfrac{kt}{\phi\overline{\mu}C_t}}$, which is used for influencing radius r_i and described in Chapter 2.

Put equivalent influence radius r_i mentioned earlier into the formula, reversely derive a shut-in time of pressure buildup to determine the shut-in pressure at that time, and that is the measured formation pressure.

If a controlled boundary of a tested well is unavailable, roughly determine its influencing area to be 1 km^2, and $r_i \approx 600$ meters is obtained; reversely, derive the shut-in time according to $r_i \approx 600$ meters, and an instant shut-in pressure is obtained, which is temporarily regarded to be the formation pressure.

As illustrated in the previous sections, for constant-volume gas reservoirs with limited volume, deliverability will decline continuously as the depletion of formation pressure during the production, so "accurate" or "absolute" AOFP (q_{AOF}) does not exist. Even in view of gas field development, it is debatable whether the so-called stable q_{AOF} index is really needed.

The following aspects are crucially important to the development of constant-volume gas reservoirs:

1. Initial q_{AOF} measured simply by single point test method
2. Understand the performance of single wells and the pressure-declining rate under different gas recovery rates by the production test
3. Effective gas supply volume or dynamic reserves controlled by single wells determined by the dynamic method
4. Determine the development strategy and economic returns

The rational flow rate can be determined with the aforementioned indices. Whether normal methods, such as the modified isochronal test method, are needed in calculating stable AOFP can be finally determined by study and analysis according to practical development and production needs.

6 STABLE POINT LAMINAR–INERTIAL–TURBULENT (LIT) DELIVERABILITY EQUATION

6.1 Background of Bringing Forward Stable Point LIT Deliverability Equation

6.1.1 Puzzles in Determining Gas Well Deliverability by Classical Methods

Classical deliverability test methods have been applied onsite for 80 years and are still considered the main methods in determining gas well deliverability so far. But various puzzles have been encountered in field applications in China for decades. Take the SB gas field in Qinghai Province for instance, the normal deliverability equation cannot be determined with deliverability test data from more than half of the gas wells because of gas and water alternating zones existing in the formations, water producing in a majority of gas wells, and the impossibility of setting a pressure gauge below the liquid level in the wellbore during the deliverability tests. In some areas, although the deliverability equation can be obtained formally from acquired data, it is thought of poor reliability due to the inconsistency of calculated AOFP from its normal range resulting from comparative analysis on formation conditions and practical production situations.

Calculating AOFP with the deliverability test method can be vividly compared to shooting at a target with a gun. The "gun" is acquisition and analysis of test data, and the "target" to be shot at is the AOFP of a gas well. The shooting will inevitably be influenced by the quality and accuracy of the gun, but a shooter can always score a hit to the target with a common gun if he aims it carefully at the target. Calculating AOFP with the deliverability equation just lacks the aiming procedure, like shooting blindfolded. Once the "gun" is set in a wrong direction due to the interference of environmental factors, it may happen that we aim at a rabbit while shooting down a bird in the sky before missing the target is discovered. If the local contour or profile of the target is seen, even though it is impossible to get a clear overall view, the hitting accuracy will be improved greatly. This is the starting point and overall idea of deducing the "stable point LIT deliverability equation."

6.1.2 Existing Problems of Classical Methods

What are the problems of classical deliverability test methods? Common problems onsite are as follow.

Abnormal Conditions of $B < 0$ or $n > 1$ in Deliverability Equation

An abnormal condition of the exponential deliverability index, n is greater than 1, occurs sometimes in the analysis with measured data. According to the original definition of the exponential equation, n means turbulent flow index. That $n = 0.5$ indicates gas flow in the formation is basically turbulent flow; and that $n = 1$ indicates laminar flow dominates the gas flow in the formation; in general, $0.5 \leq n \leq 1$, which suggests that the gas flow consists of both laminar flow and turbulent flow in the formation.

If n is greater than 1, test data are abnormal, as shown in Figure 3.26.

When $n > 1$ appears in the deliverability test analysis, the well test interpretation software used commonly, for example, FAST, will make n equal 1 instead.

The same problem appears in the LIT equation, with negative B results. It is reflected by a down–dip line in the deliverability plot, as shown in Figure 3.27. Such field examples are not rare, and Figures 3.28 and 3.29 show typical ones.

In the aforementioned field examples, deliverability plots are drawn with FAST software. The resulted values of B and n are beyond the allowable range ($n > 1$ and $B < 0$), which are unacceptable and so shown by $n = 1$ and $B = 10^{-16}$ instead by FAST.

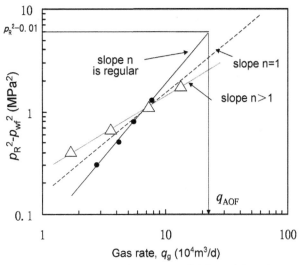

Figure 3.26 Abnormal exponential deliverability plot.

Flowing Pressures have not yet been Stabilized or Intervals of Flowing at Different Rates are not Equal in Back-Pressure Deliverability Test

These problems are usually induced by an inadequate arrangement in data acquisition, which causes $B < 0$ and $n > 1$ in the deliverability equation. For low permeability formations, if the flow duration is too short, the flowing pressure may still be in the transient decline process, especially for those low permeability formations with a no-flow boundary influence. Figure 3.30 shows this circumstance. If the durations

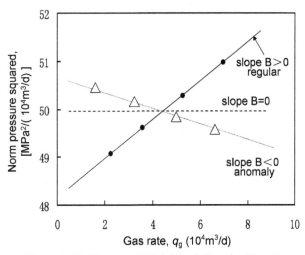

Figure 3.27 Abnormal LIT deliverability plot ($B < 0$).

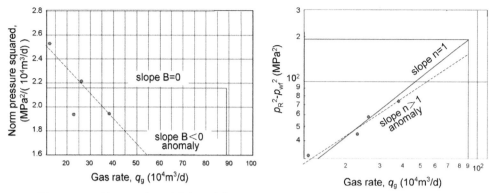

Figure 3.28 Abnormality of deliverability equation: $B < 0$ and $n > 1$ (field example 1).

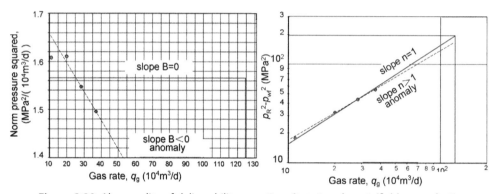

Figure 3.29 Abnormality of deliverability equation: $B < 0$ and $n > 1$ (field example 2).

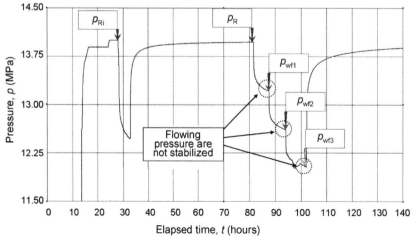

Figure 3.30 Field examples of unstable flowing pressures due to too short flowing intervals in the back-pressure test.

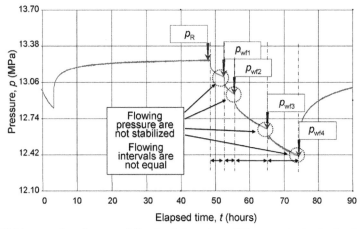

Figure 3.31 Field examples of unequal flowing intervals at different rates in the back-pressure test.

of flowing at different rates are not equal, it is likely to result in the abnormal deliverability equation (Figure 3.31).

Inverse Peak of Flowing Pressure Curve Appears due to Bottom-Hole Water Accumulated in Testing Wells

Abnormal deliverability equations, even when their B is less than 0 and n is greater than 1, often occur when bottom-hole water accumulates in the wellbore or the pressure gauges cannot be set below the accumulated liquid level because of casing program problems.

An inverse peak often appears on a flowing pressure curve when water accumulates in the bottom hole, as shown in Figure 3.32.

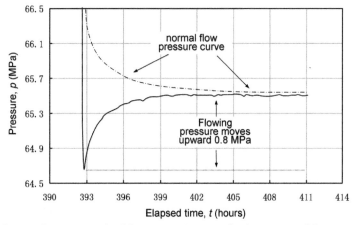

Figure 3.32 Abnormal inverse peak of flowing pressure at the beginning of flow in a gas well with bottom-hole water accumulation.

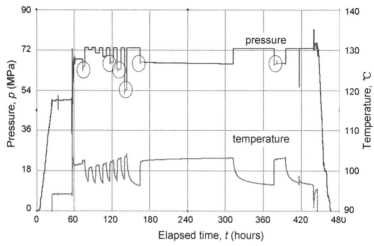

Figure 3.33 Inverse peaks on flowing pressure curve during modified isochronal test due to accumulated water in the bottom hole.

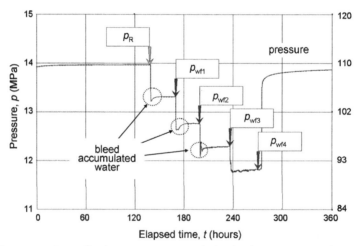

Figure 3.34 Inverse peaks on flowing pressure curve during back-pressure test due to accumulated water in the bottom hole.

Figures 3.33 and 3.34 are field examples with an inverse peak of flowing pressure.

The deviation of pressure at test points due to accumulated water in the bottom hole is analyzed as follows.

1. Conversion relationship between pressures at measured point and at the bottom hole. Figure 3.35 presents the conversion relationship between the pressure at measured point p_{cwf} and flowing pressure at bottom-hole p_{wf}, and the influence factors of pressure deviation.

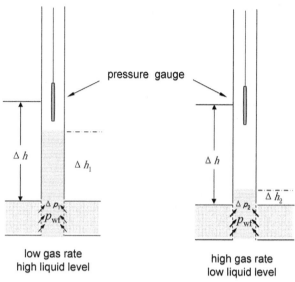

Figure 3.35 Conversion relationship between measured pressure and bottom-hole flowing pressure.

The relationship of the measured pressure at tested point p_{cwf}, bottom-hole flowing pressure p_{wf}, and deviated pressure Δp can be expressed as the following formula:

$$p_{cwf} = p_{wf} - \Delta p. \tag{3.36}$$

The height of the accumulated water column is Δh_1 at the moment of opening the well,

$$p_{cwf1} = p_{wf1} - \Delta p_1 = p_{wf1} - \rho_g g \Delta h - \left(\rho_w - \rho_g\right) g \, \Delta h_1. \tag{3.37}$$

After an interval of Δt, part of the accumulated water is produced to the surface along with gas flow, the height of the water column falls down to Δh_2, and the relationship between the measured pressure and the bottom-hole pressure presently is

$$p_{cwf2} = p_{wf2} - \Delta p_2 = p_{wf2} - \rho_g g \Delta h - \left(\rho_w - \rho_g\right) g \, \Delta h_2 \tag{3.38}$$

Pressure variation at the tested point after flowing for an interval of Δt is

$$\Delta p_{cwf} = p_{cwf2} - p_{cwf1} = \left(p_{wf2} - p_{wf1}\right) - \left(\rho_w - \rho_g\right) g \left(\Delta h_2 - \Delta h_1\right) \tag{3.39}$$

The first term in the right-hand side of Equation (3.39), $(p_{wf2} - p_{wf1})$, indicates the actual variation of flowing pressure in the bottom hole and it should be negative generally.

The second term in the right-hand side of Equation (3.39), "$-\left(\rho_w - \rho_g\right) g \left(\Delta h_2 - \Delta h_1\right)$," indicates pressure deviation due to the height change of the accumulated water, and the term is always positive because of $\Delta h_2 < \Delta h_1$.

2. Different gas wells may have absolutely different performance due to their different geological conditions.

(1) Generally, the bottom-hole flowing pressure keeps declining (reflected by negative value) after the well is opened, although accumulated water may twist the shape of the flowing pressure decline trend and make it deviate upward; however, if the deviation is not big enough, the flowing pressure is still declining continuously.

(2) If the formation deliverability coefficient kh is relatively high, the flowing pressure will become stabilized quickly after opening the well, causing the first item in the right-hand side of Equation (3.39), "$p_{wf2} - p_{wf1}$," to be very small or even approaching zero, so the effect of the second item in the right-hand side of Equation (3.39), "$- (\rho_w - \rho_g) g (\Delta h_2 - \Delta h_1)$," will appear very quickly, and the value of Δp_{cwf} will become positive, with flowing pressure rising after reaching a minimum value and forming an inverse peak as shown in Figures 3.32, 3.33, and 3.34. It is the typical influence characteristics caused by accumulated water being produced from the hole.

If inverse peaks appear in several flow periods or tested points during a deliverability test, it is known that accumulated water is produced continuously out from the well, along with gas produced with an increasing rate, and the level of the accumulated water column at different tested points is changing, which obviously brings certain distortion in the analysis results of the deliverability test.

AOFP Calculated with Different Deliverability Test Methods and/or Different Analysis Methods are Very Different

In well test industry standards, various applicable conditions are roughly recommended for a back-pressure test, isochronal test, and modified isochronal test, but there is no and impossible to be a regulation of application scope, so people are confused in the application of these methods onsite. For example, a modified isochronal test was ever performed in KL-2 gas field with very high deliverability. From the same test data, AOFP calculated by the LIT equation is two-thirds that calculated by the exponential equation (refer to Figure 3.13). Again, results obtained from the analysis of the same test data with pseudo pressure and pressure-squared methods are greatly different (refer to Figure 3.18).

Low Rate of Utilization of Classical Deliverability Test in Production Wells

Since frequently changing working systems, that is, opening the well and flowing the well with different rates, and continuous real-time monitoring of flowing and shut-in pressure variation with a downhole pressure gauge are required during classical deliverability tests, a great amount of work must be done onsite and a great deal of investment is needed; a classical deliverability test in onshore gas fields can be conducted only in a very small proportion of appraisal wells and is not permitted for production wells. The production rate programming of these production wells can only be estimated with logging data or grouped gradually after the wells are put into production.

In an offshore gas field of China, considerable technical efforts and funds were paid for performing a back-pressure test repeatedly during production in order to do production allocation. But after a period of practice, it was realized that the basic condition of stable formation pressure requested by the tests no longer existed, and so the results from such tests cannot correctly give depleted deliverability indices; sometimes they even give unbelievable deliverability indices that are much higher than the initial deliverability value.

The various problems in the classical deliverability test mentioned earlier are indeed puzzles for gas field deliverability determination.

6.2 Stable Point LIT Deliverability Equation

6.2.1 Characteristics of New-Type Deliverability Equation

It is necessary to seek a feasible new approach to solve the problems discussed previously and to evaluate successfully the deliverability of production gas wells. The new method recommended in this book is called the method of "stable point LIT deliverability equation." This method has the following advantages.

1. General applicability. It is applicable to both appraisal wells and a majority of production wells. As for gas wells from which classical deliverability test data have been acquired, such a new deliverability equation can be obtained with available data directly. The reliability of the new equation can be verified by comparison to that from the classical deliverability test. With the initial stable point LIT deliverability equation, an initial IPR curve is plotted and the initial AOFP can be calculated.

2. Rigorous theory. It can be seen from the establishment process of the stable point LIT deliverability equation introduced that the equations are deduced rigorously with related theories, in which the definition, value selection, and acquisition approach of some parameters are studied and discussed repeatedly so as to strive for meticulous deduction.

3. Tracing deliverability. As discussed before, classical deliverability test methods can only be used to determine the initial deliverability of gas wells, while the new method can give not only their initial deliverability equation, but also deduce a "dynamic deliverability equation" on the basis of the initial deliverability equation. The dynamic deliverability equation can be used to calculate dynamic AOFP in any certain production period and to obtain dynamic deliverability characteristic parameters, such as the dynamic IPR curve in that period.

4. Convenient operation. The new method does not require any additional operations, and well shutting-in or frequently changing working systems (i.e., producing with different rates) are unnecessary for normal production wells, which avoids bringing any difficulties in production planning. The new method only requires measuring the

bottom-hole flowing pressure under production at a stable rate regularly in accordance with general dynamic monitoring requirements; if there are any packers outside the production pipe string, deliverability tracing can also be accomplished by converting the casing pressure (monitored at surface) into the bottom-hole flowing pressure.

6.2.2 The New Method is Supplement and Improvement of The Original Classical Deliverability Test Method

The new method does not replace the existing deliverability test methods completely but to supplement and improve the original classical ones. The original methods are substituted with the new one only when applicable deliverability equations cannot be obtained by classical methods. Their application can be seen in Figure 3.36:

1. Exploration wells: use the single point method to calculate deliverability. The flow time in exploration wells is usually very short, flowing pressure has not been stabilized, further cleanup is needed in some wells, and the simple method like single point method is used for primarily estimating their deliverability in a transient state.

2. Appraisal wells: the single point method is also permitted for primarily estimating AOFP in the initial flow period. For some appraisal wells connected with production lines or very important appraisal wells, the test duration should be extended appropriately until entering the pseudo-steady state.

3. Deliverability test in appraisal wells: a deliverability test is needed for appraisal wells in the initial production period in accordance with industrial standard, and classical deliverability test methods are adopted: either the back-pressure test method, requiring stabilized flowing pressure for each deliverability test point, or the modified

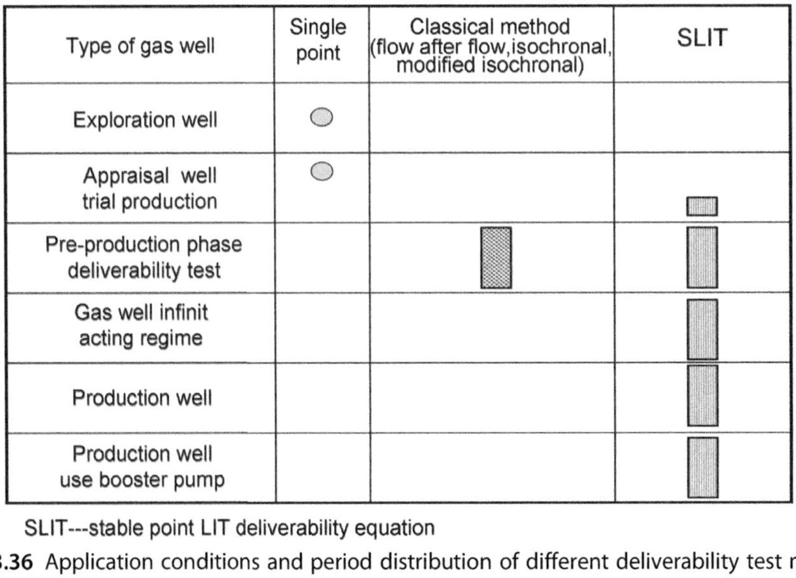

Type of gas well	Single point	Classical method (flow after flow, isochronal, modified isochronal)	SLIT
Exploration well	○		
Appraisal well trial production	○		▦
Pre-production phase deliverability test		▦	▦
Gas well infinit acting regime			▦
Production well			▦
Production well use booster pump			▦

SLIT---stable point LIT deliverability equation

Figure 3.36 Application conditions and period distribution of different deliverability test methods.

isochronal test method, requiring one stabilized deliverability point in the extended test after a set of transient point testing.

4. Application of stable point LIT deliverability equation: being a new method, the stable point LIT deliverability equation can fulfill its role in various different stages of a gas well after being put into production. Only one condition is needed in establishing the equation: there is one relatively stable point obtained during testing. Therefore, the new equation is briefly named "stable point equation" hereinafter.

    (1) In a gas well test stage, if the flow duration is sufficiently long and the gas flow rate and flowing pressure are basically stabilized, the stable point equation method can be used. For instance, in the KL–2 gas field, the flowing pressure can be stabilized several hours after the gas well is opened, deliverability equation met the requirement can be derived with a short-interval test point. It can be proved by the examples shown later (refer to section 6.4.1).

    (2) The stable point equation can be established with one stable point acquired during the deliverability test.

    The modified isochronal test contains an extended test stage. It is thought that the gas well should be characterized by a pseudo-steady state at the end of the extended test stage, and the point can be regarded as a steady deliverability point to construct a stable point equation.

    As for the back–pressure test, the flow rate and flowing pressure of each test point should be stabilized in accordance with related requirements so that any test point selected randomly can be used to establish a stable point equation. An example is given with test point selected in the back–pressure test of well-KL205 later in the deduction process of the stable point equation, and the IPR curve and AOFP from the stable point equation and the backpressure test are very close to each other.

    For establishing stable point equations of the gas wells with load water by the back–pressure test, we usually choose the final test point as the optimal point to eliminate the influence of water. The stable point equation was established by using this method in the DF gas field when the author participated in the research work of the well test there.

    (3) Production test of appraisal wells and initial producing stage of production wells. In a large or medium gas field, the number of gas wells put into production amounts to at least several hundred or even more. It is impossible to conduct a standard deliverability test on every well. Statistics show that in onshore gas fields, gas wells having undergone a standard deliverability test are less than one-tenth of the total wells. How to make a production plan for wells that have not been tested? The AOFP has to be estimated often by the single point method in a gas well test under the circumstance that there are not enough reliable data. However, practices have proved that data acquired in the early stage create too large of error, which are several times larger or smaller than the actual deliverability, and inevitably brings troubles in the plan adjustment of gas wells after being put into production.

Application of the stable point equation can deal with this problem well. In the initial stage of a production test or initial production of a gas well, as long as the average bottom-hole flowing pressure in a steady production period is measured, a reliable initial stable point deliverability equation will be established, the initial AOFP can be calculated, and the initial IPR curve can be plotted. If a permanent pressure gauge has been set at the bottom hole of a gas well, the establishment of this new type of deliverability equation will become more convenient.

(4) Dynamic deliverability equation of production wells to be established during the gas production process.

The gas well deliverability declines dramatically after several years of production. Although the production is maintained at the initial flow rate, the bottom-hole flowing pressure and wellhead pressure will decrease obviously. For gas wells with poor deliverability, the gas flow rate and pressure may begin to decline after a year or even after several months of production. How should deliverability now be evaluated? The stable point LIT deliverability equation can pay its unique role in this case. Just measure the stationary deliverability data point, substitute data of this point into the equation, calculate the dynamic formation pressure at the gas supply boundary, establish the "dynamic deliverability equation" in this period, estimate the dynamic AOFP, and plot the dynamic IPR curve.

(5) Pressure-boosting stimulation stage of production wells.

In the pressure-boosting stimulation stage, the effects of technology measures will bring a certain difficulty in monitoring the bottom-hole flowing pressure. If pressure can be measured properly, the new deliverability analysis method is also applicable.

Figure 3.37 shows the comparison of different deliverability analysis methods and their applications. It can be seen that the new method is applicable not only to vertical gas wells, but also horizontal wells; the only difference is that the equations are expressed slightly different. It will be introduced in a special section of this chapter.

6.3 Theoretical Deduction and Establishment of Stable Point LIT Deliverability Equation

According to the deduction shown in Equation (3.4), the LIT equation in the pseudo-steady flow stage is expressed as Equation (3.6) in terms of pressure-squared form, namely

analytic method	vertical well Initial deliverability analysis			vertical well dynamic deliverability analysis				horizontal well deliverability analysis		
	Initial q_{AOF}	Initial equation	Initial IPR	dynamic q_{AOF}	dynamic equation	dynamic IPR	boundary pressure	Initial q_{AOF}	dynamic q_{AOF}	dynamic IPR
flow after flow	✓ (✗ for low permibility)	✓ (B<0, n>1)	✓	✗	✗	✗	✗	✓ (✗ for low permibility)	✗	✗
modified isochronal	✓	✓	✓	✗	✗	✗	✗	✓	✗	✗
single point	✓	✗	✗	✗	✗	✗	✗	✓ (borrow)	✗	✗
SLIT	✓	✓	✓	✓	✓	✓	✓	✓	✓	✓

SLIT---stable point LIT deliverability equation   ✓---applicable   ✗---unservicable

Figure 3.37 Application ranges of different deliverability methods.

$$p_R^2 - p_{wf}^2 = Aq_g + Bq_g^2, \tag{3.40}$$

where

$$A = \frac{29.22\overline{\mu}_g\overline{ZT}}{kh}\left(\lg\frac{0.472r_e}{r_w} + \frac{S}{2.302}\right) \tag{3.41}$$

$$B = \frac{12.69\overline{\mu}_g\overline{ZT}}{kh}\cdot D \tag{3.42}$$

6.3.1 Classification of Parameters Influencing Coefficients A and B

Once the values of coefficients A and B are determined, the LIT deliverability equation is established. The parameters influencing coefficients A and B can be classified as

1. Formation deliverability coefficient kh
2. Formation pressure p_R and producing pressure differential ($p_R^2 - p_{wf}^2$)
3. Completion parameter of gas well, generally expressed as apparent skin factor S_a
4. Formation physical properties: formation temperature T_f, subsurface viscosity μ_g, and deviation factor of the gas Z
5. Gas supply radius of gas well r_e and effective wellbore radius r_w

Except for the formation deliverability coefficient kh being specially discussed, the other parameters are introduced here one by one.

Initial Formation Pressure p_R

p_R is One of Three Key Factors Influencing Gas Well Deliverability. It Should be Measured Correctly at the Right Time During the Completion of Gas Wells.

Formation Physical Properties, T_f, μ_g and Z

T_f, μ_g and Z, are Measured Onsite or in a Laboratory with Samples Acquired from the Wells. These Parameters Should be Collected at the Right Time for Each Gas Field.

Apparent Skin Factor of Gas Wells S_a

Skin factor S_a can be expressed as

$$S_a = S + q_g D,$$

where S is mechanical skin factor of gas wells and D is non-Darcy flow coefficient during subsurface percolating process of natural gas, $(10^4 \ m^3/day)^{-1}$

These two parameters are illustrated one by one as follow.

1. Mechanical skin factor S. This factor appears in coefficient A of the deliverability equation. It is generally determined by pressure buildup analysis. Mechanical skin factor S has approximately the same value in a specific completion process condition in an area or a gas field. For instance, in fractured gas wells drilled in low permeability sandstone formations of the upper Paleozoic in the Ordose Basin, the half-lengths of fractures are about 60–80 meters and the mechanical skin factor is generally about −5.5.

2. Non-Darcy flow coefficient D. Because many factors influence non-Darcy flow coefficient D, the process of determining this parameter is quite complicated.

    Above all, the D value is influenced by formation permeability k, porosity ϕ, subsurface viscosity μ_g, and relative density γ_g of the gas. Secondarily, it is influenced by well completion situations, such as effective thickness of the penetrated formation h and effective wellbore radius r_w. Therefore, some researchers established some relation formulae of non-Darcy flow coefficient versus the aforementioned parameters, for example, Jones constructed Formula 2.14 to express the D value, and this formula is written in terms of CSU units of China as follow:

$$D = \frac{1.35 \times 10^{-7} \gamma_g}{k^{0.47} \phi^{0.53} h r_w \mu_g}.$$

Other foreign scholars and some Chinese experts have also done some research on this issue and drew up similar expressions. Problems regarding this expression can be concluded as follow.

- It is very difficult to determine formation parameters (e.g., k and φ) influencing natural gas percolation.
- The influence of completion on the D value is very great, and far beyond from this simple expression only with some parameters such as effective thickness h and bottom-hole effective radius r_w. Usually, actual conditions cannot be exactly reflected by the D value calculated by Formula 2.14.

There are 3 Effective Methods to Determine the D Value Currently

1. Regression of pressure buildup analysis results. Different apparent skin factors S_a can be obtained from different pressure buildup analyses if the flow rates q_g before

shutting in for each pressure buildup are different and D (and the mechanical skin factor S as well) can be calculated by the regression analysis of S_a and q_g values (refer to Figure 2.46).

2. Changing skin factor treatment with well test interpretation software. Some commercial well test interpretation software have an option of changing the skin factor. This option enables a good match of flowing pressure variations under different gas flow rates by adjusting apparent skin factors during pressure history matching process, and so the software will give the non-Darcy flow coefficient D simultaneously.

3. Comparison of deliverability analysis results. Usually, standard deliverability tests are performed in many wells in most gas areas or gas fields. Take the SLG gas field for example, modified isochronal tests were conducted in more than several dozen wells, an initial deliverability equation was established, and an initial AOFP was calculated from each test. Then a stable point LIT deliverability equation of each well can be established using the stable point in the extended test period and the initial AOFP of each well can also be estimated. The initial AOFP values obtained from those two methods can be brought into consistency by adjusting the value of D in the stable point LIT deliverability equation; an effective D value in the area can be determined in this way. It will be introduced in the following field examples.

Gas Supply Radius r_e and Effective Wellbore Radius r_w of a Gas Well

Determination of the r_e value depends on an effective block area controlled by the tested well. But for a lithologic block with complex morphology or effective connected regions controlled by fractures, its effective area cannot usually be determined in a deliverability test so it is difficult to get the r_e value directly. However, the r_e value has only a small impact on interpretation results for it is taken logarithm in the formula. A common practice is to select an empirical value within a range of 500−1000 m according to the formations on which the gas well is drilled. The effect of value election errors will be offset during determination of the formation deliverability coefficient, as discussed specifically later.

Because the effect of its variation has been considered in the completion skin factor, the wellbore radius r_w is generally equal to the casing radius at reservoir section.

6.3.2 Determination of Deliverability Coefficient Kh and Establishment of Initial Deliverability Equation

As the most important factor influencing coefficients A and B in the deliverability equation, deliverability coefficient kh will be determined finally, and the value ultimately selected is called the "equivalent deliverability coefficient."

Once the parameters mentioned earlier in the deliverability equation are determined, and only the deliverability coefficient kh is not determined yet, rewrite the deliverability equation as

$$p_R^2 - p_{wf}^2 = \frac{A_2'}{kh}q_g + \frac{B_2'}{kh}q_g^2, \tag{3.43}$$

where

$$A'_2 = 29.22\overline{\mu}_g \overline{Z} T_f \left(\lg \frac{0.472 r_e}{r_w} + \frac{S}{2.302} \right) \tag{3.44}$$

$$B'_2 = 12.69\overline{\mu}_g \overline{Z} T_f \cdot D \tag{3.45}$$

An equivalent deliverability coefficient formula is obtained after transforming Equation (3.43):

$$kh = \frac{A'_2 q_g + B'_2 q_g^2}{p_R^2 - p_{wf}^2}. \tag{3.46}$$

Select an initial stationary deliverability point (q_{g0} p_{wf0}), and substitute it and the initial formation pressure p_{R0} into Equation (3.46); the equivalent deliverability coefficient kh of the gas well is calculated instantly.

Substitute this equivalent deliverability coefficient kh into Equation (3.43) and an initial stable point deliverability equation is established:

$$p_R^2 - p_{wf}^2 = A_2 q_g + B_2 q_g^2$$

and then the initial AOFP (q_{AOF}) and the initial IPR curve can be obtained.

It should be pointed out that the equivalent deliverability coefficient kh calculated by Equation (3.46) is only an intermediate value during the stable point LIT equation deduction process, but should not be used in parameter calculations in other situations—the reasons are given as follow.

1. The equivalent kh value reflects not only the actual permeability and effective thickness of the reservoirs around the well, but also the effects of formation heterogeneity and no-flow boundary etc. Therefore, the kh value calculated reversely from the pressure difference reflects comprehensively all formation factors mentioned previously.

2. In calculating the equivalent kh value inversely, other influence parameters, such as gas supply radius r_e, wellbore radius r_w, and mechanical skin factor S, have been determined beforehand, and errors in their determination process will be cumulated in the process of inverse calculation of the kh value, which will offset unfavorable influences due to the errors caused by selected values, but also brings the calculated value of kh a difference from its real one.

Because the deliverability equation is established based on the presupposition that the formation is homogeneous, the kh value reflects comprehensive influences of various formation factors in complicated reservoirs.

Then how to prove the reliability and applicability of deliverability equations established in this way? Simple illustrations are given as follow.

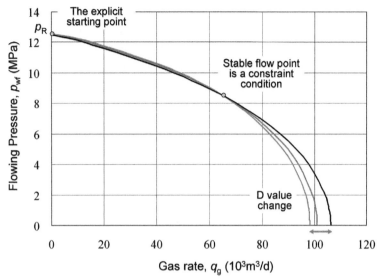

Figure 3.38 IPR curve constrained by the stable deliverability point.

IPR curves reflect best the deliverability equation characteristics. The characteristics of the IPR curve plotted on the basis of the stable point deliverability equation are shown in Figure 3.38.

There are two key points in Figure 3.38.

1. Formation pressure p_R is measured precisely onsite. It is the initial point when $q_g = 0$ and the only one explicit starting point of the IPR curves.

2. Stable deliverability point $(q_{g0}\ p_{wf0})$.

The stable deliverability point is a measured value, is acquired when the well is put into production after cleaning up and basically removing the accumulated load water, and is also a stable value in the actual production process of the well. It is a key point in constraining the trend of IPR curves. If vividly comparing the calculation of AOFP to shooting a gun at a target again, the stable deliverability point is just the "front sight" of the gun and is the front sight set between the gun and the target. The calculation of AOFP can also be compared to "shooting" when the "shooter" has partly seen the "target" and aims at it.

It can be seen from Figure 3.38 that trial changes of the D value in establishing the IPR curve bring a relatively small influence on the calculated AOFP due to constraining of the measured deliverability point.

6.4 Field Examples

6.4.1 Application of Initial Stable Point LIT Equation in Well KL-205

A short-term production test was conducted in the research of Well KL-205 before putting it into production. A back-pressure test was conducted at the beginning of production, from which a relationship between flow rate and pressure is listed in Table 3.7; flowing pressure variation during the test is plotted in Figure 3.39.

As far as Well KL-205 is concerned, physical properties and other parameters in establishing the stable point deliverability equation include

Subsurface viscosity of natural gas $\mu_g = 0.025$ mPa·s

Deviation factor of natural gas $Z = 1$

Formation temperature $T_f = 376$K

Gas supply radius of gas well $r_e = 500$ meters

Effective wellbore radius $r_w = 0.09$ meter

Mechanical skin factor $S = 0$

Non–Darcy flow coefficient $D = 0.018(10^4 \text{ m}^3/\text{day})^{-1}$ (obtained by changing skin factor treatment method with Saphir well test interpretation software)

Table 3.7 Flow Rate and Flowing Pressure Data of Back-Pressure Test in Well KL-205

Choke size, mm	Tubing pressure, MPa	Bottom-hole pressure, MPa	Gas flow rate, 10^4 m³/day
0		74.5083	0
7.94	61.95	73.3818	69.2750
11.11	59.64	72.4816	106.1368
13.04	57.95	71.6997	148.5733
14.63	56.41	70.9776	175.4265
16.34	54.21	70.1467	206.7129
17.96	51.69	69.1387	244.1955

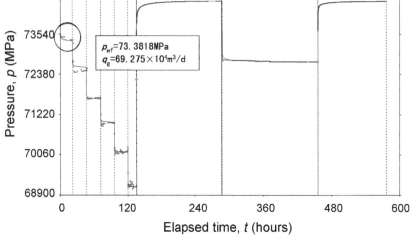

Figure 3.39 Bottom-hole pressure history of back-pressure test in Well KL205.

It is obtained by substituting the aforementioned parameters into Equations (3.44) and (3.45):

$$A' = 29.22 \times 0.025 \times 376 \left(\log \frac{0.472 \times 500}{0.09} + 0 \right) = 939.00$$

$$B' = 12.69 \times 0.025 \times 376 \times 0.018 = 2.147.$$

The first test data of the back-pressure test in Figure 3.39 and Table 3.7 are selected as the initial stable deliverability point, namely

$$p_R = 74.5083 \text{ MPa}$$

$$p_{wf} = 73.3818 \text{ MPa}$$

$$q_g = 69.2750 \times 10^{-4} \text{ m}^3/\text{day}$$

and substituted into Equation (3.36) to inversely calculate equivalent deliverability coefficient kh:

$$kh = \frac{939.0 q_g + 2.147 q_g^2}{p_R^2 - p_{wf}^2} = \frac{939.0 \times 69.275 + 2.147 \times 69.275^2}{74.508^2 - 73.3818^2} = 452.424 \text{ mD} \cdot \text{m}$$

The stable point LIT equation of Well KL205 is finally determined as

$$p_R^2 - p_{we}^2 = 2.075 q_g + 4.746 \times 10^{-3} q_g^2$$

With this deliverability equation, the deliverability curve (Figure 3.40a) and IPR curve (Figure 3.40b) are obtained consequently.

A successful back-pressure test was once conducted in this well, and the IRP curve deduced from the back-pressure test is shown in Figure 3.41. A comparison of the resulting IPR curves from the back-pressure test and stable point LIT deliverability equation is shown in Figure 3.42.

As seen from Figure 3.42, results by the two methods are very close, which prove that a deliverability equation tallying with the actual field situation and applicable to a nearly homogeneous reservoir with high permeability can absolutely be established with one early stable deliverability point.

6.4.2 LIT Equation Established in SLG Gas Field

Production tests were conducted in certain exploration wells in the early development stage in Sulige areas, and modified isochronal tests were mostly applied in the initial

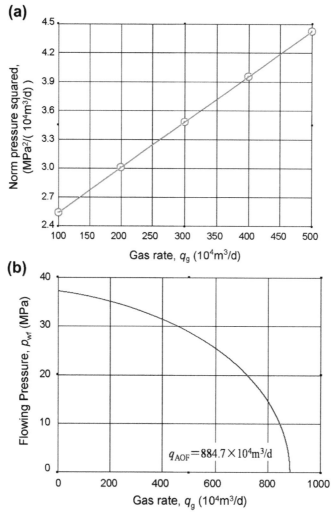

Figure 3.40 Theoretical calculated deliverability plot (a) and IPR curve (b) corrected with measured data.

product test periods. The author carried out overall deliverability test research with test data acquired from 11 wells.

Analysis of Standard Modified Isochronal Tests

Complete pressure and flow rate data of modified isochronal tests in 11 wells were acquired onsite and then analyzed carefully with well test interpretation software, initial IPR curves were plotted and initial AOFP values were calculated consequently, and the results are listed in Table 3.8.

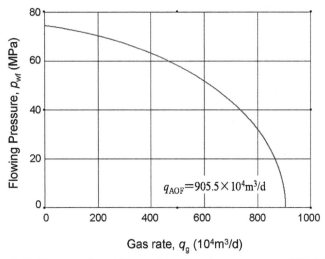

Figure 3.41 IPR curve from LIT equation of back-pressure test of Well KL205.

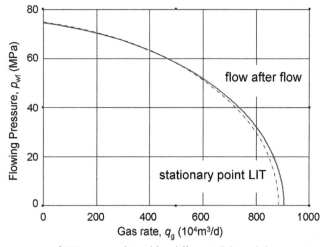

Figure 3.42 Comparison of IPR curves plotted by different deliverability equations of Well KL205.

Stable Point LIT Equation

Select mean values of the physical property and other parameters in the main producing wells of the SLG gas field as basic parameters of the established stable point LIT equation over the whole field:

Subsurface viscosity of natural gas, $\mu_g = 0.02$ mPa·s

Deviation factor of natural gas, $Z = 0.9753$

Formation temperature, $T_f = 378$ K

Mechanical skin factor, $S = -5.5$

Gas supply radius of gas well, $r_e = 500$ meters

Table 3.8 Comparison of Parameters and Analysis Results Obtained from Different Deliverability Analysis Methods

	Stable deliverability data in late extended test						Calculated AOFP, 10^4 m^3/day	
Well name	p_R MPa	p_{wf} MPa	q_g 10^4 m^3/day	Equivalent kh, Md·m	A	B	Stable point equation	Pressures squared LIT equation
S4	28.4000	15.2300	14.926	6.408	38.279	0.01460	20.9034	20.6733
S5	28.8149	19.3147	10.031	5.402	45.409	0.01732	18.1589	17.4771
S6	27.9975	15.4312	15.131	6.840	35.879	0.01368	21.6791	21.7128
S10	27.5100	10.3500	15.049	5.714	42.925	0.01637	17.5136	17.4243
S20	29.6080	16.9390	5.310	2.213	110.84	0.04227	7.8852	7.7305
S25	27.1370	21.5780	4.063	3.686	66.545	0.02538	11.0199	10.2518
T5	29.3400	9.0000	10.025	3.165	77.493	0.02955	11.0618	11.0204
S16–18	26.7142	22.1887	0.802	0.889	275.73	0.1051	2.5857	2.5793
S14	29.7615	26.2307	1.502	1.865	131.51	0.05016	6.7178	6.6909
S37–7	30.3476	25.9552	1.200	1.190	205.99	0.07857	4.4633	4.4556
S40–10	30.8743	29.0633	3.011	6.811	36.011	0.01373	26.2079	26.4358

Effective wellbore radius, $r_w = 0.07$ meters

Non-Darcy flow coefficient, $D = 0.001(10^4 \, \mathrm{m}^3/\mathrm{day})^{-1}$ (obtained from the results of comparative analysis of deliverability)

The general stable point LIT equation applicable to the SLG gas field is

$$p_R^2 - p_{wf}^2 = \frac{245.4}{kh} q_g + \frac{0.09357}{kh} q_g^2. \tag{3.47}$$

Deliverability coefficient kh in the formula varies from well to well and can be calculated with Equation (3.48) combined with measured data in the late extended test during the modified isochronal test of the tested well.

$$kh = \frac{245.4 q_{g(measured)} + 0.09357 q_{g(measured)}^2}{p_{R(measured)}^2 - p_{wf(measured)}^2}. \tag{3.48}$$

Results are listed in Table 3.8.

The aforementioned results can be expressed in a comparative column plot (Figure 3.43).

As shown in Table 3.8 and Figure 3.43, the AOFP values calculated by the stable point LIT equation and standard modified isochronal test method are very similar. And the conclusions can be made as follows:

1. Except for special needs, a modified isochronal test is not necessary to be conducted repeatedly to production wells in the SLG gas field in the future. Instead, as long as stable deliverability data are acquired, by substituting data into Equations (3.47) and (3.48), a reliable initial deliverability equation and initial AOFP are obtained.

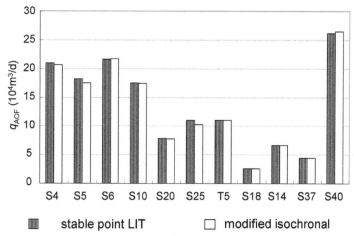

Figure 3.43 Comparative column plot of calculated AOFP resulting from different deliverability analysis methods in a SL gas field.

2. By similar methods, feasible stable point LIT equations can also be constructed for such gas fields as YL, ZZ-MZ, and WSQ, as well as other gas fields discovered in the future in the Ordose Basin, and so the deliverability test process can be simplified significantly. In fact, this job has been done or is being done there.

6.5 Methods of Establishing Dynamic Deliverability Equation

6.5.1 Initial Stable Point LIT Equation is Established First

Take Well S6 of the SLG gas field as an example. Pressure and flow rate history acquired onsite are shown in Figure 3.44.

A general initial stable point LIT equation established with physical property parameters of He-8 Formation is expressed as follows:

$$p_R^2 - p_{wf}^2 = \frac{245.4}{kh} q_g + \frac{0.09357}{kh} q_g^2. \tag{3.49}$$

End point data of the extended test during the modified isochronal test in Well S-6 acquired in 2001 are $p_R = 27.9975$ MPa, $p_{wf} = 15.4312$ MPa, and $q_g = 15.1311 \times 10^4$ m^3/day; substitute them into Equation (3.49), calculate the equivalent deliverability coefficient kh of the well inversely to be 6.84 mD·m, and establish an initial deliverability equation accordingly:

$$p_R^2 - p_{wf}^2 = 35.879 q_g + 0.01368 q_g^2. \tag{3.50}$$

The initial IPR curve of Well S-6 is plotted by this deliverability equation and is shown in Figure 3.45.

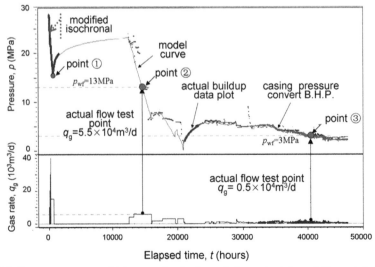

Figure 3.44 Flow rate, pressure history, and selection of dynamic deliverability points in Well S6.

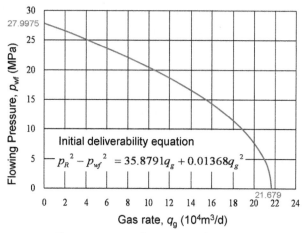

Figure 3.45 Initial IPR curve of Well S-6.

6.5.2 Establishment of Dynamic Deliverability Equation

A deliverability decline of Well S-6 occurred in December 2002 after a period of production. Substitute data of dynamic deliverability point ② (refer to Figure 3.44), p_{wf} = 13 MPa and q_g = 5.5 × 10^4 m^3/day, into Equation (3.50); the dynamic formation pressure is estimated primarily by calculating inversely as

$$p_R = \left(35.8791q_g + 0.01368q_g^2 + p_{wf}^2\right)^{0.5} = 19.2477\text{MPa}\cdot \qquad (3.51)$$

Natural gas prosperities vary along with a decrease of formation pressure. Figure 3.46 shows the relationship between subsurface viscosity μ_g and deviation factor Z of gas versus formation pressure p_R calculated on the basis of the component of the gas in Well S-6.

After formation pressure has declined, the subsurface viscosity of the gas changed to μ_g = 0.019 mPa·s and the deviation factor changed to Z = 0.8856 (obtained from

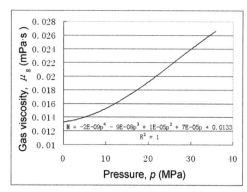
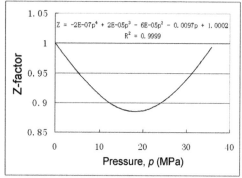

Figure 3.46 Relationship of gas properties vs pressure in Well S-6.

Figure 3.46). Substitute these new values into the expressions of A' and B', $A' = 211.7182$ and $B' = 0.8071$ are obtained; then substitute the new values of A' and B' into Equation (3.51); a modified formation pressure is obtained as $p_{R1} = 18.5147$ MPa. Sometimes the iterative process needs to be operated many times. A dynamic deliverability equation of Well S-6 in December 2002 is ultimately established as

$$p_{R1}^2 - p_{wf}^2 = 30.9502q_g + 0.1180q_g^2.$$

It should be noted that the formation pressure is substituted by p_{R1} but formation deliverability coefficient kh was unchanged in the new deliverability equation; dynamic AOFP has decreased to $q_{AOF1} = 10.64 \times 10^4$ m³/day. A dynamic IPR curve at that time is shown in Figure 3.47.

Dynamic deliverability situations of Well S-6 in April 2005 can be obtained with the same method and data of dynamic deliverability point ③ (refer to Figure 3.44): the dynamic formation pressure is $p_{R2} = 4.62$ MPa, dynamic AOFP is $q_{AOF2} = 0.86 \times 10^4$ m³/day, and dynamic IPR curve is shown in Figure 3.48.

6.5.3 Deliverability Decline Process in Gas Wells

Superposing its initial and dynamic IPR curves on one graph, the deliverability decline process of Well S-6 can be clearly seen; refer to Figure 3.49. Meanwhile, put the dynamic IPR curves at different periods on the pressure history plot of Well S-6—how the gas well tends to deplete in production can be clearly presented; refer to Figure 3.50.

The aforementioned analysis process has been compiled into a feasible Excel operation sheet to be applied onsite.

Some researchers thought that a "stress-sensitive effect" exists in low permeability formations, which means that formation permeability deceases along with a decrease of formation pressure. If research results on this subject are formed from commonly

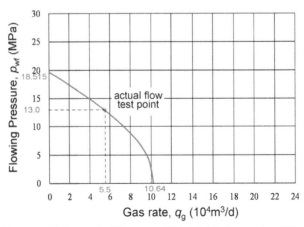

Figure 3.47 Dynamic IPR curve of Well S-6 in December 2002.

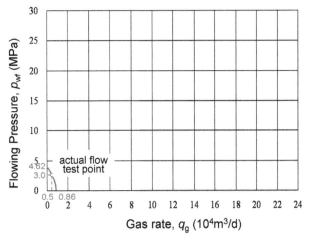

Figure 3.48 Dynamic IPR curve of Well S-6 in April 2005.

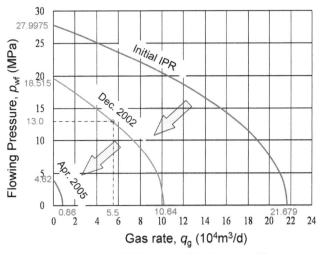

Figure 3.49 Comparison of IPR curves of Well S-6 at different periods.

acceptable numerical relations, the dynamic deliverability analysis can be brought into influencing factors of the deliverability coefficient that varies along with pressure variation without difficulties. Also, concrete operation methods are awaiting further development.

6.6 Stable Point LIT Equation of Horizontal Wells

More and more horizontal wells are drilled to exploit natural gas in both onshore and offshore gas fields in China. Unlike vertical wells, reservoir sections in horizontal wells are almost horizontally penetrated, there is a building up section existing between the straight section and the horizontal section, and either the linear distance or the vertical

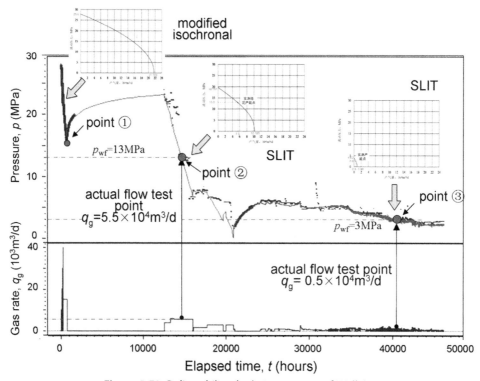

Figure 3.50 Deliverability depletion process of Well S-6.

distance between the kick-off point to the horizontal pay zone is quite long. Due to testing technologies, the downhole pressure gauge is usually set above the deflection point, as shown in Figure 3.51.

Many problems are raised in pressure analysis because the pressure gauge cannot measure directly the pay zone pressure and it is very difficult to acquire a pressure gradient between the measured point and the gas pay zone; among them, the most important one is the measured pressure shift problem due to accumulated load water. Accumulated load water is likely to occur due to a gas—water separation in a horizontal and deviated wellbore; therefore, when a deliverability test is performed in a horizontal well, a measured pressure shift caused by accumulated water effect is more possible, which makes deliverability test data abnormal, reliability of calculated deliverability indices poor, and even establishing an applicable deliverability equation impossible.

In recent years, a project of deliverability test analysis for an offshore gas field that has been put into production is performed, although plenty of back-pressure test data has been acquired with basically reliable quality, but the interpretation results of more than half of the wells are of poor reliability, and even feasible deliverability equations cannot be established.

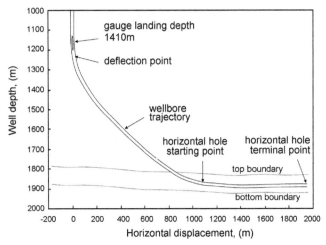

Figure 3.51 Landing depth of a pressure gauge in a horizontal well.

It is highlighted here that the application of stable point LIT equations is very necessary in the deliverability tests of horizontal wells.

6.6.1 Theoretical Deduction of Stable Point LIT Equation for Horizontal Wells

The paper (SPE 15375) by S.D. Joshi in the book of Collection of Treatise of Association of Petroleum Engineer (USA) gave an expression for calculating production under completion conditions of horizontal wells, and a deliverability equation for horizontal wells is deduced from Formula 8 in this paper [76]. Formula 8 was originally a production expression of oil wells in conditions of stable flow and without a skin effect in terms and under Darcy's units, and was rededuced and modified by Li Dang; coefficient π was added into the denominator of the expression and is now written as [77]:

$$q_{\mathrm{h}} = \frac{2\pi k_{\mathrm{h}} h \Delta p / \mu B}{\ln\left[\dfrac{a + \sqrt{a^2 - (L_{\mathrm{e}}/2)^2}}{(L_{\mathrm{e}}/2)}\right] + \dfrac{\beta h}{L_{\mathrm{e}}}\ln\left[\dfrac{\beta h}{2\pi r_{\mathrm{w}}}\right]}. \tag{3.52}$$

For gas wells, the stable flow rate expression under Darcy's units is

$$q_{\mathrm{gh}} = \frac{2\pi k_{\mathrm{h}} h T_{\mathrm{sc}} \left(p_{\mathrm{R}}^2 - p_{\mathrm{wf}}^2\right) / p_{\mathrm{sc}} \mu_{\mathrm{g}} T_{\mathrm{f}} Z}{\ln\left[\dfrac{a + \sqrt{a^2 - (L_{\mathrm{e}}/2)^2}}{(L_{\mathrm{e}}/2)}\right] + \dfrac{\beta h}{L_{\mathrm{e}}}\ln\left[\dfrac{\beta h}{2\pi r_{\mathrm{w}}}\right]} \tag{3.53}$$

After further deduction, the expression of the denominator of Equation (3.53) is changed to

$$
\ln\left[\frac{a+\sqrt{a^2-(L_e/2)^2}}{(L_e/2)}\right] + \frac{\beta h}{L_e}\ln\left[\frac{\beta h}{2\pi r_w}\right]
$$

$$
= \ln\frac{2a}{L_e}\left[1+\sqrt{1-(L_e/2)^2}\right] + \frac{\beta h}{L_e}\ln\left(\frac{\beta h}{2\pi r_w}\right)
$$

$$
= \ln\frac{2a}{L_e}\left[1+\sqrt{1-(L_e/2a)^2}\right]\left[\beta h/2\pi r_w\right]^{(\beta h/L_e)}
$$

$$
= \ln\frac{r_{eh}}{\left\{\frac{(L_e/2a)r_{eh}}{\left[1+\sqrt{1-(L_e/2a)^2}\right]\left[\beta h/2\pi r_w\right]^{(\beta h/L_e)}}\right\}} = \ln\frac{r_{eh}}{r_{wh}},
$$

(3.54)

where r_{eh} is called the equivalent gas supply radius of a horizontal well and is expressed as

$$
r_{eh} = \sqrt{\frac{A_{eh}}{\pi}} = \sqrt{\frac{\pi r_e^2 + 2r_e L_e}{\pi}},
$$

(3.55)

where A_{eh} is the drainage area of gas supply area: $A_{eh} = \pi r_e^2 + 2r_e L_e$.

The gas supply zone of a horizontal well is indicated in Figure 3.52.

Readers can refer to He Kai's paper for content concerning the calculation formula of radius of gas supply area for horizontal wells [78].

In Equation (3.54), r_{wh} is called the equivalent wellbore radius of horizontal well and is expressed as

$$
r_{wh} = \frac{r_{eh}L_e}{2a\left[1+\sqrt{1-(L_e/2a)^2}\right]\left[\beta h/2\pi r_w\right]^{(\beta h/L_e)}}.
$$

(3.56)

Variable a in Equation (3.56) can be expressed as

$$
a = (L_e/2)\left[0.5+\sqrt{0.25+(2r_{eh}/L_e)^4}\right]^{0.5}.
$$

(3.57)

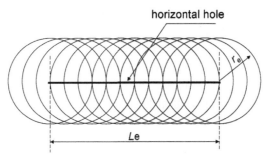

Figure 3.52 Gas supply zone of a horizontal well.

After deduction, Equation (3.53) is transformed further into an expression for gas flow rate calculation in a pseudo-steady-state flow in a gas well under the CSU unit system:

$$q = \frac{2.714 \times 10^{-5} k_h h T_{sc} \left(p_R^2 - p_{wf}^2 \right)}{p_{sc} \mu_g Z T_t \left(\ln \frac{0.472 r_{eh}}{r_{wh}} + S_a \right)}. \tag{3.58}$$

Equation (3.58) is rewritten as the following LIT equation:

$$p_R^2 - p_{wf}^2 = A_h q_g + B_h q_g^2, \tag{3.59}$$

where

$$A_h = \frac{12.69 \overline{\mu}_g \overline{ZT}}{k_h h} \left(\ln \frac{0.472 r_{eh}}{r_{wh}} + S \right) = \frac{A'_h}{k_h h} \tag{3.60}$$

$$B_h = \frac{12.69 \overline{\mu}_g \overline{ZT}}{k_h h} \square D = \frac{B'_h}{k_h h}. \tag{3.61}$$

It is the expression of the stable point LIT equation for horizontal wells.

Meanings and units of symbols in the formula are as follow:

A_h, B_h is the Coefficients of LIT equation of horizontal wells

p_R is the Formation pressure at gas supply boundary, MPa

p_{wf} is the Bottom-hole flowing pressure, MPa

k_h is the Horizontal permeability of the gas zone, mD

h is the Effective thickness of the gas zone, m

$k_h h$ is the Horizontal equivalent kh, mD·m

L_e is the Effective length of horizontal section of the well, m

μ_g is the Subsurface viscosity of the gas, mPa·s

T_{sc} is the Standard temperature condition, 293.16K

p_{sc} is the Standard pressure condition, $p_{sc} = 0.101325$ MPa

Z is the Deviation factor of the real gas, dimensionless
T_f is the Gas zone temperature, K
S_a is the Apparent (or pseudo) skin factor, dimensionless, $S_a = S + Dq_g$
S is the Mechanical skin factor of borehole wall, dimensionless
D is the Non-Darcy flow coefficient, $(10^4 \ \mathrm{m}^3/\mathrm{day})^{-1}$
r_{eh} is the Equivalent gas supply radius of horizontal well, m
r_{wh} is the Equivalent wellbore radius, m
r_w is the Wellbore radius, m
β is the Heterogeneity correction coefficient, $\beta = \sqrt{k_h/k_v}$

6.6.2 Establishment of Initial Stable Point LIT Deliverability Equation for Horizontal Wells

The method of establishing a stable point LIT deliverability equation is illustrated with a representative example. The equation can be classified into two types: initial stable point LIT deliverability equation and stable point LIT deliverability equation during the decline process; the latter is also called the "dynamic LIT deliverability equation."

Readings of Initial Steady Production Point

A permanent electronic pressure gauge is set in the representative well at the depth of the reservoirs to monitor bottom–hole pressure history real time; the result is shown in Figure 3.53.

Figure 3.53 shows that the bottom–hole flowing pressure of the well declined rapidly since being put into production in July 2003 due to initial production at a quite high flow rate; after that, the decline rate of the flowing pressure was slowed down when the flow rate was lowered; until September 2004, the well was adjusted to produce at a flow rate of

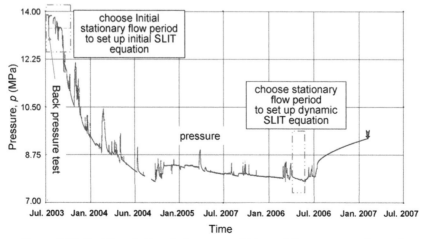

Figure 3.53 Real-time bottom-hole pressure history of the representative well.

approximate 20×10^4 m^3/day, the flowing pressure was gradually stabilized, while it has approached to 8 MPa, near the low limit of output pressure of gas wells.

The initial deliverability equation was established with a stable flowing pressure point in the initial production stage in 2003; the dynamic deliverability equation after deliverability declining was established with a stable deliverability point in 2006.

Figure 3.54 shows an enlarged flowing pressure variation in the initial production stage of the representative well, and it can be seen from Figure 3.54:

1. Before this well was opened for production, a back-pressure test had been performed and then the well was shut in for a pressure buildup test.
2. The well was produced in August 2003. At the beginning of the production, the bottom-hole pressure fluctuated due to frequent shifting working systems; a relatively stable production interval occurred in late August. The reading of bottom-hole flowing pressure during the period from August 23 to September 4 was 13.4096 MPa, which was read from well test software.
3. The reading of average gas flow rate of that period was 11.29×10^4 m^3/day.

Selection of the Parameters of the Representative Well

Based on data of logging interpretation, completion technology, natural gas property analysis, and other related reservoir comprehensive evaluation, the parameters of the representative well are selected primarily as follow. Horizontal section length $L_e = 554.2$ meters, vertical effective thickness of gas zone (from logging) $h = 11.47$ meters, ratio of horizontal and vertical permeability $k_h/k_v = 10$, equivalent wellbore radius $r_w = 0.108$

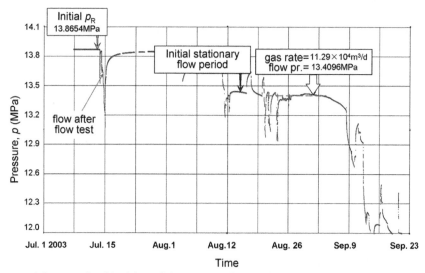

Figure 3.54 Selection of stable deliverability point in the initial production stage of the representative well.

meter, gas supply radius $r_e = 1000$ meters, mechanical skin factor of horizontal section $S = 0$, non-Darcy flow coefficient of this gas well $D = 0.01(10^4 \text{ m}^3/\text{day})^{-1}$, subsurface viscosity of the gas $\mu_g = 0.0164$ mPa·s, deviation factor of the gas $Z = 0.9358$, and formation temperature $T_f = 359.1$K.

Calculation of Deliverability Equation Coefficients of Horizontal Wells

According to Equation (3.55), the equivalent gas supply radius of a horizontal well is

$$r_{eh} = \sqrt{\frac{\pi r_e^2 + 2r_e L_e}{\pi}} = 1163.1 \text{ m}$$

From Equation (3.57), parameter α is obtained as

$$a = (L_e/2)\left[0.5 + \sqrt{0.25 + (2r_{eh}/L_e)^4}\right]^{0.5} = 1179.7 \text{ m}$$

From Equation (3.56), the equivalent wellbore radius of a horizontal well r_{wh} is

$$r_{wh} = \frac{r_{eh} L_e}{2a\left[1 + \sqrt{1 - (L_e/2a)^2}\right] [\beta h/2\pi r_w]^{(\beta h/Le)}} = 106.8 \text{ m}$$

Consequently, coefficients A_h and B_h in the deliverability equation can be calculated:

$$A_h = \frac{12.69 \times 0.0164 \times 0.9358 \times 359.1}{k_h h}\left(\ln\frac{0.472 \times 1163.1}{106.8} + 0\right) = 114.49(k_h h)^{-1}$$

$$B_h = \frac{12.69 \times 0.0164 \times 0.9358 \times 359.1 \times 0.01}{k_h h} = 0.6295(k_h h)^{-1}$$

The deliverability equation is expressed as

$$p_R^2 - p_{wf}^2 = 114.49(k_h h)^{-1}q_g + 0.6295(k_h h)^{-1}q_g^2. \tag{3.62}$$

Equation (3.62) can express the deliverability coefficient (equivalent $k_h h$) as follows:

$$k_h h = \frac{114.49q_g + 0.6295q_g^2}{p_R^2 - p_{wf}^2} \tag{3.63}$$

Establishment of Initial Stable Point LIT Equation of the Representative Well

The parameters of the initial stable point selected from Figure 3.54 are:
- Formation pressure at gas supply boundary $p_R = 13.8654$ MPa
- Flowing pressure at stable point $p_{wf} = 13.4096$ MPa
- Gas flow rate at stable point $q_g = 11.29 \times 10^4$ m^3/day

Substitute these parameters into Equation (3.63) and obtain

$$k_h h = 110.43 \text{ mD·m}$$

and the initial deliverability equation is established accordingly:

$$p_R^2 - p_{wf}^2 = 1.0368 q_g + 0.0057 q_g^2 \tag{3.64}$$

AOFP is calculated as

$$q_{AOF} = 114.0 \times 10^4 \text{ m}^3/\text{day}.$$

The deliverability plot and the IPR curve are drawn by well test interpretation software according to the aforementioned deliverability equation as shown in Figure 3.55.

It is estimated according to the calculated AOFP of 114.0×10^4 m^3/day that the well can be produced at a stable flow rate of $20-25 \times 10^4$ m^3/day; this is proved in the subsequent actual production process.

6.6.3 Method of Establishing Dynamic Deliverability Equation

1. Calculation of physical property parameters

Coefficients A_h and B_h in the initial deliverability equation established based on Equations (3.62), (3.63), and (3.64) will vary when formation pressure p_R has declined. In the expressions of these two coefficients, the main influencing factors are subsurface viscosity μ_g and deviation factor Z of the natural gas; deliverability coefficient $k_h h$ is usually considered a constant for the effect of its variation is negligible except the formation is both stress sensitive, with abnormal high pressure.

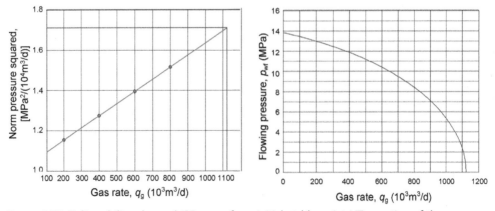

Figure 3.55 Deliverability plot and IPR curve from initial stable point LIT equation of the representative well.

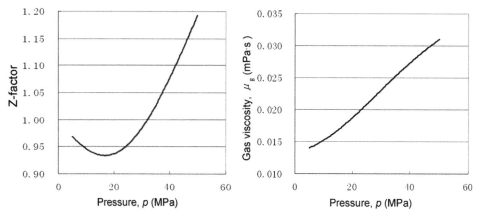
Figure 3.56 Properties of the gas of the representative well.

Properties of natural gas of the representative well vary along with formation pressure, and the results of their variation vs pressure calculated by PVT software are shown in Figure 3.56. It is suggested that the deviation factor (i.e., Z factor) and subsurface viscosity of natural gas are calculated by Standing's and Lee's methods, respectively.

2. Calculation of the coefficients of dynamic deliverability equation A_h and B_h and formation pressure p_R

In Equations (3.60) and (3.61), coefficient A_h will vary from A_{hini} to A_{hdyn} and B_h from B_{hini} to B_{hdyn} when $(Z\mu_g)$ changes from $(Z\mu_g)_{initial}$ to $(Z\mu_g)_{dynamic}$ due to a formation pressure decline:

$$A_{hdyn} = A_{hini} \times \frac{(Z\mu_g)_{dyn}}{(Z\mu_g)_{ini}} = 0.9641$$

$$B_{hdyn} = B_{hini} \times \frac{(Z\mu_g)_{dyn}}{(Z\mu_g)_{ini}} = 0.005631$$

A formula is obtained, hereby

$$p_{Rdynamic}^2 - p_{wf}^2 = 0.9641 q_g + 0.005631 q_g^2 \qquad (3.65)$$

In this formula, because formation pressure $p_{Rdynamic}$ is not the initial formation pressure any longer, a new formation pressure $p_{Rdynamic}$ should be determined prior to the application of Equation (3.65). The specific way is to select a new stable deliverability point at first, for the representative well, choose the stable deliverability point in May 2006 in Figure 3.53, and, after locally enlarging, the selected point is shown in Figure 3.57.

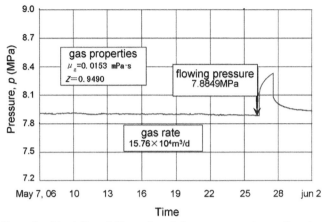

Figure 3.57 Reading of stable deliverability point of the representative well at the end of May 2006.

Select the average gas flow rate of 15.76×10^4 m³/day from May 7 to May 20; the flowing pressure during this period is 7.8849 MPa. Substitute them into Equation (3.65),

$$p_{\text{Rdynamic}} = \left(p_{\text{wf}}^2 + 0.9647 q_{\text{g}} + 0.005631 q_{\text{g}}^2\right)^{0.5} = 8.8749 \, \text{MPa}$$

Then a new dynamic deliverability equation in 2006 is obtained as

$$p_{\text{Rdynamic}}^2 - p_{\text{wf}}^2 = 0.9641 q_{\text{g}} + 0.005631 q_{\text{g}}^2 \tag{3.66}$$

and the dynamic absolute open flow potential is calculated as

$$q_{\text{AOF}} = 60.4 \times 10^4 \, \text{m}^3/\text{day}$$

It should be pointed out that only when the depleted formation pressure p_{Rdynamic} has been determined can the deviation factor of the natural gas $(Z\mu_{\text{g}})_{\text{dynamic}}$ be determined accordingly. It needs a few steps of iterative operation and is unnecessary to be described redundantly here.

3. Deliverability depletion of the representative well

Based on previous operations for the representative well, its dynamic deliverability equation has been established, its dynamic AOFP has been calculated, and at the same time the IPR curve after productivity depletion has been plotted, as shown in Figure 3.58.

The deliverability depletion process is presented demonstratively when the initial IPR curve and the IPR curve after deliverability depletion of the representative well are placed in the same graph, as seen in Figure 3.59.

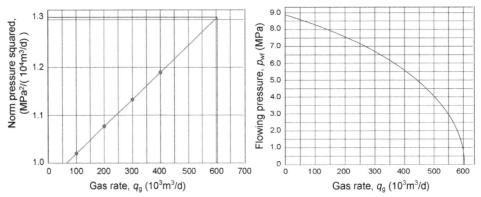

Figure 3.58 Dynamic deliverability equation and dynamic IPR curve of the representative well in May 2006.

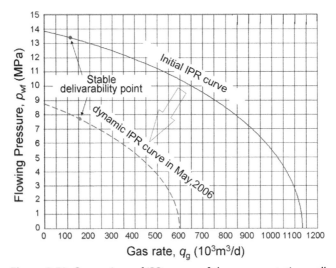

Figure 3.59 Comparison of IPR curves of the representative well.

Figure 3.59 clearly shows that after putting the well into production for 3 years, the formation pressure at the gas supply boundary has decreased from the initial value of 13.865 to 8.8758 MPa, and AOFP has declined from 114 to 60.0×10^4 m^3/day. Under the changed circumstances, the new IPR curve after deliverability depletion should be used in the production planning of this gas well.

It has also been proved that a stable point LIT equation is truly effective in determining the initial deliverability equation and in estimating the deliverability process of horizontal wells.

7 PRODUCTION PREDICTION IN DEVELOPMENT PROGRAM DESIGNING OF GAS FIELDS

Deliverability tests in gas wells aim ultimately at providing the basis of a gas flow rate designing for all planning development wells in gas fields. Therefore, after deliverability tests have been performed in some gas wells and their deliverability equations have been established, the deliverability equations applicable to the whole gas field and the corresponding IPR curves are usually required for making a production prediction of the planning production wells in development program designing. Such estimated gas flow rates are also the basis of production planning for each well in gas field numerical simulation. Common practical methods under different circumstances are elaborated here.

7.1 Deliverability Prediction of Wells with Available Well Test Data

In a newly discovered gas field, a deliverability test must be performed for all exploration wells and important development appraisal wells in accordance with the requirements of relevant regulations. The deliverability equation and IPR curve under the initial conditions are obtained by the deliverability test, as shown in Figure 3.60.

Figure 3.60 shows that the reasonable flow rate of the gas well of which the IPR curve has been obtained can be determined with one of the following methods.

7.1.1 Determining Gas Well Flow Rate with Reasonable Producing Pressure Differential

Multiple limitation factors limit producing pressure differential during gas production, for example:

1. The maximum producing pressure differential of a gas well has to be determined in order to control sand.
2. The producing pressure differential of a gas well has to be limited in order to control the coning of bottom water, etc.

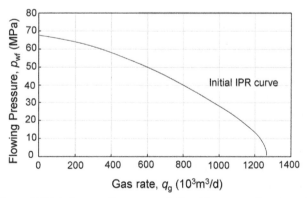

Figure 3.60 Inflow performance relationship curve of a gas well.

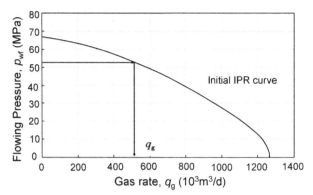

Figure 3.61 Production prediction by an IPR curve.

When the producing pressure differential is determined, bottom–hole flowing pressure is determined accordingly, and the corresponding flow rate can be obtained from the IPR curve, as shown in Figure 3.61.

It is seen from Figure 3.61 that after the flowing pressure is calculated with the given producing pressure differential and initial formation pressure, its corresponding gas flow rate can be read directly from the IPR curve. This graphic method can be replaced by direct calculation with Equations (3.2), (3.6), or (3.13), (3.14) or (3.18), (3.19) for a given flowing pressure p_{wf}.

7.1.2 Gas Flow Rate is Determined by Intersection of Inflow Performance Relationship and Outflow Performance Relationship Curves

The IPR curve mentioned earlier means the inflow performance relationship curve determined by the formation conditions. Production of the gas well depends not only on formation conditions but also on wellbore conditions; the gas flow rate is surely influenced by pressure loss due to vertical conduit flow in the wellbore, and this pressure loss will increase along with the increase of the gas flow rate, as shown in Figure 3.62.

The vertical conduit flow performance curve (or outflow performance relationship curve, i.e., OPR curve) in Figure 3.62 can be established by the related equation regarding vertical conduit flow in gas production technology. The outflow performance relationship curve is constrained by wellhead pressure. When the wellhead pressure is higher, the outflow performance relationship curve will move upward. An intersection point of inflow and outflow performance relationship curves indicates the maximum gas flow rate of the well.

Certainly, gas production at a relatively low rate can be considered in production arrangement. Surplus wellhead pressure is controllable by the choke; it is recommended controlling a relatively low gas flow rate while keeping a fairly high wellhead pressure (pressure before the choke).

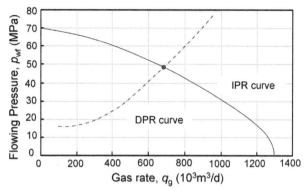

Figure 3.62 Intersection method of inflow and outflow performance relationship curves for determining the flow rate of a gas well.

7.1.3 Determining Deliverability During the Process of Formation Pressure Depletion

When the gas field is developed with natural depletion, the formation pressure declines continuously, and the deliverability curve varies accordingly. Figure 3.63 shows the variation of such a deliverability curve.

It is seen from Figure 3.63 that neither constant producing pressure differential nor constant wellhead pressure (determined by intersection of inflow and outflow performance curves) can maintain the original gas flow rate when the formation pressure has decreased. Gas deliverability at that time must be determined according to the changed dynamic deliverability curves.

It can also be seen clearly that the gas flow rate depends not only on reservoir parameters (e.g., kh, S, and D) but also on some constraint conditions, such as dynamic formation pressure and dynamic deliverability decline.

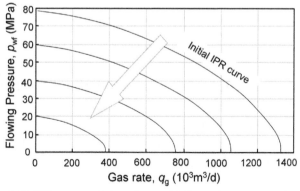

Figure 3.63 IPR curve variation along with formation pressure depletion.

7.1.4 Other Limitations for Gas Flow Rate

1. Limitations caused by erosion velocity

For wells with an extremely high gas flow rate, gas flowing with a high velocity will have an "erosion effect" on the surface of metal equipment and cause damage to the tubing, Christmas tree, and choke. The erosion phenomenon depends mainly on gas flow velocity. The higher the flow velocity is, the more possible this phenomenon occurs. In addition to flow velocity, erosion is also related to pressure, temperature, and the solid particles of sand and salt brought out by the gas flow.

The erosion phenomenon is prone to appear around the Christmas tree, especially at the choke, where the pressure is the lowest and gas flow velocity is the highest.

The maximum flow velocity limit value (e.g., $\nu_{max} = 35$ m/s) can be given on the basis of theoretical calculation or experiment research, and from which the maximum gas flow rate can be determined.

2. Flow rate limitation required for keeping stable production

At present in China, the method of determining the flow rate of gas wells with AOFP (q_{AOF}) is used commonly to meet the requirement of keeping stable production. For instance, if AOFP of a well (q_{AOF}) is 100×10^4 m^3/day, one-fifth of it, that is, $q_g = 1/5 q_{AOF} = 1/5 \times 100 \times 10^4$ m^3/day $= 20 \times 10^4$ m^3/day, is taken as the stable flow rate of the well. Based on a numerical simulation study, it is known that the stable production period of homogeneous reservoirs determined by this method is about 15–20 years.

Sometimes an important precondition is neglected when determining stable production with this method, that is, the gas flow rate must be corresponding to the gas recovery rate of 3–5%, and determination of the gas recovery rate must be related to the dynamic reserves controlled by the well. With dynamic reserves controlled by a single well determined accurately, a rational gas flow rate can be determined according to the rational gas recovery rate. As for reservoirs limited by boundaries, although AOFP of a single well may be very high, but formation pressure will decline rapidly after the well has been put into production, it is obviously impossible to keep stable production.

7.2 Deliverability Prediction of Production Wells in Development Program Designing

In a large gas field, the number of exploration wells in the early stage is limited, and sometimes some of these wells may be not suitable for production due to the limitations of initial well completion conditions; therefore, in most development programs, deliverability prediction must be done for the planned production wells. The procedures of deliverability prediction are as follow.

1. Establish deliverability equation applicable to the whole gas field based on existing data of the gas well test.
2. The main constraint of deliverability equation or IPR curve of the gas field is formation coefficient kh; other parameters, such as non-Darcy flow coefficient D, skin factor S, gas supply radius r_e, and equivalent wellbore radius r_{we}, are considered as known and usually similar.
3. Find out the distribution rule of formation coefficient kh with static methods, that is, geophysical, logging, and other geological methods, and make its distribution map.
4. Determine kh value at the position of the planned well.
5. Estimate the deliverability of the planned well in the development program with the kh value.

The following analysis tasks should be fulfilled according to the aforementioned procedures.

7.2.1 Establishing the Deliverability Equation of the Whole Gas Field

The decisive parameters in the deliverability equation for the relationship between gas flow rate and producing pressure differential are coefficients A and B in the LIT equation or coefficient C and n in the exponential equation. When the flow enters into a pseudo-steady state, the main parameter influencing coefficients A and B is kh, as seen from Equations (3.26) and (3.27)(for pressure-squared method) or Equations (3.29) and (3.30)(for pseudo-pressure method). In addition, coefficients A and B are related to completion parameters, that is, skin factor S and non-Darcy flow coefficient D to some extent. Skin factor S can be controlled by improving completion technology, whereas non-Darcy flow coefficient D is generally a constant in a gas field and can be obtained by well testing. So the main factor influencing coefficients A and B is the kh value. Therefore, as long as the relationship of A and B vs kh is established, the deliverability equation applicable to the whole gas field can be obtained.

Establishing the Deliverability Equation of a Gas Field by Calculation

Equation (3.28) can be converted into the following expression:

$$\psi_R - \psi_{wf} = \left[29.22\overline{T}\left(\log\frac{0.472r_e}{r_w} + \frac{S}{2.302} \right) \right](kh)^{-1}q_g + 12.69\overline{T}D(kh)^{-1}q_g^2$$

(3.67)

For instance, given $\overline{T} = 100\,°C = 373\,K$, $r_e = 500$ meters, $r_w = 0.09$ meter, $S = 15$, $\overline{\mu}_g = 0.025$ mPa·s, and $D = 1.8 \times 10^{-2}(10^4\ \text{m}^3/\text{day})^{-1}$, the deliverability equation expressed by pseudo-pressure is transformed into

$$\psi_R - \psi_{wf} = 1.08 \times 10^5 (kh)^{-1}q_g + 85.20(kh)^{-1}q_g^2$$

(3.68)

and that expressed by pressure squared is

$$P_R^2 - P_{wf}^2 = \left[29.22\overline{\mu}_g\overline{T}\left(\log\frac{0.472r_e}{r_w} + \frac{S}{2.302}\right)\right](kh)^{-1}q_g + 12.69\overline{\mu}_g\overline{T}D(kh)^{-1}q_g^2$$
$$= 2.707 \times 10^3(kh)^{-1}q_g + 2.130(kh)^{-1}q_g^2$$

(3.69)

Establishing Deliverability Equation of a Gas Field by Regression of Deliverability Test Data
The deliverability equation is applicable to the whole gas field derived previously is only for pseudo-steady flow conditions. This equation is suitable for gas field production planning, but the following problems are often encountered in the selection of parameters.

(1) Determining the kh value. It can be seen from the aforementioned derived formula that formation coefficient kh should be obtained from transient testing, which is the product of effective permeability and effective thickness, whereas the kh value at the position of a planned well during the development program design is obtained from the distribution map based on the air permeability from geological, geophysical, or logging analysis methods. There is a great difference between these two kinds of methods for determining the kh value and so a certain error is inevitably brought about sometimes due to application of the aforementioned formulae.

(2) Selection of S and D values. S and D values obtained from well test analysis are not very precise sometimes for some reasons, which influences the formula coefficients consequently, which is why the aforementioned theoretical formulae are seldom applied.

In order to definitely determine the deliverability equation of the gas field, the relationship between coefficients A and B vs kh obtained from statistical method can be applied. Take a gas field for instance: N wells (or intervals) have been tested normally and the deliverability equations of them are

$$\psi_R - \psi_{wf} = A_i q_g + B_i q_g^2 \quad (i = 1, 2, \ldots\ldots N)$$

(3.70)

The ith one of the N wells (or intervals) has its corresponding formation coefficient $(kh)_i$ $i = 1,2,\ldots\ldots N)$ obtained from logging analysis. Draw the relation graphs of A_i and B_i vs $(kh)_i$, respectively, both relationship of A vs kh and B vs kh can be obtained, as shown in Figure 3.64.

It can be obtained from the regression:

$$A_\psi = a_\psi(kh)^{-m(a,\psi)}$$
$$B_\psi = b_\psi(kh)^{-m(b,\psi)}$$

(3.71)

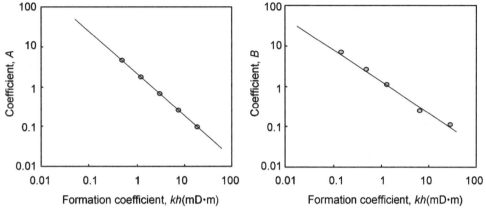

Figure 3.64 Relationship between coefficients A and B of the LIT deliverability equation vs the *kh* value obtained from regression.

When substituted into Equation (3.28), the deliverability equation of the gas field is obtained:

$$\psi_R - \psi_{wf} = a_\psi (kh)^{-m(a,\psi)} q_g + b_\psi (kh)^{-m(b,\psi)} q_g^2 \qquad (3.72)$$

Substitute the *kh* value of each planned well in the development program into Equation (3.72); the deliverability equation of that planned well can be established, from which the IPR curve of that well can be plotted, and then the deliverability prediction of that well can be performed.

(3) Similarly, coefficients of the LIT deliverability equation with the pressure-squared form can be derived with a statistical method:

$$A_2 = a_2 (kh)^{-m(a,2)}$$
$$B_2 = b_2 (kh)^{-m(b,2)} \qquad (3.73)$$

The deliverability equation applicable to the whole gas field is deduced as

$$P_R^2 - P_{wf}^2 = a_2 (kh)^{-m(a,2)} q_g + b_2 (kh)^{-m(b,2)} q_g^2. \qquad (3.74)$$

Also, the deliverability equation of each well in the gas field can be constructed for production prediction based on Equation (3.74).

The absolute value of index *m* in Equations (3.72) and (3.74) is about 1. If the value of *m* obtained from statistical analysis is far away from 1, maybe something went wrong in the analysis process.

In addition, the preconditions for applying the aforementioned statistical analysis method include:

• There are sufficient and high-quality deliverability test data available for getting a reliable deliverability equation and its coefficients A_i and B_i.

- kh values of the interval from logging analysis are representative, which means that the selected kh value is consistent with that of the interval contributing gas flow rate q_g actually, for instance, if a very thick and vertically well-connected gas zone was tested with only partial perforation: such a case is not subject to the principle.
- The completion conditions of tested intervals are basically consistent.

Do not put fracturing completed wells and normal completed wells together in statistical analysis of well test data, as otherwise serious errors are inevitable.

Take a field example for illustration. Good deliverability equations have been obtained from six wells (or intervals) in a gas field, they are

$$\psi_R - \psi_{wf} = 218.3 q_g + 5.69 \times 10^{-2} q_g^2$$

$$\psi_R - \psi_{wf} = 113.0 q_g + 2.62 \times 10^{-2} q_g^2$$

$$\psi_R - \psi_{wf} = 58.5 q_g + 1.21 \times 10^{-2} q_g^2$$

$$\psi_R - \psi_{wf} = 24.5 q_g + 4.32 \times 10^{-2} q_g^2$$

$$\psi_R - \psi_{wf} = 30.3 q_g + 5.54 \times 10^{-2} q_g^2$$

$$\psi_R - \psi_{wf} = 39.8 q_g + 7.65 \times 10^{-2} q_g^2$$

The corresponding kh values of the six wells (or intervals) are 499, 1002, 1998, 5001, 4000, and 2998 mD·m, respectively; statistical analysis described earlier is done and the graphs of coefficients A and B vs formation coefficient kh are as shown in Figure 3.65.

By regression, the deliverability equation of this gas field is established as

$$\psi_R - \psi_{wf} = 8.0 \times 10^4 (kh)^{-0.96} q_g + 59.95 (kh)^{-1.11} q_g^2$$

Establishing the Deliverability Equation of a Gas Field by Typical Well Analogy Method

If there is not enough deliverability test data provided for statistical analysis in a gas field before designing a development program, it is very difficult to perform the statistic analysis method introduced previously. However, if good deliverability test data are acquired and the deliverability equation is derived from at least one well in the field, the analogy method is applicable in establishing the deliverability equation for the whole gas field.

1. Write down the good deliverability equation of the tested well first:

$$\psi_R - \psi_{wf} = A_{0\psi} q_g + B_{0\psi} q_g^2$$

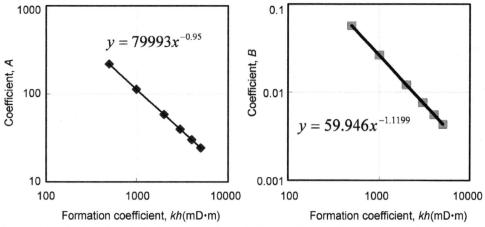

Figure 3.65 Statistical analysis for relationship of LIT equation coefficients A and B vs formation coefficient kh.

2. Find out the formation coefficient of penetrated intervals in the well $(kh)_0$ determined by logging interpretation.
3. Determine the formation coefficient $(kh)_i$ of the ith planned well in the development program from the kh value distribution map plotted with geological data.
4. The deliverability equation of the ith planned well in the development program can be expressed as

$$\psi_R - \psi_{wf} = A_{0\psi}\frac{(kh)_0}{(kh)_i}q_g + B_{0\psi}\frac{(kh)_0}{(kh)_i}q_g^2. \tag{3.75}$$

Deliverability prediction for the ith planned well in the development program can be conducted by Equation (3.75).

Similarly, a formula with the pressure-squared method can be written as

$$p_R^2 - p_{wf}^2 = A_{02}\frac{(kh)_0}{(kh)_i}q_g + B_{02}\frac{(kh)_0}{(kh)_i}q_g^2. \tag{3.76}$$

Exponential Deliverability Equation

It is generally thought that the deliverability analysis with the exponential deliverability equation is a "simple analysis," which means that it may bring a bigger error in the analysis. Therefore, the exponential equation is not usually recommended in predicting gas field deliverability, but it will be introduced here as a method anyway.

The exponential deliverability equation with pseudo-pressure is expressed as follows:

$$q_g = C_\psi(\psi_R - \psi_{wf})^n.$$

Coefficient C_ψ in the formula can further be expressed as

$$C_\psi = \frac{kh}{29.22\overline{T}\left(\log\dfrac{0.472r_e}{r_w} + \dfrac{S}{2.302}\right)} = C'_\psi kh$$

and the exponent n is thought as a constant in the whole gas field; the gas field deliverability equation can be written as

$$q_g = C'_\psi(kh)\left(\psi_R - \psi_{wf}\right)^n, \tag{3.77}$$

where

$$C'_\psi = \left[29.22\overline{T}\left(\log\frac{0.472r_e}{r_w} + \frac{S}{2.302}\right)\right]^{-1}. \tag{3.78}$$

Similarly, the exponential deliverability equation with the pressure-squared form of the whole gas field can be expressed as

$$q_g = C_2(kh)\left(p_R^2 - p_{wf}^2\right)^n \tag{3.79}$$

where

$$C_2 = \left[29.22\overline{\mu}_g\overline{ZT}\left(\log\frac{0.472r_e}{r_w} + \frac{S}{2.302}\right)\right]^{-1} \tag{3.80}$$

For a specific gas field, the gas flow rate equation can also be determined using the following methods.

1. Theoretical calculation method. With such parameters as μ_g, Z, T, r_e, r_w, and S determined, coefficient C'_ψ or C_2 can be calculated by Equation (3.78) or (3.80), and then gas field deliverability prediction performed using Equation (3.77) or (3.79).
2. Statistical method. If successful deliverability tests have been conducted in N wells (or intervals) over the whole gas field, statistical analysis can be performed for the coefficient of exponential deliverability equation C of these pay zones, as shown in Figure 3.66. It can be obtained from statistical analysis:

$$C = C_\psi(kh)^{m(\psi,c)} \tag{3.81}$$

Consequently, the deliverability equation applicable to the whole gas field with pseudo-pressure is written as

$$q_g = C_\psi(kh)^{m(\psi,c)}\left(\psi_R - \psi_{wf}\right)^n. \tag{3.82}$$

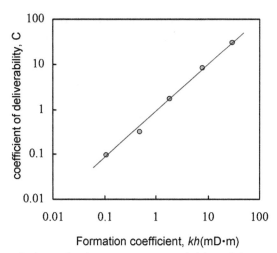

Figure 3.66 Regression of relationship between exponential deliverability equation coefficient C and formation coefficient kh.

By the same method, the deliverability equation with pressure square applicable for the whole gas field can be gained:

$$q_g = C_2(kh)^{m(2,c)}\left(p_R^2 - p_{wf}^2\right)^n \tag{3.83}$$

Deliverability prediction of the planned wells in the development program can be performed by using Equation (3.82) or (3.83) combined with the corresponding IPR curve. It should be noted that the values of exponent m (ψ,c) or m (2,c) should be around 1, as otherwise checking the statistical analysis process is necessary.

3. Typical deliverability equation analogy method. Just like in the case of the LIT deliverability equation, application of the statistical analysis method may be difficult due to lack of test data. In this case, a typical deliverability equation analogy and reasoning method can be applied.

   a. Select a typical well with high-quality deliverability test data and write down its deliverability equation:

$$q_g = C_{\psi 0}\left(\psi_R - \psi_{wf}\right)^n \tag{3.84}$$

   or

$$q_g = C_{20}\left(p_R^2 - p_{wf}^2\right)^n \tag{3.85}$$

   b. Find out the formation coefficient of penetrated intervals in the well $(kh)_0$ determined by logging interpretation.

c. Determine the formation coefficient $(kh)_i$ of the ith planned well in the development program from the kh value distribution map plotted with geological data.

d. The deliverability equation of the ith planned well in the development program can be expressed as

$$q_{gi} = C_{\psi 0}\frac{(kh)_i}{(kh)_0}\left(\psi_R - \psi_{wf}\right)^n \tag{3.86}$$

or

$$q_{gi} = C_{20}\frac{(kh)_i}{(kh)_0}\left(p_R^2 - p_{wf}^2\right)^n. \tag{3.87}$$

Perform deliverability prediction for the ith planned well in the development program using Equation (3.86) or (3.87).

7.2.2 Plotting Distribution Map of Kh Value Over Whole Gas Field and Determination of Kh Value at Well Point

This task should be fulfilled by geologists. The general way is to perform lateral reservoir prediction by using geophysical research results, proofread them by geological researching with the parameters obtained from logging analysis of the well points, plot the kh distribution map of the planar separated layering and that of the full hole, and finally determine kh values of the well point, the producing zones, and the full hole of the planned well.

7.2.3 Calculating Rational Flow Rate of Planned Wells in the Development Program by Deliverability Equation

The rational flow rate of planned wells can be calculated by LIT Equation (3.67) or (3.69), (3.72) or (3.74), (3.75) or (3.76), or by the exponential equations.

AOFP Method

The method is applied commonly in China. The basic assumption of it is that the reservoir is approximately homogeneous. When the deliverability equation is established, let $p_{wf} = 0.101$ MPa, that is, decrease the bottom-hole producing pressure down to atmospheric pressure, the maximum theoretical gas flow rate, AOFP (q_{AOF}) value is calculated.

Once the q_{AOF} value is determined, determine the stable flow rate of the well as a certain proportion of it, for example, 1/5, ¼, or other fraction of q_{AOF}, and then carry out numerical simulation to make sure that a stable production period of 15—20 years exists under this stable flow rate.

Producing Pressure Differential Method

The IPR curve can be obtained instantly by well test interpretation software after the deliverability equation of the planned well in the development program is established. The following procedures can be used if the software is unavailable.

1. Establish a relationship curve or formula between pseudo pressure ψ and pressure p.
2. Get knowledge of the relationship between q_g and p_{wf} using Equations (3.67) or (3.69), (3.72) or (3.74), and (3.75) or (3.76).
3. Plot the relationship graph of q_g and p_{wf}, that is, the IPR curve.

With the IPR curve shown in Figure 3.60, the rational flow rate can be attained with the straight line intersection method as shown in Figure 3.61.

Inflow Performance Relationship Curve and Outflow Performance Relationship (OPR) Curve Intersection Method

Based on the casing program of a production well, calculate the friction of gas when flowing from the bottom hole to the wellhead, generate the OPR curve consequently, and plot it in the coordinate graph with q_g against p_{wf}; it will intersect with the IPR curve, and the intersection point indicates the maximum initial gas flow rate of the well, shown in Figure 3.62.

If formation pressure has declined after the well has been producing for a period of time, the IPR curve of the well at that time will move downward, as shown in Figure 3.63. The outflow performance relationship curve should be adjusted in this case. A new upper threshold value of the gas flow rate can be obtained on the basis of the new inflow and new outflow performance relationship curves and their new intersection point.

All new obtained production values will be used as new production parameters in numerical simulation.

Other Flowing Velocity Control Methods

In view of technology, erosion velocity limitation conditions are often put forward. The upper limit of the flow rate is restricted by erosion velocity conditions; if the calculated gas flow rate exceeds this erosion velocity limitation, it should be adjusted lower properly. In fact, the calculated gas flow rate is very often far lower than the erosion velocity limitation. It is necessary to increase the flow rate to realize the higher economical return, but specific arrangements should be considered comprehensively.

8 DISCUSSION ON SEVERAL KEY PROBLEMS IN DELIVERABILITY TEST

As introduced at the beginning of this chapter, deliverability test methods are somewhat trial and error and were made to meet the requirement of production planning. Therefore, some abnormal phenomena are inevitable in their application in fields. This chapter analyzed specifically problems that appeared during deliverability tests in China and put forth some suggestions of the treatment measures to deal with those problems.

8.1 Design of Deliverability Test Points

8.1.1 Design of Flow Rate Sequence

In a deliverability test of a gas well, the highest and lowest gas flow rates of deliverability test points should be determined first.

1. The minimum gas flow rate in the deliverability test
   (a) The minimum gas flow rate in the deliverability test should be higher than or equal to the gas flow rate that can carry the liquid in the bottom hole out of the wellbore
   (b) The minimum gas flow rate in the deliverability test should be high enough to keep the wellhead temperature higher than the hydrate formation point
   (c) The bottom–hole flowing pressure is nearly equal to 5% of the formation pressure
   (d) The gas flow rate is about 10% of the AOFP (q_{AOF})
2. The maximum gas flow rate in the deliverability test
   (a) The maximum gas flow rate should keep the bottom–hole condition of no large amount of sand production or borehole collapse, which may cause serious well damage
   (b) For wells in which bottom water is close to the producing zones, bottom water coming into the hole must be avoided during the test to prevent serious water producing into the well
   (c) For condensate gas wells, it must be avoided that extremely too big producing pressure differential exists due to an extremely high gas flow rate, which may cause retrograde condensation in a large area in the formations and subsequently the existence of two-phase flow, well plugging, and invalid deliverability test indices due to condensate oil gathering around the bottom hole
   (d) The gas flow rate should not be higher than 50% of the AOPF. It was once advised abroad (e.g., Canada) that the maximum gas flow rate of test points be 75% of q_{AOF}, while it is proved unnecessary in further practice in China. Also, blowing off a great deal of natural gas will waste gas resources and create air pollution.

   In selecting gas flow rates for well testing, a q_{AOF} value should be estimated beforehand. The estimation is based on Equation (3.58) and uses the following formula:

$$q_{AOF} = \frac{2.714 \times 10^{-5} T_{SC} khp_R^2}{p_{SC}\bar{\mu}_g \overline{Z} T \left(\ln \dfrac{r_e}{r_w} + S_a \right)} = \frac{0.07852 khp_R^2}{\bar{\mu}_g \overline{Z} T \left(\ln \dfrac{r_e}{r_w} + S_a \right)}, \qquad (3.88)$$

where q_{AOF} is AOFP, 10^4 m³/day; k is effective formation permeability, mD; h is effective formation thickness, m; p_R is measured formation pressure, MPa; $\bar{\mu}_g$ is average gas viscosity in gas zone, mPa·s; r_e is radius of gas supply boundary, m; r_w

is equivalent well radius, m; p_{sc} is standard pressure condition; T_{sc} is standard temperature condition; $\overline{Z}$ is average Z-factor; $\overline{T}$ is average formation temperature; and S_a is pseudo skin factor of gas well.

Among the parameters in the formula, k and h can be obtained from logging data and by means of conversion, or directly refer to the test interpretation results of adjacent wells. p_R is the measured formation pressure; $\overline{\mu}_g$ and $\overline{Z}$ can be obtained from calculation based on gas component analysis or looking up in relative graphs or tables; S_a can be estimated with completion data or by referring to that of adjacent wells; and r_e can be estimated by referring to well pattern situations. A rough estimated AOFP (q_{AOF}) can finally be attained accordingly.

Of course, the best and the most common approach is to conduct a flowing test with a common choke (e.g., 8- or 10-mm choke) after well completion and cleaning up the completion fluids in a new well, measure the flowing pressure, record the gas flow rate, roughly estimate a q_{AOF} value with this data, and then, on the basis of the results, arrange the following deliverability test.

3. Selection of flow rate sequence. The deliverability test point sequence should meet the requirement of the deliverability test method, with special consideration of its application conditions; some suggestions are given here.

   (a) Usually adopt an increasing sequence, namely arrange a test point of a relatively low gas flow rate first and then higher and higher gas flow rates follow in turn.

   (b) An increasing sequence must be adopted for the modified isochronal testing method, as a reverse sequence will bring big errors in data analysis.

   (c) If an accumulated load liquid exists at the bottom hole, open the well with a big choke prior to the deliverability test, drain away the accumulated load liquid in the hole, and then begin the deliverability test according to the original plan.

   (d) If a hydrate forms in the hole, adopt a relatively high gas flow rate at first to keep the wellbore temperature at a high level and reduce the probability of a hydrate forming.

   (e) As for the back-pressure test, if both the gas flow rate and the flowing pressure can be stabilized very soon after well opening, the test results with either increasing sequence or decreasing sequence or even alternative increasing and decreasing sequence shows little difference. However, because stabilization conditions of the test onsite are not easy to grasp, the test may fail or a normal deliverability equation cannot be established in the regression analysis so it would be better to take a flow rate increasing sequence.

   (f) Extended testing in the isochronal test or modified isochronal test is usually arranged to be conducted after all transient deliverability test points, which is very necessary, particularly when the tested well is located in a block with a boundary, as the final long-term extended testing can better reflect the reservoir situation.

8.1.2 Stabilization of Gas Flow Rate

As for gas well tests in China, particularly for those performed in key wells in key areas, strict control measures are often taken for the gas flow rate during the tests. That is, install a needle valve behind the outlet valve to adjust the passing gas flow rate and install a flow meter at the wellhead to implement a timely regulation of the flow rate, and hence fluctuation of the gas flow rate will be less than 1%.

However, if the tested well is a production well, a metering device is usually installed on the gas gathering station, which brings great difficulties in flow rate control and regulation. Aiming at the unstable flow rate situation, Winestock and Colpitts derived a method. They took flow rate variation into consideration [27]. As long as the gas flow rate does not change too fast, the analysis can be done with corresponding instant values of the flow rate and pressure instead of average values during the whole duration of the test point. However, this demands not only dense recording of the bottom-hole flowing pressure but also exact and dense recording of the gas flow rate variation, as well as obtaining the relationship between them to resolve the problem of varying flow rate.

8.1.3 Selection of Duration for Each Test Point

1. Duration of test points in back-pressure test

    The back-pressure test is usually called the "steady test," meaning that each flow at a different rate reaches a stable state for every test point. "Steady" here means two things in general: stable gas flow rate and stable flowing pressure. The former is easy to realize. Many formations with high or medium permeability and without faults or other no-flow boundaries nearby can realize a stabilized gas flow rate within several hours. If the flow rate of a gas well cannot stabilize under a fixed choke, it can still be controlled by a needle valve to be stabilized. However, stable flowing pressure may not be achieved simultaneously. Theoretically, "stable flowing pressure" can never be obtained unless a supply boundary exists. Therefore, "stable flowing pressure" only means relatively stable within a certain range.

    Because flowing pressure declines absolutely, how much should be the upper limit of its fluctuation value? There is a provision in the Chinese petroleum industrial standard, *Technical Specifications for Natural Gas Well Test*, and it is not necessary to be elaborated here.

2. Duration of transient deliverability test points

    During isochronal and modified isochronal tests, transient deliverability tests are performed first. The duration of transient deliverability test points must be at least longer than the duration of wellbore storage flow, the influence area of the producing has exceeded the skin zone, and the flow has entered the radial flow stage, as shown in Figures 3.67 and 3.68.

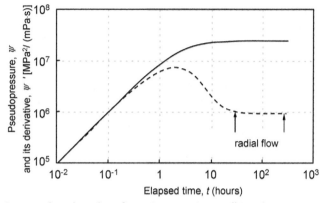

Figure 3.67 Pressure log—log plot of transient test in a well in a homogeneous formation.

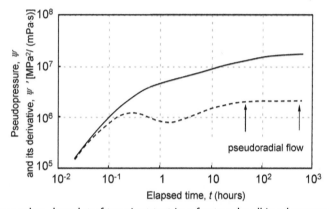

Figure 3.68 Pressure log—log plot of transient test in a fractured well in a homogeneous formation.

The positions indicated by arrows in Figures 3.67 and 3.68 are the minimum test times of each transient deliverability test point. When implementing a deliverability well test design before executing, designing a transient pressure drawdown curve should be obtained by simulation with the predicted parameters, such as k, h, μ, S, x_f, and S_f, and then the minimum transient deliverability testing duration is determined.

3. Duration of extended testing

In isochronal or modified isochronal tests, a stable deliverability point is needed and is obtained from extended testing. Stabilization in practice means that the pressure does not change obviously during a certain period of time. As for high permeability formations, it is not difficult to attain; while for tight formations, the flowing pressure cannot stabilize in a very long duration, even in such a long duration as several months or years. Particularly for formations with multiple no-flow boundaries near the well or constant-volume block gas reservoirs with a very small area, when the flow enters into the pseudo-steady state, the flowing pressure will decline sharply, and flowing pressure stabilization is impossible. Therefore, the very initial point of entering into a pseudo-steady state becomes the "mark point" of determining an extended test duration.

The radius of influence r_i is expressed as Equation (2.8) according to the definition of it in Chapter 2. For a constant-volume gas reservoir with a radius $r = r_e$, $r_i = r_e$ indicates that the flow enters into a pseudo-steady state, and the time at that point is

$$t_s = \frac{69.4\bar{\mu}_g \phi r_e^2 C_t}{k}. \tag{3.89}$$

For gas zones, the main factor influencing the total compressibility $C_t = S_g C_g + S_w C_w + C_f$ is C_g, and $C_g \approx 1/p$, so Equation (3.89) can be rewritten as

$$t_s \approx \frac{70\bar{\mu}_g \phi r_e^2}{k p_R}, \tag{3.90}$$

where t_s is time when the flow of a well in a constant-volume gas reservoir reaches the pseudo-steady state, hr; $\bar{\mu}_g$ is average gas viscosity under formation pressure, mPa·s; ϕ is porosity of the gas zone, decimal; k is effective formation permeability, mD; r_e is supply boundary radius of the gas well, m; and p_R is measured formation pressure, MPa.

Continuously distributed gas reservoirs have no natural closed boundary. As a gas field, however, a well pattern always exist, and a controlled area of single wells is defined accordingly. When the gas wells are producing simultaneously, interwell flow division lines delimit no-flow boundaries of each single well during its production, and these boundaries are equal to the closed boundaries of a constant-volume gas reservoir. Therefore, r_e is determined by the designation of the well pattern, and the corresponding t_s is the duration of designing an extended test.

However, a gas well test in a reservoir with low permeability during the early exploration stage may show no effect of a constant-volume gas reservoir but show a boundary effect; its bottom-hole flowing pressure cannot stabilize after a long time of production. It is really difficult to determine t_s. In this case the following testing and analysis methods can be adopted.

1. For gas zones without a boundary effect, a controlled area of one single well can select 1 and 2 km^2 and so on to calculate the duration of extended test t_s and corresponding AOFP (q_{AOF}).
2. For gas zones with a boundary effect, choose the shortest boundary distance L_{b1} as supply boundary r_e to calculate t_s and corresponding AOFP (q_{AOF}).

The obtained deliverability indices with an additional explanation of the test method attached can be submitted to the concerned management department for reference.

8.2 Why Calculated AOFP Sometimes is Lower than Measured Wellhead Flow Rate

AOFP (q_{AOF}) means the upper limit flow rate of a well. Theoretically, it should be higher than the actual flow rate at any condition, whereas, as described in Section 3.1, there are several different AOFP indices, such as initial transient deliverability indices and deliverability indices obtained from the extended deliverability test. In the case of initial deliverability indices, flowing pressure remains in the transient state during continuously declining, and formation pressure is its initial value, hence the calculated q_{AOF} is very high. In the case of deliverability indices obtained from an extended deliverability test, the flowing pressure has stabilized after a declining process. Furthermore, for formations with a closed boundary, formation pressure will decline along with natural gas producing, flowing pressure declines dramatically, and initial formation pressure is usually applied as its formation pressure in calculating AOFP; the calculated deliverability index this way is certainly far less than the initial transient deliverability indices or even less than the initial measured wellhead flow rate.

The field example given here is a good illustration of this phenomenon. Table 3.9 shows actual data acquired from well sites. Figures 3.69 and 3.70 are the deliverability plots drawn from aforementioned measured data.

Figure 3.69 is the deliverability analysis plot with only transient test points; based on its AOFP (q_{AOF}) it is calculated to be 30.957×10^4 m^3/day. Figure 3.70 is the deliverability analysis plot with transient and extended test points, and the AOFP (q_{AOF}) based on it is only 13.3857×10^4 m^3/day, lower than actual flow rates of the third and fourth flow during the test.

Figure 3.70 obviously shows that the deliverability line moves upward due to a dramatic decline of stable-point flowing pressure, which makes the vertical line

Table 3.9 Measured Data of Modified Isochronal Test

Testing phase	Pressure, MPa	Pseudo-pressure, MPa^2/(mPa·s)	Gas flow rate, 10^4 m^3/day
Initial pressure	29.340	54670	
Initial flowing	27.560	50042	5.0232
Initial shut in	28.570	52668	
Second flowing	25.100	43649	10.0057
Second shut in	27.330	49444	
Third flowing	22.100	35914	15.0297
Third shut in	25.620	44999	
Fourth flowing	18.220	26215	20.0182
Fourth shut in	—	—	
Extended flowing	9.000	7187	10.0250
Final shut in	19.150	28491	

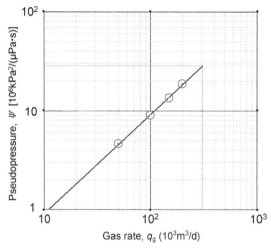

Figure 3.69 Deliverability analysis with transient test points.

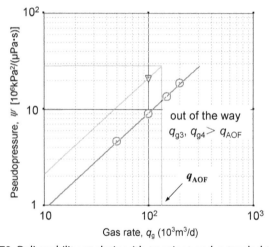

Figure 3.70 Deliverability analysis with transient and extended test points.

indicating AOFP moving leftward, and so causes a calculated AOFP lower than the actual flow rate of test points.

The example indicates that a great difference would be brought about if analysis is done with test data of different stages in different times and under different conditions. In fact, 13.3857×10^4 m^3/day is the stabilized AOFP, while 30.957×10^4 m^3/day is the transient AOFP.

8.3 Existing Problems in Calculating AOFP by Back-Pressure Test Method

The back-pressure test is the most common deliverability test method onsite for gas wells. It is generally thought that the method is applicable to gas zones with high permeability. Theoretical simulations and field practices show that the method is also applicable to gas zones with medium and even low permeability. For instance, an oil company once performed successful back-pressure tests and obtained deliverability values of the gas fields with low permeability in the central and western regions of China. The back-pressure test method does, however, show some formidable defects for some special lithologic reservoirs. The author performed special research on this problem. The research used various well test interpretation software and aimed at some typical formations, made pressure history simulation of deliverability test process at first, then made resulted deliverability analysis and comparison combined with simulation results, and finally verified the obtained understandings in field applications.

8.3.1 Back-Pressure Test for Homogeneous Formations

Reservoir simulation parameters: $k = 3$ mD, $h = 5$ meters, $S = 0$, $D = 0.1 (10^4 \, m^3/day)^{-1}$, $C = 3 \, m^3/MPa$, and $p_R = 30$ MPa.

1. Simulation of back-pressure test

    The classical back-pressure test method is applied in simulation with a flow rate of 2, 4, 6, and $8 \times 10^4 \, m^3/day$ in an increasing sequence, and flow duration for every rate is the same in a simulation but different for a different simulation. Flow durations for different simulation are 24, 72, 240, and 720 hours, respectively. Flowing pressures under each rate are measured. One of the pressure histories with a uniform flow duration of 24 hours is shown in Figure 3.71.

    Based on simulated back-pressure test flowing pressure and flow rate data, AOFPs obtained from the analysis with well test interpretation software are listed in Table 3.10.

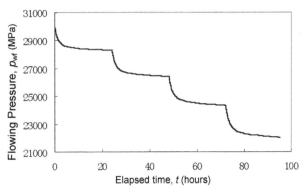

Figure 3.71 Pressure history in a back-pressure test with a uniform flow duration of 24 hours.

Table 3.10 Comparison of Calculated AOFP With Different Test Methods for a Homogeneous Formation

Test method	Flow duration of backpressure test or extended test, hours	AOFP, 10^4 m^3/day		Note
		LIT, pseudo pressure	Exponential, pseudo pressure	
Back-pressure test	24	17.5238	19.8274	$k = 3$ md
	72	16.7397	18.7042	$h = 5$ m
	240	15.9207	17.5922	$s = 0$
	720	15.2693	16.7132	$D = 0.1(10^4 \text{ m}^3/\text{day})^{-1}$
				$C = 3$ m^3/MPa
Modified isochronal test	300 (extended test)	17.6472	18.9752	$k = 3$ md
	500 (extended test)	17.4237	18.6690	$h = 5$ m
				$S = 0$
				$D = 0.1(10^4 \text{ m}^3/\text{day})^{-1}$
				$C = 3$ m^3/MPa
Back-pressure test with each flow starting at stable formation pressure	240	17.1224	18.7679	Each flow test starts at stable formation pressure

2. Simulation of modified isochronal test

   With the same formation parameters, the well was flowing and shut in with 24 hours for each transient test point and 300 and 500 hours of extended tests during the two simulated tests, respectively; the obtained AOFPs are also listed in Table 3.10.

   The simulation results of both back-pressure tests with each flow starting at stable formation pressure and different extended test durations of 300 and 500 hours, respectively, are listed in Table 3.10. In these simulations, the well is always opened for flowing when the pressure is stabilized at formation pressure every time but with a different flow rate and stable flowing pressure in each flowing, and the AOFP is calculated by regression of these data points, that is, flow rates and flowing pressures.

3. Comparative analysis of simulation results

   Table 3.10 shows:

   (1) For medium- or low-permeability homogeneous formations, AOFP can be obtained by the back-pressure test or modified isochronal test method, and the difference of the results is small.

   (2) Classical back-pressure test with 24 or 72 hours of flowing duration can provide a q_{AOF} value that is very close to the q_{AOF} resulting from the back-pressure test with each flow starting at a stable formation pressure. The longer the flowing duration, the lower the calculated q_{AOF} value, which is because formation pressure at the supply boundary declines gradually along with proceeding of the

test—based on which choke is enlarged continuously to form abnormal low flowing pressure and cause a low calculated q_{AOF} value.

(3) After 300 hours of extended testing in modified isochronal testing, the pressure has basically entered a pseudo-flow state; hereafter further lengthening extended testing does not influence the result much, and the obtained q_{AOF} value has been very close to the q_{AOF} resulting from the back-pressure test with each flow starting at a stable formation pressure, and the error is less than 3%. Therefore, the modified isochronal test is considered to be the first choice method for the deliverability test.

8.3.2 Back-Pressure Test for Fractured Wells in Channel Homogeneous Formation

In the simulation study for fractured wells in channel homogeneous formation, the applied basic formation parameters (excluding fracturing parameters and boundary conditions) are absolutely the same as described earlier: $k = 3$ mD, $h = 5$ meters, $S_f = 0$, $x_f = 60$ meters, $C = 3$ m^3/MPa, $p_{Ri} = 30$ MPa, $D = 0.01(10^4$ m^3/day$)^{-1}$, and $L_{b1} = L_{b2} = 70$ meters, and the applied flow rate sequence q_{gi} is 5, 10, 15, and 20 $\times$ 10^4 m^3/day, successively.

With these parameters and conditions, deliverability test simulations are carried out for the back-pressure test and modified isochronal test.

1. Simulation and analysis of backpressure test and modified isochronal test

Pressure histories are obtained with different test intervals of 24, 72, 240, and 720 hours. For example, the pressure history when the test interval is 24 hours is shown in Figure 3.72.

Figure 3.72 clearly shows that the flowing pressure declines rapidly due to the boundary effect.

AOFP values calculated by simulated flowing pressures can be seen in Table 3.11. Simulation of the modified isochronal well test is done with the same formation parameters and AOFP values are calculated; all the results are listed in Table 3.11.

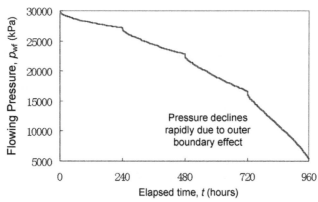

Figure 3.72 Simulated pressure history of a back-pressure test in a fractured well in a channel homogeneous formation.

Table 3.11 Deliverability Comparison of a Fractured Well in Channel Homogeneous Formation

Test method	Flow duration of back-pressure test or extended test, hours	AOFP, 10^4 m^3/day	
		LIT, pseudo pressure	Exponential, pseudo pressure
Back-pressure test	24	45.8716	54.7337
	72	31.9648	35.0461
	240	20.6596	21.1623
	720	13.8233	14.2300
Modified isochronal test	300 (extended test)	37.5658	36.9925
	500 (extended test)	31.1040	30.4647

2. Comparison and analysis of deliverability calculation results

Table 3.11 shows:

(1) Although an improved method taking of the decline of formation pressure at the supply boundary into account is applied in calculating AOFP with the modified isochronal test method, the calculated results still present a decline trend along with the prolonging of extended test duration, from 37 to $\approx 30 \times 10^4$ m^3/day, indicating that a great boundary effect is influencing the deliverability of the gas well. The longer the producing period is, the more remarkable the influence of the boundary effect is.

(2) Calculated AOFP values obtained from the back-pressure test method become lower along with a prolonging of flow duration of test points.

   • When the flow duration is selected to be 72 hours, the resulting AOFP value is equivalent to that from the modified isochronal test.
   • When the flow duration is selected to be 240 hours (10 days), the resulting AOFP value is about 20×10^4 m^3/day, two-thirds of that from the modified isochronal test.
   • When the flow duration is selected to be 720 hours (30 days), the resulting AOFP value is only about 14×10^4 m^3/day, one-half of that from the modified isochronal test.

It is mainly due to the boundary effect upon the formation pressure decline.

It should be pointed out particularly that the stable flowing pressure value is very difficult to measure in the back-pressure test because it is always continuously declining more and more quickly (refer to Figure 3.72), which makes the analysis work disoriented. It is hoped that prolonging the flow duration makes the flowing pressure more stable, but the result is just the opposite of this. This fact has been proved by field practice of a long flowing duration up to several months in a test.

Therefore, the back-pressure test method obviously is not applicable to wells in channel formations. These study results are similar to conclusions for those formations with no-flow boundaries of different configurations, especially for constant-volume gas reservoirs with a closed boundary. The field example introduced previously has proved it clearly (refer to Figure 3.25).

8.4 Method and Analysis of Single-Point Deliverability Test and its Error

8.4.1 Single Point Deliverability Test

At present, the single point deliverability test is widely used for estimating the AOFP of tested gas wells during the early stage of gas field exploration. Because this method is simple and can give important deliverability information in the very early stage of gas field exploration, it is popularly applied in many fields. It should be pointed, however, out that this method is not perfect or reliable. The reasons are given here.

1. The deliverability obtained from the single point method is often an instant and transient one. It is sometimes acceptable in cases of homogeneous formations. But for gas wells in formation with no-flow boundaries, especially those gas wells connecting with big natural fractures or having been treated by hydraulic fracturing, the instant deliverability values often offer overoptimistic information.

2. The single point method is not an independent deliverability analysis method, but a simplification of various deliverability test methods mentioned previously. If in a large gas field with plenty of reliable deliverability test data available, with which index n and coefficient C generally being applicable to the whole field are obtained from statistical analysis and so the deliverability equation for the whole field is obtained, then once only one single deliverability test point is measured, the deliverability can be calculated by that deliverability equation.

   However, calculating deliverability with a uniform single point test formula will obviously bring a certain risk for new types of formation and new wellbore conditions in new exploration areas.

8.4.2 Two Examples of AOFP Calculation Formulae for Single Point Test in Development Areas of Gas Field

Several dozen deliverability tests were conducted in the TN and SB gas fields in Qinghai Province, respectively. Statistical analyses of obtained test data were performed for exponential deliverability equations, and the results are shown in Figures 3.73 and 3.74, respectively.

The regression analyses of test points give (refer to Figures 3.73 and 3.74)

$$q_{AOF} = 1.004 q_g \left(\frac{p_R^2 - p_{wf}^2}{p_R^2} \right)^{-0.7426}$$

for the SB gas field and

$$q_{AOF} = 1.0007 q_g \left(\frac{p_R^2 - p_{wf}^2}{p_R^2} \right)^{-0.6418}$$

for the TN gas field.

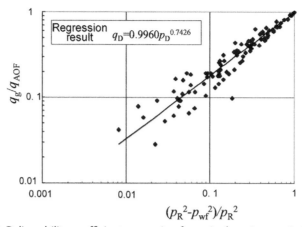

Figure 3.73 Deliverability coefficient regression for a single point test in SB gas field.

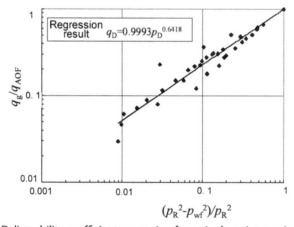

Figure 3.74 Deliverability coefficient regression for a single point test in TN gas field.

It can be seen from these formulae that once gas flow rate q_g and the corresponding flowing pressure p_{wf} of *one* test point only are obtained, they can be put into the formulas and AOFP (q_{AOF}) is calculated.

The precision of q_{AOF} values of the area calculated with the formula is rather assured because the index n of the formula is regressed from measured data in this area.

8.4.3 Some Examples of AOFP Calculation Formulae for Single Point Test Method for Exploration Wells

It is impossible that the parameters of all new wells in a new exploration area or a gas-bearing area are uniformly consistent, especially when the formation conditions vary significantly, but it is very desirable that the deliverability of newly discovered wells can be estimated roughly under the consideration of taking only the main factors into account. Therefore, some LIT and exponential deliverability calculating methods from "single point test" data, which are thought commonly applicable, are introduced in some literatures [28], and the calculation formula of the LIT single point method is

$$q_{AOF} = \frac{6q_g}{\sqrt{1 + 48 \, p_{DG}} - 1} \tag{3.91}$$

and the calculation formula of the exponential single point method is

$$q_{AOF} = q_g(p_{DG})^{-0.6594} \tag{3.92}$$

where

$$p_{DG} = \frac{p_R^2 - p_{wf}^2}{p_R^2}.$$

In Equation (3.91), coefficients A and B must meet the following assumption:

$$\alpha = \frac{A}{A + Bq_{AOF}} = 0.25.$$

The value of α is obtained by the author from statistics of deliverability test data acquired from 16 wells in the Sichuang area. In exponential Equation (3.92), n is definitely 0.6594, which also comes from statistics of deliverability test data done by the author. Certainly, these assumptions will not be fully met for common formations.

In addition, there are some more AOFP calculation formulae for the single point test method, for example, the one used in the JB gas field by the Downhole Service Sub-company of the PetroChina Changqing Oilfield Company is

$$q_{AOF} = \frac{q_g}{0.007564 + 1.2565\sqrt{0.9816 - \dfrac{p_{wf}}{p_R}}} \tag{3.93}$$

Another one used for the Upper Paleozoic Formation by the Research Institute of Petroleum Exploration & Development of PetroChina Changqing Oilfield Company is

$$q_{AOF} = \frac{q_g}{1.1613\sqrt{1.0225 - \dfrac{p^2}{p_R^2} - 0.1743}} \tag{3.94}$$

8.4.4 Errors Analysis of Single Point Deliverability Test Method

In order to verify the reliability of AOFP calculated by various single point deliverability tests, dynamic simulations are implemented by computer software with a set of representative parameters separately selected aiming at various formation conditions, and different pressure histories of different test methods in the same formation are obtained. Compare the differences between calculated deliverability values obtained from those various test methods.

Homogeneous Formation

Selected parameters include:

- Formation permeability: $k = 3$ mD
- Formation thickness: $h = 5$ meters
- Skin factor: $S = 0$
- Turbulent flow coefficient: $D = 1 \times 10^{-4}$ $(\text{m}^3/\text{day})^{-1}$
- Wellbore storage coefficient: $C = 3$ m^3/MPa
- Flow rate: $q_g = 6 \times 10^4$ m^3/day
- Initial formation pressure: $p_{\text{Ri}} = 30$ MPa

With these parameters, simulation of deliverability test processes, including the single point testing process and modified isochronal testing process, are completed with well test interpretation software.

1. Single-point deliverability test simulation

    The duration of simulation is from 0 to 500 hours for verifying the variation of AOFP values calculated by the single point deliverability test vs time, and the variation of the flowing pressure vs time is plotted in Figure 3.75.

    Flowing pressure variation vs time can be read from the pressure history shown in Figure 3.75 and listed in Table 3.12. Flowing pressure varies vs time, so calculated AOFP values change vs the time accordingly.

2. Calculating AOFP with single point deliverability test

    Four formulae of single point deliverability test methods discussed previously are applied. Because these formulae were brought forth in different periods and are based

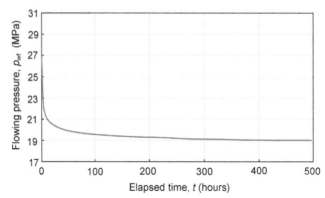

Figure 3.75 Simulated pressure history of a single-choke flow test in a well in a homogeneous formation.

Table 3.12 Flowing Pressure Variations in Single-Choke Flow Test (in a Well in Homogeneous Formation)

Time, t hours	Flowing pressure, p_{wf} MPa
10.1	21.0789
25.3	20.5143
50	20.2007
70	20.0540
100	19.9010
150	19.7300
200	19.6100
240	19.5340

on various background data, they should be rather representative. With the simulation parameters given in Table 3.12, corresponding AOFP values are calculated separately, and the results are listed in Table 3.13.

With results in Table 3.13, the relationship between calculated AOFP values obtained from different single-point deliverability formulae and test durations is plotted and shown in Figure 3.76.

3. Modified isochronal test simulation

Pressure history simulation for the modified isochronal test is fulfilled with the same reservoir parameters, and the simulated test procedure is:

a. Implement the four-point transient deliverability test, producing the well for 24 hours followed by shutting in the well for 24 hours for each transient test point, and the flow rate sequence is 2, 4, 6, and 8×10^4 m^3/day successively.

b. Perform the extended test with the flow rate of 6×10^4 m^3/day for 500 hours.

c. Shut in the well for 500 hours to measure pressure buildup for learning the reservoir pressure at the end of the test.

Table 3.13 AOFP Values Calculated by Four Different Formulae of Single Point Deliverability Tests (in a Well in Homogeneous Formation)

Test time interval (hours)	Calculated AOFP, 10^4 m^3/day			
	LIT formula 3.91	Exponential formula 3.92	Jingbian formula 3.93	Changqing formula for Upper Paleozoic formula 3.94
10.1	8.9326	9.0076	8.9390	8.9527
25.3	8.6680	8.7140	8.6550	8.6859
50	8.5331	8.5642	8.5086	8.5499
70	8.4727	8.4971	8.4426	8.4890
100	8.4114	8.4290	8.3754	8.4272
150	8.3449	8.3550	8.3022	8.3601
200	8.2994	8.3045	8.2519	8.3143
240	8.2711	8.2731	8.2205	8.2858

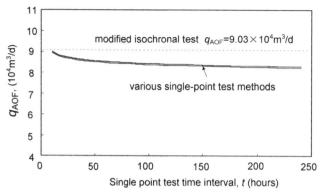

Figure 3.76 Comparison of calculated AOFP values for homogeneous formation obtained from method 1.

Figure 3.20 shows the pressure history obtained from the simulations.

The stable AOFP value calculated from the simulated pressure history curve is

$$q_{AOF(extended\ 500h)} = 9.03 \times 10^4 m^3/d$$

4. Comparison of AOFP values obtained from different methods

Deliverability values calculated by the formulae of four different single point deliverability tests and modified isochronal test are plotted in Figure 3.76 for comparative analysis.

Figure 3.76 shows:

  a. Calculated AOPF values obtained by four single point methods are approximate. Maybe the authors providing calculation formulae considered comparative relationship among them in determining formula coefficients.

  b. If the flow test lasts over 2 days (48 hours), the obtained AOFP values are basically a constant within a range of 8.3 to 8.6 $\times$ 10^4 m^3/day. This means that the time duration for data acquisition has only a small impact on AOFP.

  c. The difference of q_{AOF} values calculated by the single point test and q_{AOF} value by the modified isochronal test is also small and the difference range is within 10%, but their representative lines obviously branch off from each other, which is due to the non-Darcy flow coefficient value D of 1 $(10^4$ $m^3/day)^{-1}$ in the model, so the exponential deliverability equation obtained from the modified isochronal test is

$$q_g = 0.040(p_R - p_{wf})^{0.802}$$

In other words, index n caused by the turbulent flow effect should be 0.802, but in the formula of the single point test, index n is always equal to 0.6594 for all formations, no matter how different they are, which inevitably brings some errors in calculation results.

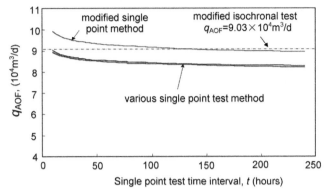

Figure 3.77 Comparison of calculated AOFP values for homogeneous formation obtained from method 2.

By changing the n value in the exponential equation of the single point test method to 0.802, the calculated results are certainly different, as shown in Figure 3.77. The results tend to be consistent with those obtained from the modified isochronal test.

Fractured Wells in Channel Homogeneous Formation

Simulating pressure variation during the flowing test period with well test interpretation software to calculate q_{AOF} and the parameters chosen in the simulation include:

Formation permeability: $k = 3$ mD, formation thickness: $h = 5$ meters, half-length of the infinite-conductivity vertical fracture: $x_f = 60$ meters, channel boundary distances: $L_{b1} = L_{b2} = 70$ meters, skin factor of fracture: $S_f = 0.1$, turbulent coefficient: $D = 0.03(10^4 \text{ m}^3/\text{day})^{-1}$, wellbore storage coefficient: $C = 3 \text{ m}^3/\text{MPa}$, flow rate: $q_g = 6 \times 10^4 \text{ m}^3/\text{day}$, and the original formation pressure: $p_R = 30$ MPa.

With these parameters, the single point deliverability test and modified isochronal deliverability test processes are stimulated with well test interpretation software.

1. Pressure history simulation of single point deliverability test

The simulation duration is from 0 to 300 hours, but AOFP calculations of the single point test for verifying the variation of obtained AOFP are only done from 24 to 300 hours, and the obtained relationship between the flowing pressure and the flow time can be seen in Figure 3.78.

The variation of flowing pressure vs time can be read from the pressure history shown in Figure 3.78 and some flowing pressures are listed in Table 3.14.

Because flowing pressure changes rapidly vs time, calculated AOFP values also vary accordingly. AOFP values calculated by the same formulae of a single point test are listed in Table 3.15.

The relationship between AOFP values calculated with different methods and test times (Table 3.15) is shown in Figure 3.79.

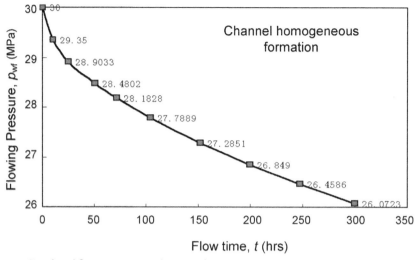

Figure 3.78 Simulated flowing pressure history of a gas well in a single choke flow test (in a fractured well in a channel homogeneous formation).

Table 3.14 Simulated Flowing Pressure History in Single-Choke Flow Test (Fractured Well in Channel Homogeneous Formation)

Test time, t (hours)	25.55	50.97	72.00	104.00	152.00	200.00	248.00	300.00
Flowing pressure, p_{wf} (MPa)	28.9033	28.4802	28.1828	27.7889	27.2851	26.8490	26.4586	26.0723

Table 3.15 AOFP Values Calculated by Different Formulae (Fractured Well in Channel Homogeneous Formation)

Test time (hours)	Calculated AOFP, 10^4 m^3/day			
	LIT formula 3.91	Exponential formula 3.92	Jingbian formula 3.93	Changqing formula for Upper Paleozoic formula 3.94
25.55	31.1156	31.4294	32.0807	31.5216
50.97	25.0275	25.7351	24.8946	25.2987
72.00	22.2774	23.0554	22.0221	22.4934
104.00	19.6557	20.4332	19.4119	19.8229
152.00	17.2963	18.0146	17.1293	17.4232
200.00	15.7958	16.4471	15.6941	15.8991
248.00	14.7304	15.3203	14.6767	14.8181
300.00	13.8620	14.3935	13.8453	13.9375

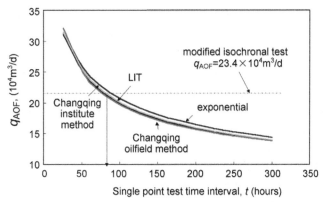

Figure 3.79 Comparison of AOFP values calculated by different methods (a fractured well in a channel homogeneous formation).

2. Modified isochronal test simulation

The pressure history simulation of the modified isochronal test process is completed with the same reservoir parameters, and the simulation procedures are as follow:

a. Implement the four-point transient deliverability test, producing the well for 24 hours, followed by shutting in the well for 24 hours for each transient test point; the flow rate sequence is 2, 4, 6, and 8×10^4 m^3/day successively.

b. Perform the extended test with the flow rate of 6×10^4 m^3/day for 500 hours.

c. Shut in the well for 500 hours to measure pressure buildup for learning the reservoir pressure at the end of the test.

Figure 3.80 shows the pressure history obtained from the simulations.

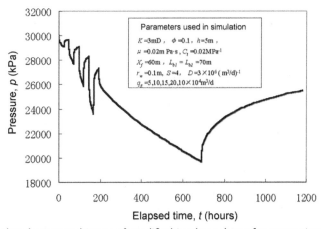

Figure 3.80 Simulated pressure history of modified isochronal test for comparison with that of the single point deliverability test.

The stable extended AOFP value of the fractured well calculated from the simulated pressure history is

$$q_{\text{AOF}(\textit{extended }500\text{h})} = 23.38 \times 10^4 \text{m}^3/\text{d}$$

3. Comparison of AOFP values obtained from different methods

The deliverability values calculated separately by different single point test methods and modified isochronal test method are plotted in Figure 3.79 for comparative analysis.

It is seen from Figure 3.79 that:

a. For a fractured well in a channel homogeneous formation, AOFP values calculated from the single point test method are decreased gradually along with prolonging of the test time due to a continuous decline of flowing pressure during producing of the well. The AOFP value is not a constant. Therefore the selection of the test time is crucial to AOFP calculation.

b. AOFP values calculated by different formulae of test methods are close to each other (refer to Figure 3.79) or the calculation results of those different tests are basically consistent.

c. Calculated AOFP of the modified isochronal test method is $23.4 \times 10^4 \text{ m}^3/\text{day}$; if this value is regarded as the judgment standard, values calculated by various formulae of the single point test at a specific moment are generally consistent to that standard.

Therefore, in specific formation conditions, if the single point test method is chosen in calculating deliverability, flowing pressure at a specific moment after the flow rate is stabilized should be selected to calculate AOFP so that the calculated result can generally be consistent with the modified isochronal test result. However, such a "specific moment" will not be the same for different formations, which is why calculating AOFP indiscriminately with the same formula of the single point test results in a certain difference, while this difference reflects different deliverability characteristics.

8.5 Deliverability Test Without any Stable Flow Points

All deliverability test methods introduced in this chapter require stable flow points:
1. Every deliverability point is required to be stabilized in back-pressure tests.
2. An extended deliverability test point is required for the modified isochronal test; during the extended test, the flow rate is constant and flowing pressure becomes stable, and this extended point is also a stable flow point.

Long-term producing is necessary for acquisition of a stable flow point, especially for tight gas zones. To reach stable flow the well may have to produce for several months or even longer so that a large amount of gas is wasted. Therefore, the test for a stable point is always undesirable in view of field application. For the back–pressure test, to rid stable points is impossible, but how about isochronal and modified isochronal tests? Is it possible to rid the stable test point of them? The answer to this question should be "yes" theoretically.

The slope of the deliverability curve can be obtained from the deliverability plot on the basis of transient test point analysis. That is, coefficient B of the LIT deliverability equation can be determined.

Also, coefficient A has been expressed explicitly by the aforementioned equations: Equation (3.26) for a gas reservoir with a round supply boundary in the form of pressured square, Equation (3.29) in the form of pseudo pressure, and Equation (3.32) for infinite homogeneous formations in the form of pseudo pressure. The following parameters in the formulae should be gained by other approaches:

k = Effective formation permeability, mD
h = Effective formation thickness, m
r_e = Radius of drainage area, m
r_w = Effective well radius, m
μ_g = Average gas viscosity in gas zones, mPa·s
S = Skin factor
Z = Average Z factor of the gas in the reservoir
T = Average formation temperature, K
p_{sc} = Pressure, standard conditions
T_{sc} = Temperature, standard conditions

Once determined, these parameters can be used to calculate coefficient A and establish the deliverability equation under stable conditions.

However, the transient test and analysis should be done for obtaining the aforementioned parameters accurately. In addition, because other parameters, for example, r_e, are usually difficult to determine, it is also difficult to gain the value of coefficient A accurately, and few people would calculate it with this method.

8.6 Discussion on Wellhead Deliverability

The gas well deliverability discussed previously is the attribute of a gas well concluded on the basis of the measured bottom-hole pressure or the production capability of a gas well only related to the formation conditions. This production capacity is considered a direct and unchanged character of the formation in gas reservoir engineering design.

Discussions about deliverability tested at the wellhead in fields are often heard. The wellhead deliverability test contains two meanings:

1. To measure pressure and gas flow rate at the wellhead simultaneously and then converse wellhead pressure into downhole pressure for analysis; the results are also bottom-hole deliverability, which is just similar with that discussed earlier, but the analysis precision will be much worse due to the limitation of pressure conversion.

2. To measure pressure and flow rate at the wellhead simultaneously, analyze the deliverability of the well with the acquired wellhead pressure and flow rate data directly and obtain the wellhead deliverability index. This is called "wellhead deliverability analysis."

It seems that wellhead deliverability can reflect the production capacity of the well more directly, but, in fact, this index is variable when being used in the development program of a gas well:

a. The coefficient n (or B) of the deliverability equation based on wellhead deliverability test data may be different from that based on test data measured at the downhole [29]. Therefore, this coefficient cannot be used directly in the gas field development program design.

b. Unless being modified, flowing temperature in the wellbore may vary along with a different gas flow rate and thus the deliverability curve is not a straight line and bends to one side consequently, which brings great difficulties in analysis [30].

c. Gas flow friction in the wellbore will change along with the variation of the average wellbore pressure so wellhead deliverability is a variable during the whole production period.

d. Sometimes there is accumulated load water and liquid in the wellbore, and because the accumulated load liquid level is different under different flow rate conditions, a normal deliverability equation may not be established.

Therefore, wellhead deliverability can only be supplied to gas reservoir production managers for reference and is seldom applied in gas reservoir engineering.

8.7 Calculating Coefficients A And B Manually in Deliverability Equation and AOFP

It is mentioned repeatedly that deliverability analysis is seldom done manually due to the popularization and application of well test interpretation software, but sometimes it is necessary for gas reservoir engineers to establish the deliverability equation and calculate AOFP directly with the acquired pressure and flow rate data by hand, combined with some simple calculating tools, for example, a calculator.

An operating sequence with a field example of the modified isochronal test is put forth here, the analysis is in pressure-squared form and with the LIT deliverability equation, and AOFP is calculated finally. It is as follows.

8.7.1 Data Acquisition

Acquired pressure and gas flow rate data in the deliverability test are listed in Table 3.16. Other basic data of the well are as follow:

Formation temperature, $T = 98.17\,°C$

Effective formation thickness, $h = 5.6$ meters

Formation porosity, $\phi = 6.93\%$

Table 3.16 Measured Pressure and Flow Rate Data of Modified Isochronal Test

Item Test point No.	Gas flow rate, q_g 10^4 m³/day	Shut-in pressure, p_{ws} MPa	Flowing pressure, p_{wf} MPa	$\Delta p^2 = p_{ws}^2 - p_{wf}^2$ MPa²	$\Delta p^2/q_g$
Transient test point 1	1.6900	29.4860	28.7814	41.055	24.2929
Transient test point 2	3.1752	29.2344	27.8455	79.278	24.9679
Transient test point 3	4.3204	28.8265	26.8747	108.718	25.1639
Transient test point 4	5.7045	28.3934	25.6999	145.700	25.5412
Extended test	2.8334	29.4860	24.1757	284.960 (with initial pressure)	100.5718 (with initial pressure)
Final shut in	—	27.600	—	177.296 (with measured shut-in pressure)	62.5736 (with measured shut-in pressure)

Relative density of natural gas, $\gamma_g = 0.581$
Gas pseudo-critical pressure, $p_{pc} = 4.6228$ MPa
Gas pseudo-critical temperature, $T_{pc} = 75.7\,°C$
Deviation factor of the gas, $Z = 0.979$
Subsurface viscosity of natural gas, $\mu_g = 0.02226$ mPa·s
Subsurface gas compressibility, $C_g = 0.02617$ MPa^{-1}
Volume factor of the gas, $B_g = 0.00433$ m³/m³
Total compressibility, $C_t = 0.02194$ MPa^{-1}

The first three columns in Table 3.16 list acquired data during the test, and the two columns after those show the calculated results that can be obtained by EXCEL or calculated with a calculator.

8.7.2 Establishment of Transient Deliverability Equation

The transient deliverability equation is obtained by the regression of data of the first column q_g and the fifth column ($\Delta p^2/q_g$) in Table 3.16 as

$$\frac{p_R^2 - p_{wf}^2}{q_g} = 23.8702 + 0.3012 q_g$$

or be written as

$$p_R^2 - p_{wf}^2 = 23.8702 q_g + 0.3012 q_g^2, \qquad (3.95)$$

where $A = 23.8702$, $B = 0.3012$, and related coefficient $\gamma = 0.9817$.

Regression Equation (3.95) can also be established via drawing a trend line of the data points in a scatter diagram by EXCEL.

8.7.3 Establishment of Stabilized Deliverability Equation

Put stable test point values into the left item $\left(\dfrac{p_R^2 - p_{wf}^2}{q_g}\right)$ of the aforementioned transient deliverability equation and regard coefficient A as a variable, then

$$A = \left(\frac{p_R^2 - p_{wf}^2}{q_g}\right)_{stable} - 0.3012 q_g$$

$$= 100.5718 - 0.3012 \times 2.8334$$

$$= 99.7184.$$

The stabilized deliverability equation is

$$p_R^2 - p_{wf}^2 = 99.7184 q_g + 0.3012 q_g^2. \tag{3.96}$$

If the measured shut-in pressure after the deliverability test is taken as the formation pressure, the deliverability equation is

$$p_R^2 - p_{wf}^2 = 61.7202 q_g + 0.3012 q_g. \tag{3.97}$$

8.7.4 Calculating AOFP

Let $p_R = p_{Ri} = 29.486$ MPa and $p_{wf} = 0.101$ MPa, the maximum value of the difference of pressure square ($p_R^2 - p_{wf}^2$) is calculated to be 869.414 MPa^2. Put it into Equation (3.96); the LIT equation about q_{AOF} is obtained as

$$0.3012 q_{AOF}^2 + 99.7184 q_{AOF} - 869.414 = 0.$$

Solve this quadratic equation; AOFP (when formation pressure is the initial pressure) is obtained as

$$q_{AOF} = \frac{-99.7184 + \sqrt{99.7184^2 + 4 \times 0.3012 \times 869.414}}{2 \times 0.3012}$$

$$= 8.500 \times 10^4 \mathrm{m}^3 / \mathrm{day}.$$

If the measured shut-in pressure after deliverability test 27.600 MPa is taken as the formation pressure, the deliverability equation is Equation (3.97) and the calculated AOFP is

$$q_{AOF} = 13.232 \times 10^4 \mathrm{m}^3 / \mathrm{day}.$$

If deliverability test data like those listed in Table 3.16 have been acquired, and there happens to be no personal computer or office software available at hand, or even if the calculator has no regression function, a method of calculating deliverability equation coefficients manually was introduced in the literature [20]. These coefficients are obtained by a regression of measured data with the least-square method and expressed as

$$A = \frac{\sum \frac{\Delta \psi}{q_g} \sum q_g^2 - \sum q_g \sum \Delta \psi}{N \sum q_g^2 - \sum q_g \sum q_g}, \tag{3.98}$$

$$B = \frac{N \sum \Delta \psi - \sum q_g \sum \frac{\Delta \psi}{q_g}}{N \sum q_g^2 - \sum q_g \sum q_g}, \tag{3.99}$$

where N is the test point number. Transient and stabilized deliverability equations can also be established in the same way.

9 SUMMARY

The deliverability test and analysis methods of gas wells were discussed in this chapter. As described in Chapter 1, deliverability analysis of gas wells is a crucial part of the gas reservoir dynamic description and needs detailed illustrations.

Basic analysis methods were introduced first:
- Three deliverability test methods
- Two deliverability equations
- Three pressure expression forms of deliverability equations
- Graphic method of deliverability equation IPR curve

Just as mentioned at the beginning of this chapter, these classical methods were put forward as early as the middle of the last century; for instance, the back-pressure test method has been applied in fields for as long as 80 years. There are some defects in these semiempirical formulae due to limitation of the technical conditions and understanding of the underground percolation process of natural gas at that time. Therefore, a "stable test point LIT equation" is particularly and emphatically recommended in this chapter, in addition to complete and detailed introduction of the existing classical methods. This expression of the equation is deduced theoretically, the establishment method of the equation is introduced, a thorough analysis on the key parameters is implemented, and the selection of them is discussed combined with field conditions. When the method is applied in fields, only one stable initial deliverability test point is required to be acquired, and if only data of the stable initial deliverability test point are required, a reliable initial deliverability equation can be constructed, which broadens significantly the application scope of the deliverability test. The expressions for both vertical wells and horizontal wells are deduced separately in this chapter; the dynamic deliverability equation can further be derived from the initial stable test point deliverability equation so that analyses of the dynamic deliverability indexes of gas wells during the production process can be conducted in time. In addition, some field examples of which the author has participated in actual analyses are listed in this chapter; these practices include application of this method in the KL-2 gas field, the Upper Paleozoic gas field of the Ordose Basin, and horizontal wells of offshore gas fields.

Based on the analysis of deliverability equations of gas wells, deliverability prediction methods in the design of gas field development program are also introduced. First, how to derive and establish the deliverability equation applicable to the whole field from the deliverability equation of one gas well is explained. Second, how to use this deliverability equation to determine the gas flow rate of each well in designing the development program and the initial gas flow rate in numerical simulations. In fact, if the stable test point LIT equation method is widely applied in a field, and the initial deliverability equation of each production well is established and is used in the design of deliverability indices in the gas field development program, the prediction of gas field production can be carried out on a totally different basis; not only the design for each well accords with the practical conditions but also the obtained deliverability equation can be applied in future dynamic deliverability analysis.

Some problems possibly encountered in deliverability tests and data analyses are also introduced in detail, combined with geological characteristics of China in this chapter. Solutions to them were discussed; these problems include:

1. How to design a deliverability test combined with Chinese geological characteristics.
2. Why calculated AOFP values are sometimes lower than the measured wellhead flow rate.
3. Why abnormal phenomena such as $n > 1$ and $B < 0$ sometimes occur in the establishment of deliverability equations with measured data.
4. In which formations is the back-pressure test method not applicable for the deliverability test.
5. The error analysis of single point deliverability test method.
6. How to analyze deliverability test data manually.

Deliverability analysis is the crucial kernel of gas reservoir dynamic description. While being able to determine the deliverability of gas wells or gas reservoirs with the methods introduced in this chapter and acquired deliverability data, engineers should also be able to (a) understand further the geological background factors influencing and determining the deliverability of the well and the reservoir through establishment of the dynamic model of the gas well and the reservoir and (b) predict the deliverability indices of the well in the future through establishment of the dynamic model for the gas well.

CHAPTER 4

Analyzing Gas Reservoir Characteristics with Pressure Gradient Method

Contents

The structural characteristics of gas reservoirs can be effectively made clear with static pressure gradient analysis. Static pressure gradient analysis methods can be generally classified into two types:

1. Static pressure gradient analysis in gas field exploration phase
2. Formation pressure gradient analysis during gas field development processes

    The initial formation pressure of a well must be measured accurately when the test of this well just starts, and this initial formation pressure is the pressure in the original state of the formation and should be measured immediately when the gas well is shut in as soon as displaced flow is finished; at that time the pressure is maintained stable, and there has not been any obvious disturbance or shortfall in the gas zone due to gas production. For a well drilled into a high-permeable formation, this is not difficult to do, but for

Dynamic Well Testing in Petroleum Exploration and Development   © 2013 Petroleum Industry Press. Published by Elsevier Inc.
ISBN 978-0-12-397161-6, http://dx.doi.org/10.1016/B978-0-12-397161-6.00004-8

a formation with low permeability, the pressure cannot be recovered to its initial value until the well has been shut in for a very long time after a certain amount of gas is produced. In particular, in cases of fractured wells in a low-permeability formation with lithologic boundaries, pressure buildup would be much slower. Therefore, seizing the right time to measure the initial formation pressure is very important.

Early stage static pressure gradient analysis can reveal which wells are located in the same connected gas-bearing block, as the measured formation pressures of those wells located in the same connected gas-bearing block should distribute in accordance with the static pressure gradient of the gas, even though the depths of their gas zones are different. In other words, they distribute on the same "gas static pressure gradient line."

If, during the exploration phase, the static pressure points of wells within a gas-bearing area do not form a "gas static pressure gradient line," there may be several possibilities; one of them is that these static pressure points are on two or three gas static pressure gradient lines, which means that these wells may possibly be distributed in two or more mutually separated independent gas reservoirs. Another possibility is that the static pressure points are on an approximate "hydrostatic pressure gradient line," which means that the gas fields used to belong to the same hydraulic system during the formation and migration process of the gas reservoir, but as gas reservoirs, the gas therein was not in the same connected gas-bearing block.

However, if the static pressures of all gas wells in the gas reservoir or gas-bearing area are really on the same "gas static pressure gradient line" during the exploration phase, is it true that these wells in the gas reservoir or gas-bearing area can be deemed as mutually connected to each other in the development phase? The answer is "not sure." In some gas fields in the Sichuan Province of China, especially some in the south of this province, their static pressures in the early stage are generally consistent, but are different after producing for a period of time. Sometimes, the measured static pressures (converted to at the same depth) of two or three dozen wells measured at about the same time are different from each other. This means that the areas controlled by the wells are actually independent systems during development, even though they may be not completely isolated from each other.

Regarding the characteristics that gas reservoir pressure may present as just described, the general methods and examples of pressure gradient analysis will be discussed below in four parts.

1 PRESSURE GRADIENT ANALYSIS OF EXPLORATION WELLS IN THE EARLY STAGE AND SOME FIELD EXAMPLES

1.1 Collection and Processing of Pressure Data

A typical field example is given here. A gas reservoir the data of 11 wells in which are distributed is shown in Figure 4.1. Seven of these 11 wells were drilled through Interval

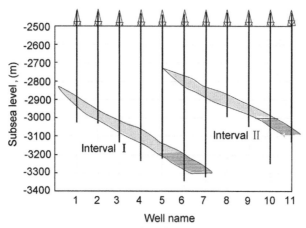

Figure 4.1 Profile of a gas reservoir.

I and the other 6 through Interval II. Each of Intervals I and II has its own gas—oil contacts: the elevation depth of the gas—oil contact of Interval I is −3200.0 meters and that of Interval II is −3000.0 meters.

Table 4.1 lists basic drilling data and the measured initial formation pressures of the wells mentioned previously.

1.2 Pressure Gradient Analysis

Plotting data of initial formation pressures and midelevation of the gas zone in Table 4.1, a pressure gradient analysis chart is obtained as shown in Figure 4.2.

Table 4.1 Relationship between Static Pressure at Middepth and Well Depth

Well name	Interval	Middepth of gas zone, m	Elevation of bushing, m	Midelevation of gas zone, m	Initial formation pressure, MPa
1	I	3410.0	500	−2910.0	36.170
2	I	3485.0	495	−2990.0	36.305
3	I	3532.0	492	−3040.0	36.390
4	I	3590.0	490	−3100.0	36.494
5	I	3660.0	486	−3180.0	36.620
6	II	3280.0	480	−2800.0	34.320
6	I	3729.0	479	−3250.0	37.300
7	II	3325.0	475	−2850.0	34.410
7	I	3777.0	477	−3300.0	37.800
8	II	3373.0	473	−2900.0	34.500
9	II	3420.0	470	−2950.0	34.590
10	II	3468.0	468	−3000.0	34.680
11	II	3540.0	470	−3070.0	35.400

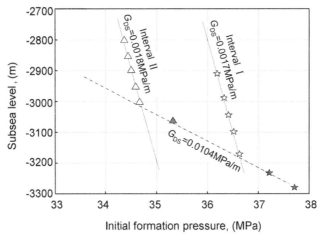

Figure 4.2 Initial formation pressure gradient analysis.

It is seen from Figure 4.2 that

Pressure points are classified into two groups; each group corresponds to an interval.

Among the seven pressure points corresponding to Interval I, five pressure points were measured in its gas zone, and the relationship between pressure and elevation depth is

$$p = 29.280 - 0.0018h.$$

This means that the static pressure gradient of the gas zone is

$$G_{DS} = 0.0018 \text{ MPa/m},$$

which is consistent with the static pressure gradient of gas and indicates that these gas wells belong to the same connected gas-bearing block.

For the other six pressure points corresponding to Interval II, the relationship between pressure and elevation depth in its gas zone is

$$p = 31.299 - 0.0017h.$$

This means that the static pressure gradient of the gas zone is

$$G_{DS} = 0.0017 \text{ MPa/m},$$

which is also consistent to the static pressure gradient of gas and indicates that these gas wells all belong to another connected gas-bearing block.

The three pressure points of the water zone are all located on another gradient line, the pressure gradient of which is

$$G_{DS} = 0.0104 \text{ MPa/m},$$

which is exactly the static pressure gradient of water. This means that the water bodies may possibly be connected or at least the reservoirs belong to the same hydraulic system during the sedimentation process.

As shown in the aforementioned example, although the bottom water pressures of the two gas zones in this gas field are both generally the hydrostatic column pressure, these two gas zones are not connected to each other and the positions of their gas–oil contacts are different. The pressure distribution of these two gas zones also clearly shows that because these two gas zones are independent gas reservoir bodies, they must be treated differently during development.

Figure 4.3 shows the analysis diagram of the measured static pressure gradient of the KL-2 gas field.

It is clearly seen from Figure 4.3 that the static pressure measured points in its gas zones are located on the same gradient line. Statistical analysis indicates that its pressure gradient is $G_{DS} = 0.00265$ MPa/m. Based on the relation expression of pressure gradient and density, $G_{DS} = \rho_g \cdot g$, the fluid density is calculated as $\rho_g = 0.2702$ g/cm^3.

In addition, according to the relative density of gas and subsurface pressure and temperature conditions in the KL-2 gas field, the average subsurface density of the gas in the KL-2 gas field is calculated as $\rho_g = 0.268$ g/cm^3 (as shown in Section 4.2 for the calculation method), which is very close to the gas density obtained from gradient conversion. This proved that reservoirs in the KL-2 gas field are connected and it is an uncompartmentalized gas field.

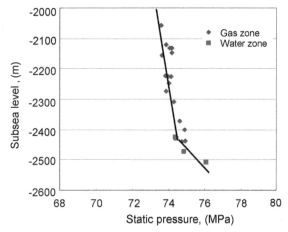

Figure 4.3 Analysis diagram of static pressure gradients at the KL-2 gas field.

2 CALCULATION OF GAS DENSITY AND PRESSURE GRADIENT UNDER FORMATION CONDITIONS

For a broadly connected gas reservoir body, the pressure gradient versus depths is expressed as

$$G_{DS} = \frac{p_2 - p_1}{h_2 - h_1}$$

$$= \frac{\Delta p}{\Delta h}$$

$$= 10^{-6}\frac{\rho \cdot \Delta h \cdot g}{\Delta h} \qquad (4.1)$$

$$= 10^{-6}\rho g$$

where G_{DS} is static pressure gradient in the formation, MPa/m; ρ is subsurface density of the gas, kg/m^3; g is acceleration of gravity, $g = 9.80665$ m/s^2; p_1 is formation pressure at depth h_1, MPa; p_2 is formation pressure at depth h_2, MPa; and h_1, h_2 is different middepths of gas zones, m.

The following formula can be used for calculation of gas density ρ [31]:

$$\rho = \frac{pM_g}{ZRT} \qquad (4.2)$$

where ρ is subsurface density of the gas, kg/m^3; p is pressure at the place where the gas is located, MPa; and $M_g = M$ is molar weight of the gas, kg/kmol; $M_{air} = 28.97$ kg/kmol is known, so

$$\gamma_g = \frac{M_g}{M_{air}} \quad \text{and} \quad M_g = \gamma_g \times M_{air} = \gamma_g \times 28.97 \text{ kg/kmol}.$$

For example, if $\gamma_g = 0.6252$ $M_g = 0.6252 \times 28.97$ kg/kmol $= 18.1120$ kg/kmol;

$Z =$ Deviation factor of the gas, it can be determined by looking up Figure 2.50 in this book according to the pseudo-reduced pressure p_{pr} and pseudo-reduced temperature T_{pr} of the real gas

$T =$ Formation temperature, K

$R =$ Universal gas constant, the unit and value of which are different in different systems of units, as shown in Table 4.2.

The pressure gradients of gas zones at different depths under formation conditions can be calculated based on Equation (4.1) or (4.2). An example is introduced as follows.

The gas reservoir conditions include:

- Reservoir pressure, $p_R = 30$ MPa
- Reservoir temperature, $T = 105\,^{\circ}C = 378$ K

Table 4.2 List of Values and Units of Universal Gas Constant

System of units	R value	Unit of R
China statutory unit	8.3143×10^{-3}	$MPa \cdot m^3 (kmol \cdot K)^{-1}$
SI unit	8.3143	$J(mol \cdot K)^{-1}$
	8.3143×10^3	$Pa \cdot m^3 (kmol \cdot K)^{-1}$
cgs unit	82.056	$atm \cdot cm^3 (mol \cdot K)^{-1}$
British oil field unit	10.732	$psi \cdot ft^3 (lbmol \cdot {}^\circ R)^{-1}$

- Relative density of gas, $\gamma_g = 0.6433$
- Pseudocritical pressure, $p_{pc} = 4.7693$ MPa
- Pseudocritical temperature, $T_{pc} = 205.567$ K
- Deviation factor of the gas, $Z = 0.962$

Equation (4.2) gives

$$\rho = \frac{30 \times 0.6433 \times 28.97}{0.962 \times 8.3143 \times 10^{-3} \times 378} = 184.92 \text{ kg/m}^3$$

Substituting it into Equation (4.1) results in

$$G_{DS} = 184.92 \text{ kg/m}^3 \times 9.80665 \text{ m/s}^2$$

$$= 1813.4 \times 10^{-6} \text{ MPa/m}$$

$$= 0.00181 \text{ MPa/m}.$$

It can be seen that because the density of the gas is very low, the gradient of the gas in the formation is also very low and is significantly different from that of the oil or water in oil-bearing or water-bearing intervals. Therefore, the pressure gradient analysis has become an important method in the identification of uncompartmentalized gas reservoirs.

Under equivalent pressure and temperature conditions, the pressure gradient in a gas reservoir is consistent with the static pressure gradient in the wellbores of gas wells. However, obviously the pressure gradient measured in the wellbore must never be used directly to perform pressure gradient analysis for formation. The formation pressure gradient must always be determined by plotting all measured static pressures of all wells with the elevation depth of the gas producing zone, just like what was done in Figure 4.2 earlier.

However, if the pressure gradient analysis shows that the static pressures of all gas wells are located on the same "gas static pressure gradient line," is it surely that the gas reservoir is an uncompartmentalized gas field? The answer to this question is that this is only a necessary condition rather than a sufficient condition; this is discussed in the following pressure analysis during the development phase of a gas field.

3 PRESSURE GRADIENT ANALYSIS DURING DEVELOPMENT OF A GAS FIELD

For a gas field featuring large area distribution and flat structure, the formation pressure can be more or less consistent or distributed according to the "natural gas static pressure gradient line" during the early development phase. Take the JB gas field of China for example, there are 44 statistic pressures with good quality measured in the early phase of exploration, the relationship between the pressure and the depth from regression is $p = 24.90 + 0.00193h$, indicating that the pressure gradient value is $G_{DS} = 0.00193$ MPa/m, which is generally the pressure gradient of dry gas. In other words, it was believed that the entire gas field was likely to be an uncompartmentalized gas field. After it was put into development, however, the situation changed and gradually showed that the gas field is divided into several development areas with different pressures of areas of Well S-45 and Well S-155 and so on. What hindered the pressure balance among the different areas of the gas field? The following factors may be to blame:

- Faults or lithologic boundaries exist among different gas-bearing areas that obstruct the gas flow, resulting in a pressure imbalance.
- Low-permeability zones or "flow barrier boundaries" exist among different gas-bearing areas that make the pressure-balancing process between different zones very slow.
- The reservoir itself is a very low permeability one that retards the pressure-balancing process.
- In fissure gas fields, the reserves controlled by individual wells are distributed within a large number of group- and/or series-distributed fissure systems in which there may possibly be connected fissures in some portions, but the permeability of these connected parts is extremely limited compared to the whole fissure system. Therefore, although the characteristic of constant-volume gas reservoirs (i.e., pressure derivative goes down rapidly during the late time on the pressure buildup curve, as shown in Chapter 5) is never shown, the formation pressures measured at different wells at about the same time are extremely different.

Each gas field with the aforementioned characteristics should be divided into different areas during development according to the different degrees of pressure dropping. As the development process advances, they will gradually disperse and form several different pressure gradient areas, as shown in Figure 4.4; therefore the gas field is also divided into different areas on the plane.

The area division of pressure will provide an important dynamic basis for the development regime of the entire gas field. Therefore, pressure analysis is an important element in development performance analysis.

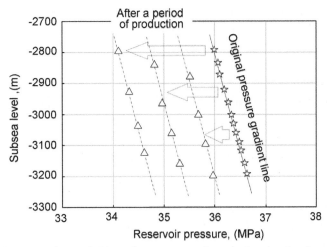

Figure 4.4 Moving and area division of pressure gradient during the development process.

4 SOME KEY POINTS IN PRESSURE GRADIENT ANALYSIS

4.1 Accuracy of Acquired Pressure Data

Accurate acquisition of formation pressure data is surely the foundation of pressure gradient analysis. However, it did happen that the difference of measured static pressure values acquired at different times in a very short duration was as large as several mega Pascals in some gas wells in the field. This would bring about many uncertain factors into the analysis. The following factors affect the accuracy of acquisition of pressure data.

- Because pressure is measured during the well test before the wellbore has been cleaned up, the measured pressure cannot reflect formation conditions accurately or is not the real formation pressure.
- Because liquid accumulated in the bottom hole after the gas well test and the pressure gauges were not set at the middepth of the gas zone but at somewhere above the liquid level, the measured value cannot represent the bottom-hole pressure.
- When pressure is measured after stimulation such as acidizing or fracturing, the pressure drop funnel has not yet restored.
- The pressure gauges are not set at the midpoint of the gas zone, and the pressure gradient below the pressure gauges was not measured, which causes inaccurate converted pressure.
- Low-precision mechanical pressure gauges are used or the calibration of the used pressure gauges has already expired before measurement.
- The pressure drop funnel created by the gas well test in an extremely low permeability formation or a special lithologic formation cannot restore in time.

Therefore, seizing the right time of pressure measurement during the gas well test process, using high-precision pressure gauges, and setting the pressure gauges at the middepth of the gas zone so that accurate static pressure data can be acquired in time are the foundations of pressure gradient and other analyses.

4.2 Pressure Gradient Analysis should be Combined Closely with Geologic Research

Pressure gradient analysis is a means for geological research of reservoirs and is designed to serve gas field development. Therefore, it must be combined closely with geologic development research.

4.2.1 The Area—Division of the Reservoir Provided by Pressure Gradient Analysis should be Supported by the Relevant Geological Basis

Producing intervals can be divided into several areas through pressure gradient analysis, but this area—division must be supported by certain geologic backgrounds, such as

- Different blocks formed by cutting of faults
- Different series of strata, each of which has its own independent gas—water relations
- Sandstone blocks with different shapes divided by the lithologic boundaries created via fluvial facies sedimentation
- Different group- and/or series-distributed fissure systems in limestone formations
- Blocks with different permeability divided by mudstone delves formed by marine sedimentations
- Massive limestone bodies formed by organic reef segments and so on.

Some special geological characteristics determine pressure distribution characteristics. Therefore, when a certain kind of pressure gradient characteristics is observed, the geological causes of their occurrence must be determined as otherwise the validity of such pressure distribution characteristics would become doubtful.

4.2.2 Analysis of Pressure Gradient Characteristics Provides Supporting Information for Validating Reserves Calculation Results

So far the major calculation method of gas reservoir reserves is still the static parameters method, that is, the so-called volumetric method. However, the reserves calculated by this method would bring about huge errors, especially for some special lithologic reservoirs such as fissured limestone reservoirs, whereas the dynamic methods, including the pressure gradient method, can discover in time the pressure-dropping characteristics of each well in the exploration area during both the early stage and a period of time after production to find out the anomalies in reserves distribution so that to timely raise questions against the risks in reserves calculation results, thereby reducing losses.

4.2.3 Analysis of Pressure Gradient Provides Basic Parameters for the Designing of Development Program

Combined with transient pressure test analysis, pressure gradient analysis can provide basic parameters for design of the gas field development program, including the division of blocks or areas, boundary configurations and permeability in each area of the field.

5 ACQUISITION OF DYNAMIC FORMATION PRESSURE AFTER A GAS FIELD HAS BEEN PUT INTO DEVELOPMENT

5.1 Dynamic Production Indices During Production of a Gas Field

All technical professionals involved in oil and gas field development know that formation pressure near an oil or a gas well is declining continuously during the production process of that well. In the 1980s, the relevant departments of the former Ministry of Petroleum Industry of China organized related experts to specially discuss the exact meaning of formation pressure and the methods of pressure measurement or acquisition in order to eliminate the errors of formation pressures acquired in various oil fields in China.

Much consideration regarding this problem has been made again and again during several dozen years of research activities in the dynamic description of gas reservoirs and then the author proposed the concept of dynamic indices of gas wells, including new concepts such as the dynamic deliverability of gas wells, dynamic inflow performance relationship (IPR) curve, and dynamic absolute open-flow potential and dynamic formation pressure, and also derived some formulae for determining these dynamic indices in gas fields. New ideas about the measurement, acquisition, and analysis of these dynamic indices regarding dynamic description research in Chapter 1 have already been discussed comprehensively. The deduction and application of the stable point laminar—inertial—turbulent deliverability equation mentioned in Section 3.6 of Chapter 3 have also provided detailed discussions about dynamic deliverability indices.

It can be seen clearly from the aforementioned studies and analyses that the gas reservoirs dealt with are never stationary. The absolute open flow potential of a gas well measured initially can only reflect the real conditions of the gas well within a short period after the well starts producing. After a short duration, for example, after 1 year, only a few months, or even only a few days, this index may already become not significant at all for understanding of this gas well, as does the formation pressure. The primary cause of a deliverability depletion of gas wells is simply the depletion of formation pressure. Once a deficit appears within the gas drainage range due to the recovery of gas, the formation pressure will decrease and the deliverability of the gas well will also decrease accordingly, especially in small constant-volume gas reservoirs. This kind of changing formation pressure is called "dynamic formation pressure" in this book.

5.2 Several Formation Pressures with Different Meanings

Formation pressure seems simple, but there are several different meanings of it in flow areas in the formation. Figure 4.5 is an illustration of formation pressure under the condition of recovering gas by elastic energy.

5.2.1 Measured Average Formation Pressure

Measured average formation pressure means the bottom-hole static pressure measured after a long-term shutting in within a limited production area. This measurement method is used commonly in oil fields. If the shut-in duration is long enough, the pressure tends to be in a balance state within the limited constant-volume area controlled by the well, and the measured static pressure can represent the average formation pressure in the block.

In the oil field, however, the shut-in duration is usually limited; for example, a shut-in duration of 1 month is already very long, but for low-permeability gas reservoirs, especially for extremely low-permeability reservoirs, it is not long enough; for example, in a long and narrow reservoir trapped by lithologic boundaries in the SLG gas field, the formation pressure is still building up after 1 year of shut-in duration; the pressure measured at that moment still fails to reach the value of average formation pressure.

If the area controlled by an individual well is divided by the well pattern, the formation pressure obtained by measurement of the shut-in pressure is also affected by the production history of all wells in adjacent areas.

5.2.2 Formation Pressure Determined by Deduction Based on Dynamic Model

It is known from the aforementioned discussions that the method of measuring pressure after a long shut-in duration has always been used to acquire the average formation pressure of gas wells in the field, but in fact the average formation pressure value cannot be measured successfully due to the restriction of field conditions.

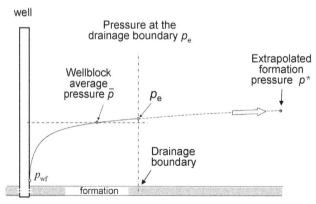

Figure 4.5 Illustration of formation pressure under the condition of recovering gas by elastic energy.

If a gas well is located in a limited constant-volume reservoir, the dynamic description and analysis of the reservoir have been carried out, and a perfect dynamic model of it has been built, the exact value of average formation pressure can be determined any time—an example of which is shown in Figure 4.6.

In gas reservoirs such as the SLG gas field, YL gas field, and DF gas field described in Chapter 8 of this book, the average formation pressures of those gas wells located in a constant volume block have already been deducted.

5.2.3 Calculation of Formation Pressure at Gas Drainage Boundary p_e

For any gas well, all energy that drives the gas to flow to the bottom hole comes from the producing pressure differential. The producing pressure differential is usually expressed as the pressure form, the pressure-squared form, or the pseudo-pressure form written as $(p_R - p_{wf})$, $(p_R^2 - p_{wf}^2)$, or $(\psi_R - \psi_{wf})$, respectively, in which formation pressure p_R is the formation pressure at the gas drainage boundary. For heterogeneous gas drainage areas with complicated configurations, the specific geometric position corresponding to this concept is actually not easy to determine. However, the proportional relationship of producing a pressure differential to the gas flow rate would generally and basically maintain unchanged during the production process. The flowing pressure can be measured at any moment so that the formation pressure can be calculated accordingly, and the formation pressure at the gas drainage boundary can therefore be determined.

The dynamic deliverability equation established in Section 3.6 of Chapter 3 of this book provided a practical and feasible method for determining this kind of formation pressure at the dynamic gas drainage boundary. Figure 4.7 shows dynamic IPR curves

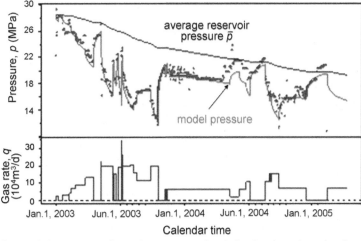

Figure 4.6 Determining average formation pressure by deduction based on the dynamic model.

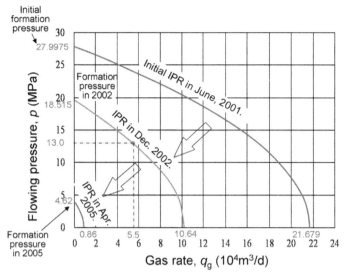

Figure 4.7 Determining dynamic formation pressure at gas drainage boundary by dynamic IPR curves.

generated by the dynamic deliverability equations in different periods; the starting points of them simply correspond to the dynamic formation pressures at the corresponding periods.

For gas wells, as long as the stable points during the production process are tracked and monitored, the dynamic deliverability equation and the dynamic formation pressure can simply be obtained at any time. This is really an easy and convenient method for acquiring formation pressure applicable to any gas well.

5.2.4 Other Frequently Used Formation Pressure Concepts

1. Extrapolated formation pressure p^*

For exploration wells in large homogeneous sandstone formations, the pressure buildup tests must last long enough to obtain data of the radial flow stage; then based on Horner's method, the straight line segment of the pressure buildup curve on the semilog plot can be used to extrapolate the formation pressure p^* (as shown in Figure 4.5). This formation pressure value, p^*, means pressure at the infinity point of the reservoir. The extrapolation method can be found in conference literature [2].

If there are impermeable or heterogeneous boundaries near the tested well, this extrapolation method is usually invalid, not only because the straight line segment of the radial flow is usually hard to identify, but also because the value of formation pressure p^* itself at the infinity point does not exist; therefore, this extrapolation become meaningless.

2. Matched initial formation pressure p_{R0}

A matched initial formation pressure p_{R0} can be obtained during well test data analysis with well test interpretation software. It is wrong that some well test interpreters regard this matched formation pressure as the current formation pressure. In fact, the preliminary dynamic model built by well test interpretation software based on the pressure buildup curve must be verified by the pressure history match. The criterion whether the interpretation results can pass verification is that the pressure history generated by the used theoretical model in interpretation must be consistent with the measured pressure history, wherein the most important consistency is that the initial pressure of the model must be exactly consistent with the actual measured initial pressure. Therefore, the interpretation results include specially the matched initial pressure p_{R0}.

5.3 Performing Gas Reservoir Analysis with Dynamic Formation Pressures

Once the formation pressure values have been obtained, the following analysis can be performed.

5.3.1 Research on Reservoir Division

Geologically, a large gas field may be divided into several different areas or blocks by impermeable boundaries such as faults. For a gas field with special lithologic features it is even possible that one single well controls an entire gas reservoir. After producing for a period of time, the formation pressure may become different as a result of different depletion due to different cumulative withdrawals from various well blocks. The gas field can be divided into subareas according to the differences of dynamic formation pressure.

5.3.2 Dynamic Variation Analysis of Pressure Gradient Line

If the permeability in the gas reservoir is high and the gas reservoir is clearly divided into some subareas, the initial pressures of different gas wells located on the same initial pressure gradient line initially may decline differently after a period of time; they are likely to move onto several different pressure gradient lines, as shown in Figure 4.4. This kind of dynamic variation of a pressure gradient line can help identify the division of gas reservoirs.

CHAPTER 5

Gas Reservoir Dynamic Model and Well Test

Contents

Dynamic Well Testing in Petroleum Exploration and Development   © 2013 Petroleum Industry Press. Published by Elsevier Inc.
ISBN 978-0-12-397161-6, http://dx.doi.org/10.1016/B978-0-12-397161-6.00005-X

This chapter is an essential part of this book because:

1. The content of this chapter is transient well test analysis, especially pressure buildup analysis, which is the main content of all professional books entitled "well test" or well test specialized courses in petroleum institutes or universities. It is also considered as the main connotation of "modern well test analysis" being developed in recent 30 years.

2. As described in Chapter 1, a preliminary "dynamic model" of a gas well can be established from pressure buildup analysis. This dynamic model contains much information about the reservoir, including information about the area near the wellbore and well completion (e.g., kh and S), and that about the boundaries near the well. After verifying with production test data of the well, the model is further developed, and the configuration and property of all outer boundaries can be identified; based on these results, a rather perfect dynamic model that can overall describe the reservoirs near the well is formed.

Perfect dynamic modeling can reappear the whole pressure history of the gas well and the reservoir, and further than that, it can also display the variation of gas output and flowing pressure of the well in the future with the aid of well test interpretation software, and therefore it can undoubtedly and greatly help the realization of scientific and procedural management of gas fields.

1 INTRODUCTION

1.1 Static and Dynamic Models of Gas Reservoir

At present, gas field studies place an emphasis on the numerical simulation of gas reservoirs, especially numerical simulation before its development. Nowadays every development program design contains a set of production indices obtained from numerical simulation. A numerical simulation study is usually done with following steps:

- To complete geological modeling
- To establish an initial parameter field and to simulate the pressure variation of the gas field with a given flow rate variation
- To carry out a match verification of pressure history in the early development stage
- To predict production indices in the middle and late development stages

1.1.1 Geological Modeling of Gas Reservoirs

The basis of geological modeling is:

- Structural characteristics maps drawn from geophysical prospecting results and confirmed by geological research
- Planar distribution maps of kh value drawn mostly on the basis of reservoir lateral prediction, corrected by the geophysical method and calibrated by logging data
- Initial pressure and initial temperature data of gas wells
- Analysis results of physical properties of fluid

Before development of a gas field, the pressure history of the whole field has not formed yet and match verification of pressure history cannot be done then so a geological model established mainly on the basis of static information is used in the development program design, and various development indices are predicted on the basis of this model.

However, many problems of static model have been exposed during its application in gas field development practices.

Uncertainty of Sealing of Faults Marked on Structural Map

Some gas fields are not only trapped by some faults with big downthrows at their periphery, but are also cut by many faults with small downthrows inside the field. The Kela-2 gas field in west China is a typical representative. There are more than 70 faults of class II or III inside the field (refer to Figure 5.1); if these faults play a role in separating gas flow, their positions and related influence on field development are considered carefully in well spacing and development indice prediction; however, if the effect of these faults is negligible, the gas field can be developed overall. Therefore, a model established only by static information cannot provide reliable prediction for gas field development in the early stage and so brings various uncertain factors to the numerical simulation study.

Then it has been confirmed based on the subsequent well tests that all small faults inside the gas reservoir do not have any effects on separating gas flow.

There are Some Lithologic Buffers in Homogeneously Distributed Sandstone Reservoirs

In contrast, in view of lateral prediction by geophysical exploration, some sandstone reservoirs developed from fluvial sedimentation are considered as integrally widespread and steadily extended. Therefore, a set of optimistic and comprehensive indices are obtained in reserves calculations. However, there is a prevision that, based on geological research, effective permeable sublayers controlling gas flows are staggered, distributed, separated, and disconnected from each other, as shown in Figure 5.2.

A gas well may only control a few elongated effective zones along the river channel and controls very limited dynamic reserves due to the buffer restriction along the direction vertical to the channel. The problem is that a geological study can only provide

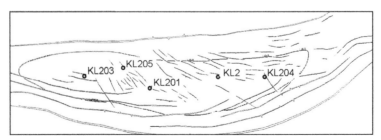

Figure 5.1 Distribution of faults inside a gas field development area.

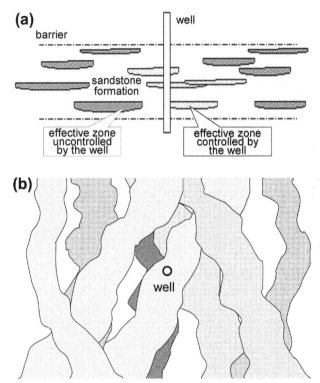

Figure 5.2 (a) Schematic section of fluvial sandstone subzones. (b) Planar distribution of fluvial sandstone subzones.

a possible reservoir pattern, but cannot confirm its actual position, width, and length. However, the buffer configuration can be made clear by well test studies.

Reserves and Physical Properties Distribution of Buried Hill Fissured Limestone Reservoirs are Uncertain

Plenty of buried hill limestone reservoirs have been discovered in China. One of them, the Wumishan reservoir in the Renqiu oil field, was once a major part of the Huabei oil field. But there are still many reservoirs of this kind with significant planar distribution variation, and their reserves calculated by the static method are quite uncertain. Take the Qianmiqiao gas field discovered in recent years as a typical example: in the early stage, a very high initial gas flow rate of a few wells in this field encouraged everyone greatly, but due attentions have not been paid to its complicated geological conditions. In fact, for such a reservoir whose major storage space for fluid is group- and/or series-distributed fractures, reserves calculated by the static method, which is valid for large sandstone reservoirs, is absolutely much larger than its real value.

If a geological model is established on the basis of such geological understandings, a related reservoir prediction made by numerical simulation will bring extremely optimistic results and potential risk for the implementation of a development program. This fact discloses sufficiently the uncertainty of static geological modeling for complicated gas reservoirs.

Configuration Description of Special Lithologic Reservoir Such as Volcanic Rock

Some special lithologic gas reservoirs have been found in recent years in China; some among them are volcanic reservoirs. The distribution rules of volcanic reservoirs are very different from those of common sandstone reservoirs, but no reliable geological description methods have been formulated for them yet.

It seems that there are some uncertainties in the static model, which is the basis of geological modeling, especially in models for special lithologic reservoirs. This will bring great risk to the prediction of development indices and the next implementation of the development program.

1.1.2 Dynamic Model of Gas Reservoirs and Gas Wells

As seen from the aforementioned discussion, present geological modeling can only describe gas reservoir characteristics from one aspect. Visually speaking, it is just like a "static photo" with a few clear points (well points) only. But in view of a gas field study, what are needed best and urgently are "alive and dynamic images" of a gas reservoir, the dynamic performance and the dynamic model of it.

What aspects does a dynamic model of a gas field include in general?

Pressure History and Related Production History of Gas Field

The pressure history of a gas field is characterized by the pressure histories of gas wells in the field, including:

- Pressure histories of observation wells. An observation well only reflects a "point" in the flow process of gas zones. Such a well is drilled not for gas production, but reflects actual pressure performance of gas zones in the percolation process. When an interference test betweens wells or a pulse test is run, it is in observation wells where pressure data are measured and then used to study the gas reservoir.
- Pressure histories of production wells. A production well represents a "convergent point" in flow equations, for it is drilled for gas producing from the reservoir, and it records bottom-hole flowing pressure during production and pressure buildup during well shutting in. It is these transient pressure variation processes, that is, pressure drawdown during production and pressure build up during shutting in, that reflect the properties and distribution of reservoir parameters such as permeability and information such as whether reservoir boundaries exist or not, the distance between the boundary and the well, and boundary properties.

Well Test Model of Reservoir Established on the Basis of Pressure Analysis

The core of modern well test analysis is model analysis. Model analysis establishes a "well test model" based on transient well test analysis, and the model should be consistent with reservoir dynamic performance. Actually, the well test model here is still a geological model, but it contains more content than a static model: in addition to a related description of reservoir static parameter distribution, it contains a description of gas flow in the reservoir as well. Specifically, it includes:

1. Description of reservoir parameter distribution
   - Permeable space types of the reservoir: homogeneous medium, double porosity medium, or double permeability medium, etc.
   - Reservoir parameters: permeability k, effective thickness h, flow coefficient kh/μ, storativity $\phi h C_t$, storativity ratio ω and interporosity flow coefficient λ of double porosity medium, and formation coefficient ratio κ of double permeability medium, etc.
   - Reservoir planar distribution and related parameters: the configuration and distance of impermeable boundaries around the well, variation and parameter ratio of flow coefficient kh/μ and storativity $\phi h C_t$ for composite reservoir, and other features of the boundary
   - Completion status and related parameters: damage during drilling and completion and skin factor S; fracture configuration and related parameters: half-length of fracture x_f, flow conductivity F_{CD}, and fracture skin S_f, etc.
2. Description of fluid flow in reservoir with focus on the well
   - Description of flow status in chronological order and also in order of distance from the well, such as wellbore storage flow, radial flow, linear flow, bilinear flow, and pseudo radial flow
   - Description of flow status is often completed on the basis of curve characteristic analysis, and establishment of relationsships between transient test curves and formation flow status is done by the graphics analysis method
   - Pressure curve analysis is usually based on pressure derivative curve analysis mainly and is set up strictly on the basis of the solution of percolation mechanic equations

Reservoir Dynamic Model Formed by Integrating Individual Well Models

Well test models of individual wells are the basis of the reservoir dynamic model. The well test models of all individual wells will delineate the overall configuration of the whole reservoir. The following is an example.

1. Determination of well test model of an individual well

   A gas field developed by a petroleum company is composed of fluvial sandstone reservoirs. In the early stage, a transient test was run in an exploration well of this field, and the pressure buildup curve is shown in Figure 5.3.

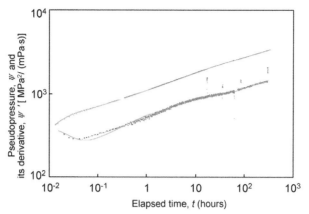

Figure 5.3 Typical single well test model.

The dynamic model of this well from its individual well analysis can be described as follows.

- There is a long fracture connecting with the well; the half-length of the fracture is $x_f = 87$ meters. It is verified that the analysis result is consistent to completion condition: this well had really been stimulated by sand fracturing before testing.
- There are rectangular impermeable boundaries very close to the well; the distances between the well and the boundaries are, respectively, $L_{b1} = 198$ meters, $L_{b3} = 250$ meters, $L_{b2} = 2000$ meters, and $L_{b4} = 2000$ meters. It is also verified that formations of this well are composed of meandering river sandstones, and the effective thickness of a single layer is about 4 meters. From the general rules of width-thickness ratio for such sandstones, the width of an effective single channel formation is approximately 400−500 meters; analysis results are in line with formation features.
- From well test interpretation, formation permeability around the well is $k = 1.5$ mD, and the reservoir is considered to be a low permeable sandstone gas zone.
- Dynamic reserves controlled by this well are calculated as about 3000×10^4 m^3 from the boundary conditions and pressure decline rate of this individual well.
  A schematic view to present formation conditions made from well test analysis results is shown in Figure 5.4.

2. Determination of gas field dynamic model

   Some common and consistent understanding on this gas field is obtained on the basis of pressure buildup analysis results of nearly 10 wells. Also, dynamic model features of this gas field are summarized as follow based on the general understanding of all wells.

   - The reservoirs of this gas field were cut by lithologic boundaries developed from fluvial sediments, striped, and separated local regions formed.

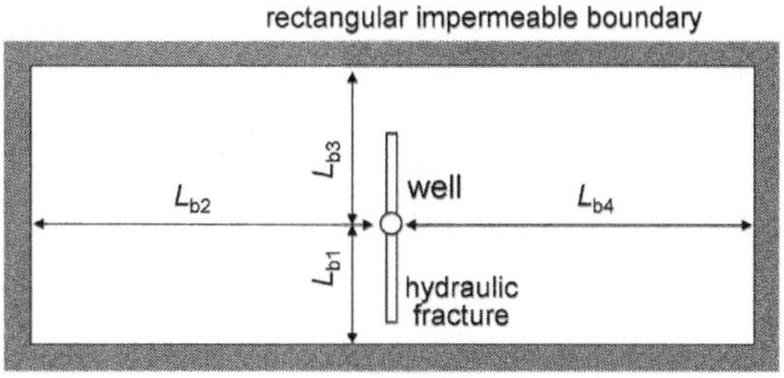

rectangular impermeable boundary

Figure 5.4 Schematic plot of individual well dynamic model.

- The local region controlled by an individual well is 400–500 meters wide and 2000–3000 meters long. A single well controlled reserve is about $3000-5000 \times 10^4$ m³.
- A single well controlled reserve is consistent with static pressure test data of the other six wells.
- The boundaries separating blocks have extremely low permeability.
- The reservoir permeability ranges from 0.5 to 3 mD and varies from block to block.
- After stimulation, the quality of well completion is very good: the skin factor is as low as $-3 \sim -5$.
- The non-Darcy flow coefficient (D) during gas well production is about 0.5 $(10^4 \text{ m}^3/\text{day})^{-1}$.

The reservoir dynamic model established from the aforementioned well test analysis makes geological understandings much richer, more concrete, and can be used for numerical simulation analysis. By integrating geological modeling with dynamic modeling, this gas bearing area in the field is subdivided into many small long strips, which are isolated from each other by many buffers, as shown in Figure 5.5.

As can be seen from this example, development indices resulting from the rebuilt model based on reservoir dynamic characteristics are much worse than those without boundaries, but are more consistent to the actual conditions:

- Because the single well controlled reserve is limited, the rate-maintenance capability indices are much lower than those without boundaries. The period of stabilized production of an individual well is short and the cumulative recovery of it is small.
- Buffer distribution characteristics should be sufficiently considered in designing of well spacing density so that the whole reservoir can be controlled by gas wells.
- If well spacing is too sparse, a large quantity of reserves cannot be controlled effectively, and therefore a very low recovery percent of reserves will result. More infill wells should be drilled during the development to enhance the recovery.

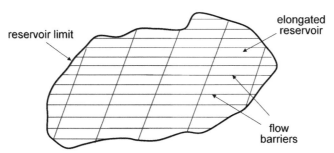

Figure 5.5 Schematic plot of gas reservoir synthetic modeling.

1.2 Pressure History of a Gas Well Symbolizes the Life History of it

A gas well put into production can be compared to that of a baby being born. The pressure history of the well during its production is just like its life history. The productivity of a well is vigorous when the pressure is still its initial value, but with pressure declining, the well ages gradually; and when pressure is totally depleted, it dies. Therefore, future performance prediction of a gas well can be done by studying its pressure history and the law of the pressure variation.

1.2.1 Different Pressure Histories Exist Under Different Reservoirs and/or Different Well Completion Conditions

Several kinds of pressure histories of some production testing wells have been introduced in Chapter 3 of this book. These pressure histories reflected the reservoir and well completion conditions. This chapter shows a similar pressure history plot (Figure 5.6) for a more thorough understanding of gas well life history from the viewpoint of the reservoir dynamic model.

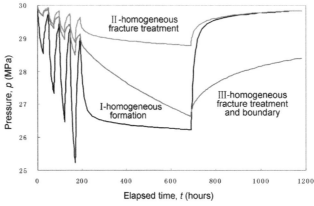

Figure 5.6 Pressure history of a gas well symbolizes the life history of it.

In Figure 5.6, Pressure History Curves of Three Production Testing Wells Reflect three Different Kinds of Formation and Well Completions Conditions

Case I

- Homogeneous sandstone formations with permeability $k = 3$ mD and effective thickness $h = 5$ meters.
- No boundary effects around the well.
- Normal completion and skin factor $S = 0$.
- Well production history: in the early stage, the modified isochronal test was run with flow rates of 4, 6, 8, and 10×10^4 m^3/day successively, and an extended test was run with the last rate of 8×10^4 m^3/day, and then the well was shut in.

Case II

- Just the same as case I, homogeneous sandstone formations have permeability $k = 3$ mD and effective thickness $h = 5$ meters.
- Just the same as case I too: No boundary effects around the well.
- Well completion is improved by fracturing, $x_f = 70$ meters and $S_f = 0.1$.
- Same production history with case I.

Case III

- Just the same as cases I and II: homogeneous sandstone formations with permeability $k = 3$ mD and effective thickness $h = 5$ meters.
- Two parallel impermeable boundaries exist around the well; distances to the well are the same: $L_{b1} = L_{b2} = 70$ meters.
- Same as case II, well completion is improved by fracturing, $x_f = 70$ meters and $S_f = 0.1$.
- Same production history with cases I and II.

Phenomenon Seen from Figure 5.6

1. Producing pressure differentials are different

    Although formation permeability and thickness in the three cases are just the same, they display different producing pressure differentials during short-term producing. In case I, the well was normally completed and a relatively big producing pressure differential appears; whereas in cases II and III, because stimulation of fracturing treatment has improved the production conditions, a relatively much smaller producing pressure differential for the same rate is observed.

2. Pressure decline trends during flowing are different

    Producing pressure differentials in cases I and II are very different, but for quite a long time after opening the well the flowing pressures are quite stable in both cases, which indicate that the formation supply conditions are similar in both cases and forecast that the wells can keep stabilized production in both cases if producing with reasonable flow rates.

    Case III is quite different. Its flowing pressure declines continuously at a certain rate after well opening and shows no signs of gradual stabilization.

3. Pressure buildups after shut in are different

In cases I and II because shut-in pressure can nearly build up to the initial pressure, the well may have a long period of stabilized production. However, in case III, the formation pressure drops obviously and significantly and will deplete gradually as cumulative production increases.

Summarily, in case I, the gas production rate of the well is not very high but stabilized; in case II, the gas well has the potential to enhance its production; if it produces with adequate rates, it can withdraw more gas in a duration as short as possible and keep producing with a stabilized flow rate; and case III is very different from cases I and II because the gas drainage area is limited, gas productivity is declining gradually, and no long-term stabilized production can be maintained. This is the life history of a gas well symbolized by its pressure history.

1.2.2 Pressure History Trend of Gas Well is Determined by Reservoir Conditions

Various pressure history trends of different gas wells introduced earlier are determined by related formation conditions of the wells, mainly including:

- Dynamic reserves controlled by the well.
- Distribution of dynamic reserves—natural gas is stored in concentrative masses or thin beds; in homogenous sandstone or double porosity medium or even in the system with group- and/or series-distributed fractures; the reservoir is approximately round, square, or very long striped, etc..

Different reservoir dynamic models depend on different formation conditions, and the long-term trend of formation pressure depends on the dynamic reservoir model.

It should be especially pointed out that a dynamic model contains completion conditions of gas wells (e.g., S, x_f, and S_f); these completion conditions can influence the pattern of pressure history, but their influence exists only in a short period after well opening, that is, only flowing pressure or producing pressure differential is affected. A long-term trend of the pressure is beyond the influence of completion conditions. In other words, stimulation such as acidizing and/or fracturing treatments can only enhance production of the treated wells, and so quicken the capital recovery; but with the current technical conditions, reservoir reconstruction is impossible, reservoir configuration cannot be changed, dynamic reserves controlled by individual well cannot be increased, and the overall pattern of the gas reservoir dynamic model cannot be changed accordingly. It is very crucial to understand dynamic characteristics of gas fields.

1.2.3 Main Approach to Confirm Reservoir Dynamic Model is Pressure History Match Verification

Main Approach to Establish Reservoir Dynamic Model is Mainly Transient Well Test Analysis
As discussed earlier, the main basis of establishing a reservoir dynamic model is transient well test analysis. Main analysis methods of a transient well test include:

- Pressure buildup analysis
- Pressure drawdown analysis

- Interference test or pulse test analysis
- Comprehensive analysis of well test data combined with geological information

Type Curve Match Analysis is Major Approach of Transient Well Test Analysis

Transient well test analysis has been developed for half a century. The first breakthrough of the technique took place in the early 1950s. Miller and colleagues [6] and Horner [7] found an obvious straight-line portion on the curve of pressure vs logarithm of time, and this straight line corresponds to the radial flow of the fluid in formation, which laid the theoretical foundation for the semilog straight-line analysis method, or semilog analysis method for short.

However, aiming merely at the simplest homogenous sandstone reservoirs, this semilog analysis can only provide a few parameters of the formations, such as permeability (k) and skin factor (S) near the wellbore.

As study objects of the well test become more and more extensive and complex, the semilog method can neither provide required interpretation one to one correspondingly nor even find out the straight-line portion corresponding to radial flow in some cases.

In order to resolve these problems, from the early 1970s to the middle 1980s, a number of percolation mechanics and well test specialists, headed by Ramey, Agarwal, Gringarten, and Bourdet, successfully developed the "modern well test analysis method" whose core is the "log—log type curve match method." This method is featured by the establishment of various well test models, which contain the type and parameters of the reservoir, the type and parameters of the boundaries, and completion parameters of oil or gas wells. Different type curves correspond to different models, and different model characteristics are expressed by different type curve shapes. And so an accurate description of reservoir dynamic characteristics is obtained by well test analysis.

Ambiguity of Type Curve Match Analysis

However, the ambiguity of type curve match analysis is always a puzzling problem in modern well test interpretation. Different reservoirs may sometimes display analogous pressure behavior in short-term performance. Take a well in a double porosity reservoir with transient interporosity flow, for example; its pressure and pressure derivative curves are shown in Figure 5.7.

The curve characteristics are shown in Figure 5.7, after the horizontal straight line of transition flow period (0.25 line), the derivative curve goes up and becomes another horizontal straight line of radial flow in the whole double porosity system, that is, fissure and matrix system (0.5 line). In the log—log plot, the pressure derivative rises by 0.25 and the logarithmic coordinate absolute value rises by 0.301 (lg0.5—lg0.25) accordingly.

However, if there is a single no-flow boundary near a well in the homogeneous reservoir, the shape of its well test curve is totally consistent to that of the aforementioned case: on the log—log plot, the pressure derivative curve rises from the horizontal straight line (0.5 line) to another one (1.0 line), and the logarithmic coordinate absolute value

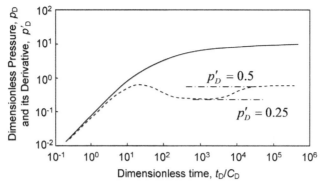

Figure 5.7 A type curve of a well in double porosity medium with transient interporosity flow.

also rises by 0.301 (lg1.0—lg0.5) accordingly, absolutely the same as the aforementioned case, as shown in Figure 5.8.

It tells us that two absolutely different formation models may be obtained from the interpretation of a curve depending only on the curve shape. It is a typical puzzling problem of ambiguity in well test analysis.

Pressure History Match is the Final Verification for Removing Ambiguity

As mentioned previously, an ambiguity problem exists in well test interpretation. The only effective method of removing the ambiguity is to compare the calculated pressure history of the model, which is established by type curve match analysis by well test interpretation software, with the pressure history actually measured on site: if the two pressure histories are very consistent, the reliability of the dynamic model is conformed; if they show any inconsistencies, model parameters need to be adjusted—even the used model itself needs to be changed sometimes, for further improving understanding of the model, until perfect consistency is obtained.

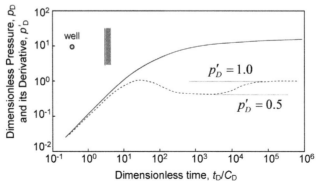

Figure 5.8 A type curve of a well in homogenous formation with a single no-flow boundary nearby.

To illustrate the importance of pressure history match, real practice in a gas well is given as follows.

Well X-6 in west China was tested, the drilled formations of which are composed of sandstones with low permeability developed by fluvial sediment. It was known from geological analysis that if a well is drilled in the area of river channel sandstones, its gas output would be fairly high, and there probably are some areas with extremely low permeability around Well X-6 indicated by either serious worsening of formation or buffers. The measured pressure history of this well is shown in Figure 5.9. The pressure buildup curve of the final shut in is analyzed with the type curve match method, and the log–log match plot of pressure and pressure derivatives is shown in Figure 5.10.

It can be seen from Figure 5.10 that the type curve match is satisfactory, and the following dynamic model is established based on the analysis results:

- This gas well is located in a round composite formation; the flow resistance in inner zone is less than that in outer one
- Formation permeability around the well is $k_1 = 7$ mD; that of the outer zone is $k_2 = 0.49$ mD
- Flow coefficient ratio of outer to inner zones is $(kh/\mu)_2/(kh/\mu)_1 = 0.07$
- Storativity ratio of outer to inner zones is $(\phi h\ C_t)_2/(\phi h\ C_t)_1 = 0.06$
- Radius of the interface of inner and outer zones is $r_{1,2} = 101$ meters
- Skin factor of the well is $S = -5.9$ (hydraulic fracture exists)
- Non-Darcy flow coefficient is $D = 0.028(10^4\ \mathrm{m}^3/\mathrm{day})^{-1}$
- Wellbore storage coefficient is $C = 2.2\ \mathrm{m}^3/\mathrm{MPa}$

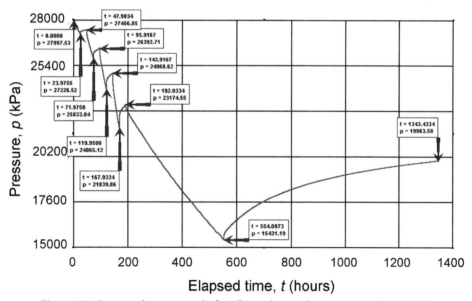

Figure 5.9 Pressure history trend of Well X-6 during short-term production test.

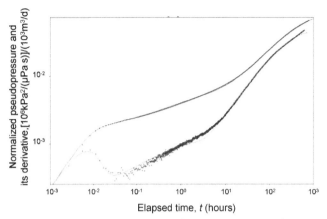

Figure 5.10 Type curve match plot of pressure buildup of Well X-6.

Semilog match verification shows that the match result seems favorable (Figure 5.11).

The whole process of well test interpretation finishes hereto in common practices in most of the well test analysis. However, it was discovered by further pressure history match verification that the model obtained from aforementioned analysis results was inadequate, and especially the understanding about reservoir boundaries was incorrect. A related pressure history match verification graph can be seen in Figure 5.12.

It can be seen from Figure 5.12 that the calculated theoretical pressure of the model declines much slower than the actual measured one. It indicates that the real conditions of the reservoir are far worse than those of the theoretical model.

After further modification of the theoretical model, the improved analysis results are obtained and are as follow.

• The reservoir is a triple porosity composite formation, the most inner zone of which is best and the most outer zone the worst

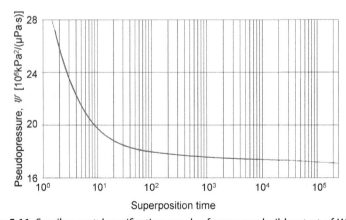

Figure 5.11 Semilog match verification graph of press ure buildup test of Well X-6.

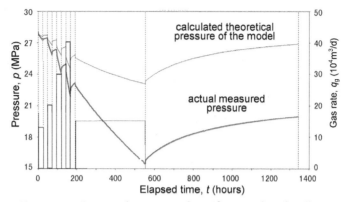

Figure 5.12 Pressure history match verification plot of Well X-6.

- The permeability of the most inner zone, middle zone and the most outer zone are $k_1 = 7$ mD, $k_2 = 0.15$ mD, and $k_3 = 0.03$ mD, respectively
- Radius of the interface of the most inner and middle zones is $r_{12} = 71$ meters and that of the middle and most outer zones is $r_{23} = 240$ meters
- Flow coefficient ratios are $(kh/\mu)_2/(kh/\mu)_1 = 0.02$, $(kh/\mu)_3/(kh/\mu)_1 = 0.004$
- Storativity ratios are $(\phi h\, C_t)_2/(\phi h\, C_t)_1 = 0.12$, $(\phi h\, C_t)_3/(\phi h\, C_t)_1 = 0.06$
- Skin factor is $S = -5.9$, non-Darcy flow coefficient is $D = 0.028(10^4\,\text{m}^3/\text{day})^{-1}$, and wellbore storage coefficient is $C = 2.2$ m^3/MPa

The improved model is conspicuously characterized by a much lower permeability of outer zones and much less gas supply in the reservoir than the former model, and it truly represents the characteristics of the reservoir.

Pressure history match verification of the improved model is shown in Figure 5.13.

It can be seen from Figure 5.13 that the trends of calculated theoretical pressure of the improved dynamic model and the actual measured pressure have been very similar.

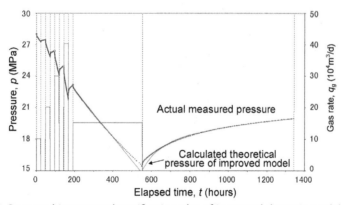

Figure 5.13 Pressure history match verification plot of improved dynamic model of Well X-6.

Thanks to pressure history match verification, this more accurate dynamic model was thus established.

However, the following special attention must be paid to:

1. To conduct pressure history match verification, reliable and long-term pressure history information must be acquired on site. Such pressure information can be obtained, for example, with a continuous-recording memory electrical pressure gauge laid at the bottom hole of the well.
2. The pressure history should contain not only buildup pressure histories of shut-in periods, but also, and more importantly, drawdown pressure histories of all flowing periods.
3. Enough pressure drawdown history can more truly reflect the characteristics of the formation and its boundaries and plays a key role in verifying the correctness of the reservoir dynamic model.

Pressure derivative analysis is sometimes very difficult because pressure buildup data are not accurate enough or the reservoir is extremely complex. In this case, a "principle model" analysis method can be considered. That is, make the calculated pressure history of the model basically consistent to actual measured pressure history by adjusting model parameters. Such a dynamic model established this way should be able to reflect the real formation conditions more accurately, despite no very satisfactory type curve match results obtained.

1.3 Study Characteristics of Reservoir Dynamic Model Based on Characteristics of Transient Well Test Curves

1.3.1 Different Portions of Transient Pressure Curve Reflect Characteristics of Different Zones of the Reservoir

As mentioned in Chapter 2, a radius of influence of pressure exists when a gas well (or an oil well) is producing. In other words, a pressure-drop funnel forms in the reservoir after opening a well; this pressure-drop funnel spreads outward continuously as the production time goes on (refer to Figure 2.33), and the radius of influence is calculated by Equation (2.8).

The formula shows that the radius of influence r_i expands continuously as the production time t goes on and is direct proportional to the square root of t. In addition, r_i is also directly proportional to the square root of permeability k, while inversely proportional to the square root of $(\phi \mu C_t)$.

If any one of the aforementioned parameters is different for a gas well or an oil well, not only will the pressure vary differently around the well, but also the depth of the pressure-drop funnels; for gas wells, pseudo-pressure instead of pressure is used in analysis. This means that once the process of bottom hole pressure drawdown is measured accurately, reservoir parameters such as formation permeability can be calculated from it. This idea is explained by a schematic plot (Figure 5.14).

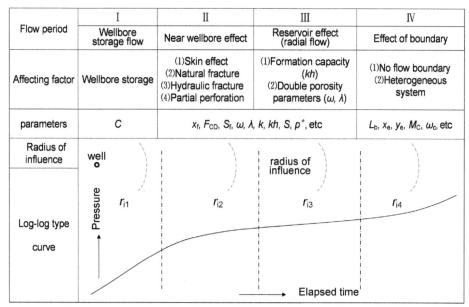

Flow period	I	II	III	IV
	Wellbore storage flow	Near wellbore effect	Reservoir effect (radial flow)	Effect of boundary
Affecting factor	Wellbore storage	(1)Skin effect (2)Natural fracture (3)Hydraulic fracture (4)Partial perforation	(1)Formation capacity (kh) (2)Double porosity parameters (ω, λ)	(1)No flow boundary (2)Heterogeneous system
parameters	C	x_f, F_{CD}, S_f, ω, λ, k, kh, S, p^*, etc		L_b, x_e, y_e, M_C, ω_c, etc

Figure 5.14 Schematic view of transient flow periods.

Early Wellbore Storage Flow Period I

The wellbore storage portion is the first one of pressure drawdown or buildup curves. In cases of pressure buildup, wellbore storage is also called "afterflow." The cause and measurement of wellbore storage were introduced in Chapter 2. The characteristic of wellbore storage stage is that the pressure curve on the log–log plot is a unit-slope straight line.

$$\frac{\mathrm{d}\lg\Delta p}{\mathrm{d}\lg\Delta t} = 1$$

The wellbore storage coefficient (C) can be calculated from this portion of the transient pressure curve, that is, pressure drawdown curve or pressure buildup curve.

Near Wellbore Formation Influence Period II

As time goes on, the area, the factors of which influence pressure-drop funnel extension, extends outward. As discussed earlier, the meaning of extension contains the following two aspects:

- The radius of influence r_i expands, that is, the pressure-drop funnel becomes larger and larger
- The pressure-drop funnel becomes deeper and deeper, namely the flowing pressure (p_{wf}) declines with time gradually during well producing

The pressure drawdown curve given in Figure 5.14 is just the flowing pressure measured at the bottom hole of the well.

Factors influencing this portion of the curve are as follow:

1. When a large fracture is created in the formation by artificial fracturing, the fracture effects will reach the bottom hole earlier than formation effects so that linear flow is formed around the fracture and delays declining of bottom-hole flowing pressure.

2. When there is a primary fissure system in formation, these fissures delay the pressure decline process in forms of its high conductivity, its group and/or series distribution, or its fissure system in dual porous media.

3. If the reservoir is partial perforated or seriously damaged during and/or completion, gas flow resistance near the wellbore increases and the flowing pressure will decline more quickly.

It is known that this regime is affected by many factors near the wellbore such as skin, partial perforation, fracture, and fissures, but unfortunately, depending only on the curve characteristics of this regime, one can only discover the existence of these factors but cannot find out the seriousness of them or the value of these factors. What causes this problem is simple: for instance, to estimate skin factor (S) caused by damage, parameters such as formation permeability before formation damage must be determined first, but formation permeability can only be determined by data in a later period of the transient well test curve (i.e., radial flow period or reservoir influencing period III in Figure 5.14 when the skin effect has finished).

Reservoir Influencing Period III (Radial Flow Regime)

This portion is the most important part in a transient well test curve because

1. This portion has the most obvious characteristics, shown as a straight line with slope m on the semilog plot and a horizontal straight line on the pressure derivative plot

2. The most important parameters of the formation and the well are all calculated from it:
   - Formation permeability (k) and flow coefficient (kh/μ) are calculated from semilog straight line slope (m) or log–log type curve match
   - Skin factor (S) indicating completion quality is also calculated from semilog straight line slope (m) or log–log type curve match
   - Half-length of fracture (x_f), fracture conductivity (F_{CD}), and fracture skin factor (S_f) are calculated from it
   - Storativity ratio (ω) and interporosity flow coefficient (λ) of double porosity reservoir are calculated from it
   - Reservoir pressure ($p^\star$) is calculated from it, etc.

If this portion has been acquired during the transient well test, the test is thought basically successful; otherwise the test is judged unsuccessful generally.

In the 1970s–1980s, many researchers attempted to gain the needed parameters by a well test terminated at period II, but those parameters were calculated commonly from analysis of period III. However, it has been recognized that those efforts produce very little effect.

Nowadays testing tools have been improved a lot. If downhole shut-in tools are used in well testing, the wellbore storage effect will be reduced significantly, with the wellbore storage flow period shortened by 2—3 orders of magnitude consequently. It is very favorable in cases of hardly identifying the radial flow period due to too great a wellbore storage effect, but because this technique brings certain difficulties in testing the string structure, it is sometimes considered not feasible for gas wells whose pipe string cannot or is very difficult to move.

Boundary Influence Period IV

In this period, curve shape variation reflects whether the flow resistance of fluid increases or decreases in formation. This period is very important, but very difficult to acquire and analyze.

1. This portion of the curve is extremely important in establishing the dynamic model of a gas well and gas field.

   As mentioned at the beginning of this chapter, a dynamic method can validate parameters such as effective permeability of the reservoir, but the most important ones are as follow.

   - It can identify whether boundaries such as a fault identified in the static reservoir model separate the reservoir or not
   - It can identify the position, distance to the well, and influence on the gas well production of lithologic boundaries that have not been identified yet in the static reservoir model
   - It can identify the gas storage pattern in reservoirs with group- and/or series-distributed fractures in buried hill limestone
   - It can identify the gas storage characteristics in some other special lithologic formations

2. To acquire this portion of the curve, the test must last a sufficient time.

   Especially for low permeable gas zones, such a long-term test is often abandoned due to its difficult support.

3. The ambiguity effect is more serious and brings more difficulties in data analysis.

Pressure history match verification can usually compensate for these deficiencies.

1.3.2 Pressure Derivative Curve is the Main Basis in Identifying Reservoir Characteristics

Since developed by Bourdet and colleagues [32, 33] in the early 1980s, the pressure derivative analysis method has given a totally new meaning to the "modern well test." Pressure curves reflect the bottom hole pressure variation with time, and pressure derivative curves reflect the pressure variation rate with time. Such a derivative is defined as "differential of pressure with respect to logarithm of time" and its expression was introduced in Equation (2.4) in Chapter 2.

Table 5.1 Relationship of Flow Type and Character of Pressure Derivative Type Curve

Flow Type	Typical Character of Derivative Curve
Wellbore storage flow in early period	Early unit-slope straight line
Radial flow in homogeneous reservoir	Horizontal straight line
Radial flow in fissures of double porosity system	Early horizontal straight line
Radial flow in total double porosity system	Late horizontal straight line
Local radial flow and radial flow in whole formation in partial perforated formation	Two horizontal straight lines; later one locates lower than earlier one
Vertical radial flow of horizontal well	Early horizontal straight line
Pseudo radial flow in reservoir with vertical or horizontal fracture	Late horizontal straight line
Pseudo-steady flow portion on pressure drawdown curve in closed reservoir	Late unit-slope straight line
Pseudo-steady flow portion on pressure buildup curve in closed reservoir	Late steep down dip line
Linear flow in a well with infinite-conductivity vertical fracture	Early half-unit slope straight line
Late linear flow in channeled reservoir	Late half-unit slope straight line
Bilinear flow in a well with finite-conductivity vertical fractures	Early quarter-unit slope straight line
Spherical flow in partial perforated formation	Straight line of slope -0.5
Flow resistance increases	Derivative curve goes upward
Flow resistance decreases	Derivative curve goes downward
Transitional flow of double porosity reservoir	Derivative curve becomes downward concave

The characteristics of formation, or dynamic reservoir model, have their corresponding special and typical characteristics on transient pressure derivative curves and can be described in detail in Table 5.1.

The aforementioned characteristics are shown in Figure 5.15. Figures inside parentheses in the plot are the slope of that special portion of pressure derivative curve.

1. Wellbore storage flow

   It appears at the beginning of producing or shut in of a well and is indicated by a unit-slope straight line on the derivative curve in the early period.

   Such a special unit-slope straight line appears in most transient gas well tests, but possibly no data of this period are measured if the tested reservoir has particularly high permeability and a downhole shut-in tool is used in well testing.

2. Early linear flow and bilinear flow

   Early linear flow and bilinear flow are indicated by a half-unit slope straight line and a quarter-unit slope straight line, respectively.

   Such special typical curves describe the near-wellbore formation that has experienced massive sand fracturing. Because a long vertical fracture connected with the well has been formed by fracturing treatment, linear flow perpendicular to the fracture

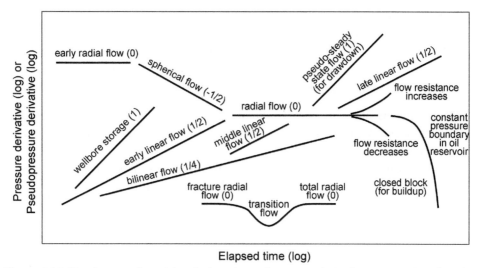

Elapsed time (log)

Figure 5.15 Sketch map of special typical portions of pressure derivative curve on log–log plot.

surface results around the fracture (and bilinear flow in and around the fracture, respectively, in some cases); the state of this flow is shown in Figure 2.20.

3. Middle-time linear flow

The special typical curve of middle-time linear flow is also a half-unit slope straight line, but it appears later than the early one.

Such a special typical curve appears in the period of linear flow perpendicular to the horizontal section in a horizontal well. It is also in the near-wellbore formation influencing period, and its flow pattern can be seen in Figure 2.22b.

4. Late linear flow

Its special typical curve is also a half-unit slope straight line, but it appears much later, usually extends to the end of the test, and is in the late boundary influencing period.

Such a special typical curve is induced by parallel impermeable boundaries. Geologically, they may result from a graben originated from a fault (Figure 2.16), by river channel boundaries made up of fluvial deposits (Figure 2.18), or by an axial-striped fractured zone made up of folding of limestone formation and so on.

5. Radial flow

The special typical curve of radial flow is a horizontal straight line on a pressure derivative curve, that is, a straight line with a slope of 0.

A radial flow regime is the reservoir influencing period. It can be classified into various types: radial flow in homogenous formation, radial flows in fissure system and total system of double porosity reservoir, radial flows in inner zone and outer zone of composite reservoir, radial flows in local and whole formation of partial perforated formation, pseudo radial flow in fractured reservoir, and vertical radial flow and pseudo radial flow in horizontal well.

Flow patterns of radial flow can be seen in Figures 2.7, 2.15, 2.22, 2.24, and 2.25.

The following parameters and information of the radial flow regime and its adjacent flow regimes can be obtained through radial flow analysis:

(1) Parameters of the reservoir influencing period, such as permeability k, flow coefficient kh/μ, formation capacity kh, and fissure permeability of double porosity system k_f

(2) Skin factor reflecting the seriousness of damage in near-wellbore formation S, the end point of linear flow, and half-length of fracture x_f

(3) To identify boundary influencing period and determine boundary property and distance of the well by the variation trend of the derivative curve

Therefore, this portion of radial flow period is the most important characteristic portion in well test curves.

6. Spherical flow

The special typical curve of spherical flow is a straight line with slope -0.5.

Spherical flow occurs when the formation is partially perforated. Its flow pattern is shown in Figure 2.13. Such a flow state is a transitional one. Sometimes its special typical curve is just a transitory downward line dropping from the horizontal straight line reflecting radial flow in a perforated part of the formation to another horizontal line reflecting radial flow in the whole formation.

The special typical curve of spherical flow can be used to not only identify the general flow characteristics, but also infer the degree of partial perforation from the distance between the two horizontal straight lines.

7. Pseudo steady flow

The special typical curve of pseudo steady flow is a unit-slope straight line.

Pseudo steady flow means the flow state of a gas well in a closed block, when the pressure everywhere in this block declines uniformly after flowing for a certain period of time. It is the final flow state of pressure drawdown and will continue to the end of the flow. The flow state can be seen in Figure 2.10.

8. Stable state line of pressure buildup curves

Stable state flow during pressure buildup is the flow state corresponding to pseudo steady flow during producing of the well in a closed gas block but is totally different from that during well producing; the special typical derivative curve of pressure buildup drops down abruptly and sharply.

After shutting in the well in a closed block, pressure in the whole block equilibrates soon and so the pressure derivative (i.e., pressure variation rate) trends to zero quickly. Correspondingly, the curve shows a sharp decline in the log–log plot.

For an oil well, if a broad water zone or injection well around it forms a boundary maintaining approximately constant pressure, the pressure derivative curve also drops. This is called "constant pressure boundary" in well test analysis. But unlike oil wells, a gas well has no such flow areas where the flow resistance is less than that in the gas region so there is no "constant pressure boundary" there. An important point that needs to be emphasized is that the pressure derivative curve drop can only occur during the pressure equilibration process of the closed block.

9. Transition flow in double porosity reservoir

This is a special flow pattern, a representation of transition from radial flow in the fissure system to radial flow in the whole reservoir, that is, in both fissure and matrix systems. The pressure derivative curve shows a down concave curve, two horizontal straight lines of the same height lie before and after this transition flow concave curve separately, and these two horizontal straight lines indicate those two radial flows in different homogenous media, that is, the fissure system and the whole reservoir.

In general, the appearance of such a special and typical concave curve means the existence of characteristics of a double porosity reservoir. However, it sometimes may be covered by the wellbore storage effect.

10. Flow resistance increases or decreases

Various different distributions of formation heterogeneity are shown in Figures 2.26–2.30 in Chapter 2. These heterogeneity distributions will cause gas flow resistance increasing [e.g., when permeability and/or thickness of the formation becomes lower or viscosity of fluid becomes higher, i.e., the flow coefficient (kh/μ) becomes lower] or decreasing [e.g., when permeability and/or thickness of the formation becomes higher or viscosity of fluid becomes lower, i.e., the flow coefficient (kh/μ) becomes higher]. If gas flow resistance increases, the pressure derivative curve will tilt upward from the radial flow horizontal straight line; however, the pressure derivative curve will tilt downward.

The causes making flow coefficient (kh/μ) higher or lower are various, and so such curves should usually be analyzed by combining geological research.

1.3.3 "Graphics Analytical Method" used to Identify Reservoir Dynamic Model
Principles of Graphics Analysis

As described earlier, dynamic models of different reservoirs have different flow behaviors and constitutions, and different flow behaviors are shown by different special typical pressure derivative curves. Therefore, accurate understanding of reservoir dynamic models can be achieved on the basis of interpretation of those special typical pressure derivative curves [48].

The core of this chapter is discussion focused on this subject. All discussions in this chapter are a crucial content of this book. This subject is discussed further in detail combined with application of well test interpretation software in Section 5.4.

What should be made certain here is that if a complete pressure buildup curve is acquired on site and a log–log plot of pressure and pressure derivative is completed accordingly, what important information can be obtained instantly or what semiquantitative judgments can be concluded instantly from the characteristics of the curve? In other words, is it possible to get a quick-look interpretation of the pressure derivative curve? The answer to this question is YES.

The following simple examples illustrate this subject.

1. A log–log plot of pressure and its derivative vs time is obtained for a well after the pressure buildup test, as shown in Figure 5.16a. For the same well, the pressure buildup curve turns into the pattern of Figure 5.16b after stimulation treatment. Graphic characteristics clearly indicate that the treatment is very effective, judging from the decrease of distance between the pressure curve and derivative curve A, namely $A_2 < A_1$ (Figure 5.16).

   Another question is whether the value of skin factor (S) can be judged by value A? The answer is also YES.

2. A gas well is stimulated by massive sand fracturing treatment, and the log–log plots of pressure buildup before and after fracturing treatment are as shown in Figures 5.17a and 5.17b, respectively. It can be clearly seen that a long fracture has been formed by fracturing treatment, indicated by that in Figure 5.17a, the formation displays common homogeneous formation characteristics, but Figure 5.17b shows a very long

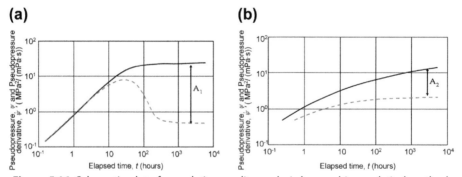

Figure 5.16 Schematic plot of completion quality analysis by graphics analytical method.

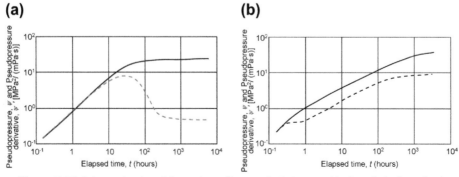

Figure 5.17 Schematic plot of fracturing effect analysis by graphical analytical method.

linear flow regime. As is known, the longer the linear flow regime, the longer the propped fracture (Figure 2.21).

Certainly, after inputting the aforementioned test data into any well test interpretation software, the answers can be obtained from correct calculations. However, some questions still remain:

a. Can some certain important information or semiquantitative answers to the well and the reservoir be obtained immediately from the shape of the curve itself?

b. Even though certain quantitative results have been obtained from interpretation by software, can it be assured that those results are surely correct?

c. If a production management personnel or a supervisor onsite has no well test interpretation software available or no applicable software for repetitive inspections, how can he or she identify the reliability of interpretation results rapidly and immediately make decisions about the next measures to be taken onsite?

These problems can be resolved by some simple graphical analysis methods.

Some Simple Graphical Analysis Methods

1. Graphic characteristics of homogenous reservoir and skin factor (S) estimation

A special typical log–log curve of a well in a homogeneous reservoir is shown in Figure 5.18.

In Figure 5.18, the upper solid curve is the pressure difference (Δp)(pseudo pressure difference $\Delta \psi$ for gas wells) curve and the lower dash curve is the pressure derivative ($\Delta p'$) (pseudo pressure derivative $\Delta \psi'$ for gas wells). Combining the pressure and pressure derivative curves is just like a fork with two teeth. It can be divided into three sections.

- Section I is "the handle" of "the fork," a unit-slope straight line (i.e., the slope angle of the straight line is 45°, if the lengths of one log cycle of both coordinates on the log–log

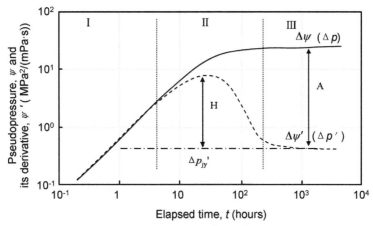

Figure 5.18 Special typical log–log curve of a well in homogeneous reservoir.

plot are the same). Here pressure and its derivative curves merge together, indicating wellbore storage flow period, or afterflow regime in cases of pressure buildup.

- Section II is the transition period; the pressure derivative curve reaches its maximum gradually and then goes downward, and the maximum value is expressed by H (mm; see Figure 5.18). The H value is determined by the value of parameter group $(C_D e^{2S})$, which is called the "shape parameter of curve."
- Section III is the radial flow regime.

The pressure derivative curve in this regime is a horizontal straight line. If the curves are plotted in a dimensionless coordinate system, that is, in p_D, $p_D{}'\sim t_D/C_D$ coordinate system, the ordinate of a radial flow horizontal straight line is 0.5.

It can be seen from Figure 5.18 that, in addition to H, there is another characteristic value, A, indicating the distance between the pressure curve and the derivative horizontal straight line.

H means the difference between the maximum value of the pressure derivative and the value of the radial flow horizontal straight line, and A means the difference between the values of pressure and its derivative in the radial flow period (mm)(see Figure 5.18).

$$H = \log\Delta p' - \log\Delta p'_{j\gamma} \ (\Delta p'_{j\gamma} \text{ is the ordinate of radial flow horizontal straight line of}$$
$$\text{pressure derivative curve}) \tag{5.1}$$

$$A = \log\Delta p - \log\Delta p' \ (\text{Both } \Delta p \text{ and } \Delta p' \text{ are measured in radial flow regime}) \tag{5.2}$$

It is unnecessary to calculate accurately from tested data for obtaining the values of H or A. What is needed is to just measure the approximate lengths of them with a ruler on a printed log–log plot and then estimate their dimensionless values:

$$H_D = H/L_C \tag{5.3}$$

$$A_D = A/L_C \tag{5.4}$$

where L_C is the length of one log cycle in ordinate (mm)
and skin factor (S) can be estimated as

$$S = 10^{\left(\sqrt{8.65H_D+6.14}-2.75\right)} - 0.5\ln C_D \tag{5.5}$$

or

$$S = 10^{\left(\sqrt{5.53A_D-3.37}-1.12\right)} - 0.5\ln C_D \tag{5.6}$$

Something needs to be explained here:

a. Equation (5.5) or (5.6) is just an approximate expression from statistical analysis on the basis of type curve characteristics, and H and A are approximations measured by ruler, so the calculated S value may not be very accurate;

b. C_D in Equation (5.5) or (5.6) is the dimensionless wellbore storage coefficient, and its definition is

$$C_D = 0.15916 \frac{C}{\phi h C_t r_w^2} \tag{5.7}$$

where

C = wellbore storage coefficient, m^3/MPa

ϕ = porosity, fraction

h = effective formation thickness, m

C_t = total formation compressibility, MPa^{-1}

r_w = effective well radius, m

and the wellbore storage coefficient (C) can be estimated by

- Equation (2.5) in Chapter 2 or
- Table 2.1 in Chapter 2 or
- Equation (2.5a)

In Equation (2.5a), $\Delta p/\Delta t$ can be calculated directly by a Cartesian plot drawn from data of the wellbore storage flow (afterflow in cases of buildup) period in early time.

It needs to be noted that the error of estimated S value due to the inaccuracy of C_D is very limited because C_D is under logarithm operator in Equations (5.5) and (5.6).

2. Graphics characteristics of double porosity reservoir and estimation of ω value

Theoretically, where there are two different storage spaces of fissure system and matrix rocks in the reservoir simultaneously, there will be a special flow state of the double porosity reservoir, and the log—log plot of the pressure and its derivative is shown in Figure 5.19. After a gas well is opened and the wellbore storage flow period I is over, the fluids will flow from the fissures to the wellbore, and

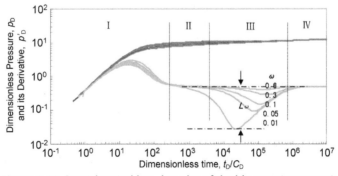

Figure 5.19 Special typical log—log plot of double porosity reservoir.

this is flow period II; when the pressure inside the fissure system decreases, transition flow from the matrix system to the fissure system occurs, and this is flow period III; and as the pressures in the fissure and matrix systems are equilibrating, the flow state turns into radial flow of the total system and this is flow period IV.

It can be seen from Figure 5.19 that double porosity reservoirs have the most prominent characteristic in its special typical curve in the transition flow period—this special typical curve is determined by a special parameter, storativity ratio ω, and its definition is

$$\omega = \frac{(\phi h C_t)_f}{(\phi h C_t)_f + (\phi h C_t)_m} \tag{5.8}$$

This storativity ratio ω indicates the ratio of fluids stored in the fissure system to those stored in the total reservoir (i.e., in both fissure and matrix systems), and it is the most important parameter of a double porosity reservoir. More detailed discussions about it are given in Section 5.4.

It should be pointed out that the sunken depth of the transition regime increases as the ω value decreases (Figure 5.19); there seems to be a simple numerical relationship between them, and in fact the expression of this relationship is

$$\omega = 10^{-2L_D} \tag{5.9}$$

where
$L_D = L_\omega / L_C$
L_ω = absolute sunken depth of transitional regime, mm
L_C = length of one log cycle in pressure coordinate, mm
For instance, in the log–log plot of a real example, $L_\omega = 12$ mm is obtained by measuring and the length of one log cycle in the pressure coordinate is $L_C = 15.0$ mm, then $L_D = 12/15 = 0.8$ and $\omega = 10^{-2 \times 0.8} = 10^{-1.6} = 0.025$.

3. Graphics characteristics of composite reservoir and estimation of M_C value

A composite reservoir forms when the flow coefficient value of the zone near the well, $(kh/\mu)_{near}$, is quite different from that of the zone far from the well, $(kh/\mu)_{far}$. The most important parameter indicating the existence of a composite reservoir is the ratio of the flow coefficient in the near-well zone to that in the far-well zone, M_C:

$$M_C = \left(\frac{kh}{\mu}\right)_{near} \Big/ \left(\frac{kh}{\mu}\right)_{far} \tag{5.10}$$

The pressure and its derivative log–log plot of a round composite reservoir with different M_C values are shown in Figure 5.20. If the flow resistance in the outer zone is greater than that in the inner zone (i.e., $M_C > 1$), pressure derivative curve will tilt

(a)   **(b)**

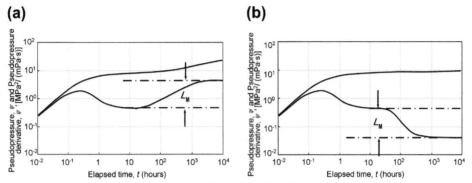

Figure 5.20 Special typical log–log plot of composite reservoir.

up and form another horizontal straight line reflecting radial flow in the outer zone. Contrarily, if the flow resistance in the outer zone is less than that in the inner zone (i.e., $M_C < 1$), the pressure derivative curve will tilt down and form another horizontal straight line reflecting radial flow in the outer zone.

It can be seen from the graphics characteristics, based on the height difference of the two horizontal straight lines of outer and inner zones, the variation trend from inner to outer zones of the reservoir can be identified, and the flow coefficient ratio of the two zones can also be estimated, the expression is

$$M_C = 10^{L_{MD}} \tag{5.11}$$

where
$L_{MD} = L_M / L_C$
L_C = the length of one log cycle on ordinate (pressure coordinate), mm
L_M = height difference of two horizontal straight lines reflecting radial flows in outer and inner zones, respectively, on the derivative curve, mm
L_M is positive if pressure derivative curve tilts up and is negative if tilts down.
For instance, if permeability of the inner zone is $k_{inner} = 10$ mD, and reservoir thickness and fluid physical properties are the same in inner and outer zones, it is seen from the log–log plot that the later pressure derivative horizontal straight line is 5 mm higher than the earlier horizontal straight line, and if the length of one log cycle is $L_C = 15$ mm, then
$L_{MD} = L_M / L_C = 5\text{mm}/15\text{mm} = 0.333$
so
$M_C = 10^{0.333} = 2.15$,
and the permeability of outer zone is obtained as
$k_{outer} = 10\text{mD}/2.15 = 4.65\text{mD}.$

It can be seen from the aforementioned analysis that the analysis method intro-duced earlier, that is, from graphics characteristics analysis to flow characteristics analysis to the reservoir dynamic model step by step, is very effective and powerful for interpreting many problems during gas field development. More detailed descrip-tions for various types of formation and field examples are given one by one in the following parts of this chapter.

2 PRESSURE CARTESIAN PLOT—PRESSURE HISTORY PLOT

2.1 Content and Drawing of Gas Well Pressure History Plot

2.1.1 Preprocessing and Data Examinination of Gas Well Pressure History Records

Many authors engaged in well test theory and analysis method research usually do not care how pressure data are acquired onsite and do not pay any attention to the accuracy of data, as they think that data acquisition is a routine operation, and acquired data should certainly be correct. Unfortunately, sometimes the facts are reversed.

It is quite true that if pressure data are acquired onsite precisely and timely, the majority of the well test research task has been fulfilled.

The correct operation to acquire pressure data of a well is to run pressure gauges down to the bottom hole and place them just at the middepth of gas pay zones and close to the sand face so as to measure and record pressure variation continuously during the well flow and shut-in periods. A pressure history plot, that is, the plot of pressure vs time in Cartesian coordinates, can be drawn, and on the same plot the gas flow rate and its changes, and measures taken on the ground during the test should also be marked or shown in chronological order.

The recorded pressure history must be standardized by "preprocessing" because it may be influenced by some factors as follow.

- Usually, pressure gauges cannot be set at the middepth of tested gas pay zones so acquired pressure variations are not the real ones; particularly, the measured shut-in static pressure is very often lower than the real static pressure so data correction is very necessary.
- Because the measurement point is not the middepth of gas pay zones, pressure measured at the measured depth has to be converted into that at the middepth of gas pay zones according to the pressure gradient in the wellbore. Some new problems may arise in pressure conversion: flowing pressure gradient is different from shut-in static pressure gradient; flowing pressure gradient itself will vary along with the gas flow rate; generally, the shut-in static pressure gradient should be constant, but the measured pressure data of a tested well showed that the pressure gradient was still varying continuously due to phase transition effects in the wellbore and other reasons even after the well had been shut in. If such a conversion method commonly used is adopted, many uncertainties will be brought out in practice, and a proper approach to resolve this problem needs to be found out by further studies.

- If pressure history monitoring is not fulfilled with one trip of the pressure gauge, there are some connection problems such as time connection, pressure data connection, and, if the measuring depths in two or more trips are different, pressure conversion connection etc., and the original records should be checked to achieve a reasonable connection.
- In order to improve interpretation, the pressure sample rate is usually higher during the periods before and after the scheduled flow rate changes than others. Usually, very many, sometimes several hundred thousand even more than one million, pressure data are acquired. On the precondition that important and effective records must be retained, overfull data should be deleted by the filtration program to ensure smooth operation of well test interpretation software.

Due to the aforementioned reasons, original pressure data brought back from the field cannot be used directly for analysis or interpretation by well test interpretation software, and it is necessary to conduct preprocessing for making them reflect the pressure variation of the reservoir truly and accurately.

A "Data examine" should be conducted for abnormal pressure history records. The Data examine means evaluation and analysis of original test data, usually including:

- The start and end points and the connection parts of measured pressure data, as well as abrupt rise or drop of the pressure, always indicate the trip [i.e., running in hole (RIH) and pulling out of hole (POOH)] of pressure gauges, the corresponding time should be found out from the onsite well test operation report, and data in the portions of RIH and POOH should be deleted. Please refer to Figure 5.21.
- Identify any intervals when abnormal measured pressure changes appear during testing, and find out the causes of these abnormalities.

For instance (Figure 5.22a), during a pressure buildup test, the pressure abruptly went down; it was believed that this resulted due to gas leakage at the wellhead and so

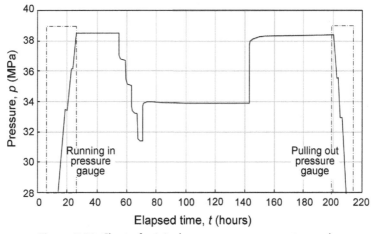

Figure 5.21 Chart of original pressure measurement records.

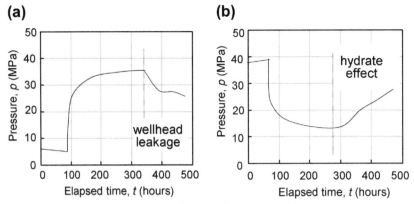

Figure 5.22 Test operation problems influence normal pressure variation.

eliminated the possibility of influence of changes of reservoir properties. Another example was when the pressure went up abruptly during the pressure drawdown test (Figure 5.22b); by investigation and verification of the well test operation report, the cause of which was some hydrate accumulated at the wellhead and made the gas production rate decline.

- Some water accumulates at the bottom hole of the gas well, the pressure gauge is not run at the middepth of gas pay zones, and is located above the liquid level. The measured pressure may become abnormal when the well is just opened or during the pressure buildup period: a downward spike appears when the well is just opened (Figure 5.23a); a downward rather than an upward trend appears in the later stage of the pressure buildup period (Figure 5.23b). The causes can be discovered by verifying water cut records or even sometimes the static pressure gradient after shutting in the well. Please refer to Figure 5.23.

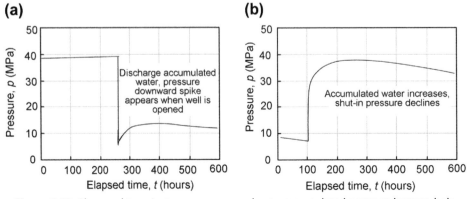

Figure 5.23 Abnormal transient pressure curve due to accumulated water at bottom hole.

- During the injection-pressure drawdown test process in a coalbed gas well, an abrupt rise of a pressure derivative may occur sometimes; it is found based on a thorough investigation that always happens when the wellhead pressure (p_H) drops down to zero; and it is caused by variable wellbore storage effects: the variation process from compression of the fluids in the closed coalbed to building up of the liquid level. Please refer to Figures 5.24a and 5.24b; a detailed discussion of this process is given in the related parts of Chapter 7.
- Some abnormal changes of pressure originate from the operations or measures, such as flushing circulation, packer failure, or abnormal opening of circulation valve.
- Some more very common pressure features, although seen quite often, cannot be understood if the process conditions during the test are not known. Actually, some well test analysts deal with these special kind of problems just with routine methods, resulting in more troubles.

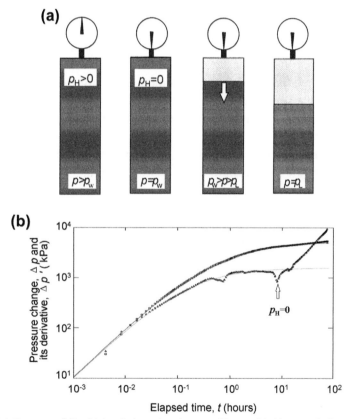

Figure 5.24 (a) Process of liquid level dropping in a wellbore. (b) Abnormal change of pressure derivative curve in a well.

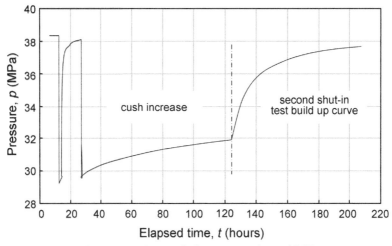

Figure 5.25 Bottom-hole pressure chart of DST.

Figure 5.25 gives a common dual-flow—dual shut-in test curve in drill stem testing (DST). For the flow periods, especially the second flow period, the flowing pressure does not decline but rises gradually, reflecting a continuous discharge of completion fluid from formation to the wellbore during this period, which causes the liquid cushion level in the test string to rise continuously. It is the pressure of the fluid column that pressure gauges measure and record.

If analysts have no idea of the process and operation of the well test, they are not able to learn and identify the truth behind the pressure variation.

2.1.2 Pressure History Plot of Gas Well

A pressure history plot applicable to well test interpretation is finally obtained on the basis of preprocessing and evaluation analysis of the pressure history records mentioned earlier. A typical pressure history plot of a normal gas well is as shown in Figure 5.26.

It is seen from Figure 5.26 that the pressure history of gas well testing can be divided into four periods.

1. Period I: The primary test of gas flow rate

    As for the well completion and gas well test technique developed up until now, the perforation and test are often joint operated to open up a gas well. After testing tools are run down the hole and settled at a planned position, perforation under negative pressure conditions is conducted. For gas zones with a certain deliverability, natural gas flows into the wellhole immediately, and the pressure measured at that moment is formation pressure.

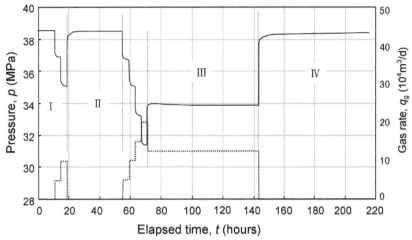

Figure 5.26 Pressure history plot of gas well test.

In order to learn primarily about gas deliverability of the tested well, an initial gas flow measurement is performed with one or two chokes in order to better arrange the coming deliverability test.

2. Period II: Deliverability test

Gas well deliverability is the most important index in gas field development. Chapter 3 introduced various deliverability test methods of gas wells. A deliverability test is very necessary for a new drilled gas well. In Figure 5.26, a back–pressure test was arranged in the second period. Four flow rates, that is, four chokes, from small to large, are selected for the test, and flowing pressure variation during flowing is monitored simultaneously. With the four stable gas flow rates (q_g) and four corresponding stable flowing pressures (p_{wf}), absolute open flow potential (AOFP) of the gas well can be obtained from deliverability analysis or IPR curve.

For some simple homogeneous reservoirs, one-quarter or one-fifth of the AOFP is often taken as the design index of a reasonable flow rate in the development plan of a gas field. The method is proven lack of necessary theoretical basis for various complicated gas zones. A more thorough study shows that both pressure history and pressure prediction of a gas well are the most important bases for production arrangement of the gas well.

3. Period III: Short-term production test

As discussed at the beginning of this chapter and emphasized repetitively in this book, the purpose of a well test is to find out and identify the dynamic model of a gas well eventually and then to fulfill a further dynamic prediction based on this dynamic model. A pressure drawdown can best reflect the dynamic features of a reservoir and so is crucial to confirming the dynamic model. Therefore, the flowing pressure variation in the pressure drawdown period must be included in monitoring the pressure history of a gas well. At present, some people do not attach much importance to it and think that every problem can be resolved if only a pressure buildup curve is acquired; unfortunately and out of their expectation, many pressure buildup data have been obtained but the reservoir model cannot be identified reliably: either related interpretation results are far from static data and cannot be accepted by the production planning department or interpretation results cannot be conformed due to the ambiguity problem, whereas the pressure history, especially pressure drawdown history, which is the only thing able to verify the reliability of the model, has not been acquired, resulting falling short of success for lack of a final effort.

4. Period IV: Pressure buildup

Both pressure drawdown and pressure buildup are originally basic test periods in a transient well test. Actually, all type curves for well test interpretation are designed on the basis of the pressure drawdown test. But gas flow rate is often fluctuating during producing, bringing great difficulties to the data analysis, especially when using the pressure derivative type curve match method. Eventually people gradually paid more attention to the analysis of pressure buildup curves. As the gas flow rate is 0 during the period of pressure buildup, the interference of the flow rate fluctuating to pressure data can be minimized, and this period becomes the most important one in data analysis. The coming analysis of the dynamic model is often conducted on the basis of shut-in pressure buildup curves.

Tests of different periods of gas well pressure history can be implemented repeatedly depending on the practical conditions onsite. In fact, a well may be opened and shut in several times; all periods in its pressure history are of great importance and should be recorded if possible and would play an active role in reservoir dynamic analysis.

2.2 Information About Formation and Well Shown in Pressure History Plot

2.2.1 Pressure History Plot During DST Of Natural Flow Gas Well

A pressure history plot obtained from DST is shown in Figure 5.27. It is a gas well producing gas by natural flow after testing. It is seen from Figure 5.27 that the liquid cushion, which was put into the wellbore at the very beginning of testing, is promptly discharged at the beginning of the second flow. After that, flowing pressure tends to be stable very soon. Then the well is shut in for the pressure buildup test.

Various formation conditions are displayed in the shut-in period.

1. Normal gas well

If the gas well is located in a reservoir with relatively high or medium permeability, the shut-in pressure can build up quickly and reach its formation pressure. Further judgments cannot be made directly based on the pressure history Cartesian plot, but if data of this period are plotted in a log—log plot of pressure and its derivative and

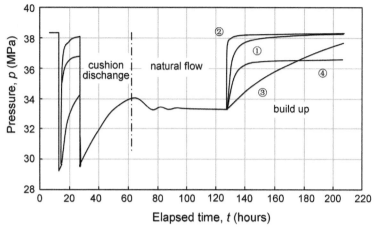

Figure 5.27 Pressure history plot of natural flow gas well test.

analyzed, it would be found that the well has a relatively small skin factor and a rather high permeability in the order of mD or even higher.

2. Gas well with high permeability and serious damage

This curve is typically characterized by a very rapid pressure buildup after shutting in and approximate shaping a right angle, indicating that the producing pressure differential is mainly expended around the well hole.

3. Gas well with low permeability

Pressure curve rises slowly.

4. Gas well with very limited energy

Judged by the curve shape, a very short duration of production has caused an obvious decline of formation pressure, and so the formation pressure drops after each production for a very short time. If a short-term production test in this kind of gas well can be conducted, the reservoir pressure loss can be observed more clearly and dynamic reserves controlled by the tested well can be determined accordingly.

2.2.2 Pressure History Plot During DST of Low Production Rate Gas Well

If a well can only produce several hundred cubic meters of gas or even less than that during testing, it cannot discharge a liquid cushion, which was put into the wellbore at the beginning of the test; the pressure history plot in this case is shown in Figure 5.28.

As shown in Figure 5.28, because the deliverability of the tested well is low, the liquid cushion has not discharged yet, although some gas can go through the cushion and flow out of the wellhead, the pressure gauge mainly records the pressure of the liquid column and reflects the rising liquid level. Therefore, the pressure rising recorded slowly during the second flow cannot truly reflect the flowing pressure of the gas well, while the later shut-in pressure buildup records somewhat the status of gas zones and gas well.

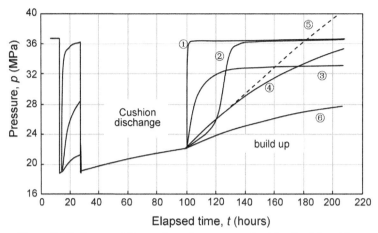

Figure 5.28 Pressure history plot of gas test well with a liquid cushion.

1. Gas well locating in a seriously damaged reservoir with medium permeability

   The feature of the pressure curve is the rapid rise up to formation pressure after shutting in with shaping almost a right angle. The location where the pressure differential is consumed most is around the sand face. The fissured limestone of the HTH gas field in the Tarim Basin behaves just like this in the early stage of well completion. After stimulated by acidizing treatment, the wells there can produce with a higher rate.

2. Gas well locating in a damaged reservoir with low permeability

   The producing pressure differential is very large in this case. It is different from the case of gas wells locating in the reservoir with high permeability, as the pressure buildup cannot quickly reach the formation pressure after shutting in, and the producing pressure differential is mainly consumed around the wellbore. Its pressure buildup process is retarded due to low permeability. In the South YL area of the Tarim Basin, formation permeability is very low, pressure coefficient is about 2, and producing pressure differential in Well Yinan 2 is as big as 80 MPa. In order to make the downhole shut-in valve work normally, the well was shut in at the wellhead first to delay downhole shut in—an alike curve is obtained.

3. Gas well locating in a reservoir with very limited reservoir energy

   Once a certain amount of gas is produced from such a well, its formation pressure measured after shutting in would be much lower than that measured before producing.

4. Gas well locating in a reservoir with low permeability

   After shutting in such a well, the pressure rises gradually and continuously and shows afterflow behavior basically.

5. Gas well locating in a reservoir with high pressure and low permeability

Similar to ④, but sometimes the shut-in pressure increases to a value exceeding the initially measured static liquid column pressure, and the pressure buildup curve is basically like afterflow behavior.

6. Dry sand

Dry sand is usually defined as a gas zone that can only produce very little gas after the well locating in which is opened and a pressure depletion signal appears after shutting in the well.

3 PRESSURE SEMILOG PLOT

3.1 Several Semilog Plots used for Parameter Calculation with their Graphic Characteristics

The ordinate and abscissa of a semilog plot in well test interpretation are pressure and (logarithm of) time, respectively.

There are several kinds of semilog plots, and the commonly used coordinates of them are listed in Table 5.2.

The drawings and characteristics of various plots are introduced here one by one.

3.1.1 Pressure Drawdown Analysis Plot

A pressure drawdown analysis plot is plotted aiming at transient pressure in the pressure drawdown test. The ordinate is pseudo pressure ($\psi(p_{wf})$) transformed from flowing pressure (p_{wf}), and the abscissa is flow time (t), as shown in Figure 5.29.

The plot shows that the pseudo pressure declines gradually. Some people take pseudo pressure differential $[\psi(p_i)-\psi(p_{wf})]$ as the ordinate (Figure 5.30), and the curve shape is consistent with the MDH plot or Horner plot, which is introduced later.

The ordinate of pseudo pressure in Figure 5.29 and pseudo pressure difference in Figure 5.30 can be replaced by pressure square and difference of pressure square if pressures are fairly low.

The most important portion in the pressure drawdown curve is the radial flow straight line with a slope of $m_{\psi d}$ (pseudo pressure plot) or m_{2d} (pressure square plot). Once determined, $m_{\psi d}$ or m_{2d} can be used to calculate the parameters of the reservoir, such as permeability k and skin factor S, with the following equation:

Pseudo pressure form:

$$k = 42.42 \times 10^3 \frac{q_g T_f}{m_{\psi d} h} \cdot \frac{p_{sc}}{T_{sc}} = 14.67 \frac{q_g T_f}{m_{\psi d} h} \tag{5.12}$$

$$S_a = 1.151 \left\{ \frac{\psi\left(p_i\right) - \psi\left(p_{wf}\left(1h\right)\right)}{m_{\psi d}} - \lg \frac{k}{\phi \mu_g C_t r_w^2} - 0.9077 \right\} \tag{5.13}$$

Table 5.2 Comparison of Coordinate Expression of Semilog Plots

Method	Ordinate		Abscissa			Application
	Coordinate type	Expression	Coordinate type	Expression	Coordinate type	
Pressure drawdown analysis method	Cartesian	p_{wf} (dimension) $$p_D = \frac{2.714 \times 10^{-5}\,khT_{sc}[\psi(p_i) - \psi(p_{wf})]}{q_g T_f p_{sc}}$$ (dimensionless) $$p_D = \frac{2.714 \times 10^{-5}\,khT_{sc}\left(p_i^2 - p_{wf}^2\right)}{q_g \mu_g ZT_f p_{sc}}$$ (dimensionless)	Cartesian	t (dimension) $$t_D = \frac{3.6 \times 10^{-3} k}{\phi \mu C_t r_w^2} \cdot \Delta t$$ (dimensionless)	logarithm	To analyze well pressure drawdown and calculate related parameters (e.g., k and S)
Horner method	Cartesian	p_{ws} (dimension) $p_D = p_D[(\Delta t)_D] - p_D[(t_p)_D]$ (dimensionless)	Cartesian	$\dfrac{t_p + \Delta t}{\Delta t}$	logarithm	To calculate related parameters (e.g., k, S, and $p*$)
MDH method	Cartesian	p_{ws} (dimension) $p_D = p_D[(\Delta t)_D] - p_D[(t_p)_D]$ (dimensionless)	Cartesian	Δt (dimension) $$t_D = \frac{3.6 \times 10^{-3} k}{\phi \mu_g C_t r_w^2} \cdot t$$ (dimensionless)	logarithm	To calculate related parameters (e.g., k, S, and $p*$)

Method	Equation	Coordinate	Purpose
Superposition function method (SUPF)	p_{ws} (dimension) $$p_D = p_D[(\Delta t)_D] + \sum_{i=1}^{n}\frac{q_i - q_{i-1}}{q_{n-1} - q_n}$$ $$\left\{ p_D\left[\sum_{j=i}^{n}(\Delta t_j)_D\right] - p_D\left[\left(\sum_{j=i}^{n-1}\Delta t_j + \Delta t\right)_D\right] \right\}$$ (dimensionless)	Cartesian $$\sum_{i=1}^{n-1}(q_i - q_{i-1})\lg\left(\sum_{j=1}^{n-1}\Delta t_j + \Delta t\right)$$ $$+(q_n - q_{n-1})\lg\Delta t$$	Cartesian	To be applied in multiflow operations, and to calculate related parameters (e.g., k, S, and p^*)
MBH method	$$p_{DMBH} = \frac{1085.7(p^* - \bar{p})}{qB\mu}$$ $(\bar{p} - p_{ws})$ (dimension)	Cartesian	logarithmic	To calculate pressure, p^*
Musket method	$$\frac{0.54287kh}{qB\mu}(\bar{p} - p_{ws}) \text{ (dimensionless)}$$	logarithm $$t_{pDA} = \frac{3.6\times10^{-3}kt_p}{\phi\mu C_t A}$$ Δt (dimension) Δt_{DA} (dimensionless)	Cartesian	To calculate $\bar{p}$

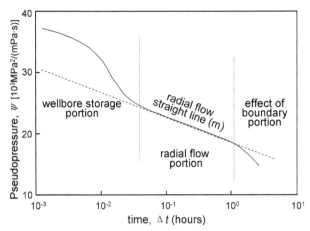

Figure 5.29 Semilog pressure drawdown curve.

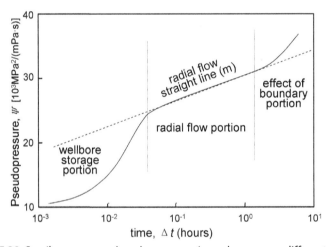

Figure 5.30 Semilog pressure drawdown curve (pseudo pressure differential form).

Pressure square form:

$$k = 42.42 \times 10^3 \frac{\overline{\mu}_g \overline{Z} q_g T_f p_{sc}}{m_{2d} h T_{sc}} = 14.67 \frac{\overline{\mu}_g \overline{Z} q_g T_f}{m_{2d} h} \tag{5.14}$$

$$S_a = 1.151 \left\{ \frac{p_i^2 - p_{wf}^2 \left(1h\right)}{m_{2d}} - \lg \frac{k}{\phi \mu_g C_t r_w^2} - 0.9077 \right\} \tag{5.15}$$

3.1.2 Horner Plot

If the well is tested with one-flow—one-shut in, the pseudo shut-in pressure is usually plotted with the Horner plot. This method was created by Horner in 1951[7]. The abscissa of this plot is "Horner time" $\dfrac{t_p + \Delta t}{\Delta t}$, and the ordinate is pseudo shut-in pressure p_{ws}. The most important portion of this plot is also a radial flow straight line, and its slope is m, as shown in Figures 5.31 and 5.32.

There usually are two drawing methods of the Horner plot:

- The first one is that its scale of abscissa $\dfrac{t_p + \Delta t}{\Delta t}$ decreases from left to right, as shown in Figure 5.31. If such a method is used, the early pressure point is located at the left,

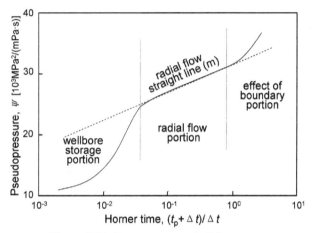

Figure 5.31 Horner pressure buildup curve.

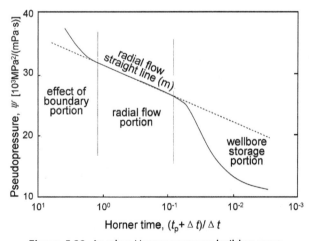

Figure 5.32 Another Horner pressure buildup curve.

while the late pressure point is located at the right, in line with general practice. Such a method is popular in early well test books published in China.

- The order one is that its scale of abscissa $\dfrac{t_p + \Delta t}{\Delta t}$ increases from right to left, as shown in Figure 5.32. It is widely applied in foreign literatures and imported computer software (e.g., SSI software).

Similar to pressure drawdown curves, the ordinate of pseudo pressure in Figures 5.31 and 5.32 can be replaced by pressure square, p^2_{ws} (MP$_a^2$), if pressures are fairly low

The slope of radial flow straight line m, once obtained from the figure, can be used to calculate reservoir permeability k and skin factor S, with the formulae

Pseudo pressure form:

$$k = 42.42 \times 10^3 \frac{q_g T_f}{m_{\psi b} h} \cdot \frac{p_{sc}}{T_{sc}} = 14.67 \frac{q_g T_f}{m_{\psi b} h} \tag{5.16}$$

$$S_a = 1.151 \left\{ \frac{\psi\left(p_{ws}\left(1h\right)\right) - \psi\left(p_{wf}\right)}{m_{\psi b}} - \lg \frac{k}{\phi \bar{\mu}_g C_t r_w^2} - 0.9077 \right\} \tag{5.17}$$

Pressure square form:

$$k = 42.42 \times 10^3 \frac{\bar{\mu}_g \overline{Z} q_g T_f}{m_{2b} h} \cdot \frac{p_{sc}}{T_{sc}} = 14.67 \frac{\bar{\mu}_g \overline{Z} q_g T_f}{m_{2b} h} \tag{5.18}$$

$$S_a = 1.151 \left\{ \frac{p^2_{ws}\left(1h\right) - p^2_{wf}}{m_{2b}} - \lg \frac{k}{\phi \bar{\mu}_g C_t r_w^2} - 0.9077 \right\} \tag{5.19}$$

3.1.3 MDH Plot

The plots are named after three authors—Miller, Davis, and Hutchinson. This plot is applicable to shut in pressure buildup in this case: the flow lasts quite a long duration t_p and the flow rate q is quite stable before shutting in. The ordinate is pseudo shut-in pressure ($\psi(p_{ws})$), and the abscissa is a logarithm of shut-in time log Δt, as shown in Figure 5.33.

Such a plot is very similar to the Horner plot. The important portion of the plot is also the radial flow straight line, whose slope m can be used to calculate reservoir parameters, and the calculation formulas are the same as those applied in the Horner plot [Equations (5.16)–(5.19)].

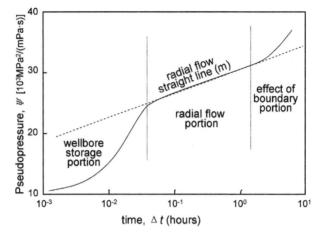

Figure 5.33 MDH pressure buildup plot.

3.1.4 Superposition Function Plot

The multiflow test means opening and shutting in the well or the flow rate changes a lot several times; each operation would influence the variation of reservoir pressure and the bottom-hole pressure, no matter if it is shut-in pressure buildup or pressure drawdown. The semilog plot in this case takes the same ordinate of p_{ws} (suppose it is buildup), but its abscissa is changed into SUPF:

$$SUPF = \sum_{i=1}^{n-1}\left(q_i - q_{i-1}\right)\lg\left(\sum_{j=i}^{n-1}\Delta t_j + \Delta t\right) + \left(q_n - q_{n-1}\right)\lg\Delta t \qquad (5.20)$$

where

n = times of working system shifting form the first flow up to the flow period to be analyzed (e.g., if tested well has three flows and three shut ins, and the final shut-in pressure buildup is to be analyzed, $n = 6$)

q_i = flow rate in ith period (e.g., if the final period, or the sixth one, is shut in, $n = 6$, $q_n = q_6 = 0$), m^3/day

Δt_i = duration of the ith period, h

Δt = time variable in the analyzed period, h

SUPF = time superposition function

The abscissa is SUPF function values in Equation (5.20) instead of simple time t or Horner time $\dfrac{t_p + \Delta t}{\Delta t}$; SUPF is also called "superposition time," and the obtained curve is shown in Figure 5.34. Due to time superposition function SUPF is adopted and the logarithm of time $\lg \Delta t$ has been taken inside the expression, Cartesian coordinates are applied in Figure 5.34. Although no logarithmic scales are shown in the coordinates, it is essentially a semilog plot.

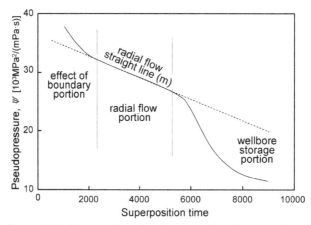

Figure 5.34 Superposition function plot of multirate well test.

It can be noted that factors of various flow rates q_i have been contained in superposition time. It is different from general time coordinate in dimension, and the time has been taken logarithm. The plot is mostly characterized by its straight line portion. The appearance of the straight line can be proved theoretically, which reflects a radial flow of elastic fluids during transient pressure variation in infinite formation.

It is important and should be pointed out that since well test interpretation software is widely used nowadays, people rarely calculate the parameters, relying on the aforementioned equations by hand. Providing related formulas in this book are just for helping readers to understand the parameter calculation process done by well test interpretation software. Special attention should be paid to data acquisition of radial flow straight lines; otherwise nothing could be brought to a conclusion!

Even though the semilog straight line analysis method used for parameter calculation has been applied for nearly 50 years, correct identification of straight lines in a semilog curve has been always being a puzzling problem. For those formations with abnormal permeability (either ultrahigh or ultralow), heterogeneous formations or oil wells with large S and C values, the straight line portion is often determined and applied in error, resulting in interpretation mistakes. The problem was not solved properly until identification by using the pressure derivative plot in 1983.

In addition to the semilog plots introduced previously, there still are some other semilog plots in the development of petroleum industry, for example, the MBH plot [34], named after three creators—Matthews, Brons, and Hazebroek. With the MBH plot, an average formation pressure of oil bearing areas with different boundary configurations can be calculated, but it is not widely used in gas field development.

Another semilog plot is the Maskat plot. Its ordinate is a logarithm of pressure differential $\log(\bar{p} - p_{ws})$, or expressed as dimensionless pressure of $\log \dfrac{0.54287kh}{qB\mu}(\bar{p} - p_{ws})$, and its abscissa is dimensionless time (Δt_{DA}). The average formation pressure is determined by adjusting the average formation pressure in the formula to change the curve to a straight line. The Maskat plot was used to determine average formation pressure during oil field development in the United States around the 1950s, but is seldom used currently, and no application examples of it in gas field development are found.

3.2 Semilog Plot used in Analysis by Well Test Interpretation Software

It is thought in view of modern well test analysis theory that the well test analysis process is essentially a process of identifying and confirming the well test model. Pressure behavior of the well test model is mostly indicated by its pressure log–log plot, and a type curve match method is formed. Semilog plot analysis methods developed in the 1950s are collectively called "conventional analysis methods," which are no longer used as an independent method in the well test analysis process. Its roles are mainly in the following two aspects.

3.2.1 Model Diagnosis in Early Interpretation Process

In some advanced well test interpretation software such as EPS, WorkBench, Saphir, and FAST, pressure history graph, semilog plot, and log–log plot are drawn simultaneously with preprocessed transient test data, and primary model diagnosis starts. Model type and related parameters (e.g., permeability k, skin factor S, and wellbore storage coefficient C) can be determined primarily.

During model diagnosis, radial flow and wellbore storage flow regimes can be identified or conformed by the radial flow horizontal straight line and the initial unit-slope (45°) straight line on the log–log pressure derivative curve, respectively. The correspondent portions on the semilog plot are so determined definitely at the same time, and the values of parameters such as k, S, and C are calculated primarily. Then these parameter calculations can be applied easily by applying the slope m of radial flow straight line in the semilog plot.

However, model parameters calculated in this diagnosis process are usually primary, rough, and uncertain; particularly for those complex types of formations, the diagnosis results only provide initial parameters of a near-wellbore area. In cases of composite formations, formations with boundaries, and double porosity systems, more burdensome analysis tasks are waiting to be fulfilled.

3.2.2 Verification of Match Analysis Results of Well Test Model

A complete dynamic model of a tested well and a complete set of parameters of various different zones such as wellbore, near-wellbore area, and reservoir outer boundaries are obtained by further well test analysis study. Interpretation results centering on the

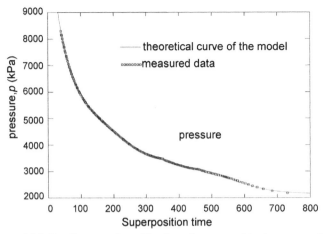

Figure 5.35 Semilog match verification of measured test pressure data.

log−log match method are an overall description for the well test model. However, whether these results are consistent with the real conditions of a tested well, the reservoir should be verified further by various match verification of specialized plots, including pressure semilog match verification.

For verification, a pressure semilog plot of the well test model is drawn, generally in dimensionless coordinates (or in dimensioned pseudo pressure and time transformed from them); this semilog plot represents the pressure behavior of the theoretical model; meanwhile, measured data are drawn in the same plot as well. If both curves are consistent, interpretation results are in line with the real conditions, at least both are consistent in the portion of data in which they are interpreted, as shown in Figure 5.35.

As seen from Figure 5.35, the abscissa is superposition time, and a complete curve of the theoretical model (solid line) and measured data (dotted line) are closely matched, proving that model verification is good.

In addition, Figure 5.35 shows a complete semilog curve, the different flow regimes are not necessary to be distinguished, and the consistency of the theoretical model and measured data is judged by the whole curve.

4 LOG−LOG PLOT AND MODEL GRAPH OF PRESSURE AND ITS DERIVATIVE

4.1 Log−log Plots and Type Curves for Modern Well Test Interpretation

4.1.1 Type Curve Analysis is the Core of Modern Well Test Interpretation

Since the end of the 1970s, the well performance study method named "modern well test" has been developed and improved in the international petroleum industry. The so-called "modern well test" includes the following three aspects:

- A complete package of theories and methods for pressure data analyses centralized on the type curve match method
- Pressure data acquired using high-accuracy electronic pressure gauges
- Development and application of well test interpretation software

Development of the type curve match method paved the foundation for the modern well test. There is no modern well test without the type curve match method; it is because of the following three reasons.

Features of Type Curves May Fully Reflect Characteristics of the Reservoir Dynamic Model

As discussed in Section 5.1 of this chapter, from understanding of the features of transient pressure curves to that of percolation of underground fluids and that of dynamic models, an effective way is formed for understanding the reservoir. The features of the curve here refer to the features of "type curves" used in modern well test interpretation.

Modern well test interpretation is based mainly on the type curve match analysis method. Almost all recognized reservoir structures can be modeled statically and described by corresponding mathematical models and so a dynamic model representing their performances is formed. In fact, every standardized dynamic model may be regarded as one type curve.

Type curves have the following features:

1. Characteristic features

   Figure 5.15 clearly shows the characteristic features of type curves. A different flow status or different flow periods present its own unique feature lines on pressure derivative type curves. Therefore, underground flow states can be deduced, and the structural configurations of a reservoir can be recognized.

2. Completeness of whole course

   The completeness of a whole course during studying of a reservoir by the type curve match method means that the entire scanning covering from wellbore to bottom hole, to near-wellbore area, and to far-away area can be performed in different time periods of the transient pressure test (see Figure 5.14). In this way, all conditions encountered during scanning can be reflected and so the overall structure of the dynamic model can be described.

   As for the other methods, say, semilog straight-line analysis method (e.g., Horner plot analysis, MDH plot analysis, pressure drawdown analysis), only the radial flow straight line portion may be used to determine certain reservoir parameters (k, S, etc.) in the vicinity of the wellbore.

3. Geologic correlation

   During the theory research of modern well tests, significant efforts have been devoted to developing new well test models.

   - As soon as a new type of reservoir was discovered by geologists and penetrated by wells, a corresponding mathematical model would be constructed and corresponding type curves were generated promptly. For example, development of river channel sandstone reservoirs formed by meandering rivers led to the construction of well test models for channel formation and corresponding type curves; and development of limestone reservoirs with fissures led to the construction of well test models and type curves for double porosity reservoirs.

- Relevant researchers will develop new well test models corresponding to every new drilling and completion operations, such as horizontal wells, highly deviated wells, multilateral wells, large-scale sand fracturing, and other stimulation measures.
- Development of numerical well test software greatly enhances a correlation between well test interpretation type curves and geologic models. Theoretically, various different well test models and corresponding type curves can be constructed for all reservoir structures, all well structures, and all possible fluid distributions.

It can be seen that the type curve match method has great potential for reservoir dynamic modeling and related studies. If used properly, the method may play a very important role in understanding the characteristics of a reservoir.

Type Curve Match Analysis Method

Features of type curves discussed previously can qualitatively describe the features of reservoirs. As for understanding of dynamic features, it is needed to know both the structural configurations and the parameters of the reservoirs. Log—log coordinates of type curves can be used to resolve this problem.

1. Dimensionless expressions of type curve coordinates

Nearly all type curves are expressed in dimensionless formats. For example, for infinity homogeneous formations, one set of type curves, as shown as Figure 5.36, can describe both oil wells and gas wells, no matter if the

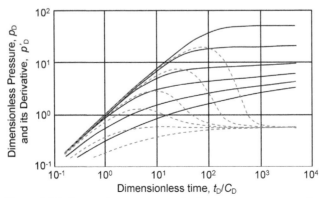

Figure 5.36 Composite type curves for homogeneous reservoirs.

permeability of the formation the wells locale is high or low and no matter what their parameters are.

These type curves are obtained from a solution of the differential equation (2.64). As for the gas layer, the ordinate is expressed as

$$p_D = \frac{2.714 \times 10^{-5} kh T_{sc} \Delta\psi}{q_g T_f p_{sc}} \tag{5.21}$$

$$\frac{t_D}{C_D} = 2.262 \times 10^{-2} \frac{kh \Delta t}{\mu_g C} \tag{5.22}$$

2. Consistency between type curves and measured curves

In the logarithmic coordinate system, the aforementioned coordinates are expressed as

$$lg p_D = lg\Delta\psi + lg\left\{2.7143 \times 10^{-5}\frac{kh}{q_g T_f} \cdot \frac{T_{sc}}{p_{sc}}\right\}$$

$$= lg\Delta\psi + A \tag{5.23}$$

$$lg\frac{t_D}{C_D} = lg\Delta t + lg\left\{2.262 \times 10^{-2}\frac{kh}{\mu_g C}\right\}$$

$$= lg\Delta t + B \tag{5.24}$$

Where,

$$A = lg\left\{2.7143 \times 10^{-5}\frac{kh}{q_g T_f} \cdot \frac{T_{sc}}{p_{sc}}\right\} \tag{5.25}$$

$$B = lg\left\{2.262 \times 10^{-2}\frac{kh}{\mu_g C}\right\} \tag{5.26}$$

It can be seen from Equations (5.23) and (5.24) that the same formation or reservoir models should have identical configurations in transient pressure curves, no matter whether under dimensionless coordinate (p_D–t_D/C_D) or dimensional coordinate ($\Delta\psi$–Δt) systems. The only difference is a displacement of A along the ordinate direction and B along the abscissa direction existing between them.

3. Determining parameters by type curve match method

  The procedures can be summarized as follow.

  • If transient pressure data of a gas well in a homogeneous reservoir were obtained onsite, draw Δp and $\Delta p'$ vs. Δt curves can be drawn on log–log paper, as shown in Figure 5.37.

  • Superimpose the aforementioned measured curves on type curves with identical scales (i.e., same log cycle length); it can be found that one type curve coincides perfectly with the measured curve. This procedure is called "type curve match," and this type curve is the "matched curve," the corresponding parameter group of which is denoted as $(C_D e^{2S})_M$, as shown in Figure 5.38.

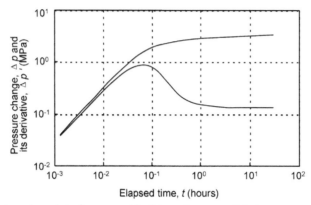

Figure 5.37 Log–log plot of measured buildup of a gas well in homogeneous reservoir.

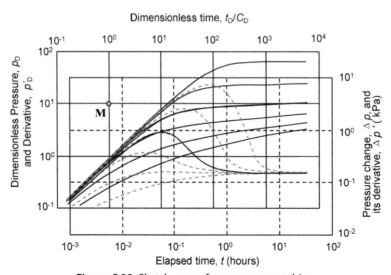

Figure 5.38 Sketch map of type curve matching.

Note: both ordinate and abscissa axes in a measured curve plot and those in type curves must be kept parallel accordingly and respectively during the whole process of type curve matching.

• Through matching procedures, differences A and B on abscissas and ordinates of the two coordinate systems can be found out and will be used for the calculation of parameters.

The method of parameter calculation can be summarized as follows: One match point, M, can be taken randomly on the figure. The point has a pair of ordinate readings and another pair of abscissa readings in the two coordinate systems:

On the measured coordinates: $\Delta\psi_M$, Δt_M

On the type curve coordinates: p_{DM}, $(t_D/C_D)_M$

Once point M is known, reservoir parameters can be calculated promptly:

$$k = 12.741 \frac{q_g T_f}{h} \cdot \frac{p_{DM}}{\Delta\psi_M} \tag{5.27}$$

$$C = 2.262 \times 10^{-2} \frac{kh}{\mu_g} \cdot \frac{t_M}{\left(t_D/C_D\right)_M} \tag{5.28}$$

$$S = \frac{1}{2}\ln \frac{\left(C_D e^{2S}\right)_M}{C_D} \tag{5.29}$$

Where

$$C_D = \frac{0.1592C}{\phi h C_t r_w^2} \tag{5.30}$$

It is by these formulae that key parameters such as k, S, and C are calculated when analyzing data by the type curve match method.

As mentioned repeatedly in this book, nowadays printed type curves and Equations (5.27)–(5.30) are rarely used for manual well test analyses due to development and common application of well test interpretation software. But the following two points should be noted:

a. It is necessary for an experienced reservoir engineer or a professional well test specialist to fathom techniques and operation procedures of the type curve match analysis. If possible, it is recommended to try the interpretation manually following these procedures at least once. Many engineers being trained for well test software started from here.

b. Some traces of manual type curve match procedures can be found on interfaces of well test interpretation software developed in early stages, but these traces have disappeared gradually as development goes on or there is no longer any grouped type curves being seen on software interfaces now. However, this does not mean that these type curves are no longer useful; in fact, they are replaced by "type curves for the targeted theoretic model." The generation procedures of these curves can be summarized as follow:

- Selection of reservoir type, or well test model
- Type curve diagnosis to determine initial parameters (k, S, C, etc.)
- The type curve for the theoretic model with obtained specific parameters, that is, specific type curve, is generated by software
- Log–log type curve match analyzing and rectifying the model and its parameters
- Match verification to confirm the final generated model and then the final interpretation conclusions

During the whole interpretation procedure, the log–log type curve match method has been the dominant one. Accordingly, it is also the core of modern well test analyses.

4.1.2 Some Common Log–Log Type Curves
Dimensionless Parameters in Various Type Curves

The log–log type curve match analysis method was developed by Ramey and colleagues [8,9,10] in the early 1970s. Since then, countless type curves have been developed. It is worth mentioning that the pressure derivative type curves created by Bourdet [32, 33, 80] in 1983 have made the type curve match analysis method a great leap forward in the conformation of reservoir dynamic models. To help readers understand some common type curves, the expressions of their coordinates, their characteristics, and applications are introduced.

Except for a few special cases (e.g., Mckinley's type curves), most type curves take pressure as the ordinate and time as the abscissa; logarithms of both coordinates are taken for drawing (see Table 5.3).

It should be pointed out that most of the type curves listed in Table 5.3 are used for homogeneous reservoirs and double porosity reservoirs and that the expressions of coordinates in the same kind of type curves may vary in accordance with different reservoirs or even different authors so attention should be paid for identification when using them. In addition, because all type curves have their coordinates in the dimensionless mode, they can be used for both gas and oil wells, and there is neither well test analysis type curves specifically for gas wells only nor those specifically for oil wells only.

Introduction of Commonly Used Type Curves for Gas Well Test Interpretation

1. Composite type curves resulted from combining pressure and pressure derivative type curves for homogeneous reservoir

    This type curve consists of two sets of type curves:
    - Gringarten's pressure type curves [9]
    - Bourdet's pressure derivative type curves [32, 33]

Table 5.3 Dimensionless Expressions for Common Log–log Coordinates

Name of Type Curve	Ordinate	Abscissa	Parameter
Gringarten's type curves for homogeneous reservoir	$p_D = \dfrac{0.5428 kh\Delta p}{q\mu B}$ (for oil well) $\quad$ $p_D = \dfrac{2.714 \times 10^{-5} kh T_{sc}\Delta\psi}{q_g T_f p_{sc}}$ (for gas well)	$\dfrac{t_D}{C_D} = 2.262 \times 10^{-2}\dfrac{kh\Delta t}{\mu C}$	$C_D e^{2S}$
Bourdet's derivative type curves for homogeneous reservoir	$p'_D = \dfrac{0.5428 kh\Delta p'}{q\mu B}$ (for oil well) $\quad$ $p'_D = \dfrac{2.714 \times 10^{-5} kh T_{SC}\Delta\psi'}{q_g T_f p_{SC}}$ (for gas well)	$\dfrac{t_D}{C_D} = 2.262 \times 10^{-2}\dfrac{kh\Delta t}{\mu C}$	$C_D e^{2S}$
Agarwal and Remy's type curves for homogeneous reservoir	$p_D = \dfrac{0.5428 kh\Delta p}{q\mu B}$ (for oil wells) $\quad$ $p_D = \dfrac{2.714 \times 10^{-5} kh T_{sc}\Delta\psi}{q_g T_f p_{sc}}$ (for gas wells)	$t_D = \dfrac{3.6 \times 10^{-3} k}{\phi\mu C_t r_w^2}\cdot\Delta t$	C_D S

(*Continued*)

Table 5.3 Dimensionless Expressions for Common Log–log Coordinates—cont'd

Name of Type Curve	Ordinate	Abscissa	Parameter
Interference test type curves for homogeneous reservoir	$p_D = \dfrac{0.5428 k h \Delta p}{q \mu B}$ (for oil wells) $p_D = \dfrac{2.714 \times 10^{-5} k h T_{sc} \Delta \psi}{q_g T_f p_{sc}}$ (for gas wells)	$\dfrac{t_D}{r_D^2} = \dfrac{3.6 \times 10^{-3} k}{\phi \mu C_t r^2} \cdot \Delta t$	C_D (when wellbore storage effect exists)
Gringarten's interference test type curves for double porosity media	$p_D = \dfrac{0.5428 k h \Delta p}{q \mu B}$ (for oil well) $p_D = \dfrac{2.714 \times 10^{-5} k h T_{sc} \Delta \psi}{q_g T_f p_{sc}}$ (for gas wells)	$\dfrac{t_D}{r_D^2} = \dfrac{3.6 \times 10^{-3} k}{\mu (\phi C_t)_f r^2} \cdot \Delta t$	ω λr_D^2
Zhuang-Zhu's interference test type curves for double porosity media	$p_D = \dfrac{0.5428 k h \Delta p}{q \mu B}$ (for oil wells) $p_D = \dfrac{2.714 \times 10^{-5} k h T_{sc} \Delta \psi}{q_g T_f p_{sc}}$ (for gas wells)	$\dfrac{t_D}{r_D^2} = \dfrac{3.6 \times 10^{-3} k}{\mu [(\phi C_t)_f + (\phi C_t)_m] r^2} \cdot \Delta t$	ω λr_D^2

Type curves for a well with an infinite conductivity vertical fracture or uniform flow vertical fracture in homogeneous reservoir	$p_D = \dfrac{0.5428 kh \Delta p}{q\mu B}$ (for oil wells) $p_D = \dfrac{2.714 \times 10^{-5} kh T_{sc} \Delta \psi}{q_g T_f p_{sc}}$ (for gas wells)	$t_D = \dfrac{3.6 \times 10^{-3} k}{\phi \mu C_t x_f^2} \cdot \Delta t$	t_{pDf} (for correction of pressure buildup) $x_f / \sqrt{A}$ (for closed block)
Type curves for a well with a finite conductivity vertical fracture in homogeneous reservoir	$p_D = \dfrac{0.5428 kh \Delta p}{q\mu B}$ (for oil wells) $p_D = \dfrac{2.714 \times 10^{-5} kh T_{sc} \Delta \psi}{q_g T_f p_{sc}}$ (for gas wells)	$t_D = \dfrac{3.6 \times 10^{-3} k}{\phi \mu C_t x_f^2} \cdot \Delta t$ or $t_{Dre} = \dfrac{3.6 \times 10^{-3} k}{\phi \mu C_t r_{we}} \cdot \Delta t$	$F_{CD} = \dfrac{k_f \cdot W}{k \cdot x_f} x_f / \sqrt{A}$ (for closed block)
Type curves for a well with horizontal fracture and uniform flow in homogeneous reservoir	$p_D \cdot F_{CD}$ $\dfrac{p_D}{h_D}$	$t_{Dxf} \cdot F_{CD}^2$ $t_{Dxf} = \dfrac{3.6 \times 10^{-3} k}{\phi \mu C_t x_f^2} \cdot \Delta t$	$h_D = \dfrac{h}{r_f} \cdot \sqrt{k_r / k_v}$

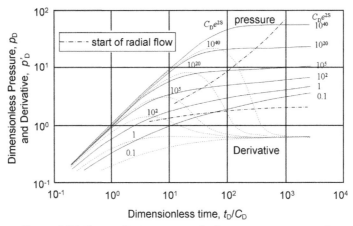

Figure 5.39 Composite type curves for homogeneous reservoir.

Although generated in different times, they are seldom used separately. These type curves have been presented again and again in this book, but as being most widely and commonly adopted, it is presented here again and as shown in Figure 5.39.

Definitions of dimensionless pressure and dimensionless time in type curves are given in Table 5.3, and pressure derivative p'_D is the derivative of pressure to the logarithm of time.

$$p'_D = \frac{dp_D}{d \ln\left(\frac{t_D}{C_D}\right)} = \frac{dp_D}{d\left(\frac{t_D}{C_D}\right)} \cdot \left(\frac{t_D}{C_D}\right)$$ (5.31)

Accordingly, it is essentially multiplying the pressure derivation to time. The dashed line in Figure 5.39 indicates the start of radial flow.

By type curve match analysis, formation permeability k, skin factor S, and wellbore storage coefficient C can be calculated with Equations (5.27)−(5.30).

2. Composite type curves for double porosity reservoir

Similar to composite type curves for homogeneous reservoirs, this type curve also consists of two sets of type curves: pressure and pressure derivative type curves. Pressure type curves contain two groups of curves: curves for the flow in a homogeneous reservoir (homogeneous flow for short) and curves for the transition flow [10]; correspondingly, derivative type curves also contain two groups of curves: homogeneous flow curves and transition flow curves [32].

Figure 5.40a shows part of the composite type curves: pressure type curves.

It is not difficult to see that the homogeneous flow in pressure type curves presented in Figure 5.40a is just the same as those in Figure 5.39, but type curves of transition flow are added. More details about the flow features of double porosity reservoirs are introduced in the following sections.

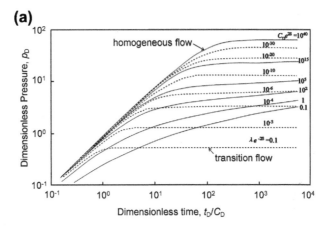

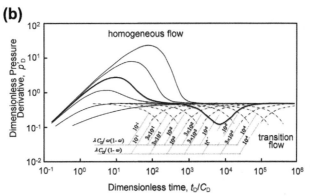

Figure 5.40 (a) Pressure type curves for double porosity reservoirs (pseudo-steady transition flow). (b) Pressure derivative type curves for double porosity reservoir (pseudo-steady transitional flow).

Composite type curves for double porosity media include pressure derivative type curves in addition to pressure type curves. To be shown more clearly, they are displayed in Figure 5.40b. Figure 5.40a and Figure 5.40b are combined together as one composite type curve.

It can be seen from Figure 5.40b that in addition to the identical pressure derivative type curves with homogeneous flow shown in Figure 5.39, transition flow pressure derivative type curves are also added, and these transition flow pressure derivative type curves correspond to the transition flow pressure type curves shown in Figure 5.40a.

By type curve matching, the permeability of fissure system k_f, total skin factor S, wellbore storage coefficient C, storativity ratio ω, and interporosity flow coefficient λ can be calculated.

3. Agarwal and Remy's type curves for homogeneous reservoir

As pioneer of the type curve match method [8], this type curve (Figure 5.41) was developed quite early and was published in SPEJ in 1970.

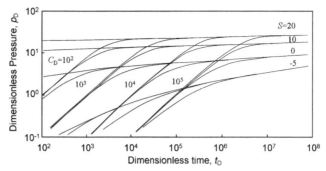

Figure 5.41 Agarwal and Remy's type curves for homogeneous reservoir with C_D and S as parameters.

Different from the type curves for homogeneous reservoirs discussed earlier, Figure 5.41 takes both dimensionless wellbore storage coefficient C_D and skin factor S as parameters and so forms multiple families of curves. As a tool for well test interpretation, it increases uncertainty in match analysis. For type curve analysis, however, it shows different influences of C_D and S, respectively, and more clearly. Therefore, it is often cited in some books, but is hardly used in well test interpretation software.

By type curve match, formation permeability k, skin factor S, and wellbore storage coefficient C can be calculated.

4. Interference test type curves for homogeneous formation

Figure 5.42 shows the log—log type curve—interference test type curve for homogeneous formation, which has been used for the longest time. This type curve can be generated by exponential integral function (Ei function for short) and used for interference test analysis.

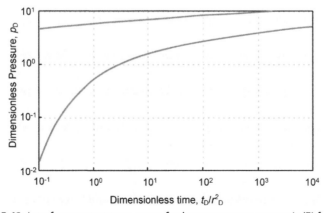

Figure 5.42 Interference test type curve for homogeneous reservoir (Ei function).

The flow coefficient of the reservoir between tested wells kh/μ and storativity $\phi h C_t$ can be calculated by type curve match analysis. Connection permeability k and some other parameters can be determined through further decomposition of them, and diffusivity coefficient η and some other parameters can be determined through further combination of them.

5. Deruyck's interference test type curve for double porosity media

This type curve can be used for interference test interpretation for double porosity reservoirs (see Figure 5.43).

It can be seen from Figure 5.43 that the type curves are composed of two parts: homogeneous flow (solid lines) and transition flow (dashed lines), and they have ω and λr_D^2 as their parameters, respectively. Flows change from homogeneous flow in fissures to transition flow and then change once more to homogeneous flow in the total system (fractures plus matrix). Accordingly, the measured data curve can be separated into three portions, and each portion matches one of these type curves respectively and sequentially (refer to the dot–dashed line in Figure 5.43).

Fracture permeability of the area between the wells k_f, storativity coefficient $(\phi \ hC_t)_f$, storativity ratio ω, interporosity flow coefficient λ, and some other parameters can be obtained by type curve match analysis.

6. Zhuang–Zhu's interference test type curves for double porosity media [35]

Another type curve with a combination of "homogeneous flow" and "transition flow" (Figure 5.44) was developed by Zhuang Huinong and Zhu Yadong for match analysis of the measured data curve with the theoretical type curve more visually.

It can be seen clearly from Figure 5.44 that the change of interference pressure starts with rapid increasing of the pressure in the fissure system, switches to a relatively flat transition flow interval, and then increases again rapidly for the flow in the total system.

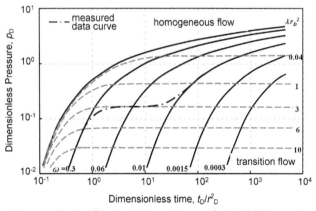

Figure 5.43 Deruyck's interference test type curves for double porosity reservoirs.

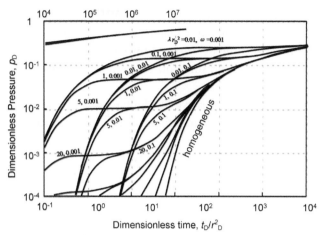

Figure 5.44 Zhuang–Zhu's interference test type curves for double porosity reservoir (by Zhuang Huinong and Zhu Yadong).

The abscissas of both Figures 5.44 and 5.43 seem the same (t_D/r_D^2), but their definitions are different (refer to Table 5.3). Consequently, these two type curves cannot coincide with each other.

The type curves shown in Figure 5.44 can be used to determine the same parameters as Gringarten's type curves; in addition, it can also be used for determining permeability k_f, storativity coefficient of the fissure system $(\phi h C_t)_f$, storativity ratio ω, and interporosity flow coefficient λ of the reservoir between the tested wells.

7. Pressure derivative type curves for interference tests in double porosity media

Differentiated dimensionless pressures displayed in Figure 5.43 or 5.44, pressure derivative type curves for interference tests in double porosity media are generated as shown in Figure 5.45. These derivative type curves have ω and λ (or λr_D^2) as their parameters and change from rising up to transition flow concave portion and then to radial flow horizontal straight-line portion.

This type curve is generally used jointly with interference test pressure type curves to determine the parameters: fracture permeability k_f, storativity coefficient $(\phi h C_t)_f$, storativity ratio ω, and interporosity flow coefficient λ.

8. Type curves for wells with vertical fracture in homogeneous reservoir [36]

The type curve for this situation is shown in Figure 5.46.

Formation permeability k, half-length of fracture x_f, wellbore storage coefficient C, conductivity F_{CD}, and some other parameters can be determined by match analysis with this type curve.

9. Type curves for wells with finite conductivity vertical fracture in homogeneous reservoir [37]

The type curve for this situation is shown in Figure 5.47.

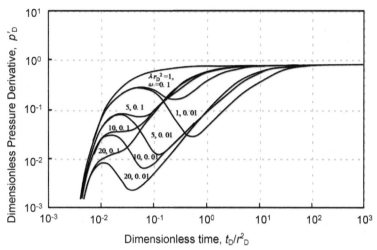

Figure 5.45 Pressure derivative type curves for interference tests in double porosity reservoir.

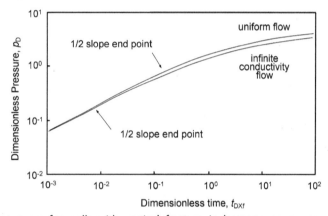

Figure 5.46 Type curve for wells with vertical fracture in homogeneous reservoir (with infinite conductivity flow and uniform flow, respectively).

Formation permeability k, fracture half-length x_f, fracture conductivity F_{CD}, and some other parameters can be determined by type curve match analysis.

Some Infrequently Used Type Curves in Gas Well Test Analysis

Although dimensionless log—log type curves can be used in gas wells, oil wells, or water wells, some type curves are rarely used in gas well test analyses due to some historical background of their generation; such type curves may not be included in the type curve databases of common well test interpretation software, including those developed specially for oil well tests or water well tests. Some of these type curves are presented here for reference.

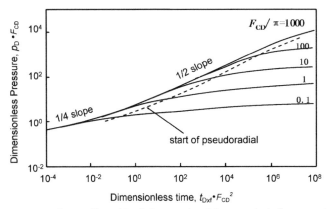

Figure 5.47 Type curves for wells with finite-conductivity vertical fracture in homogeneous reservoir.

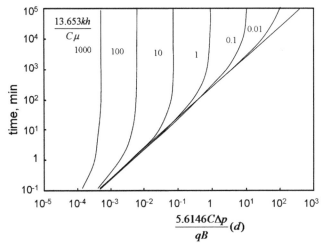

Figure 5.48 Mckinley's type curves.

1. Mckinley's type curve

   The Mckinley's type curve is a log–log type curve, but it is quite different from commonly used log–log type curves (Figure 5.48):

   • Type curve coordinates are not dimensionless
   • Ordinate is time t, in minutes, whereas the abscissa is a pressure-related parameter combination of $\dfrac{5.6146C \cdot \Delta p}{qB}$ in days
   • Parameter is $\dfrac{13.653kh}{C\mu}$ (CSU measurement unit)

- Influence of skin factor S has not been taken into account.

Because its coordinates are not dimensionless, Mckinley's type curves cannot be applied directly in gas well test analysis when pseudo pressure is used. Under such circumstances, the type curve must be modified accordingly

2. Slug flow type curves

During DST in a gas well with low deliverability, fluid may stop flowing before the fluid column in the wellbore reaches ground and so "slug flow" forms. Under such circumstances, the flow features of the tested well may not be reflected. If formation permeability estimation is required, the "slug flow type curve" can be used.

Ramey and colleagues [38] have committed themselves to research of this type curve. However, most well test interpretation software, especially gas well test interpretation software, does not contain this kind of type curve so far. Figure 5.49 shows one form of this kind of type curve.

Some New Type Curves

1. Normalized pressure analysis method and normalized pressure analysis curves

Type curves introduced earlier are commonly applied as "standard analysis plots" at present for onsite analyses and in well test interpretation software. New progress has been made in well test graph analyses from the late 1980s to early 1990s. Agarwal, Onur, Duong, Reynolds, and others first recombined pressure, time, and pressure derivative to generate new coordinate variables, created many new graph modes consequently, and used them to analyze pressure curve features and interpretation of parameters [39,40,41].

In these new type curves, a new pressure variable, p_{DG}, is introduced.

$$p_{DG} = \frac{1}{2} \cdot \frac{p_D}{p_D'} \tag{5.32}$$

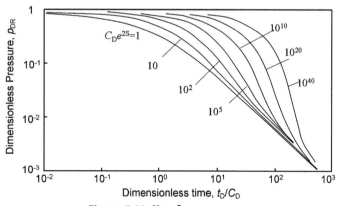

Figure 5.49 Slug flow type curve.

It is also known as "normalized pressure," where definitions of p_D and p'_D for oil wells are

$$p_D = \frac{0.54287kh\left(p_i - p_{wf}\right)}{qB\mu} = \frac{0.54287kh\Delta p}{qB\mu} \tag{5.33}$$

$$p'_D = \frac{dp_D}{dt_D} \cdot t_D = \frac{0.54287kh}{qB\mu} \cdot \left(\frac{d\Delta p}{dt} \cdot t\right) = \frac{0.54287kh}{qB\mu} \cdot \Delta p' \tag{5.34}$$

and

$$\Delta p' = \frac{d\Delta p}{d\ln t} = \frac{d\Delta p}{dt} \cdot t$$

It can be seen that p_{DG} may also be expressed as

$$p_{DG} = \frac{1}{2} \cdot \frac{\Delta p}{\Delta p'} \tag{5.35}$$

During the radial flow period, the ordinate values of the new type curve are identical to those of the measured data curve; when performing the curve match, only moving one plot along the direction of the abscissa is necessary. This is the key feature of such a graph mode.

As for gas wells, definitions of dimensionless pressure and its derivative are similar:

$$p_D = 2.7143 \times 10^{-5}\frac{kh}{q_g T_f} \cdot \frac{T_{sc}}{p_{sc}} \cdot \Delta\psi(p) \tag{5.36}$$

$$p'_D = 2.7143 \times 10^{-5}\frac{kh}{q_g T_f} \cdot \frac{T_{sc}}{p_{sc}} \cdot \Delta\psi'(p) \tag{5.37}$$

Accordingly,

$$p_{DG} = \frac{1}{2} \cdot \frac{p_D}{p'_D} = \frac{1}{2} \cdot \frac{\Delta\psi(p)}{\Delta\psi'(p)} \tag{5.38}$$

Similarly, ordinate values of the new type curves are identical to those on measured pseudo pressure curves in the radial flow period; when performing the curve match, moving one plot along the direction of the abscissa is enough.

A new time variable for homogeneous reservoir t_{DG} is introduced as

$$t_{DG} = \frac{1}{2} \cdot \frac{t_D}{p'_D} \tag{5.39}$$

Therefore,

$$t_{DG} = 3.3157 \times 10^{-3} \frac{qB}{\phi h C_t r_w^2} \cdot \frac{t}{\Delta p'} \tag{5.40}$$

As for a homogeneous formation with fracture, the definition of p_{DG} is the same, but the time variable should be t_{DxfG}:

$$t_{DxfG} = \frac{t_{Dxf}}{2p_D'} = 3.3157 \times 10^{-3} \frac{qB}{\phi h C_t x_f^2} \cdot \frac{t}{\Delta p'} \tag{5.41}$$

In which definitions of t_{Dxf} and other parameters are identical to those specified previously. In addition, the wellbore storage coefficient can be defined as

$$C_{Dxf} = \frac{0.1592C}{\phi h C_t x_f^2} \tag{5.42}$$

Coordinate variables and curve features of new type curves that have been published are listed in Table 5.4.

Different denotations for variables may be used by different authors, but their meanings are basically the same. To ensure consistency in the document, denotations of Agarwal were used. In addition, some authors also made more tries with regards to the combination of different parameters and performed analyses of heterogeneous formations. They are not discussed in detail here.

Figure 5.50 shows type curves specifically for infinity homogeneous formations corresponding to graph model 1 in Table 5.4.

It can be seen from Figure 5.50 that the wellbore storage flow period (the unit-slope straight-line portion in standard type curves) concentrating to a point (0.5, 0.5); the right-hand side of the dashed line is the radial flow period, which totally coincides with the standard type curves.

The abscissa and ordinate of the measured data curve are $\frac{t}{\Delta p'}$ and $\frac{\Delta p}{2\Delta p'}$ respectively. It can be seen from Equations (5.32) and (5.35) that p_{DG} can be expressed as $\frac{p_D}{2p_D'}$ or $\frac{\Delta p}{2\Delta p'}$, so the ordinate values of corresponding points on measured curves and theoretical type curve are equal. Therefore, when performing a type curve match, it is necessary only to move the curve horizontally. C_D can be calculated by coordinates of the match point M_1:

$$C_D = 3.3157 \times 10^{-3} \frac{qB}{\phi h C_t r_w^2} \cdot \frac{\left(t/\Delta p'\right)_{M_1}}{\left(t_{DG}/C_D\right)_{M_1}} \tag{5.43}$$

Table 5.4 Coordinates and Features of New Type Curves

Graph Model No.	Plot Type	Ordinate Expression	Abscissa Expression	Parameter	Targeted Reservoir	Curve Feature
1	Theoretical type curves	p_{DG}	t_{DG}/C_D	$C_D e^{2S}$	Homogeneous reservoir	1. Coincide with standard type curves in radial flow period 2. Whole wellbore storage flow period concentrating to a point of (0.5,0.5) 3. C_D can be determined from the abscissa of match point M
	Measured curves	$\dfrac{\Delta p}{2\Delta p'}$	$\dfrac{t}{\Delta p'}$			
2	Theoretical plot	p_{DG}	$\dfrac{t_D}{C_D}$	$C_D e^{2S}$	Homogeneous reservoir	1. Coincide with standard type curves in radial flow period 2. Wellbore storage flow period is the horizontal line $p_{DG}=0.5$ 3. Permeability k and flow coefficient kh/μ can be calculated from the abscissa of match point
	Measured curves	$\dfrac{\Delta p}{2\Delta p'}$	t			
3	Theoretical plot	p_{DG}	$\dfrac{t_D}{C_D E}$ $E=\lg(C_D e^{2S})$	$C_D e^{2S}$	Homogeneous reservoir	1. Wellbore storage flow period is the horizontal line $p_{DG}=0.5$ 2. Radial flow period starts basically from the same point that abscissa is $(t_D/C_D \cdot E)$. This can be used to identify the time when radial flow starts
	Measured curves	$\dfrac{\Delta p}{2\Delta p'}$	t			

No.	Curve				Model	Notes
4	Theoretical plot	p_{DG}	t_{DxfG}	C_{Dxf}	Homogeneous reservoir with infinite conductivity vertical fracture or uniform-flow fracture	1. Linear flow period is the horizontal line $p_{DG} = 1$ 2. Wellbore storage flow period starts at point E and goes to the horizontal line $p_{DG} = 1$. E point has ordinate = 0.5, abscissa = $(C_{Dxf}/2)$ 3. C_{Dxf} and x_f can be determined through abscissa match
	Measured curves	$\dfrac{\Delta p}{2\Delta p'}$	$\dfrac{t}{\Delta p'}$			
5	Theoretical plot	p_{DG}	t_{Dxf}	C_{Dxf}	Homogeneous reservoir with infinite conductivity vertical fracture or uniform flow fracture	1. Linear flow period is the horizontal line $p_{DG} = 1$ 2. Permeability k and flow coefficient kh/μ can be determined from the abscissa of match point M
	Measured curves	$\dfrac{\Delta p}{2\Delta p'}$	t			
6	Theoretical plot	p_{DG}	t_{DxfG}	F_{CD}	Homogeneous reservoir with finite conductivity vertical fracture	1. When F_{CD} is low, early time is bilinear flow horizontal line $p_{DG} = 2$ 2. When F_{CD} is higher, it may change to linear flow horizontal line $p_{DG} = 1$ 3. F_{CD} and x_f can be determined through horizontal match
	Measured curves	$\dfrac{\Delta p}{2\Delta p'}$	$\dfrac{t}{\Delta p'}$			
7	Theoretical plot	p_{DG}	$t_{DxfG} \cdot F_{CD}$	F_{CD}	Homogeneous reservoir with finite conductivity vertical fracture	1. Bilinear flow period: straight line $p_{DG} = 2$; linear flow period: straight line $p_{DG} = 1$ 2. Early transition period: merge into one curve 3. F_{CD} and x_f can be determined from the abscissa match
	Measured curves	$\dfrac{\Delta p}{2\Delta p'}$	$\dfrac{t}{\Delta p'}$			

(Continued)

Table 5.4 Coordinates and Features of New Type Curves—cont'd

Graph Model No.	Plot Type	Ordinate Expression	Abscissa Expression	Parameter	Targeted Reservoir	Curve Feature
8	Theoretical plot Measured curves	p_{DG} $\dfrac{\Delta p}{2\Delta p'}$	$t_{DxG} \cdot F_{CD}$ $\dfrac{t}{\Delta p'}$	F_{CD} C_{Dxf}	Homogeneous reservoir with finite conductivity vertical fracture	1. Bilinear flow period: straight line $p_{DG} = 2$; radial flow period: straight line $p_{DG} = 1$ 2. Wellbore storage flow period starts at point E ($C_{Dxf}/2, 1$) 3. F_{CD}, C_{Dxf}, and x_f can be determined from the abscissa match
9	Theoretical plot Measured curves	p_{DG} $\dfrac{\Delta p}{2\Delta p'}$	t_{Dxf} $\dfrac{t}{\Delta p'}$	F_{CD}	Homogeneous reservoir with finite conductivity vertical fracture	1. Bilinear flow period: straight line $p_{DG} = 2$; linear flow period: straight line $p_{DG} = 1$ 2. Permeability k and flow coefficient kh/μ can be determined from the abscissa match
10	Theoretical plot Measured curves	p_{DG} $\dfrac{\Delta p}{2\Delta p'}$	t_{Dxf} t	$\dfrac{x_e}{x_f}$	Homogeneous reservoir with infinite conductivity vertical fracture (or uniform flow) with square boundaries	1. Linear flow period: straight line $p_{DG} = 1$ 2. Boundary reflection period: straight line $p_{DG} = 0.5$ 3. $\dfrac{x_e}{x_f}$ can be determined from the abscissa match
11	Theoretical plot Measured curves	p_{DG} Δp	t_D t	L_D	Horizontal well	1. Initial radial flow period: coincides with straight line $L_D = 0.5$ of the type curve 2. Linear flow period matches corresponding pressure type curve 3. Pseudo radial flow period matches measured pressure curve
12	Theoretical plot Measured curves	$p_{DRG} = \dfrac{p_D}{p_D'}$ $\dfrac{I(\Delta p)}{t \cdot \Delta p}$	$\dfrac{t_D}{p_D}$ $\dfrac{tC_{FD}(p_i - p)}{I(\Delta p)}$	C_D S	Slug flow	1. Early period concentrating to a point of (C_D, 0.5) 2. Match analysis gives S 3. Additional type curves is required for determination of kh/μ

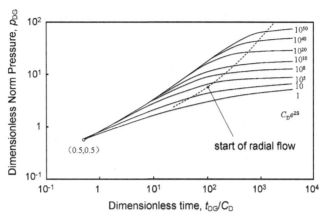

Figure 5.50 Agarwal's type curve for combined parameters in infinity homogeneous reservoir (graph model No. 1 in Table 5.4).

It is impossible to calculate formation permeability by this type curve.

Figure 5.51 shows another type of curve with other combined parameters proposed by Duong [41]. It is model graph No. 2 in Table 5.4.

As seen from Figure 5.51, the wellbore storage flow period is the horizontal straight line $p_{DG} = 0.5$. In the transition from the horizontal line to the radial flow portion of the curves, the higher the C_{De}^{2S} value is, the more rightward the curve locates, thus forming a set of intersecting curves. The radial flow portion lies on the right-hand side of the dashed line, and they coincide with the standard type curves.

Match point M_2, together with the match value of abscissa t_{M2} and $(t_D/C_D)_{M2}$, can be found out through moving the measured data curve horizontally and

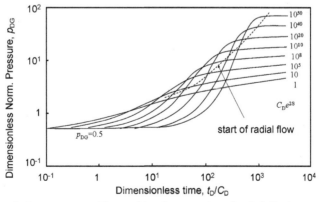

Figure 5.51 Duong's type curves with combined parameters for infinity homogeneous reservoir (graph model No. 2 in Table 5.4).

matching it with the theoretical type curve. These values can be used to calculate the value of k:

$$k = 0.2778 \frac{\mu \phi C_t r_w^2 C_D}{t_{M2}} \cdot \left(\frac{t_D}{C_D}\right)_{M2} \quad (5.44)$$

where C_D is determined by match point M_1 in Figure 5.50 [refer to Equation (5.43)]. Substituting Equation (5.43) into Equation (5.44) results in

$$\frac{kh}{\mu} = 9.21 \times 10^{-4} qB \frac{(t/\Delta p')_{M1}}{(t_{DG}/C_D)_{M1}} \cdot \frac{(t_D/C_D)_{M2}}{t_{M2}} \quad (5.45)$$

Figure 5.52 presents a type curve specifically for wells connecting a vertical fracture with infinite-conductivity or uniform flow.

It can be seen from Figure 5.52 that the linear flow portion is the horizontal straight line $p_{DG} = 1$ (a–b), whereas the transition portion is (b–c); the pseudo-radial flow portion starts at the point marked with an "$\times$".

Wellbore storage flow starts at point E ($C_{Dxf}/2$, 0.5).

By horizontal moving the measured data plot horizontally to get a good match, the fracture half-length x_f can be calculated from the match values of match point M_1,

$$x_f = \left[3.3175 \times 10^{-3} \frac{qB}{\phi C_t h} \cdot \frac{\left(t/\Delta p'\right)_{M1}}{\left(t_{DxfG}\right)_{M1}} \right]^{1/2} \quad (5.46)$$

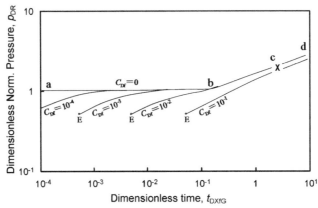

Figure 5.52 Type curve with combined parameters of well connecting a vertical fracture with uniform flow (graph model No. 4 in Table 5.4).

Permeability k can be calculated from either the slope of the semilog straight line in the pseudo-radial flow portion or the abscissa of match point M_2 in graph model No. 5 in Table 5.4:

$$\frac{kh}{\mu} = 9.21 \times 10^{-3} qB \frac{\left(t/\Delta p'\right)_{M1}}{\left(t_{DxfG}\right)_{M1}} \cdot \frac{\left(t_{DxfG}\right)_{M2}}{t_{M2}} \tag{5.47}$$

Plots with combined parameters may highlight certain features of well test curves. When comprehensive analysis is performed by a combination of these curves and standard pressure and pressure derivative curves, pressure features of the formation can be identified more accurately.

2. Integral pressure analysis method and integral pressure analysis plots

Conventional multiflow evaluator tools and some mechanical pressure gauges are still in use during operations of formation tests in China. Data acquired by mechanical pressure gauges are characterized by low accuracy; consequently, calculated pressure derivative points distribute disorderly and identification of the horizontal straight line becomes very difficult. Improvement of differentiation methods, for example, smoothing, may lead to distortion of the derivative curve. Consequently, it is impossible to apply the modern well test type curve match method to analyze data acquired by mechanical pressure gauges.

Blasingame and co-workers [42] developed an analysis method that can be used to solve the problem to a certain degree. The method is known as the integral pressure analysis method, or "average pressure" method.

Define dimensionless integral pressure

$$p_{Di} = \frac{1}{t_D} \int_0^{t_i} p_D(T) dT \tag{5.48}$$

dimensionless integral pressure differential

$$p_{Did} = p_D - p_{Di} \tag{5.49}$$

and derivative of dimensionless integral pressure:

$$p'_{Di} = t_D \cdot \frac{dp_{Di}}{dt_D} \tag{5.50}$$

It can be proved that $p_{Did} = p'_{Di}$. Therefore, these two expressions are changeable from each other theoretically. However, with different numerical calculation methods, some different results may be obtained, and p'_{Di} is mostly used.

In addition, define the first class normalized integral pressure

$$p_{Dir1} = \frac{p_{Di}}{2p'_{Di}} \tag{5.51}$$

and the second class normalized integral pressure

$$p_{Dir2} = \frac{p'_{Di}}{p'_D} \qquad (5.52)$$

where p'_D is the commonly used dimensionless pressure derivative

$$p'_D = t_D \cdot \frac{dp_D}{dt_D} \qquad (5.53)$$

With integral pressure, pressure and the pressure derivative of homogeneous formation are shown in Figure 5.53.

As can be seen from Figure 5.53:

- Integral pressure type curves and pressure type curves are similar in shape: it is a unit-slope straight line during the pure wellbore storage flow period, and the pressure curve becomes gentler in the radial flow period. Integral pressure derivative type curves and Bourdet's type curves are also similar in shape: they both start with a unit-slope straight line, followed by a maximum value; and in the radial flow period they both show a horizontal straight line $p'_{Di} = 0.5$.

- Different from standard pressure type curves and derivative type curves, integral pressure type curves may look like horizontally relocate the composite type curves of Gringarten's and Bourdet's to the right.

- The biggest difference between integral pressure derivative type curves and Bourdet's type curves is that integral pressure derivative type curves have "diffusion" features named by the author himself. Accordingly, the starting points of the horizontal straight line of those derivative type curves whose C_De^{2S} is greater than 10^{10} do not totally correspond to the starting point of radial flow; instead, they are slightly on the right side of the starting point of the radial flow straight line. In Figure 5.53, the starting point of the radial flow is J_1, but the location on which this derivative type

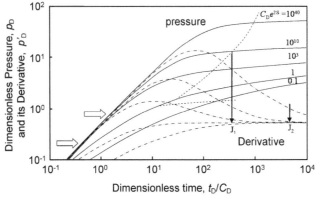

Figure 5.53 Type curves of dimensionless integral pressure and its derivative for homogeneous reservoir.

curve becomes horizontal lies at point J$_2$. Consequently, it may fail to indicate the starting point of radial flow.

- Measured integral pressure and integral pressure derivative can be calculated with Equations (5.48) and (5.50) and plotted and matched with the type curves shown in Figure 5.53; formation parameters can be calculated similarly. During integration, the pressure is changed into average pressure; consequently, fluctuation of pressure and pressure derivative curves is removed and evened. In this way, the type curve match may become easier.

  Wide coverage in the final part of this chapter is given to the introduction of analysis plots with parameter groups generated by recombining some parameters. This is because such analysis plots attracted the attention of many researchers for quite a while in the early 1990s and many papers have been published. But more than 10 years later, when looking back at the application of these research results, it was found that those analysis plots were not included in commonly used and successful well test interpretation software. Reasons for this include the following.

- The new method *cannot* provide new parameters or new understanding to the dynamic model of reservoir.
- In case software is compiled and applied, data interpretation becomes more complicated instead of simplified because some parameters that used to be determined by one type curve match must be determined by two type curve matches. In addition, sometimes figure conversions may also lead to confusion in concepts.
- Due to widely adopting high-accuracy electronic pressure gauges, integral pressure methods adopted originally for compensating the accuracy of data acquired by low-accuracy pressure gauges have become redundant.

The purpose of summarizing and introducing these methods in this book is to remind readers, especially those engaged in studies of gas field performances, to avoid excessive reliance on the software based on those new methods and expecting to obtain new understanding of gas reservoir dynamic performance research from which.

4.2 Typical Characteristic Curves—Model Graphs for Well Test Analyses

As emphasized repeatedly in this book, graph analysis is the core of modern well test analyses. With sufficient consideration to boundary conditions of reservoir, characteristic log—log curves generated for specific reservoirs can reflect flow features of gas (or oil, water) in the reservoir. These curves are called "model graphs." Parameters of the model graphs can be classified into the following categories:

1. Basic media in reservoir: homogeneous reservoir, double porosity reservoir, and double permeability reservoir
2. External boundary conditions of reservoir:
   - Different boundary configurations—linear, composite straight lines, circular, rectangular, and complex configurations

- Different boundary properties—no-flow boundary, constant-pressure boundary (for oil wells), semipermeable boundary, etc.

3. Horizontal distribution of reservoir:
- Distribution of *kh* values of reservoir
- Distribution of fluids (gas, oil, and water) in the reservoir

4. Different bottom-hole conditions:
- Bottom hole with wellbore storage coefficient C and skin factor S
- Bottom hole connecting to a vertical fracture with infinite conductivity
- Bottom hole connecting to a vertical fracture with finite conductivity
- Bottom hole connecting to a horizontal fracture
- Bottom hole connecting with fissure system
- Bottom hole partially perforated, etc.

5. Different well structures:
- Vertical wells
- Horizontal wells
- Highly deviated wells and multibranch wells, etc.

Some typical conditions were selected to generate the model graphs, and these typical conditions have been met in gas fields (or in oil fields).

It should be pointed out that theoretic derivative characteristic curves for some complicated reservoirs may have multiple turns, and at least 10 log cycles are required to fully display all of these characteristics. With consideration to the capacities of currently available pressure gauges, the first measured point can be recorded 1 second after starting the test, the entire test may consume over 300 years. Consequently, such theoretic curves can only be used for theoretic study because they are meaningless in field applications. For example, some type curves involving characteristic curves for horizontal wells in a double porosity reservoir contained in some well test interpretation software may take over 12 log cycles—it may take 30,000 years to obtain the desired pressure data. Therefore, this book will not introduce those plots.

Table 5.5 contains model graphs that have commonly been met onsite during gas field development.

5 CHARACTERISTIC DIAGRAM AND FIELD EXAMPLES OF TRANSIENT WELL TEST IN DIFFERENT TYPES OF RESERVOIRS

5.1 Characteristic Diagram (Model Graph M-1) and Field Examples of Homogeneous Formations

5.1.1 Homogeneous Formations in Gas Fields
Various Types of Reservoirs in China

There are many types of reservoirs in the gas fields in China, for example:

1. Large porous sandstone reservoirs, for example, TN field and SB field in the Chaidamu Basin

Table 5.5-1 Model Graphs Met Commonly in Gas Well Test and Their Features

Model Graph No.	Reservoir and Boundary Conditions	Representation of Reservoir Conditions	Model Graph	Parameters and Features
M-1	Homogeneous reservoir, wellbore storage coefficient C, skin factor S and infinite boundary	⊙		a-b-c: wellbore storage portion c-d: radial flow portion
M-2	Double porosity reservoir, wellbore storage coefficient C, skin factor S and two radial flows in fissure and total systems	⊙		a-b: wellbore storage portion b-c: radial flow in fissure system c-d: transition flow portion d-e: Radial flow in total system
M-3	Double porosity reservoir, wellbore storage coefficient C, Skin factor S & with only one radial flow of the total system	⊙		a-b-c: wellbore storage portion c-d: transition flow portion d-e: radial flow in total system
M-4	Homogeneous reservoir, wellbore storage coefficient C, skin factor of fracture S_f and infinite conductivity vertical fracture			a-b: wellbore storage portion b-c: linear flow portion c-d: transition flow portion d-e: pseudo-radial flow portion

(Continued)

Table 5.5-1 Model Graphs Met Commonly in Gas Well Test and Their Features—cont'd

Model Graph No.	Reservoir and Boundary Conditions	Representation of Reservoir Conditions	Model Graph	Parameters and Features
M–5	Homogeneous reservoir, wellbore storage coefficient C, fracture skin factor S_f and finite conductivity vertical fracture			a–b: wellbore storage portion b–c: bilinear flow portion c–d: transition flow portion d–e: pseudo-radial flow portion
M–6	Homogeneous reservoir, wellbore storage coefficient C, skin factor S and reservoir partially perforated			a–b: wellbore storage portion b–c: partial radial flow portion c–d: transition portion d–e: reservoir radial flow portion
M–7	Homogeneous reservoir, wellbore storage coefficient C, skin factor S and composite reservoir (flow condition of inner zone is better than that of outer zone)			a–b: wellbore storage portion b–c: radial flow of inner zone c–d: transition flow portion d–e: radial flow of outer zone
M–8	Homogeneous reservoir, wellbore storage coefficient C, skin factor S and, composite reservoir (flow condition of inner zone is worse than that of outer zone)			a–b: wellbore storage portion b–c: radial flow of inner zone c–d: transition flow portion d–e: radial flow of outer zone

Table 5.5-2 Model Graphs Met Commonly in Gas Well Test and Their Features

Model Graph No.	Formation and Boundary Conditions	Representation of Formation Conditions	Model Graph	Parameters and Features
M-9	Homogeneous reservoir, well storage coefficient C, Skin factor S, Singular straight-line impermeable boundary			a–b: wellbore storage portion b–c: radial flow portion c–d: transition flow portion d–e: fault reflection portion
M-10	Homogeneous reservoir, well storage coefficient C, Skin factor S, Angular impermeable boundary composed by straight lines			a–b: wellbore storage portion b–c: radial flow portion c–d: transition flow portion d–e: fault reflection portion
M-11	Homogeneous reservoir, well storage coefficient C, Skin factor S, Enclosure with approximately square configuration			a–b–c: wellbore storage portion c–d: radial flow portion d–e$_1$: boundary reflection portion (pressure drawdown) d–e$_2$: boundary reflection portion (pressure buildup)
M-12	Homogeneous reservoir, well storage coefficient C, Skin factor S, Impermeable boundary in strip			a–b–c: wellbore storage portion c–d: radial flow portion d–e: boundary linear flow portion

(*Continued*)

Table 5.5-2 Model Graphs Met Commonly in Gas Well Test and Their Features—cont'd

Model Graph No.	Formation and Boundary Conditions	Representation of Formation Conditions	Model Graph	Parameters and Features
M-13	Homogeneous reservoir, well storage coefficient C Fractured well Channel impermeable boundaries			a–b: wellbore storage portion b–c: radial flow portion c–d: pseudo-radial flow portion e–f: boundary linear flow portion
M-14	Well storage coefficient C Skin factor S group- and/or series-distributed channel fractures			a–b: radial flow in area A b–c: boundary–reflecting and transition flow in area A c–d–e: flow in area C
M-15	Well storage coefficient C Skin factor S Complicated group- and/or series-distributed fractures			a–b: wellbore storage portion b–c: radial flow in area A c–d: boundary–reflecting portion d–e: gas supplied from areas B, C, and others
M-16	Homogeneous reservoir, well storage coefficient C Skin factor S horizontal wells			a–b: wellbore storage portion b–c: vertical radial flow portion c–d: linear flow portion d–e: pseudo-radial flow portion

2. Common sandstone formations cut by faults, for example, gas fields in the four coastal oil provinces of East Bohai Bay and Hutubi field in the Zhungaer Basin
3. Large Ordovician fissured carbonate reservoirs, for example, JB gas field in the Ordos Basin
4. Permian—Carboniferous carbonate reservoirs, for example, majority of gas fields in the Sichuan Basin
5. Meandering rivers formed by fluvial deposits and braided stream strip-like sandstone reservoirs, for example, Permian—Carboniferous sandstone reservoirs in the Ordos Basin
6. Very thick fluvial sandstone reservoirs, for example, KL2 gas field in Tarim Basin, whose major pay zones are in the Cretaceous system

The characteristics of these gas fields are very different not only in the scale or size of their single reservoir bodies, but also in the accumulation conditions of their single reservoir bodies. These problems undoubtedly bring many difficulties in field development. However, based on thorough studies on reservoir dynamic characteristics, it is discovered, based on thorough studies on reservoir dynamic characteristics, that although the different research targets differ in thousands of ways from the viewpoint of geology, there is one common feature of them, that is, the distribution pattern of most reservoirs is uniform in a certain limited area.

Concept of "Homogenous Formation"

The concept of "homogenous formation" was proposed in well test research on the basis of the aforementioned understandings. Although it has not been defined strictly so far, the concept is widely accepted and used by reservoir engineers.

There are some common views on how to describe the homogeneous formation:

1. Generally, reservoir parameters (e.g., k and ϕ) are approximately uniform within the controlled area of the well, no matter if the reservoir is a porous or fissured medium.
2. As far as the flow characteristics reflected by well test curves are concerned, radial flow (or pseudo radial flow) occurs in a certain limited area without a very obvious reflection of increasing or decreasing flow resistance.

Even though the description of homogenous formations does not involve, and should not involve, any limitation of the fluids in the formations, as a gas zone, if a gas—oil or gas—water interface exists in it, they will have some impact on gas flow and make it deviate from the flow characteristics of homogeneous formations.

Examples of Homogenous Formation

There are no absolute homogeneous geological conditions naturally, but dynamic characteristics with "homogeneous medium phenomena" are seen very often in gas fields. For example:

1. In the TN and SB gas fields, homogeneous characteristic curves are obtained most commonly in the well test, and most of the test results indicate that there is a large homogeneous area around the tested well.

2. In the KL-2 gas field, all transient well test curves show homogeneous medium characteristics, even though great differences exist in its very thick Cretaceous sandstones vertically, and there are a great number of small throw faults developed in its gas reservoirs.

3. The Ordovician fissured limestone formations penetrated in the JB gas field are described as double porosity systems or multiple porosity systems by geologists. However, most of the wells, except some in a few wellblocks, do not show characteristics of a double porosity system, and the variations in reservoir parameters displayed on the tests seem to occur as far as several hundred or even thousands of meters away from the wells.

4. The geological environments where gas wells are drilled in the Sichuan Basin are more complicated. The carbonate reservoirs in this field are featured by well-developed fissures, pores, and caves and great effects of complex structures, but quite a few gas wells still show characteristics of homogenous formation.

Therefore, analysis of transient well test curves of homogenous formations plays an important role in gas field dynamic research.

5.1.2 Positioning Analysis

If two pressure buildup curves are acquired in a gas field, one of which shown in Figure 5.54a and the other in Figure 5.54b, it may be concluded that they represent totally two different formations.

In fact, Figures 5.54a and 5.54b were drawn based on the same well, between which the only difference is that curves in Figure 5.54a were acquired by wellhead shut in, whereas those in Figure 5.54b were acquired by downhole shut in. For a gas well deeper than 3000 meters, its wellbore storage coefficient C can be as large as

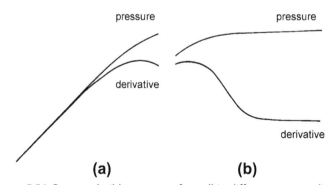

Figure 5.54 Pressure buildup curves of a well in different test conditions.

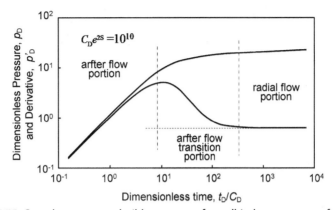

Figure 5.55 Complete pressure buildup curves of a well in homogeneous formation.

$1-3$ m^3/MPa; if the formation permeability k is low, the afterflow duration will be very long, even the measured curve is useless. In pressure curves measured by downhole shut in shown in Figure 5.54b, the afterflow duration is very short and radial flow duration is very long, from which formation parameters such as k and S can be calculated accurately.

A complete pressure buildup curve is plotted in Figure 5.55. It contains three portions: afterflow, transition flow, and radial flow.

The curves in Figure 5.55 are in dimensionless coordinates with $C_D e^{2S}$ as the parametric variable.

The length of the curve in dimension is limited because of the following aspects:

1. Acquiring time reaches its lower bound. The initial data point or the first data point is at one or several seconds (electronic pressure gauge) or minute (mechanical pressure gauge).

2. Acquiring time reaches its upper bound. A mechanical pressure gauge can uninterruptedly work only for several hours or dozens of hours; the continuous working time of an electronic pressure gauge can be prolonged significantly to several hundred hours due to the limitations of the operation conditions.

3. Pressure acquisition reaches its lower bound, which is about 0.0001 MPa for electronic pressure gauges and $0.001-0.01$ MPa for mechanical pressure gauges.

Hence the starting and ending time of the measured curves is

$$\left(\frac{t_D}{C_D}\right)_{\text{start/end}} = 2.262 \times 10^{-2} \frac{kh}{\mu C} \cdot \Delta t_{\text{start/end}} \tag{5.54}$$

The lower pressure limit is

$$p_D = 0.54287 \frac{kh}{qB\mu} \cdot \Delta p_{\text{start}} \tag{5.55}$$

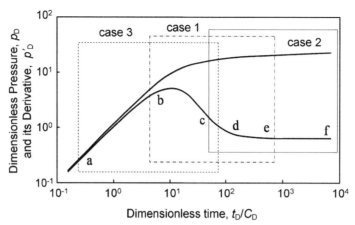

Figure 5.56 Positioning analysis examples (log—log plot).

If $\Delta t_{start} = 1$ minute, $\Delta t_{end} = 4$ hours, and $\Delta p_{start} = 0.01$ MPa, the position of measured curves can be traced in Figure 5.55. The parameters of several real field cases are listed in Table 5.6. The positioning of the curves can be seen in Figure 5.56. Note: the parameters listed in Table 5.6 are from oil wells.

Figure 5.56 shows that case 1 is with a complete curve consisting of afterflow portion (b—c) and radial flow portion (d—e); case 2 is only with radial flow portion (d—e—f) but without afterflow portion, and case 3 is with afterflow portion and transition flow portion (a—b—c) but without radial flow portion.

Such positioning analysis is also applicable to a semilog curve, referring to Figure 5.57.

5.1.3 Classified Model Graphs for Positioning Analysis of Homogeneous Formations

Two parameter groups are formed from parameters influencing the shape of the transient well test curve, including

$C_D e^{2S}$ = graphic parameter, dimensionless

$\dfrac{kh}{\mu C}$ = position parameter, $[\text{mD} \cdot \text{m}/(\text{mPa} \cdot \text{s})]/(\text{m}^3/\text{MPa})$

and each of the two parameter groups is classified into seven levels, that is,

- For graphic parameter $C_D e^{2S}$, intervals of 0.1, 1, 10, 10^4, 10^{10}, 10^{20}, and 10^{30} are given.
- For position parameter $\dfrac{kh}{\mu C}$, intervals of 100, 300, 800, 2000, 6000, 20,000, and 70,000 are given.

There are 49 model graphs obtained based on the positioning analysis of the type curves and the classification can be seen in Table 5.7. The model graphs are divided into left and right regions by thick solid lines.

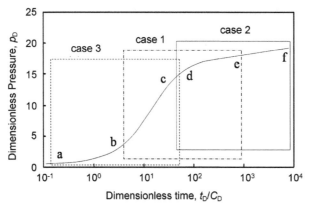

Figure 5.57 Positioning analysis examples (semilog plot).

Table 5.6 Some Real Field Cases of Positioning Analysis

Case No.	Parameters of the Case							Position of Curves		
	k/μ $\dfrac{mD}{mPa\cdot s}$	h m	C $\dfrac{m^3}{MPa}$	Δt_{start} min	Δt_{end} h	Δp_{start} MPa	qB $\dfrac{m^3}{d}$	$\left(\dfrac{t_D}{C_D}\right)_{start}$	$\left(\dfrac{t_D}{C_D}\right)_{end}$	p_D
1	100	10	0.1	1	4	0.01	10	3.8	905	0.453
2	1500	30	0.5	1	4	0.01	100	33.9	8140	2.44
3	50	1	0.1	1	4	0.01	10	0.19	45.2	0.027

a. For the right region under general circumstances, curves with a radial flow straight-line regime can be acquired by either a mechanical pressure gauge or an electronic pressure gauge and can be used successfully in graphic analysis and parameter calculation by well test interpretation software.

b. For the left region, the radial flow straight-line regime cannot be acquired from tests lasting several hours.

It is seen from Table 5.7 that if a pressure buildup curve of the well is obtained during a wellhead shut-in test, the log—log plot will be similar to model graph M-1-11 in Table 5.7, that is, $C_D e^{2S} = 10^4$ and $kh/(\mu C) = 300$ almost hold. Such curves are only with an afterflow segment and cannot be used to interpret the parameter calculation.

If a gas well is tested with wellhead shut in, and its wellbore storage coefficient C is about 3 m^3/MPa, the flow coefficient kh/μ of the well is estimated to be nearly 900 $mD\cdot m/(mPa\cdot s)$; meanwhile, if tested downhole shut in, it can be predicted that its wellbore storage coefficient C would decrease to 0.05 m^3/MPa, and thus the position parameter would change to

$$kh/(\mu C) = 18,000\left[mD\cdot m/(mPa\cdot s)\right]/(m^3/MPa).$$

Table 5.7 Classification of Positioning Analysis Model Graphs for Homogeneous Formations

$\dfrac{kh}{\mu C}$ / $C_D e^{2S}$	0.1 (very low)	0.3 (low)	0.8 (Fairly low)	2 (middle)	6 (Fairly high)	20 (high)	70 (very high)
0.1 (very low)	M-1-1	M-1-8	M-1-15	M-1-22	M-1-29	M-1-36	M-1-43
1 (low)	M-1-2	M-1-9	M-1-16	M-1-23	M-1-30	M-1-37	M-1-44
10 (Fairly low)	M-1-3	M-1-10	M-1-17	M-1-24	M-1-31	M-1-38	M-1-45
10^4 (middle)	M-1-4	M-1-11	M-1-18	M-1-25	M-1-32	M-1-39	M-1-46
10^{10} (Fairly high)	M-1-5	M-1-12	M-1-19	M-1-26	M-1-33	M-1-40	M-1-47
10^{20} (high)	M-1-6	M-1-13	M-1-20	M-1-27	M-1-34	M-1-41	M-1-48
10^{30} (very high)	M-1-7	M-1-14	M-1-21	M-1-28	M-1-35	M-1-42	M-1-49

It can be seen from the variation of the position parameter that when the test method is improved, the curve will approach to the pattern of model graph M-1-39, and its shape is improved dramatically, that is, a little afterflow portion is absent but a complete radial flow regime is obtained, from which formation parameters can be calculated accurately.

5.1.4 Field Examples

There are plenty of field examples of gas well tests in homogenous formations, including common sandstone and fissured carbonate formations. They are introduced separately here.

Large-Area Homogeneous Sandstone Formation

The SB gas field covers a gas-bearing area of 190 km^2. It can be divided into three anticlines with a wide and gentle structure and low amplitude. The main gas zone is the SB formation sandstone of the Quaternary system. Refer to Figure 5.58 for its gas-bearing area and structural conditions.

Based on well tests it is found that a majority of pressure buildup curves represent the characteristics of large-area homogeneous formations, two examples of which are shown in Figure 5.59.

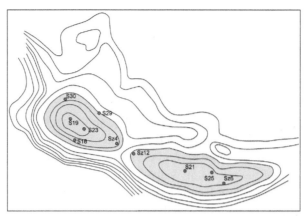

Figure 5.58 Structure and gas-bearing area in SB gas field.

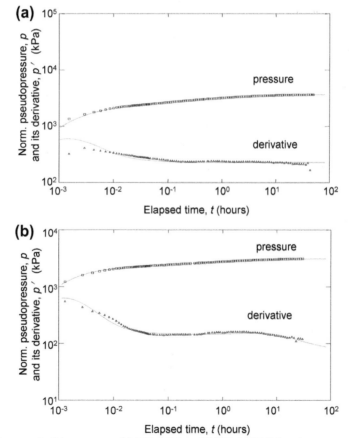

Figure 5.59 Pressure buildup curves of (a) Well T-5 and (b) Well SB4-3 (in a large-area homogeneous formation).

Parameters obtained from the analysis by well test interpretation software are as follow.

1. For Well T-5, permeability $k = 4.5$ mD, skin factor $S = 1.5$, and wellbore storage coefficient $C = 1.73 \times 10^{-2}$ m^3/MPa

2. For Well SB4-3, permeability $k = 5.9$ mD, skin factor $S = 4.2$, and wellbore storage coefficient $C = 4.99 \times 10^{-3}$ m^3/MPa.

The graphic shape of the curves of Well T-5 is analogous to that of model graph M-1-38. $kh/(\mu C) \approx 13,000$ and $C_D e^{2S} \approx 10^2$, which were calculated from interpreted formation parameters of Well T-5, under similar conditions of model graph M-1-38.

The graphic shape of the curves of Well SB4-3 is analogous to that of model graph M-1-39. $kh/(\mu C) \approx 18,000$ and $C_D e^{2S} \approx 3.3 \times 10^4$, which were calculated from interpreted formation parameters of Well SB4-3, under similar conditions of model graph M-1-39.

Carbonate Formation Displayed as Homogeneous Reservoir

There are many fissured gas fields with carbonate medium in China. Among them is the GB gas field, a limestone field mainly composed of Ordovician marine deposits. The field covers a fairly large area and is a gentle structural monocline. Its gas storage space is mainly fissure—porous medium. The field is described as a double or multiple porosity system by geologists, but obvious characteristics of a double porosity system, judged from the shape of the pressure buildup curve, only appear in a local zone (around Well L-5), and in another large area show homogeneous medium characteristics. The existence of chases made partial formations absent and formed no—flow boundaries or obvious heterogeneous variations. Uniform reservoir distribution is still seen in the regional blocks among the chases. Figure 5.60 displays a log—log plot of Well S-155 obtained

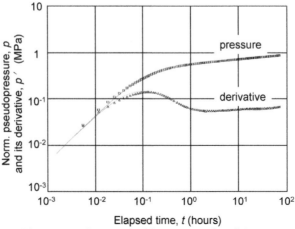

Figure 5.60 Pressure buildup curves of Well S-155.

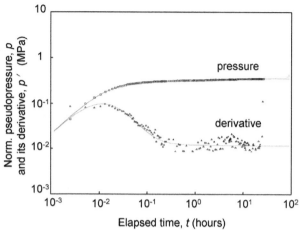

Figure 5.61 Log–log pressure buildup plot of Well M-4 (with homogeneous characteristic).

from a pressure buildup test lasting 90 hours, which reflects homogeneous characteristics obviously.

The graphic characteristic of Well S-155 is that its first half portion is basically similar to that of model graph M-1-25 in Table 5.7; the radial flow regime is extended dramatically due to a relatively long testing duration of about 100 hours. It also implies that a good radial flow regime can be acquired that is able to be used for parameter calculations even in a well having a large wellbore storage coefficient and tested by the wellhead shut-in method, if the pressure shut-in test lasts long enough.

The HTH gas field is another typical carbonate field. During the development of the field, the Carboniferous bioclastic limestone and Ordovician limestone were drilled, and most of the gas wells showed no commercial flow in the initial well completion stage, but they met the industrial gas recovery standards after acidizing stimulation. For example, before acidizing stimulation, Well M-4 has a gas flow rate of 1158 m^3/day, and the measured buildup curves were only with the afterflow portion. After acidizing, its flow rate was increased to 13.4×10^4 m^3/day and its pressure buildup test curves showed obvious homogeneous formation characteristics, as seen in Figure 5.61. The interpretation with well test interpretation software gives permeability $k = 88.7$ mD, suggesting good physical properties of the reservoirs, and $S = 8.5$, indicating existence of a certain damage.

The calculated graphic parameter and position parameter of Well M-4 are $C_D e^{2S} \approx 2 \times 10^{10}$ and $kh/(\mu C) \approx 30,000$, generally in line with the conditions of model graph M-1-40. When compared with the model graphs in Table 5.7, the graphic shape of it is really basically consistent with that of model graph M-1-40.

Condensate Gas Wells During Initial Flowing Periods

The YH gas field is a condensate gas field, and its main gas reservoir is Tertiary and Cretaceous sandstone. In a condensate gas field, before the formation pressure declines

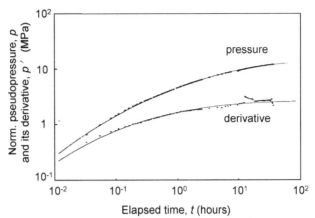

Figure 5.62 Log–log plot of pressure buildup of Well YH-6.

below the critical condensate pressure, there is no condensate oil generated there, and a single gas phase flow is maintained, just like the gas flow regime in common dry gas reservoirs. Figure 5.62 is the pressure buildup curve of Well YH-6 in this gas field. The Tertiary Jidike Formation was penetrated during the drilling of Well YH-6; the tested interval was 4995–4998 meters. The curve shows typical homogenous formation characteristics.

It can be seen from Figure 5.62 that the curve shape is the same as that of model graph M-1-22 in Table 5.7, indicating good well completion quality and skin factor to be negative in this well. Also, permeability $k = 5.94$ mD and skin factor $S = -1$ can be obtained from well test interpretation software.

5.2 Characteristic Graph of Double Porosity System (Model Graphs M-2 and M-3) and Field Examples

5.2.1 Composition and Flow Characteristics of Double Porosity System

The "double porosity medium system" is usually called "double medium" for short in China. It is the one being discussed here. Actually, "double medium" should cover "double permeability medium system" as well, often indicating multilayer oil zones with different permeability. The double permeability medium system is called the "double permeability system" in this book. It is generally thought that formations with a double porosity system develop in fissured carbonate reservoirs, such as the Paleozoic buried hills in East China, the Carboniferous–Permian carbonate gas zones in Sichuan Basin, and the Ordovician limestone gas zones in the Ordos Basin. This concept is so accepted by geologists that they define the formations as double medium whenever they see various fissures in cores or rocks isolated by fissures in the reservoir during their core observations; they even calculate the reserves of the fissure system and the matrix system, respectively, with data of logging analysis and core analysis statistics.

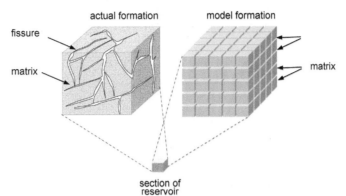

Figure 5.63 Cell body composition of formation with a double porosity system.

However, thorough investigations are rarely involved on what meanings of calculated reserves with these static methods are and what influences they bring to future development of the fields. In fact, it did happen in such a fissured gas field discovered in East China in the past few years, where undesirable errors were caused because the structural characteristics of the reservoirs had not been made clear, and reserves estimation was totally worked out with static data with misapplication of the concept of a double porosity system.

Composition of Formation with Double Porosity System

The cell of a reservoir with a double porosity system is shown in Figure 5.63. The definition of a "cell" can refer to Figure 2.47 in Chapter 2, and its volumetric size tallies with the definition of continuous medium.

The cell body shown in Figure 5.63 contains both fissure and matrix rock. Their volumes are respectively as follow.

1. Fissure volume per unit reservoir volume

$$V_{\mathrm{f}} = \frac{\text{Volume of fissure system in the cell}}{\text{Total volume of the cell body}}$$

2. Matrix volume per unit reservoir volume

$$V_{\mathrm{m}} = \frac{\text{Volume of the matrix system in the cell}}{\text{Total volume of the cell body}}$$

3. Total unit volume: $V_f + V_m = 1$

4. Pore volume of fissure system per unit reservoir volume: $V_f \phi_f$

Pore volume of matrix system per unit reservoir volume: $V_m \phi_m$

6. Storativity coefficient of fissure system: $V_f \phi_f \ C_{tf} \ (usually \ (V\phi C_t)_f)$
7. Storativity coefficient of matrix system: $V_m \phi_m \ C_{tm} \ (usually \ (V\phi C_t)_m)$
The storativity ratio (ω) is defined as

$$\omega = \frac{(V\phi C_t)_f}{(V\phi C_t)_f + (V\phi C_t)_m} \tag{5.56}$$

The definition of ω reflects the ratio of the fluid volume stored in two kinds of different storage spaces. The larger the value ω, the greater proportion of fluids stored in the fissure system, and vice versa.

Fluid Flow in Formation with Double Porosity System

In reservoirs with a double porosity system, only the fissure system is connected with the wellbore and has relatively high permeability; matrix rocks have very low permeability, and thus gas (or oil) can flow into the wellbore only via fissures. The flow process can be seen in Figure 5.64.

a. Fissure flow

Figure 5.64a shows that when the well is opened, the pressure of the fissures connected with the wellbore begins to decline, and the gas flows toward the bottom hole along the pathway of the fissure and is produced from the well, while no fluids flow out from the matrix rocks due to very poor permeability and thus no sufficient pressure difference.

b. Transition flow

When the pressure of the fissure system declines due to the production of gas, while the pressure of the matrix does not decline at all, there is a pressure difference formed between the matrix system and the fissure system. As a result, fluid in the matrix system flows into the fissure system to replenish it and buffer the pressure declining process in it.

c. The total system flow

When the pressure of the fissure system and the matrix system becomes balanced, both fissure and matrix systems participate in fluid supply to the well. This stage is called "the total system flow regime."

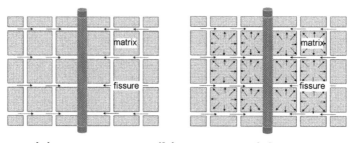

(a) flow in fissures   **(b)** transition flow & **(c)** flow in total system
Figure 5.64 Fluid flow process in formation with double porosity system.

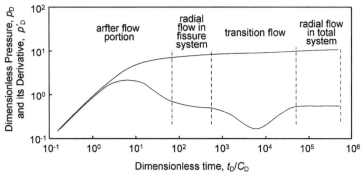

Figure 5.65 Model graph of double porosity formation with radial flows in fissure system and total system.

This flow regime was put forth by Balenblatt, a flow mechanics specialist in the Soviet Union. He also created a mathematic model for it and obtained the solution of the differential equation describing it [4]. It is the origin of the dynamic model concept of the double porosity system. Therefore, the "double porosity system" model is a dynamic model focusing on flow characteristics rather than a static reservoir description.

Flow Characteristic Graph of the Formation with Double Porosity System
A typical log—log plot of the formation with double porosity system is shown in Figures 5.65 and 5.66.
1. Figure 5.65 shows characteristics of the curve corresponding to the flow process in Figure 5.64, including
   • Radial flow can occur during the stage of flowing in a fissure system.

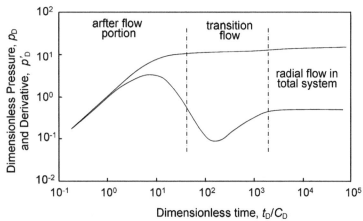

Figure 5.66 Model graph of double porosity system formation with radial flow in total system only.

- After radial flow in the fissure system, hydrocarbon in the matrix rock starts flowing to the fissure system so that pressure declining in the fissure system is hindered and transition flow is formed, which is indicated by the pressure derivative becoming a down concave. This pressure derivative transition concave is the most important characteristic of the model graph of double porosity system formation.
- When pressures in both the fissure system and the matrix system become dynamically balanced, pressures of the fissure system and the matrix system decline simultaneously and radial flow in the total system is formed. The pressure derivative curve becomes a horizontal straight line once again.

These are flow characteristics of a double porosity system with two radial flows.

2. Figure 5.66 shows characteristics of the double porosity system curve without the regime of radial flow in the fissure system:

- The afterflow segment extends rightward due to large wellbore storage coefficient C in the tested well or the transition flow segment moves leftward due to large interporosity flow coefficient λ, with the horizontal straight line indicating that radial flow in the fissure system on the pressure derivative curve is overlapped, causing it to be absent.

  Figure 5.67 reflects the transformation process. Originally, dimensionless wellbore coefficient $C_D = 10^3$ and the curve is with two radial flow segments. When the wellbore storage coefficient increases to $C_D = 10^5$, radial flow in the fissure system is overlapped by the wellbore storage effect, and only radial flow in the total system remains existing on the curve.

- Down concave of transition flow on the pressure derivative curve representing the characteristics of the double porosity system is still the main characteristics of the curve.
- Total system radial flow regime appears after the transition flow regime.

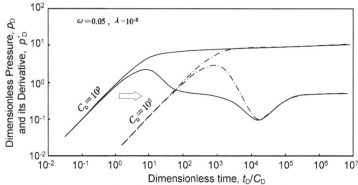

Figure 5.67 Curve characteristics of double porosity system being influenced by wellbore storage effect.

Difference of Formation with Double Porosity System and Common Homogeneous Formation

The performance of formations with the double porosity system is absolutely different from that of common homogeneous formations due to its special flow characteristics.

1. Relationship between high stabilized production rate conditions and reservoir structure

    Regarding all fissured carbonate reservoirs as reservoirs with the double porosity system as defined by geologists, some reservoirs among them with big fissures and caves often show a high initial production rate but cannot maintain it afterward. Some gas wells have an initial gas production rate as high as $100 \times 10^4 m^3$/day, but its bottom-hole flowing pressure declines more than 1 MPa or even 9 MPa per day during production, and the wells stop producing within 10 to 15 days after being put into production.

    What causes these phenomena? They are mainly the factors of the reservoir structure, that is,

    a. Foundation of high production rate. The gas wells mentioned earlier are often connected to big fissures and caves of the reservoirs, causing drilling break phenomenon, and are reflected by obvious big fissures in downhole imaging. A high production rate is due to good well completion conditions and high initial formation pressure. People are excited by a high production rate but very often neglect the stabilized production conditions.

    b. Foundation of stabilized production maintenance. Stabilized production maintenance relies mainly on dynamic reserves connected with the well. The fluids (gas and oil) stored in big fissures are very limited, so long-term stabilized production relies on those stored in matrix rocks. If the porosity ϕ of the matrix is not big enough and there is not enough fluids stored there or its porosity is big enough but its permeability is very low, for instance, as low as 10^{-5}–10^{-8} mD, the fluids cannot flow from matrix rocks to the fissure system under the general producing pressure differential within a certain production period, making stabilized production therefore impossible.

    Due to large differences among various fissured limestone reservoirs in their structures, the understanding from general logging data can only be very limited; core analysis data are often lacking representativeness, which result in more uncertain factors in static data.

2. Analysis of reserves distribution in the formation with double porosity system by parameter ω

    As mentioned previously, the concept of a double porosity system concerns the dynamic reservoir model. It is initially brought forward in well test analysis by percolation mechanic specialists. Their numerical range can be determined only

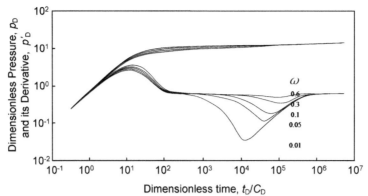

Figure 5.68 Well test model graph of double porosity system with different parameters ω.

by well test methods, and then accurate analysis on reservoir dynamic performance and stabilized high production rate conditions can be performed.

Several complete pressure buildup (or pressure drawdown) curves of the formations with different ω values are acquired from fields are shown in Figure 5.68.

Figure 5.68 shows that if the ω value is small, for example, $\omega = 0.01$, the down concave depth of the transition flow segment of the curve is relatively deep, and according to Equation (5.9) discussed previously in this chapter, there is an explicit relationship between down concave depth L_D and the ω value, that is, $\omega = 10^{-2LD}$. The ω can be obtained from the type curve matching by well test interpretation software

$$\omega = \frac{\left(C_D e^{2S}\right)_{f+m}}{\left(C_D e^{2S}\right)_f} \tag{5.57}$$

where

$\left(C_D e^{2S}\right)_f$ = graphic parameter of fissure flow segment (one of the curve match values)

$\left(C_D e^{2S}\right)_{f+m}$ = graphic parameter of total system flow segment (another one of the curve match values)

The calculation result of parameter ω is given directly by the software during its running.

Which kind of ω values will lead to better results—higher or lower? The answer to this question is under the conditions of the same gas production rate: the lower the ω value, the better. Reasons include

- The lower the ω, the richer the backup reserves, and the more possible the gas wells maintain stabilized production.

- However, if the ω value is higher, for example, increased to 0.3–0.5, although more than half of the reserves are stored in matrix rocks theoretically, reserves stored in the matrix often have a very low recovery factor in a very long recovery period, which are unfavorable for stabilized production.

  Obviously the parameter ω is very crucial to formation with the double porosity system and is very important on the prospect and destiny of gas fields and deserves to be treated seriously. Unfortunately, it has not yet been paid enough attention by reservoir engineers in the field and some phenomena exist, which must be noted:

- One phenomenon is not to understand the meaning of the ω value thoroughly, not to attach sufficient importance to the acquisition of parameter ω, and not to study and predict the future trend of oil or gas fields with ω values.
- The determination of ω value is very hasty. Some results are even very absurd and listed in well test analysis reports, but few people distinguish the truth from falsehood.

3. Analysis difficulty of recovery from the matrix system of double porosity reservoir with parameter λ

Parameter λ can be obtained from the analysis of well test curves, and the λ value is dimensionless and called the interporosity-flow coefficient. Its expression is

$$\lambda = \alpha r_{\mathrm{w}}^2 \frac{k_{\mathrm{m}}}{k_{\mathrm{f}}} \tag{5.58}$$

where

α = shape factor of matrix rocks, m^{-2}. Its values vary slightly according to the different assumptions of the matrix rock shape

r_{w} = well radius, m

k_{m} = permeability of matrix system, mD

k_{f} = permeability of fissure system, mD

Equation (5.58) shows that the λ value reflects the hydrocarbon supply capacity of the matrix system to the fissure system. Figure 5.69 presents the effect of the λ value upon flow performance.

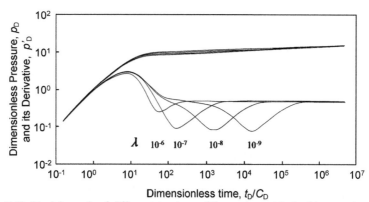

Figure 5.69 Model graph of different λ values of formation with double porosity system.

The graph shows that if the λ value decreases to one-tenth the original value, the occurrence time of transition flow will be 10 times later than the original one. Transition flow lasts about 100 days if $\lambda = 10^{-9}$ and 3 years if $\lambda = 10^{-10}$. It means

- A large amount of gas being supplied by matrix rocks to the fissure system will not start until the gas well has been opened for 3 years; it is obviously too late for a gas field under normal development.
- The curve characteristic of the transition flow segment acquired in the well test does take a fairly long time, and it is obviously very hard to test a well for such a long duration in field practice.

5.2.2 Several Influencing Factors in Acquiring Parameters of Double Porosity System

The aforementioned analysis shows that identification of the formation with a double porosity system depends on not only necessary geological conditions of the reservoir, but also whether dynamic characteristics or flow performance of the double porosity system really exists. Also, the key element to confirm double porosity system flow is the typical curve similar to the model graph shown in Table 5.7.

The following aspects can influence acquisition of a typical curve:

Pressure Buildup test Must Last Sufficiently Long

A typical model graph shows that the typical characteristic of a double porosity system on the model graph is a downward concave, which represents the transition flow regime between the two radial flow segments on the pressure derivative curve; if the λ value of the tested formation is very low, the test duration will be very long. Sometimes it is an insufficient test time that leads to a missed opportunity of a characteristic curve acquisition.

Effect of Wellbore Storage Coefficient on Test Data

Another factor influencing acquisition of a typical curve is the wellbore storage effect. Figure 5.69 shows that if the λ value is high, the downward concave of the curve will be deformed obviously; if the λ value is high enough, it will totally cover up the features of the transition flow. A great wellbore storage effect is very likely to occur in gas wells, especially in the case of wellhead shut in, and the C value can be as high as $3-5$ m^3/MPa, which will hamper seriously the acquisition of double porosity system typical curves.

It should be noted particularly that if condensate gas is produced from the well, the curve shape will be disturbed greatly due to the "hump" formed by variation of the wellbore storage coefficient.

Disturbance of Complex Fissure System to Double Porosity System Characteristic

Shapes of the pressure buildup curve that probably appear in complex fissure systems are displayed in graphs M-15 and M-16 in Table 5.5.2. Field experience tell us that there exist few successful acquisitions of typical curves of formations with the double porosity system that can be used for interpretation and that there are inevitable risks in interpreting ω and λ values from confusing curves.

Pressure Buildup Curve Cannot Show Characteristics of the Transition Flow Regime Due to too Short Flowing Duration Before Shutting In

In a paper about typical curves of the double porosity system, Gringarten explains the effect of flowing duration on the followed shut-in pressure buildup curve in detail.

As discussed before, the flow process during the pressure drawdown test in double porosity system reservoirs consists of three stages, that is, flow in the fissure system, transition flow, and flow in the total system occurring successively. If the pressure drawdown test undergoes all of the three stages, all of them will be repeated occurring in the pressure buildup test. The pressure buildup curve acquired in this case will show all curve characteristics of the double porosity system reservoir.

Contrarily, if the flow duration is very short and the shut-in pressure buildup test begins before transition flow occurs, it could be inferred that the transition flow regime is absolutely impossible to appear on the pressure buildup curve. Taking Well Wei-34 in the WY gas field in China as an example: when it was just completed, the acquired pressure buildup curve indicated that the reservoir was homogeneous, while the pressure buildup curve acquired after the well had been put into production for a period of time indicated that the reservoir was a double porosity system, which could be good evidence for that.

Effect of Reservoir Boundary

Edge water around oil reservoirs and reservoir permeability becoming better will cause the pressure derivative to incline downward, whereas a no-flow reservoir boundary will make the pressure derivative tilt upward. The inclining up and down will create a false impression of the double porosity system on the one hand and disturb the acquisition of the real double porosity system characteristic curve on the other hand, resulting in analysis results inconsistent with real formation conditions.

Effect of Pressure Measurement Precision

The quality of some pressure data acquired in the early years was quite poor due to low-precision mechanical pressure gauges. Data points jumping up and down occur after differentiation, and it is difficult to discuss the analysis of double porosity system characteristics.

5.2.3 Conditions for High-Quality Data Acquisition and Some Field Examples

As introduced previously, parameters ω and λ are very important for the development of gas fields with a double porosity system, but some difficulties indeed exist in acquiring good pressure buildup curves. It is due to its significance that such data must be acquired and analyzed well by all means for fissured limestone gas fields.

Conditions of Acquiring High-Quality Pressure Buildup Curves in Double Porosity System

A well test design must be well done prior to testing, and the following points must be brought to attention during the design, including

1. Sufficient production time prior to the shut-in pressure buildup test must be required, and flowing pressure history during the whole test must be acquired
2. Sufficient pressure buildup test time must be required
3. A high-precision electronic pressure gauge should be used
4. A pressure gauge must be set at the middepth of gas zones or below the liquid level in the case loaded water or liquids exist so as to obtain a real bottom-hole pressure variation
5. Wells located in areas free from the boundary effect must be selected to be tested to learn about double porosity system characteristics
6. Parameter ω and λ analysis may not be conducted necessarily if the tested reservoir is a very complicated fissure system.

Field Examples

The main pay zone of the GB gas field is Ordovician fissured limestone. Geologically, the formation has a physical basis of a reservoir with a double porosity system.

Interference tests between wells were performed as one part of a feasibility study prior to field development in the middle areas of the GB gas field. A well group composed of Wells L-5 (active well), L-1, and SC-1 (observation wells) were selected for the inter-ference test and pressure buildup test study, very typical curves of a double porosity system were obtained from Wells L-1 and SC-1, and the test result of Well L-1 can be seen in Figure 5.70.

Figure 5.70 shows clearly that the transition flow regime follows the afterflow segment and then radial flow in the total system appears. It is too bad that the total system radial flow segment was not fully acquired. Because a pressure buildup test lasts nearly 1000 hours, acquiring a curve is not easy for field operation.

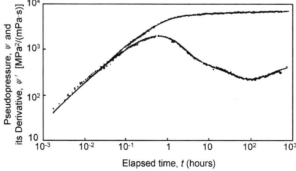

Figure 5.70 Log–log plot of pressure buildup of Well L-1.

Parameter results based on well test software analysis are shown as follow:

Permeability of fissures $k_f = 1.0$ mD

Storativity ratio $\omega = 0.285$

Interporosity-flow coefficient $\lambda = 1.4 \times 10^{-8}$

It can be seen that nearly one-third of natural gas in a reservoir is stored in the fissure system, while the remaining part is stored in the matrix rock. It can be said that, due to the high value of ω in the formation, the gas reserve supply in matrix rock is insufficient. But it should also be noted that netted fissures are dominant in this area and that the dense fissure system serves not only the flowing channel but also storage space of the gas. In particular, from the pressure history of Well L-1, the pressure did not decline quickly during production, which means that stabilized production is still possible under certain conditions.

5.3 Characteristic Graph of Homogenous Formation with Hydraulic Fractures (Model Graphs M-4 and M-5) and Field Examples

Some gas fields with a large area and low permeability were discovered in the central and western areas of China. For instance, formation permeability of the upper Paleozoic gas zones in the Ordos Basin is less than 1 mD, the best one is less than 10 mD, and the flow rates of the wells there cannot meet the industrial gas flow conditions. In order to develop these low deliverability gas fields, such stimulation treatments as sand acidizing must be taken to improve the bottom-hole flowing conditions and enhance the production rate of each well.

5.3.1 Creation and Retention Mechanism of Hydraulic Fracture

A vertical fracture (in deep wells) or horizontal fracture (in shallow wells) connected to wells is usually formed by sand fracturing. A fracture creation occurs when the pressure of the fracturing fluid in the bottom-hole exceeds the minimum principle stress of formation rocks, and so the fractures always extend outward along the direction of the maximum formation principle stress, as shown in Figure 5.71.

For example, in the Ordos Basin mentioned earlier, the direction of the maximum principle stress of the formation is approximately east–west, so is that of the created fractures consequently, just perpendicular to the direction of the river channel.

Ahead fluids for fracturing treatment continue to advance toward deep formation after the fracture has been created. Subsequently, the sand carrier carries sands to enter the formation. The fracturing front margins of the fracture formed in the zone where fracturing fluids invade always extend further than fracturing sands do, as liquid advancement are much easier than migration of solid matter.

Once pumps are stopped and a flowback operation is conducted, fractures will be closed again due to formation stress, and only those filled with sufficient fracturing sands can remain and are called "propped fractures." It is this propped fracture that makes flow more expedite and improves the production of oil and/or gas, as shown in Figure 5.72.

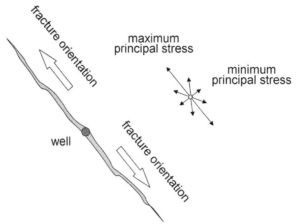

Figure 5.71 Creation and development of hydraulic fracture.

If fracturing sands added into the fluid are in good grading, the permeability of the propped fracture will be very high. Define a dimensionless parameter named fracture conductivity F_{CD} (according to Agarwal):

$$F_{CD} = \frac{k_f \cdot W}{k \cdot x_f} \qquad (5.59)$$

where
k = formation permeability, mD
k_f = permeability of filled proppant in fracture, mD
x_f = half length of the propped fracture, m
W = average width of the propped fracture, m

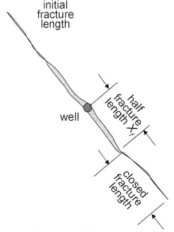

Figure 5.72 Retention of propped fractures after fracturing operation.

If F_{CD} is larger than 100, the fracture is considered a high conductivity one and is called an infinite conductivity fracture. If F_{CD} is small (e.g., $F_{CD} = 5$), it is a low conductivity fracture and is called a finite conductivity fracture.

In the case of carbonate reservoirs with primarily developed high-angle fractures, the invasion of drilling fluid and completion fluid into the reservoirs occurs very often during drilling and completion operation, causing damage of the reservoir. For a well drilled on this kind of reservoir, stimulation treatment of acid fracturing is usually taken after well completion. Acid fracturing will make natural fractures in the formation unobstructed and connecting each other or even extend them further to form a fracture zone, as shown in Figure 5.73.

The well in such a fracture zone will show performance similar to that in an infinite conductivity fracture regime in the bottom hole when producing, which is called uniform flow. In a uniform flow regime, flow velocity perpendicular to the fracture zone surface is considered uniform.

Since 1937, many researchers have studied wells connecting with a big fracture at the bottom hole and developed more than 40 theoretical models. Such fractures are summarized as follow:

a. Infinite conductivity vertical fracture created by hydraulic fracturing treatment

b. Natural vertical fissure with uniform flow in formation

c. Finite conductivity vertical fracture created by hydraulic fracturing operation with sand infilling and proper grain size ratio

d. Horizontal fracture formed by fracturing in shallow oil or gas zones

e. Vertical fracture with certain damage in skin of fracture created by hydraulic fracturing treatment

They are illustrated separately.

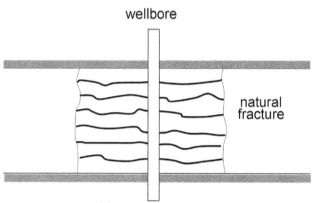

Figure 5.73 Natural fracture zone improved by acid fracturing.

5.3.2 Curve Characteristics of Well Connecting with a High Conductivity Vertical Fracture [36]

Curve Characteristics

Figure 5.74 shows typical log–log transient pressure curves of a fractured well connecting with a high conductivity hydraulic fracture in a homogeneous formation.

The parameters of typical curves in Figure 5.74 are as follow:

- Formation permeability $k = 1$ mD, formation thickness $h = 10$ meters
- Half length of hydraulic fracture $x_f = 200$ meters
- Fracture conductivity $F_{CD} = 150$, $k_f W = 30,000$ mD·m
- Fracture skin factor $S_f = 0$

High conductivity capacity is usually called infinite conductivity, which means that F_{CD} is generally higher than 100; even higher than 500 as defined by some authors. The transient pressure curves of the wells with high conductivity capacity have very prominent features: the whole curve can be divided into four segments, including

- Wellbore storage flow segment
- Linear flow segment
- Transition flow segment
- Pseudo radial flow segment

1. Wellbore storage flow segment

A well test with a wellhead shut in is usually performed in some gas wells, causing an afterflow (i.e., wellbore storage flow in buildup) segment appearing at the beginning of the pressure buildup curve of the fractured gas wells. However, this segment is often absent in fractured oil wells, especially in the cases of bottom–hole shutting in.

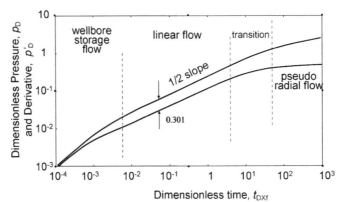

Figure 5.74 Log–log model graph of a well connecting a high conductivity hydraulic fracture ($kh = 10$ mD·m, $x_f = 200$ meters, $F_{CD} = 150$, $S_f = 0$).

Be especially careful of the characteristics of an afterflow segment in the graph: a low skin appearance is met sometimes due to the existence of a hydraulic fracture, but not all high conductivity fractured wells show this low skin appearance. When the fracture surface is damaged, that is, $S_f \neq 0$, great variations of the shape of the afterflow segment will appear. It is specifically discussed later.

2. Linear flow segment

Refer to Figure 2.20 in Chapter 2 for information on the linear flow regime. A linear flow segment is the most typical segment reflecting the characteristics of fractured wells. Both pressure and pressure derivative curves display a half-unit slope straight line, and the ratio of the distance between these two lines to the ordinate scale (the length of 1 log cycle) is 0.301.

The following demonstration is given to prove these characteristics:

Dimensionless pressure in linear flow regime can be expressed as follows:

$$p_D\left(t_{Dxf}\right) = \sqrt{\pi t_{Dxf}} \tag{5.60}$$

Logarithm transformation of both sides gives

$$\lg p_D(t_{Dxf}) = \frac{1}{2}\lg t_{Dxf} + 0.24857 \tag{5.61}$$

This formula shows that this segment of the curve is a half-unit slope straight line in a log–log plot.

The time limit of this straight-line segment is $t_{DXf} \leq 0.016$. But actually, when t_{DXf} is slightly higher than 0.016, the aforementioned characteristics still approximately exist.

It is obtained from the derivation of Equation (5.60):

$$\frac{dp_{eD}}{dt_{Dxf}} \cdot t_{Dxf} = \frac{\sqrt{\pi}}{2}\sqrt{t_{Dxf}}. \tag{5.62}$$

Logarithm transformation of both sides gives

$$\lg\left[\frac{dp_{eDxf}}{dt_{Dxf}} \cdot t_{Dxf}\right] = \frac{1}{2}\lg t_{Dxf} - 0.05246. \tag{5.63}$$

It can be seen that the derivative curve is also a half-unit slope straight line in a log–log plot and the difference between it and the pressure straight line is $\lg 2 = 0.301$ of the ordinate scale (the length of 1 log cycle).

3. Transition flow segment

The abscissa interval of the transition flow segment is $0.016 \leq t_{DXf} < 3$, and both pressure and its derivative curves incline upward and approximately keep in parallel in this segment.

4. Pseudo radial flow segment

The pressure disturbance transmits further as time goes on, the effect of the fracture becomes slight, when $t_{Dxf} > 3$, pseudo radial flow starts, and the pressure derivative curve presents a horizontal straight line, as shown in Figure 5.74.

In a pseudo radial flow regime,

$$p_D\left(t_{Dxf}\right) = \frac{1}{2}\lg\left(t_{Dxf}\right) + 1.100 \tag{5.64}$$

and

$$\frac{dp_D}{dt_{Dxf}} \cdot t_{Dxf} = 0.5 \tag{5.65}$$

Equation (5.65) shows that the pseudo radial flow segment of the derivative curve displays a horizontal straight line with a numeric value of 0.5 in dimensionless coordinates.

Dimensionless time t_{DXf} in Equation (5.65) can be expressed as

$$t_{Dxf} = \frac{3.6 \times 10^{-3} \, kt}{\mu \phi C_t x_f^2} \tag{5.66}$$

The dimensionless time t_{DXf} value is usually very small because the x_f value is relatively large, referring to Equation (5.66).

Most researchers express dimensionless wellbore storage coefficient in fractured wells as

$$C_{Dxf} = \frac{0.15916 \, C}{\phi h C_t x_f^2}. \tag{5.67}$$

If the hydraulic fracture is relatively short, for example, the x_f value decreases to 100 meters in Figure 5.74, the log–log plot looks like that of an ordinal stimulated well (with negative skin factor values) and loses the characteristics of a linear flow segment, as shown in Figure 5.75.

It is very necessary to learn about the well test operation procedure when interpreting curves like the aforementioned plot; if the fracturing operation or acidizing treatment for the fissured formation has really been performed, the analysis result of the artificial fracture can be concluded.

In addition, as for a hydraulic-fractured well with an infinite conductivity fracture, the skin factor of the whole well can be calculated with a half-length of fracture x_f with the following formula:

$$S_t = \ln \frac{2r_w}{x_f} \tag{5.68}$$

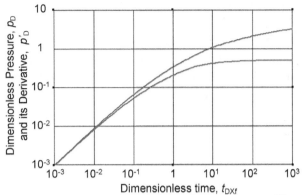

Figure 5.75 Model graph of a well with a short high conductivity fracture ($kh = 10\,\text{m D·m}$, $x_f = 100$ meters, $S_f = 0$, $F_{CD} = 150$).

For instance, if $x_f = 70$ meters and $r_w = 0.07$ meters, $S_t = -6.2$ can be obtained. It can also be seen that the value of the negative skin factor of the whole well is limited, even if $x_f = 500$ meters and the S_t value is merely -8.2.

Field Examples

The fracturing and completion operation is done very commonly for low permeability sandstone reservoirs in the Ordos Basin; such kinds of curves are acquired many times, and two of them are given here.

a. Well S-178

Figure 5.76 shows the measured curves of this well.

The well penetrates the Shanxi Formation in Permian and is completed after fracturing treatment. The graph shows that the pressure and pressure derivative

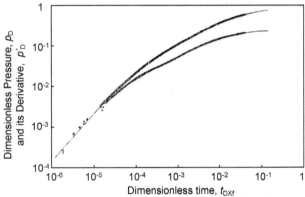

Figure 5.76 Log–log buildup plot of Well S-178 after fracturing.

curves clearly display half-unit slope straight-line segments following the afterflow segment, indicating that a high conductivity vertical fracture has been created from the fracturing operation; a pseudo radial flow is acquired afterward. The following reservoir parameters are obtained from the analysis with well test interpretation software: permeability $k = 0.024$ mD, indicating that the formation is with ultra-low permeability; half length of fractures $x_f = 59$ meters, suggesting dramatic improvement of flow around the bottom hole has been obtained; the effective total skin factor $S = -5.88$.

b. Well S-173

Figure 5.77 shows the measured curves of this well.

This well is located in the WSQ area of the CQ oil field, penetrated low permeability sandstone reservoir of the Shihezi formation in Permian and was completed after fracturing treatment. Total injected fracturing fluid was 186 m^3 with 24 m^3 of sands and ceramsites in it. After fracturing treatment and unloading fracturing and drilling fluids, the well was tested and a gas flow rate of 6.38×10^4 m^3/day was obtained. The test procedure was composed of a modified isochronal deliverability test, a short-term production test, and then a shut-in pressure buildup test in succession. Figure 5.77 shows the log–log plot of the pressure buildup test.

It is thought according to well test interpretation results that the fracturing treatment had achieved obvious effects: half-length of the fracture was $x_f = 70$ meters, formation permeability was $k = 3.5$ mD, and channel boundaries exist.

5.3.3 Flow Characteristics of in Fracture with Uniform Flow

Uniform flow means fluid flowing into different parts of the fracture surface with equal flow velocity per unit area. The literature [36] put forth an opinion that uniform flow type curves applicable to those wells connect a natural vertical fissure.

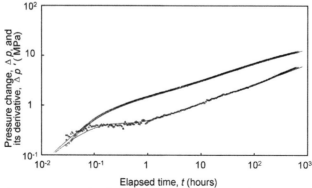

Figure 5.77 Log–log buildup plot of Well S-173 after fracturing.

Different from an infinite conductivity vertical fracture, the early half-unit slope straight line is longer, as long as t_{DXf} is smaller than or equal to 0.16. The aforementioned conditions for uniform flow are satisfied and then both pressure and pressure derivative curves in the log—log plot are half-unit slope straight lines and the ordinate difference between them is still 0.301 (of the log cycle). At a later time the curve is also a pseudo radial flow segment, the derivative value is 0.5, but the limitation of its start point is $t_{DXf} \geq 0.2$.

The curves of this case are also shown in Figure 5.74.

5.3.4 Vertical Fracture with Finite Conductivity [37]

If the hydraulic fracture is filled with sand and the granularity ratio of it is proper, the conductivity of the fracture becomes comparative with the formation permeability, and the fracture is of finite conductivity.

Curve Shape for Low Conductivity Fracture

If F_{CD} is small, for example, $F_{CD} = 1$, the curve shape is shown in Figure 5.78.

Figure 5.78 shows:

1. Afterflow segment. For a test of gas wells with the wellhead shut-in method, an afterflow segment often exists in the early stage, while it will be missed if downhole shut-in tools are used.

2. Bilinear flow segment. It is the characteristic line segment that reflects the existence of the finite conductivity fracture and expressed by quarter-unit slope straight lines. The so-called bilinear flow means:

   - Transient linear flow toward the bottom hole of the well inside the fracture
   - Formation linear flow perpendicular to the surface of the vertical fracture, as shown in Figure 5.79

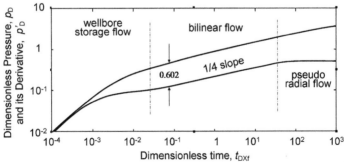

Figure 5.78 Log—log model graph of a well connecting a finite conductivity fracture ($kh = 10$ mD·m, $x_f = 200$ meters, $F_{CD} = 1$, $S_f = 0$).

well

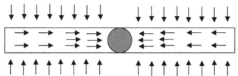

Figure 5.79 Schematic diagram of bilinear flow.

During the bilinear flow period,

$$p_D(t_{Dxf}) = \frac{\pi}{0.906\sqrt{2F_{CD}}} \cdot \sqrt[4]{t_{Dxf}}. \tag{5.69}$$

Logarithm transformation gives

$$\lg p_D = \frac{1}{4}\lg t_{Dxt} + \lg\frac{\pi}{0.906\sqrt{2F_{CD}}} \tag{5.70}$$

It is a quarter-unit slope straight line of p_D and t_{Dxf} on a log–log plot.
After logarithm transformation of the derivative of Equation (5.69), the following equation is obtained:

$$\lg(p_D \cdot t_{Dxf}) = \frac{1}{4}\lg t_{Dxf} - \lg 4 + \lg \frac{\pi}{0.906\sqrt{2F_{CD}}}. \tag{5.71}$$

Subtracting Equation (5.69) from Equation (5.71) results in

$$\lg p_D - \lg(p_D \cdot t_{Dxf}) = \lg 4 = 0.602. \tag{5.72}$$

It can be seen that both pressure and pressure derivative curves on log–log plot in a bilinear flow segment are straight lines, and the ordinate difference between them is 0.602 (of log cycle).

3. Pseudo radial flow segment. This segment is similar to the infinite conductivity vertical fracture mentioned earler, the pressure derivative curve is a horizontal straight line $\frac{dp_D}{dt_{Dxf}} \cdot t_{Dxf} = 0.5$ in a log–log plot, and is a straight line in a semilog plot.

Curve Shape in the Case of Large Conductivity

If F_{CD} is large, for example, $F_{CD} = 15$, the half-unit slope straight-line segment representing pure formation linear flow will appear after the quarter-unit slope line segment representing bilinear flow, as shown in Figure 5.80.

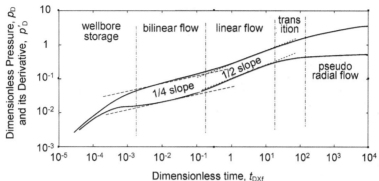

Figure 5.80 Log–log model graph of a well connecting a finite conductivity fracture ($kh =10$ mD·m, $x_f = 250$ meters, $F_{CD} = 15$, $S_f = 0$).

Figure 5.80 shows that the flow regime can be divided into four stages, including

- Afterflow (wellbore storage flow in buildup) segment
- Bilinear flow segment—both pressure and pressure derivative curves are quarter-unit slope straight lines and the distance between them is 0.602 (of log cycle)
- Linear flow segment—both pressure and pressure derivative curves are half-unit slope straight lines and the distance between them is 0.301 (of log cycle)
- Pseudo radial flow segment

5.3.5 Fracture Skin Factor and its Effect
Damage Mechanism of Fracture Skin Zone

During the fracturing operation, especially large-scale fracturing, several hundred or even a thousand cubic meters of fracturing fluid will often be injected into the formation with a high pumping rate and high pressure. The fracturing fluid will infiltrate into the fracture surface to cause formation pollution and damage simultaneously when the formation is fractured and a big fracture is created, as shown in Figure 5.81.

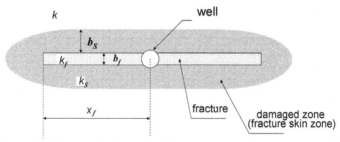

Figure 5.81 Schematic diagram of damage mechanism on fracture skin.

Fracture skin factor S_f is defined and expressed by the following formula:

$$S_f = \frac{\pi b_s}{2X_f}\left(\frac{k}{k_s} - 1\right),$$ (5.73)

where

S_f = skin factor of fracture, dimensionless
k = formation permeability, mD
k_S = permeability of skin zone, mD
b_S = thickness of skin zone, m
x_f = half length of hydraulic fractures, m

In Equation (5.73), the main factors influencing the S_f value include skin zone permeability k_S and skin zone thickness b_S; the more serious fracture plugging, the smaller k_S will be. Table 5.8 shows the relationship between k_S and S_f when $b_S = 0.1$ meter, $k = 1$ mD, and $x_f = 50$ meters.

Table 5.8 shows that when k_S decreases to one-tenth or one-hundredth of the original permeability, S_f will be 0.03 or 0.3, respectively. Therefore, even though the absolute value of skin S_f does not change much, the damage of the fracture skin zone is tremendous.

Typical Curves and Field Examples of Existence of Fracture Skin

1. Effect of fracture skin on the shape of the curves

The damage of the fracture skin zone will undoubtedly lead to the increase of flow resistance and producing pressure differential, as well as the decrease of gas deliverability of gas wells consequently. Although the existence of a hydraulic fracture makes the bottom-hole flow become better, total skin still increases to some extent due to the existence of fracture skin. The pressure derivative curve becomes a down concave obviously early due to the increase of S_f; absolutely different from the characteristics reflected in Figures 5.74, 5.75, 5.78, and 5.80, as shown in Figure 5.82.

Figure 5.82 shows that the pressure derivative inclines downward, for linear flow segment appearing originally next to afterflow segment, the flow is hindered due to the existence of fracture skin. This is the most significant characteristic of the existence of fracture skin.

Table 5.8 Relationship Between Fracture Skin Factor s_f and Permeability of Skin Zone k_S

k_S(mD)	1	0.5	0.2	0.1	0.05	0.01
S_f	0	0.003	0.013	0.03	0.06	0.3

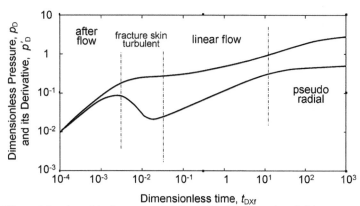

Figure 5.82 Effect of fracture skin S_f upon model graph of fractured well ($kh = 10$ mD·m, $x_f = 200$ meters, $F_{CD} = 15$, $S_f = 0.3$).

2. Field examples

Field examples of displaying fracture skin S_f after fracturing operation occur very universally.

a. Well S117

Figure 5.83 shows a measured pressure buildup curve of Well S117 in the CB gas field. This well penetrated the Shanxi formation sandstone in Permian and was completed after fracturing treatment. The gas flow rate was 9×10^4 m³/day before shutting in, and the following parameters are calculated in the analysis with well test interpretation software:

- Permeability $k = 0.45$ mD
- Half-length of hydraulic fractures $x_f = 75.8$ meters
- Fracture skin factor $S_f = 0.90$
- Non–Darcy flow coefficient $D = 0$

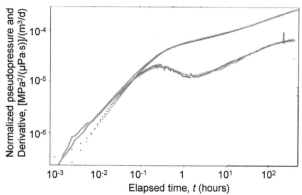

Figure 5.83 Pressure buildup log–log plot of Well S117 (quoted from Shell's report on Changbei gas field).

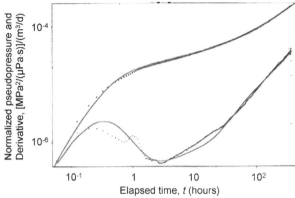

Figure 5.84 Pressure buildup log–log plot of Well Y27-11 (quoted from Shell's report on Changbei gas field).

b. Well Y27-11

Figure 5.84 shows the measured pressure buildup curves of Well Y-27-11 in the CB gas field. The well penetrated the Shanxi formation sandstone in Permian and was completed after fracturing treatment. Parameters of the formation and the well obtained from interpretation are as follow:

- Permeability $k = 0.65$ mD
- Half-length of hydraulic fractures $x_f = 74.1$ meters
- Fracture skin factor $S_f = 1.02$
- Non-Darcy flow coefficient $D = 0$

These field examples suggest that some technique defects exist in fracturing operations and that further improvements are necessary.

5.4 Characteristic Diagram of Wells with Partial Perforation (Model Graph M-6) and Field Examples

5.4.1 Geological Background of Well Completion with Partial Perforation

Whether the partial perforation method is selected for oil or gas wells mainly depends on the special geological conditions.

Partial Perforation for Gas Well Completion

Partial perforation is mostly applied in gas wells that exploit thick reservoirs with bottom water in order to preventing coning of the bottom water and causing reservoir water-out.

In China, partial perforation for gas wells was earliest applied in the WY gas field in the Sichuan Basin. The main gas pay zone of the field is the massive gas reservoirs with bottom water in Sinian. The thickness of the gas reservoir was 244 meters, and the elevation of the initial gas—water interface was −2434 meters. Fissures developed very well in the reservoir, and most of the development wells were completed above the gas—water interface and partially perforated the upper part of the formation in order to prevent coning of the bottom water. These wells displayed partial perforation characteristics in tests.

Th new found KL-2 gas field is also a very thick faulted anticline field with bottom water. The main gas-bearing horizon is the sandstone formation in Cretaceous. The thickness of the gas reservoir is about 350 meters and the gas—water interface is at −2468 meters. Only the upper part of the gas zone was perforated in order to prevent coning of the bottom water in the development plan.

In addition, the layering, perforating, and testing method was applied for a well test in the KL-2 gas field, which means that the layers of the reservoir are tested one by one. Test results show obvious partial perforation characteristics in the pressure buildup curves. Some intervals being perforated in the tests were at the top of the layer, whereas some were in the middle of the layer. Therefore, the models discussed here aim at various different possible perforated parts in formation.

Partial Perforation for Oil Well Completion

Similar to gas wells, the wells in some thick massive oil reservoirs will be completed with partial perforation and then be tested. The difference from gas wells is that it is not only for avoiding bottom water coning upward but also for avoiding the gas cap coning downward and gas channeling. Therefore, partial perforation is usually at the middle part of the oil reservoirs.

5.4.2 Flow Model in Cases of Partial Perforation

Partial perforation of gas zones will generate the flow pattern of a spherical flow or a semispherical flow. Refer to Figures 2.12, 2.13, and 2.15 in Chapter 2. Either spherical or semispherical flows present the same model graph as shown in Figure 5.85.

It can be seen from Figure 5.85 that the curve can be divided into four segments:

1. Afterflow segment. This segment is similar to conventional wells in homogenous formations, while the skin factor S in shape parameter $C_D e^{2S}$ reflects the damage of the perforated part of the formation.

2. Partial radial flow segment. It is often seen in most of the layered strata that some thin interbeds exist inside the thick layer. Even though these thin interbeds cannot absolutely stop vertical flow, they can make the vertical permeability k_V of the gas layer far smaller than lateral permeability k_H, thus retarding the occurrence of vertical flow. Therefore, a short-term radial flow in partially perforated intervals may sometimes occur in the early stage next to the afterflow segment.

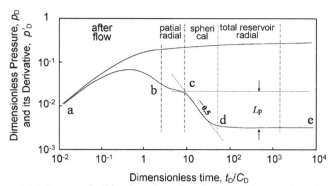

Figure 5.85 Pressure buildup log—log plot of wells with partial perforation.

If this flow segment really exists in the test curves, parameter values of perforated intervals can be calculated by interpretation software:

- Flow coefficient, $(kh/\mu)_{partial}$
- Permeability, $k_{partial}$
- Skin factor, $S_{partial}$, etc.

However, this flow segment does not necessarily exist for reservoirs without an obvious barrier effect and with fairly good vertical permeability among intervals. Therefore, parameters of different parts of a thick formation cannot be obtained by partial perforation and then testing of the well.

3. Spherical flow segment. Planar radial flow is transformed into spherical flow when flow also occurs in the intervals outside the perforated part (besides this perforated part) in the thick formation. Refer to Figure 2.15 for this flow. Spherical flow displays an inclined downward straight line with a negative half-unit slope in the pressure derivative curve as shown in Figure 5.85. It is the main graphic characteristic of spherical flow.

4. Radial flow of total formation. Generally, radial flow of the whole formation will appear after spherical flow if the test duration is long enough.

The following parameters of whole formation can be calculated from the radial flow segment of whole formation, including the overall flow coefficient $(kh/\mu)_{whole\ formation}$, overall permeability $k_{whole\ formation}$, and skin factor of whole formation $S_{whole\ formation}$.

Two points should be noted, including the following.

a. The real thickness of the whole formation must be made clear before calculating the k value of the whole formation; generally, the thickness of the reservoirs around perforated parts in which gas flow occurs is unknown; perhaps the flow occurs in thicker intervals as the flow time goes on, resulting in difficulties in defining the meaning of calculated $k_{whole\ formation}$.

b. The meaning of skin factor $S_{\text{whole formation}}$. $S_{\text{whole formation}}$ has two meanings: one is to indicate formation damage due to well drilling and completion and the other is to indicate an additional skin factor induced by bottom–hole concentrating flow in the perforation part of the formation. The influence of the latter is always much more serious, causing the skin factor to be much higher, even as high as over 100.

Figure 5.85 shows that there is a height difference between horizontal straight lines of partial radial flow and whole formation radial flow on a derivative curve, which is symbolized by L_p. L_p divided by the length of a log cycle of ordinate gives the dimensionless height difference $L_{pD} = L_p/L_C$. The ratio of flow coefficient of the whole formation to perforated intervals can be expressed as

$$M_p = \frac{(kh/\mu)_{\text{whole formation}}}{(kh/\mu)_{\text{partial}}} = 10^{L_{\text{PD}}}. \tag{5.74}$$

This equation shows that the larger L_{PD}, namely the larger height difference of derivative horizontal straight lines, the larger the flow coefficient ratio of the whole formation to perforated parts M_P will be, with M_P always exceeding 1.

The shape of the model graph will change in consideration of effect of partial perforation parameters, as shown in Figure 5.86.

Figure 5.86 shows the following.

- The effect of the ratio of thickness of perforated intervals to that of the whole formation

The thickness of the whole formation is 200 meters and that of perforated intervals h_P is 10 meters, accounting for 5% of the total thickness of the formation, reflected by a big height difference between horizontal straight lines of partial radial flow and whole formation radial flow on the derivative curve.

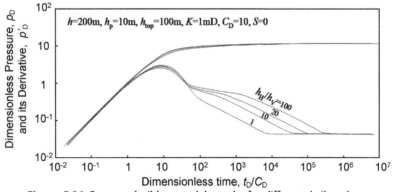

Figure 5.86 Pressure buildup model graphs for different k_h/k_v values.

- The effect of k_H/k_V

    Another important factor influencing curve shape is the ratio of lateral permeability k_H to vertical permeability k_V. The curve shows that when $k_H/k_V = 100$, two obvious radial flow segment exist, whereas when $k_H/k_V = 1$, the partial radial flow segment is missing, and the spherical flow segment appears as soon as the afterflow segment (negative half-unit slope straight line on derivative curve) ends.

5.4.3 Field Examples

Well KL-201

Well KL-201 is a gas well in the KL-2 gas field. Nine subintervals in intervals of 3600−4021 meters were selected to be tested, among which the longest and the shortest subintervals were 30 and 2 meters, respectively, accounting for 1/10th and one 1/150th of effective thickness of the whole formation.

Most curves display the partial perforation effect clearly, two of which are shown in Figures 5.87 and 5.88. For the main pay zone of Well Kela-201, the Bashijiqike formation in Cretaceous is a continuous thick one without obvious barriers inside.

The aforementioned test leads to useful understandings of the reservoir, yet some defects exist and need to be further improved.

1. Understandings of the reservoir

    Test data are acquired with an electronic pressure gauge. From pressure with high precision and good quality, the following understandings for the reservoirs are obtained.

    - Test curves contain a good radial flow segment, indicating that the main pay zones, the Bashijiqike formation in Cretaceous, are with homogeneous characteristics obviously, and that the small faults around the wells do not serve as an effective barrier.

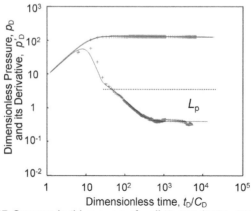

Figure 5.87 Pressure buildup curve of well KL-201 (3926–3930 meters).

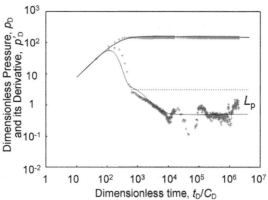

Figure 5.88 Pressure buildup curve of well KL-201 (3770–3795 meters).

- Test curves show typical partial perforation characteristics. The height difference between derivative horizontal straight lines of partial and whole formation radial flows ranges between 1 and 1.4 or so. M_p is roughly calculated to be approximate 10–20 by Equation (5.74), suggesting that the thickness of the effective pay zone in gas well production in the formation is more than 100 meters. Dynamically, main pay zones of the well are in Cretaceous, which are connected vertically as a whole and can be developed integrally. Very good development effects can be obtained no matter which intervals in this formation are perforated.
- Comparing the curve shape with that in Figure 5.86, it is found that the spherical flow segment appears quite early. It means that the k_H/k_V value is quite low, which implies that the vertical connectivity of the formation is good. This provides a further basis for integrated development of the gas field.

2. Exiting defects

The gas test arrangement for this well was originally used to understand the formation parameters and deliverability contribution of every gas zone by means of testing zone by zone. However, test results show neither interpreted parameters nor deliverability indices for each zone. Moreover, the deliverability index of the whole well cannot be calculated exactly because of interlacing of intervals caused by vertical flow in the test.

Well KL-2

This is a discovery well in the KL-2 gas field. Testing zone by zone was also conducted in this well. The gas-bearing intervals were 3567–4071 meters and were divided into 14 test zones for zone-by-zone tests. The perforation thickness varies from 1 to 10 meters, averaging 5 meters. Most test curves show partial perforation characteristics, as shown in Figure 5.89.

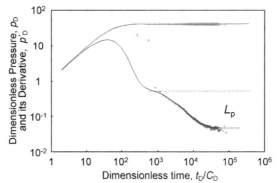

Figure 5.89 Pressure buildup log–log curves of Well KL-2 (3593.5–3595.5 meters).

The test of this well was performed after well completion. The perforated intervals were thinner and more obvious characteristics of the partial perforation effect appeared more than other wells. Expected results were not obtained from interpretation due to the fact that acquired data were regarded abnormal. In fact, the same information as that of Well KL-201 introduced earlier can also be extracted from the test curves. Moreover, Well KL-2 was tested before Well KL-201. The lesson of this well should be taken in the following tests, but unfortunately what happened in the test of Well KL-2 occurred again in the test of Well KL-201. This shows the importance of good well test design before test operation.

Well KL-204

Only one zone was tested in this well, and the test results can be seen in Figure 5.90.

The perforation thickness of this well is 5 meters and its gas flow rate is 21.84×10^4 m^3/day in the test. Partial perforation characteristics are also shown obviously.

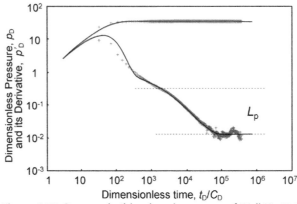

Figure 5.90 Pressure buildup log–log curves of Well KL-204.

5.5 Characteristic Diagram and Field Examples of Composite Formation (Model Graphs M-7 and M-8)

5.5.1 Principles for Evaluation of Type of Reservoir Boundary

Among various typical gas zone conditions and gas well model graphs listed in Table 5.5-1 and 5.5-2 in this chapter, those prior to model graph M-6 have not taken outer boundary conditions into account, while various outer boundary effects have been taken into account in model graphs from M-7 to M-16, and being reflected usually by tilting up or inclination down of pressure derivative curves at a later time. After a gas well starts flowing, the pressure change occurs and diffuses outward. Also, the pressure derivative curve will tilt up when the flow is obstructed and will incline downward when the flow becomes smoother. Readers can refer to Figures 2.26—2.30 in Chapter 2 in which common situations causing flow to be obstructed or smoother are listed.

However, ambiguous problem will be encountered if the reservoir structure is judged only by the shape of the pressure derivative curve. Several principles for analyzing reservoir dynamic models with well test curves are put forward here.

On the Basis of Geological Conditions

First of all, it is very necessary to understand and analyze reservoir conditions carefully and is very necessary first in both well test analysis and dynamic model validation. For example,

1. To determine if it is a sandstone reservoir or a carbonate reservoir
2. To make clear whether there are faults, the property and the throw of the faults, and their isolation to the formation in structural maps
3. For sandstone reservoirs, make clear the formation correlation for effective sandstones, and development and heterogeneity variations of the sandstone
4. For carbonate reservoirs, make clear the rules of fissure development, whether group- and/or series-distributed fractures exist, and whether the fissures developed in certain directions
5. For marine sandstone reservoirs, make clear whether there is tight tidal mud areas, that is, so-called delves exist
6. For thin fluvial sandstone reservoirs, make clear the width, length, and orientation of the river channels
7. For buried hill carbonate reservoirs, make clear the geological evaluation about fissure zones
8. For gas zones with special lithologic properties such as volcanic rock, massive reef limestone, and oolitic limestone gas zones, make clear the characteristic of them on the basis of geological research fruits and analyze its influence on gas percolation

A correct direction of reservoir dynamic model validation can be found only when sufficient understanding of geological conditions of gas reservoirs has been attained.

Model Graphs Obtained from Transient Well Test Analysis are the Main Analysis Basis

As mentioned several times in this book, the characteristics of pressure curves obtained from transient well tests describe the characteristics of hydrocarbon flow in the reservoirs.

As long as the well test design is well done and precise pressure data in sufficiently long test intervals are acquired successfully, various information of the reservoir and the well has been faithfully contained in the tested data. The next step is to analyze data with proper well test interpretation software to extract this information.

Production Performance of Gas Wells Will Verify the Correctness of the Model

Outer boundaries far away from the tested well cannot be reflected sufficiently due to the limited transient well test duration. Therefore, production performance during the production test process, that is, the pressure history, can be used to judge the reliability of the dynamic model recognized preliminarily from transient testing. Pressure history matching verification is thus considered as the most effective approach to verify dynamic models of gas wells and reservoirs.

5.5.2 Geological Conditions of Composite Formations

Widespread gas zones are usually heterogeneous due to the effects of a complicated sedimentary process. Geologically, the basis or conditions to speculate a reservoir as a composite formation are summarized as follow.

1. For porous sandstone reservoirs, a planar distribution map of an effective individual sand body of main pay zones has been obtained based on the interwell correlation analysis of subzones, and these individual sand bodies should be widely developed.
2. There are no faults segregating production zones around the wells on the structural map.
3. For fluvial sandstone formation identified by sedimentary facies research, there are no flow barriers in the effective sandstone bodies.
4. For fissured limestone formations in which there should be widespread netted fissures, a structural configuration may result in heterogeneity and orientation of fissure development, but it fails to shape a local development trend.

For a system with large fissures causing drilling break phenomenon and/or serious circulation loss, although the connection among fissures seems to be very good, it features local development from the past experience. Therefore, its distribution area may not be necessarily extensive.

Based on the aforementioned sufficient analysis, it can be identified preliminarily that the gas field may feature connected and continuously widespread distribution with planar heterogeneity.

5.5.3 Model Graph of Composite Formation

The composite formation is a kind of description for formation heterogeneity; formation heterogeneity is too complex to be described exactly. One of the simplest models is circular composite formation.

A circular composite formation means (as shown in Figure 5.91):

- There is a circular area around the well called the inner region with a flow coefficient of $(kh/\mu)_{inner}$ and a storability coefficient of $(\phi h C_t)_{inner}$; the radius of which is r_M
- Another area outside the inner region is called the outer region with a flow coefficient of $(kh/\mu)_{outer}$ and a storability coefficient of $(\phi h C_t)_{outer}$

In Figure 5.91a, the flow condition in the inner region is better than that in the outer region, that is,

$(kh/\mu)_{inner} > (kh/\mu)_{outer}$; whereas in Figure 5.91b., the flow condition in the inner region is worse than that in the outer region, that is, $(kh/\mu)_{inner} < (kh/\mu)_{outer}$.

Two Parameters Defined for Composite Formation Model

Two parameters defined for composite formation model include ratio of flow coefficient M_C

$$M_C = \frac{(k/h\mu)_{inner}}{(kh/\mu)_{outer}} \tag{5.75}$$

and ratio of storability coefficient ω_c

$$\omega_C = \frac{(\phi h C_t)_{inner}}{(\phi h C_t)_{outer}}. \tag{5.76}$$

M_C is greater than 1 in the case of Figure 5.91a and less than 1 in the case of Figure 5.91b.

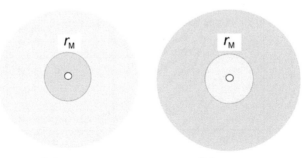

(a) $M_c > 1$ **(b)** $M_c < 1$
Figure 5.91 Circular composite formation.

Geologic Condition and Model Graph for Pressure Buildup Curves

Reasons why composite formation forms are varied:

- It forms due to the difference between formation coefficients of inner and outer regions. ω_c will be influenced by M_C in this case.
- It forms due to the difference between flow properties of inner and outer regions. For instance, if an oil ring or edge/bottom water exists in a gas reservoir where the tested gas well locates, the outer region will show the effects of the oil ring or edge/bottom water and form a composite formation with the situation that the flow condition in the inner region is better than that in the outer region, and the differences of M_C and ω_c will be slightly different from the former.

Transient well test type curves of composite formation are shown in Figure 5.92 and M-7 and M-8 in Table 5.5-1.

It can be seen from Figures 5.92a and 5.92b that the curves are divided into four segments:

1. Afterflow segment
2. Segment of radial flow in inner region, indicating the formation around the well is basically homogeneous
3. Transition segment: The pressure derivative curve tilts up due to flow hindered by a bad formation condition (Figure 5.92a) or inclines downward due to the flow condition becoming better (Figure 5.92b)
4. Segment of radial flow in outer region: Pressure derivative horizontal straight line appears again, suggesting that formation in the outer region is also basically homogeneous.

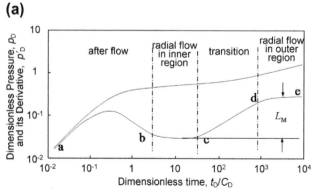

Figure 5.92a Model graph for pressure buildup curves of circular composite formation (flow condition in inner region is better than in outer region; M-7 in Table 5.5-1).

(b)

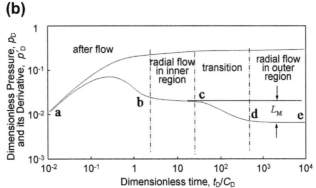

Figure 5.92b Model graph for pressure buildup curves of circular composite formation (flow condition in inner region is worse than in outer region; M-8 in Table 5.5-1).

Attention must be paid to the following points in the model graph of the composite formation shown in Figure 5.92.

a. There is a height difference between the two horizontal straight lines of radial flows in inner and outer regions of the derivative curve, expressed as L_M. A dimensionless value is defined as the ratio of L_M to L_C (one log cycle length of ordinate), that is,

$$L_{MD} = \frac{L_M}{L_C}. \tag{5.77}$$

The flow coefficient ratio of inner and outer regions can be calculated with the L_{MD} value:

$$M_C = 10^{L_{MD}}, \tag{5.78}$$

where L_{MD} is positive if the pressure derivative curve tilts upward (flow condition in inner region is better than that in outer region) and L_{MD} is negative if the derivative curve inclines downward (flow condition in inner region is worse than that in outer region).

According to Equation (5.78), the bigger the height difference (L_{MD}) of horizontal straight lines of inner and outer regions on the derivative curve, the bigger the flow coefficient ratio of inner and outer regions will be; therefore, the more abruptly the derivative curve tilts upward, the more sharply the outer region becomes worse.

b. Features difference between composite formation and formation partial perforation
  * These are two absolutely different situations in well completion conditions. Partial perforation is created by well completion conditions, and its related basis can be found from well completion records; if there is no such problem in well completion, the reason can be found from formation variation, that is, outer region of the formation making the flow condition better.

- In addition, an inclined downward segment in the two cases occurs at different times: a derivative curve inclining downward occurs in the early stage due to partial perforated formations, reflecting an influence of the well completion skin effect, whereas a derivative curve inclining downward appears in the middle or late time due to composite formations (the flow condition of the outer region becoming better).

5.5.4 Analysis of Field Examples

Field Example of Composite Formation in GB Gas Field

The Ordovician fissured limestone reservoirs are penetrated in the GB gas field. Most of the pressure buildup derivative curves acquired when the field was initially discovered in the 1990s tilt upward. Due to this fact, some experts doubt whether the reservoirs are cut by no-flow boundaries and local small blocks are formed consequently. If it is true, the reserves of the gas field will be reduced greatly. Such reservoir understandings bring a critical problem to hindering examination and approval of gas field reserves and subsequent development.

An expert team involved in thorough analysis on the gas field dynamic model denied the opinion of "gas fields distribute as various small blocks" on the basis of thorough research based on following three bases.

1. Geological basis

    The Ordovician limestone reservoir is a marine sedimentary formation developed in a large area and a gentle monocline. Neither faults nor other erosive boundaries or lithologic boundaries develop in the region. Therefore, it is estimated that the geological basis of that many closed blocks were formed is insufficient. But formation heterogeneity is still serious from the analysis of formation parameters of the existing exploration wells.

2. Characteristics analysis of pressure buildup curves

    Pressure buildup curves should be like model graph M-11 in Table 5.5-1 if the reservoirs are dominated by isolated local small blocks as predicted by some experts, and the pressure derivative curves will incline downward greatly at a later time. Except for Well S6, however, a well located at the gas zone margin that penetrated the formation of Mw-4, which is not the major pay zone, all pressure buildup curves of more than 30 wells are not like model graph M-11 at all.

    Moreover, characteristics of composite formation speared in pressure derivative curves of many wells. Most pressure derivative curves of the wells, for example, those of Well S5, Well S155, Well S52, and Well S175, show characteristics of composite formation in which the flow condition in the inner region is better than in the outer region; some others, one among which is Well S13, show characteristics of composite formation in which the flow condition in the inner region is worse than in the outer region. This phenomenon is mainly generated by the fact that the tested wells were usually selected among those with high deliverability. For wells with poor deliverability, some are abandoned because they cannot meet the commercial production specification, and some are not taken for the later production test, even though they meet the commercial production specification, causing a scarcity of data acquisition of those wells where the flow condition in the inner region is worse than in the outer region.

3. Pressure history performance

The pressure of most wells is relatively steady during the long-term production test, and pressure history matching verifications are also consistent with the trend of the model of composite formation.

It is confirmed according to the aforementioned analysis that the Ordovician reservoirs are well connected and widespread over large areas with a certain degree of heterogeneity, providing strong support for the reliability of reserves calculation. This judgment has been proved to be correct by nearly 10 years of production practice. What follows are some field examples.

1. Well S5

Well S5 is located in the southwest side of the central district of the JB gas field. Its kh is high, as seen from the contour diagram of the main pay zone, Mawu-1 Formation. The gas flow rate is as high as $8 \times 10^4 \, \text{m}^3/\text{day}$ in the deliverability test; a short-term production test was conducted following the deliverability test, the well was shut in to test pressure buildup curves, and the measured log−log plot is shown in Figure 5.93.

The plot shows that the pressure derivative curve tilts upward dramatically at a late time. The buildup test lasted nearly 500 hours but did not reach the radial flow horizontal straight line of the outer region yet.

A pressure history matching verification was performed to verify the reliability of the interpretation result about composite formation, and the pressure history matching plot is shown in Figure 5.94. The plot shows a good match of the

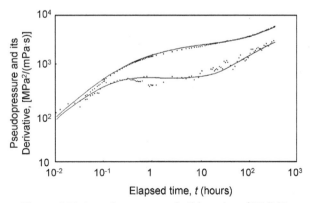

Figure 5.93 Log−log pressure buildup plot of Well S5.

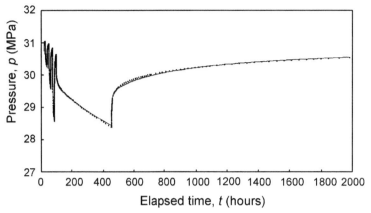

Figure 5.94 Pressure history matching verification of Well S5.

theoretical model and measured pressure data for either pressure drawdown sections or pressure buildup sections, which indicate that the interpretation results are reliable.

Parameters obtained from interpretations with related software include

Flow coefficient of inner region $(kh/\mu)_1 = 128.1$ mD·m/(mPa·s)

Permeability of inner region $k_1 = 2.7$ mD

Skin factor $S = -4.60$

Flow coefficient ratio of inner to outer regions $M_C = 16$

Storability coefficient ratio of inner to outer regions $\omega_C = 16$

Permeability of outer region $k_2 = 0.168$ mD

Radius of inner region $r_M = 225$ meters

Wellbore storage coefficient $C = 4.2$ m³/MPa

2. Well S155

   This well is located in the north district of the JB gas field, and its kh is also high, as seen from the contour diagram. Mawu-1 is the main pay zone, and the gas flow rate was 9.84×10^4 m³/day in the test. Similarly, a short-term production test was conducted after the modified isochronal test, and then the well was shut in for the pressure buildup test. Its log–log plot is shown in Figure 5.95.

   It can be seen from Figure 5.95 that the pressure and pressure derivative curves show composite formation characteristics. Also, the radial horizontal straight-line segment of radial flow in the outer region was not acquired in the test lasted around 1000 hours.

   Parameters obtained from interpretations with related software include

   Flow coefficient of inner region $(kh/\mu)_1 = 1229.1$ (mD·m/mPa·s)

   Permeability of inner region $k_1 = 25.6$ mD

   Skin factor $S = -1.7$

   Flow coefficient ratio of inner to outer regions $M_C = 9.3$

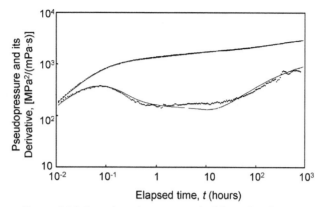

Figure 5.95 Log–log plot pressure buildup of Well S155.

Storability coefficient ratio of inner to outer regions $\omega_C = 9.3$
Permeability of outer region $k_2 = 2.75$ mD
Radius of inner region $r_M = 383$ meters
Wellbore storage coefficient $C = 2.58$ m^3/MPa

Figure 5.96 shows the pressure history matching verification of Well S155. Good matching of the theoretical model and measured data points indicates that the interpretation results are reliable.

3. Well S13

This well is located in the south district of the JB gas field and is a low deliverability well. The gas flow rate is 2×10^4 m^3/day in the test. Its pressure buildup curves are

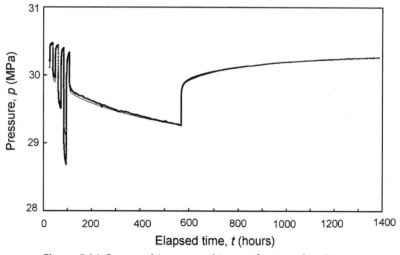

Figure 5.96 Pressure history matching verification of Well S155.

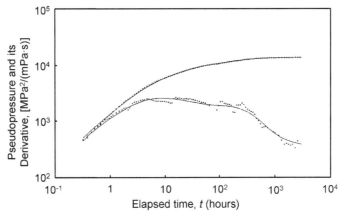

Figure 5.97 Log–log pressure buildup plot of Well S13.

acquired using the same test methods as that for tests in Well S5 and Well S155, as shown in Figure 5.97.

The log–log plot of Well S13 is completely consistent with the model graph of composite formation where the flow condition in the inner region is better than in the outer region (as shown in Figure 5.92).

Parameters obtained from interpretations with related software include

Flow coefficient of inner region $(kh/\mu)_1 = 51.97$ mD·m/(mPa·s)

Permeability of inner region $k_1 = 1.09$ mD

Skin factor $S = -4.24$

Flow coefficient ratio of inner to outer regions $M_C = 0.2$

Storability coefficient ratio of inner to outer regions $\omega_C = 0.2$

Permeability of outer region $k_2 = 5.45$ mD

Radius of inner region $r_M = 191$ meters

Good match verification of pressure history (see Figure 5.98) confirms the reliability of the interpretation results.

KL-2 Gas Field: Well KL205

The KL-2 gas field is a very high yield field with abundant reserves. Its main pay zone is Cretaceous sandstone. Well KL205 is one of the tested wells with the best acquired pressure data. After well deliverability was derived from the back-pressure test, a pressure buildup test was performed to learn about the reservoirs. The log–log plot of the well is shown in Figure 5.99.

It can be seen from Figure 5.99 that there are two different inclining sections on the pressure derivative curve after radial flow finishing. This implies a three-zone composite formation where the flow condition in the middle region is better than that in the inner region and that in the outer region is better than that in the middle region.

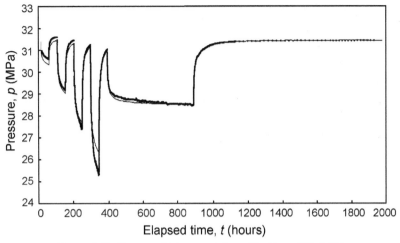

Figure 5.98 Pressure history match verification of Well S13.

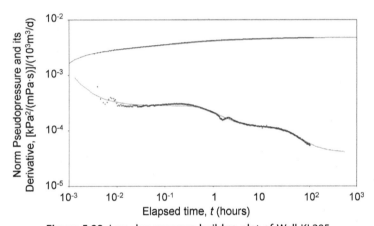

Figure 5.99 Log–log pressure buildup plot of Well KL205.

Parameters obtained from interpretations with related software include
Permeability of inner region $k_1 = 5.55$ mD
Permeability of middle region $k_2 = 14.1$ mD
Permeability of outer region $k_3 = 40$ mD
Radius of inner region $r_{1,2} = 40$ meters
Radius of middle zone $r_{2,3} = 400$ meters
Skin factor $S = 0$
Interpretation results of this well and those of adjacent Well KL203 confirm each other. Well KL203 is in the west side of Well KL205, and the calculated permeability is 70 mD in the interpretation, suggesting that permeability in the peripheral formations of Well KL205 become better westward.

The aforementioned results have been confirmed further by pressure history match verification; refer to Chapter 8.

An opinion of "there is a constant pressure boundary" was proposed in the well test analysis of Well KL-205. It is a conceptual misunderstanding for gas well test analysis. If it is interpreted that "there is a constant pressure boundary" and that there is no geological basis found in the gas field supporting this interpretation conclusion, it cannot be used to verify the rational dynamic model. It is a misapplication of well test interpretation methods for oil wells in gas well interpretation.

5.6 Characteristic Graph of Formations with No-Flow Boundaries (Model Graphs M-9—M-13) and Field Examples

5.6.1 Geological Background

As discussed in Chapter 2, pressure diffusivity is hindered if there is a no-flow boundary around the tested wells. The most common geological backgrounds of no-flow boundaries are described here.

1. Fault

    Not all faults shown in structural maps are closed ones; only those with quite high throw and being very tight can be real no-flow boundaries because there is a large amount of argillaceous content in the restraining materials in the fault face.

2. Lithologic boundary

    Fluvial sedimentary formations, especially shallow channels, undergoing active swinging during the sedimentary process will form thin and no-flow lithologic boundaries.

3. Groove formed by tide

    The gas pay zone of the JB gas field is in the marine Ordovician sedimentary formation. A great deal of "grooves" developed around the buried platform in which tidal flat dolomite gas reservoirs developed under a tidal erosion effect. Main gas reservoirs are missing due to incision of the troughs, and simultaneously argillaceous contents deposited in the troughs and formed impermeable barriers. Some gas wells located near the troughs are high-yielding ones but their dynamic performance is limited by the troughs.

4. Boundaries of fissured zones

    The main gas-bearing reservoirs of some carbonate fissured formations, for example, fissured reservoirs of buried hills, are in the systems of fissure-pore-solution cavity. These systems of fissure-pore-solution cavity were developed gradually during the geologic weathering and leaching process. They were formed and distributed within a certain range, alike sheet by sheet or string by string, and with obvious boundaries, just as observed in karst caves. Although some parts of the boundaries may connect to the outside via some kind of channels, these channels were probably closed by late tectonic movements over very long geologic time.

5.6.2 Flow Model Graph of a Well With No-Flow Outer Boundary

The velocity of flow along the direction being perpendicular to a linear no-flow boundary is zero. Assuming there is an "image well" of the tested well on the symmetric location of the linear boundary, this image well produces at the same flow rate and is closed at the same time with the tested well; the flow pattern of this system is shown in Figure 5.100.

The same principles can be applied to simulation of formations with multiple boundaries with multiple image wells. For instance, a producer drilled in a strip–like formation can be simulated by "well array" made up of infinite wells (Figure 5.101). Also, simulation of a well drilled in an angular formed by two interacting faults can also be accomplished by a serious of image wells (Figure 5.102), etc.

Angular No-Flow Boundaries (M-9, M-10)

If a well is located in the middle of an angular reservoir formed by two interacting faults, its model graph of the transient well test is as shown in Figure 5.103.

It can be seen from this graph that the flow pattern can be divided into four segments, including:

1. Wellbore storage flow (i.e., afterflow in buildup case) segment

    It is the same as that of common homogeneous formations.

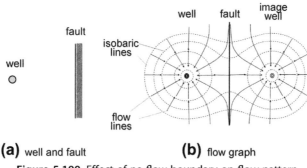

(a) well and fault   **(b)** flow graph

Figure 5.100 Effect of no-flow boundary on flow pattern.

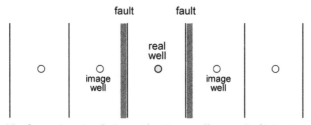

Figure 5.101 Strip-like formation simulation with mirror well array. (Left) Position of the tested well. (Right) Flow pattern of the well with no-flow boundary and the fault.

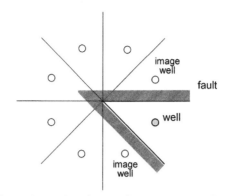

Figure 5.102 Simulation of two interacting faults.

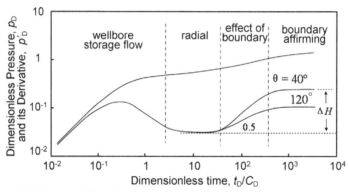

Figure 5.103 Flow model graph of a well on the bisector of two angular linear faults.

2. Radial flow segment

When the effect of transient pressure variation does not reach any boundaries, pressure still reflects features of the common homogeneous formation, and the pressure derivative curve in the radial flow segment is a horizontal straight line.

3. Boundary reflecting segment

The pressure derivative curve tilts upward in this segment. Many test data are judged reflecting no-flow boundaries from the fact of acquisition of the pressure derivative curve's tilting upward, but it is not enough to confirm the existence and properties of no-flow boundaries. As illustrated in Chapter 2, the pressure derivative curve's tilting upward may be induced by flow hindered geologically. Therefore, in order to make clear the existence and distance of no-flow boundaries such as faults, it is necessary to acquire data of the boundary confirming segment as next describing.

4. Boundary confirming segment

The pressure derivative curve shows a horizontal straight line again in this segment. The height of the second horizontal straight line, or the height difference between the second and the first horizontal straight lines, is expressed by symbol ΔH. A dimensionless height difference, ΔH_D, is obtained from dividing ΔH by L_C (i.e., the length of one log cycle of ordinate), that is, $\Delta H_D = \Delta H / L_C$, and then the angle between the intersecting faults θ can be expressed as

$$\theta = 360° \times 10^{-\Delta H_D} \tag{5.79}$$

It means that if a test well is located on the bisector of angle θ and the characteristic curves are acquired from the test as shown in Figure 5.103, the angle between the intersecting faults θ can be estimated instantly according to the height difference between the two derivative horizontal straight lines ΔH_D.

Certainly, exact formation parameters, including boundary angle θ, should be obtained by analysis with well test interpretation software.

Field example: Well S8

Test of Well S8 in the GB gas field is a typical one reflecting an angular boundary. It is located in the northern side of the middle district of the field, and near HHJ buried delve (Figure 5.104).

After a short time production test, a pressure buildup test was performed. Its log–log plot is shown in Figure 5.105.

It can be seen from Figure 5.105 that the curve shape is completely consistent with the model graph of Figure 5.103. Reservoir parameters are obtained from the interpretation with well test interpretation software, including:

- Flow coefficient $kh/\mu = 160.58$ mD·m/(mPa·s)
- Formation permeability $k = 3.22$ mD
- Angle of the groove $\theta = 32°$
- Skin factor $S = -3.9$
- Wellbore storage coefficient $C = 3.35$ m³/MPa

The angle of delve θ estimated with Equation (5.79) is about $30°$.

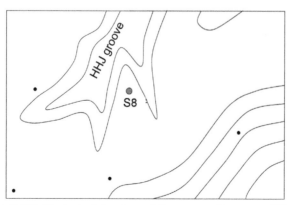

Figure 5.104 Geological structure of the surrounding area of Well S8.

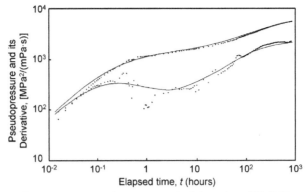

Figure 5.105 Log–log pressure buildup plot of Well S8.

To verify the reliability of interpretation results, pressure history matching verification was conducted, and the result is shown in Figure 5.106.

It can be seen that the simulated pressure history and measured pressure history match each other very well, proving the reliability of interpretation results.

No-Flow Boundaries Shaped Approximately Like a Circle

1. Geological background

    If a well is in a small reservoir with closed boundaries shaped like a circle or approximately a circle, a square, or even an irregular polygon, its production behavior is very similar. The shapes of the reservoir are shown in Figure 5.107.

2. Flow model graph

    At the late time the curve shape is as follows:

    • A straight line with a unit slope (45°) for pressure drawdown derivative curve.

    • When the effect of all boundaries in different directions reaches the bottom hole of the tested well, the pressure derivative curve goes down rapidly in a pressure

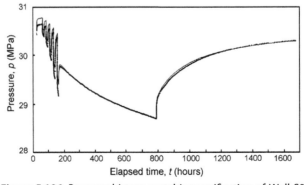

Figure 5.106 Pressure history matching verification of Well S8.

Figure 5.107 Reservoir structures of closed no-flow boundaries.

buildup plot. This is because, for a closed block, after the well is shut in, the pressure in the block tends to be balanced very quickly, the pressure drawdown funnel disappears, and the pressure in the block will reach the average pressure ($\bar{p}$). The derivative of pressure that tends to be constant is close to zero, and the derivative curve will drop downward rapidly in log coordinates, as shown in Figure 5.108.

Many closed blocks that seem to be limited geologically do not show the dynamic characteristics of a closed block. The reasons are as follow.

- Some closed blocks are shaped as a rectangular whose length—width ratio may be over 10 and among which are commonly seen fluvial sedimentary formations. If formation permeability is relatively low, the farthest boundary effect will be unable to reach the bottom hole of the well for a long time and thus the characteristics of the limited reservoir will not show for a long time, possibly dozens of days or even several months later.
- Some blocks confined by no-flow boundaries are connected to the outside in a certain direction or a certain part. Even if the connected area is very limited, the pressure derivative curve will not drop.
- Some boundaries are so-called "flow barriers" with ultralow permeability but not completely closed. It often happens to lithologic formations with fluvial sedimentation.

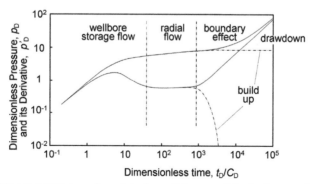

Figure 5.108 Log—log plot of a well in a small limited reservoir.

- Some wells in a reservoir with group- and/or series-distributed fractures, after production for a certain period of time, do not seem to be in the same pressure system. However, they may be connected to each other by some channels judged from their formation pressures. This situation is met commonly in the south of the Sichuan Basin. There is a rapid decline of a pressure buildup derivative curve of the wells, which is rarely seen until later.

Field example: Well S6

Well S6 is one of the few test wells with typical closed boundary characteristics, whose test curves are as shown in Figure 5.109.

This well is located in the western side of the middle district of the JB gas field. The main pay zone is the Mawu-4 reservoir of the Majiagou formation in Ordovician. The reservoir developed only in local areas and formed isolated local small blocks. The test interval of Well Shan-6 is 3578.21−3621.31 meters, and the gas flow rate was 10.26×10^4 m^3/day during the well test and 37.77×10^4 m^3/day after acidizing treatment. Pressure buildup curves showed closed formation features clearly both before and after acidizing operations. It is the only tested well being observed with closed reservoir characteristics in the GB gas field so far.

The local distribution characteristic does not bring any impacts on reserves estimation of the whole field because the Ordovician formation penetrated in Well S-6 is not the main pay zone, whose reserves account for a very small proportion in the entire gas field.

Band-Shaped No-Flow Boundary

1. Geological background
   a. A long band-shaped no-flow boundary is met commonly in a meandering formation formed by fluvial sedimentary, especially in low permeability sandstone

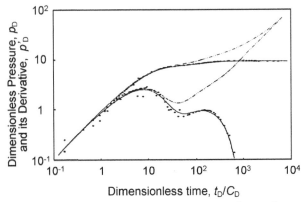

Figure 5.109 Log−log pressure buildup plot of Well S6 after acidizing treatment.

with thin layers. The reason for this is that, according to geologists, the effective thickness of the single layer relates to the width of the river channel, that is, the thinner the single layer, the narrower the river channel, as shown in Figure 5.110.

It indicates that if the single sand layer is 3 meters thick, the corresponding river channel will be as narrow as 300 meters. Therefore, it is very likely to form a long band-shaped sandstone reservoir. Its structural map is shown in Figure 5.111.

   b. The reservoir with band-shaped boundaries may locate in a graben formed by faults. Please refer to Figures 2.16 and 2.17 in Chapter 2 for a detailed description of its geological backgrounds.

2. Flow model graph
   a. Conventional completed gas wells. Linear gas flow often forms in the formations confined by band-shaped boundaries. Please refer to Figure 2.18 for linear flow and Figure 5.112 for the flow characteristic graph.

   Figure 5.112 shows that the radial flow segment is usually very short for narrow band-shaped formation, and it will be extended as widening of the band-shaped formation. The main typical segment is a half-unit slope straight line of linear flow.

If both ends of the band-shaped formation are closed, closed formation characteristics (pressure derivative curve inclines down sharply) will occur. However, practical

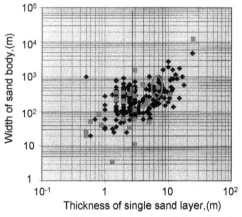

Figure 5.110 Relationship between width and thickness of fluvial sedimentary single sand layer (quoted from a Shell research report on the CB gas field).

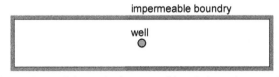

Figure 5.111 Schematic diagram of band-shaped reservoirs.

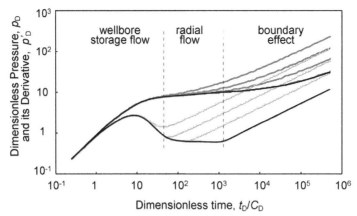

Figure 5.112 Log—log characteristic graph of gas flow in a band-shaped formation.

experience indicates that it seldom occurs in fact. It is due to either this kind of boundary is too far away to be reflected within the limited testing time or it is a partially permeable boundary, that is, the "flow barrier" discussed in Chapter 2.

Field example: Well Hu2

Well Hu2 is located in the HTB gas field, the southern margin of the Zhungaer Basin, Xingjiang Autonomous Region. Its main gas-bearing zone is the sandstone of Ziniquanzi formation in Tertiary. Tests were performed separately for two main gas-bearing intervals at 3561—3575 and 3594—3614 meters. Figure 5.113 shows the structural map of the HTB gas field.

To the south of Well Hu2 is a fault, as shown in Figure 5.113. This fault plays a role of trap in gas storage, so it could be considered as totally impermeable. To the north of the well is the gas—water contact, bringing a great effect on gas flow resistance due to a big gas—water mobility ratio; it is similar to the dynamic influences of a no-flow boundary in a short time. This situation makes Well Hu2 alike in a nearly band-shaped closed

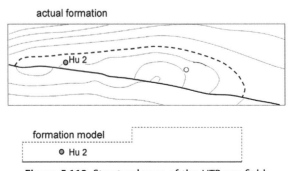

Figure 5.113 Structural map of the HTB gas field.

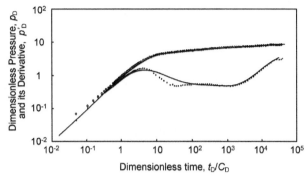

Figure 5.114 Log—log pressure buildup plot of Well Hu2 (3561—3575 meters).

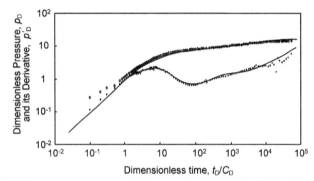

Figure 5.115 Log—log pressure buildup plot of Well Hu2 (3594—3614 meters).

reservoir. The aforementioned understanding is proved later by the curve characteristics of measured data (Figures 5.114 and 5.115).

It can be seen from the curve shape in Figure 5.115 that the wellbore at the lower part of Well Hu2 is closer to the fault boundary than the upper one, so the boundary effect appears earlier; thus the duration of radial flow is shorter.

Fractured Gas Wells in Band-Shaped Formation

1. Geological background

In the gas fields of the middle and western regions of China, the low permeability fluvial sedimentary sandstone reservoirs are usually companied with band-shaped lithologic boundaries. Due to low deliverability, well completion with sand fracturing is often adopted, and formation and bottom-hole structures are somewhat different from the situations in Figure 5.111. A boundary schematic diagram is shown in Figure 5.116 in which the direction of the man-made fracture is perpendicular to that of the river channel due to controlling by the maximum principal stress in the formations. Of course, the direction of the man-made fracture may be different because

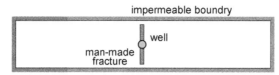

Figure 5.116 Schematic diagram for formation model of fractured wells with band-like formations.

of other actual formation conditions, but it can be proved that the different directions will have little impact on their flow model graph.

2. Flow model graph

The characteristics plot of pressure buildup obtained from wells in this kind of formation is shown in Figure 5.117.

Figure 5.117 shows that curves of a well in homogeneous medium appear prior to fracturing treatment; after fracturing, the early segments of the curves display a dramatic decrease of producing pressure differential and the whole curves shift downward, which can be divided into four segments as follows:

- Afterflow segment. This often occurs in low permeability gas wells with wellhead shut in after fracturing.
- Fracture linear flow segment. The characteristic curve of this segment is a half-unit slope straight line early on. This was introduced in detail in Section 5.3.
- Pseudo radial flow segment. Three different conditions exist in this segment, including:
  a. If the band-shaped formation is fairly wide, the pressure derivative curve will show a horizontal straight-line segment of pseudo radial flow.

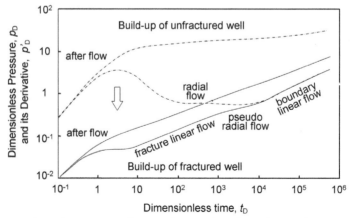

Figure 5.117 Log–log plot variation from unfractured well to fractured well in band-shaped formations.

b. If the formation is quite narrow, the pseudo radial flow segment disappears, leading to the half-unit slope straight lines of linear flow in both fracture and reservoir connecting together, making it difficult to tell them from curve characteristics.

c. As shown in Figure 5.117, the formation width is between the situations of I and II. There is a slight appearance of pseudo radial flow, by which the fracture linear flow segment and reservoir linear flow segment are demarcated.

The three conditions just mentioned are observed in the field examples given here.

- Reservoir linear flow segment. A half-unit slope straight line of linear flow appears again later. This segment will extend very long if the length of the band-shaped formation is very long.

Field examples:

Such test data are abundant in China due to the special geological conditions of its gas field. Some examples are given here.

a. Well T5

The Permian formation was perforated in Well T5. It is a low permeability sandstone gas reservoir formed by fluvial deposition. The perforation thickness was 3.5 meters, and fracturing treatment was performed during completion. The gas flow rate of the well was $5 \times 10^4 \, \mathrm{m^3}$/day before shutting in. The measured pressure buildup curve of the well, as shown in Figure 5.118, is basically consistent with Figure 5.117. The pseudo radial flow segment is missing because the reservoir is very narrow, just similar to condition b mentioned earlier.

Analysis with well test interpretation software results, including:

- Formation permeability $k = 0.78$ mD
- Half-length of fractured fracture $x_f = 43$ meters
- Skin factor of fracture $S_f = 0.0345$
- Total skin factor $S = -5.7$
- Boundary distances: $L_1 = 33$ meters, $L_2 = 700$ meters, $L_3 = 66$ meters, and $L_4 = 1890$ meters

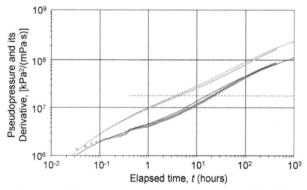

Figure 5.118 Log–log pressure buildup plot of Well T5.

The interpretation results indicate that the permeability of the formation is ultralow, and the well drilled in the band-shaped formation whose width is as small as less than 100 meters and whose length is over 2500 meters. The pressure history verification of the test results, as shown in Figure 5.119, shows a good match between theoretical curve and measured data, confirming the reliability of the interpretation results.

b. Well S20

The formation in Permian was perforated in Well S20. It is a sandstone gas reservoir with low permeability developed from fluvial deposition. The perforation thickness is 11.8 meters, and the fracturing treatment was performed during completion. The well was produced at a gas flow rate of 5.3×10^4 m^3/day before shutting in. Figure 5.120 shows pressure buildup curves acquired in the test, which are also basically consistent with Figure 5.117. The pseudo radial flow segment is missing because of the same reasons as the last example and similar to condition b mentioned earlier as well.

Parameters resulting from the analysis with well test interpretation software include:
- Formation permeability $k = 0.491$ mD
- Half-length of fracture $x_f = 43$ meters
- Skin factor of fracture $S_f = 0.05$
- Total skin factor $S = -5.59$
- Boundary distances: $L_1 = 46.7$ meters, $L_2 = 1000$ meters, $L_3 = 19.7$ meters, and $L_4 = 1200$ meters

The analysis results indicate that the formation permeability is low and that the gas well is located in the band-shaped formation whose width is less than 70 meters. Figure 5.121 is the pressure history match verification result of the interpretation. It can

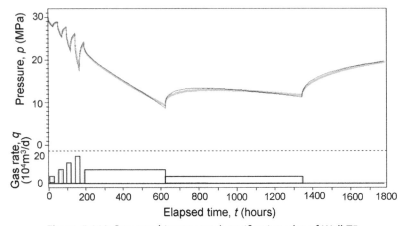

Figure 5.119 Pressure history match verification plot of Well T5.

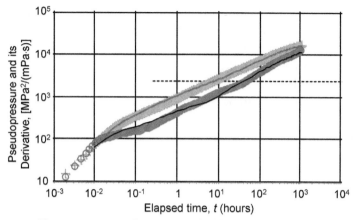

Figure 5.120 Log–log pressure buildup plot of Well S20.

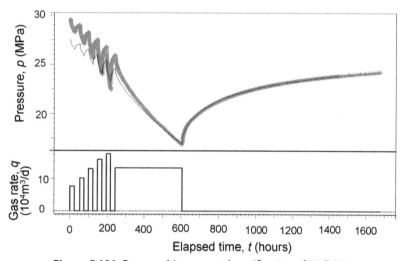

Figure 5.121 Pressure history match verification of Well S20.

be seen from Figure 5.121 that the match of the pressure buildup segment is very good, and the measured data curve declines a little bit more quickly than the curve of the theoretical model during the flow period. This implies that the scale of the real formation may be slightly smaller than the interpretation result, or the theoretical model.

c. Well S211

Well S211 perforated Shanxi Formations in Permian, a low permeability fluvial sandstone gas reservoir. The perforation thickness is 37.0 meters, and the fracturing treatment was performed during completion. The well was produced at a gas flow rate of 26.2×10^4 m^3/day before shutting in. Pressure buildup curves plotted and analyzed by the Shell Co. are shown in Figure 5.122. The shape of the curves is basically similar to that of Figure 5.117. The pseudo radial flow segment is also missing in the curves because the reservoir is very narrow, similar to condition b mentioned previously.

Parameter values obtained from the analysis with well test interpretation software include:

- Formation permeability $k = 2.0$ mD
- Half-length of fracture $x_f = 44.3$ meters
- Skin factor of fracture $S_f = 0.17$
- Total skin factor $S = -5.2$
- Boundary distances: $L_1 = 48.0$ meters, $L_2 = 10,000$ meters, $L_3 = 79$ meters, and L_4, 10,000 meters

The interpretation results indicate that the well is located in a long band-shaped formation with a width of about 130 meters and a length of 20 km. It can be said that this band-shaped formation could be very long because the boundary furthest away from the well has not been reflected yet in the curves, and thus its real length cannot be determined from test data.

The pressure history match verification of the interpretation results can be seen in Figure 5.123.

Figure 5.123 shows that the measured bottom-hole pressure declines more slowly than the theoretical model does during flowing periods, indicating that the real formation conditions are better than those of interpretation results with the theoretical model. The researchers of Shell Co. believe that this fact suggests that the boundaries of the river channels are not completely sealed but have a somewhat semipermeable barrier, which plays a very important role in balancing pressure stabilization during gas production.

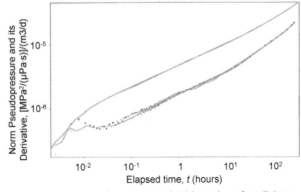

Figure 5.122 Log–log pressure buildup plot of Well S211.

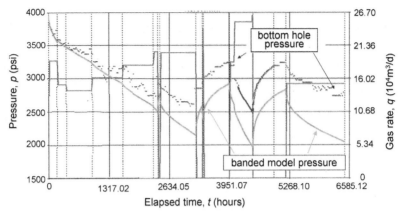

Figure 5.123 Pressure history match verification of Well S211 (quoted from interpretation results of Shell Co.).

d. Well CH 141

Well CH141 perforated Shanxi Formation in Permian, a sandstone gas reservoir with low permeability originating from fluvial deposition. The perforation thickness is 24.75 meters, and the fracturing treatment was performed during completion. The well was produced at a gas flow rate of 36.3×10^4 m³/day before shutting in. Pressure buildup curves plotted and analyzed by Shell Co. are shown in Figure 5.124. The characteristics of the curves are basically similar to those of Figure 5.117. The pseudo radial flow segment appearing in the curves for the reservoir is quite wide, similar to condition c mentioned earlier.

Parameters obtained from analysis with well test interpretation software are:

- Formation permeability $k = 3.45$ mD
- Half-length of fractured fractures $x_f = 45$ meters

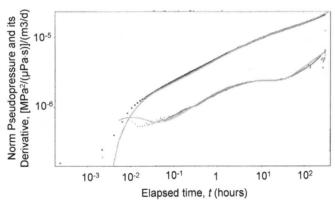

Figure 5.124 Log–log pressure buildup plot of Well CH 141 (quoted from interpretation results of Shell Co.).

- Skin factor of fracture $S_f = 0.21$
- Total skin factor $S = -5.2$
- Boundary distances: $L_1 = 205$ meters, $L_2 = 1700$ meters, $L_3 = 205$ meters, and $L_4 = 1700$ meters

The interpretation results show that the well is located in a long band formation with a width of about 400 meters and a length of 3500 meters. The pressure derivative curve contains a pseudo radial flow segment next to the fracture linear flow segment because the band formation is wide enough.

The pressure history verification of the interpretation results is similar to Well S211 and shows semipermeable boundary effects as well.

e. Well S10

Well S10 is located in the northwest section of the JB gas field in the Ordos Basin. The Shihezi formation in Permian was penetrated. The reservoir is thin with low permeability fluvial sandstone. Perforation thickness is 9 meters. A modified isochronal test was performed in which transient test points were selected, and their flow rates were 7, 12, 20, and 30×10^4 m^3/day, respectively; an extended test was conducted with a flow rate of 15×10^4 m^3/day; and the cumulative gas production was 445×10^4 m^3. A pressure buildup test was then performed, and the log—log pressure buildup plot is shown in Figure 5.125.

The curves in Figure 5.125 show that

- Skin effect in the man-made fracture exists in the early segment (as shown in Figure 5.82)
- Obvious linear flow characteristics of man-made fracture appear in the middle segment
- Pseudo radial flow characteristics appear for a short duration in the late segment
- Boundary effect is seen in the final segment

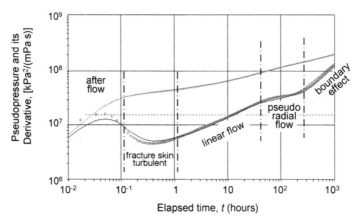

Figure 5.125 Log—log pressure buildup plot of Well S10.

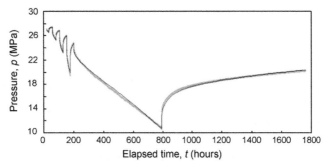

Figure 5.126 Pressure history match verification of Well S10.

Parameters are calculated via analysis with EPS well test interpretation software, including
- Formation permeability $k = 2.08$ mD
- Half-length of the fracture $x_f = 49$ meters
- Skin factor of fracture $S_f = 0.5$
- Total skin factor $S = -5.85$
- Boundary distances: $L_{b1} = 205$ meters, $L_{b2} = 281$ meters, $L_{b3} = 59$ meters, and $L_{b4} = 1500$ meters
- Wellbore storage coefficient $C = 1.21$ m^3/MPa

The interpretation results are confirmed reliably by pressure history match verification, as shown in Figure 5.126.

5.7 Characteristic Graph and Field Examples of Fissured Zone with Boundaries (Model Graphs M-14 and M-15)

As discussed when describing no-flow boundary properties in this chapter, the development morphology of the permeable parts of some fissured carbonate formations is extremely complicated and it is impossible to describe the fissured carbonate formations. However, a thorough discussion about the most common basic types of them can be done by means of geological research combined with a dynamic feature analysis of gas wells; then dynamic models of gas wells and reservoirs can be identified, and the prediction can be offered for the production performance of gas wells and reservoirs consequently. The approach is proved to be effective by field practices.

Some primary understandings for the following types of formation have been obtained.

5.7.1 Strip-Like Fissured Zone with Directional Permeability
Geological Background

Open fractures parallel to its axis in the uplifted axial area were formed in some fold belts of gentle formations. These fracture development areas are mainly restricted within the uplift in axial parts and shaped along its structural configuration and strike, as shown in Figure 5.127.

Ordovician limestone, the main pay zone in the GB gas field, is a monocline inclining westward in which a series of nose-like structures developed from north to south. Geological conditions are favorable for fracture development described earlier in the western margins, for example, the surrounding areas of Wells S–181 and S–71. This judgment is confirmed by logging data concerning fracture strike.

In addition, some fissured zones often develop in carbonate formations surrounding faults, and the strike of them is usually consistent with the fault strike due to the effect of the faulting process. This has been proven in the KL buried hill reservoirs of the JY depression, where interference tests and pulse tests of over 30 well groups were performed in more than 10 wells. This further confirms that the permeability of the formation around the faults and along the development direction of the main fissures is super high, whereas the permeability of that along the vertical direction of the strike of fissures is ultralow [43, 44].

Flow Modal Graph

Flow modal graphs of the aforementioned formations are shown in Figure 5.128.

At first glance, there are some resemblances between Figure 5.128 and model graphs of a fractured well in common formation or strip-shaped formation discussed previously. They are similar actually in flow characteristics but in fact their formations and completion conditions are totally different as shown later.

1. Ithologic boundaries are formed during fluvial deposition in strip-shaped sandstone formations, while there are no such boundaries with explicit geological background in a fissured band

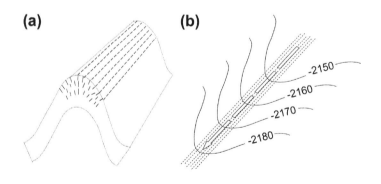

cross profile   plan view

Figure 5.127 Schematic diagram for fracture development in folded area.

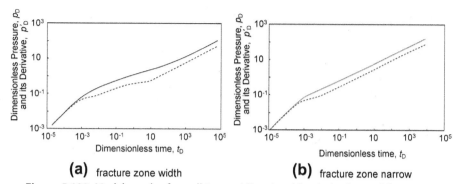

(a) fracture zone width   **(b)** fracture zone narrow

Figure 5.128 Modal graph of a well in a unidirectional fascicular fissured formation.

2. The performance of fractured wells introduced earlier is aimed at those treated by large-scale sand fracturing treatment. It is known from technique design that fractures can extend several hundred meters deep into the formation, and the propped fractures can also reach formations as deep as dozens of meters.

In the GB gas field, only the original fissures were connected, for the gas wells were only treated by acidizing or acid fracturing stimulation. Therefore, such curves as shown in Figure 5.128 seem impossible at first glance.

Field Examples
a. Well S181

Well S181 is located in the west side of the central district of the GB gas field, a relatively independent small gas block. From the original structural map, the well is drilled on the axial part of a nose structure. Mawu-4 formation was perforated at 3549.0—3553.4 meters, and deliverability and pressure buildup tests were performed after acidizing treatment. The well was produced at a gas flow rate of $7 \times 10^4 \text{ m}^3$/day prior to well shut in. The pressure buildup curves acquired in the test are shown in Figure 5.129.

Figure 5.129 shows that the shape of the curves is consistent with that shown in Figure 5.128a, from which it is concluded that the well is located in a narrow and long fissure system and that the rate-maintenance capability is quite poor. The analysis results are proved by the following production performance.

Parameters are obtained from the analysis with EPS well test interpretation software, including:

- Permeability $k = 4.79$ mD
- Width of fissure zone $d \approx 500$ meters
- Skin factor $S = -5.9$

b. Well S71

Well S71 is located in the western margin of the south district of the JB gas field, also on the nose lift of the nose structure. It perforated Mawu-1 and Mawu-2 formations with

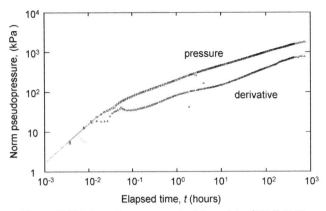

Figure 5.129 Log–log pressure buildup plot of Well S181.

a thickness of 6.4 meters. The well was produced at a gas flow rate of 2×10^4 m³/day before shutting in. Pressure buildup curves are shown in Figure 5.130.

Figure 5.130 shows that the curve shape is similar to that shown in Figure 5.128b. Parameters are obtained from the analysis with well test interpretation software, including

- Formation permeability $k = 0.0753$ mD
- Width of fissure zone $d \approx 260$ meters

Figure 5.131 shows the pressure history match verification of interpretation results, which has very good consistency of the theoretical model to measured pressure data and proves the reliability of interpretation results.

Figure 5.131 shows that Well S71 is produced at a low rate but is accompanied with a rapid pressure drop, further confirming that the gas supply area is small.

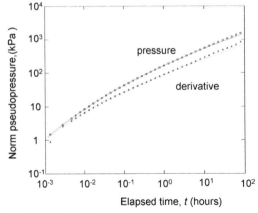

Figure 5.130 Log–log pressure buildup plot of Well S71.

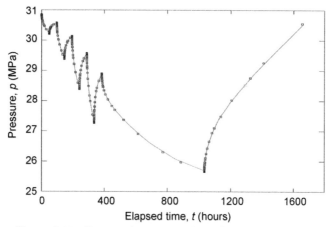

Figure 5.131 Pressure history match verification of Well S71.

c. Well Q12-18

Well Q12-18 is an appraisal well in the west side of the Banshen-7 fault block in the QMQ gas field. It is located near a fault and drilled in two intervals of 4217.0–4250.0 and 4180.0–4202.0 meters in the Ordovician limestone formation. It flowed in the middle and late part of July and then shut in for the pressure buildup test separately. The obtained pressure buildup curves are shown in Figures 5.132 and 5.133.

These two plots show that the well is probably located in a narrow and long fissured band, making the pressure derivative curve an approximate half-unit slope straight line. Due to the extremely irregular development of the fissured band, the pressure derivative curves are with present frequent fluctuation accordingly.

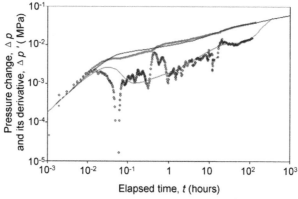

Figure 5.132 Log–log pressure buildup plot of Well Q12-18 (4217.0–4250.0 meters).

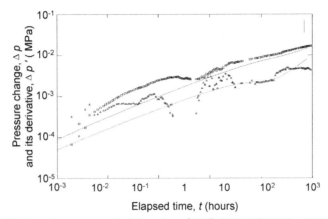

Figure 5.133 Log—log pressure buildup plot of Well Q12-18 (4180.0—4202.0 meters).

5.7.2 Beaded Fissured Bands
Geological Background

Some geological observations show that the development band of fissures, pores, and caves in carbonate formation sometimes presents local sheeted distribution. These local sheets are connected by very permeable but extremely narrow channels and form beaded bands, as shown in Figure 5.134.

Great risk will be brought to well drilling on such formations due to very serious heterogeneity, as listed here.

- Although there is a certain relationship between fissured band and geological structural characteristics, it may not be fully recognized by geologists at the beginning. Therefore, the success rate of well drilling will be low.
- If an exploration well was drilled and encountered a big fissure system, for example, zone A or zone B in Figure 5.134, its flow rate may be very high, by which the operators will be encouraged greatly. However, the fissure system usually has a limited

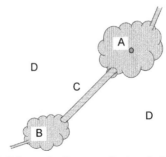

Figure 5.134 Schematic diagram of a beaded fissured band.

development area. As a result, a big problem about rate maintenance will be posed when the well is put into production.

- If a well was drilled in the channel between fissure zones or in the underdeveloped fissure zone formed by stimulation of acid fracturing, for example, zone C in Figure 5.134, its conduction will be very low. However, if the well can be made to connect the fissure system(s) by stimulation, it will be able to keep a certain stabilized flow rate, even though the rate may not be very high. Of course, the precondition is that the well located in the underdeveloped zone is really connected to the fissured zone(s) and that this connection has been verified by production performance of the well.
- If the well is drilled in area D of the formation, where no fissured zones can be encountered at all, the well will fail to possess any industrial flow.

Well Test Model Graphs

It is impossible to plot model graphs of this kind of formation by common analytical methods for well test interpretation; only by a numerical well test involving some advanced well test interpretation software can the model graphs be obtained. The typical plots are shown as follow.

1. Wells located in zone A (of Figure 5.134)

If a well is located in zone A of Figure 5.134, it is usually with a very high deliverability. Its flow characteristic plot is shown in Figure 5.135.

Figure 5.135 shows that the flow can be roughly divided into four segments, including

a. Afterflow segment

b. Segment of flow in near-wellbore area

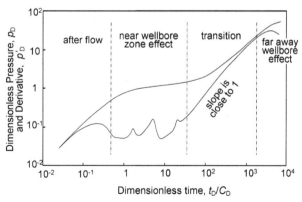

Figure 5.135 Log–log pressure buildup plot of beaded fissured band (Case 1: well located in zone A of Figure 5.134).

Because the well is located in a large big-fissure zone, its deliverability may be quite high and the flow in the near-wellbore area seems to be radial flow. However, there are frequently irregular fluctuations on its pressure derivative curve. The longer the segment, the larger the near-wellbore fissure zone.

Not much attention should be paid to analysis on fluctuation of the derivative curve when interpreting data because such analysis is difficult and unnecessary. What is important for the production of an oil or gas field is to judge whether the gas or oil supply area is broad or narrow from the curve characteristics, which is an important and significant guide for oil or gas field development in the future.

This segment can be used to calculate the reservoir parameters, for example, permeability k; radius of gas drainage area r_e; and skin factor S. To obtain more explicit information about the reservoir, analysis and interpretation should be implemented with numerical well test software combined with geological research.

Because core observation indicates fissures developed in the reservoir, the reservoir was defined as a double medium formation geologically, and some well test analysts insist on extracting double medium parameters such as storativity ratio ω and interporosity-flow coefficient λ. From these data, it is very difficult to obtain convincing results because of irregular fluctuations of pressure derivative curve; even if interpretation results are obtained, they are not significant for practical guidance of gas field production.

c. Transition flow segment

When the flow effect reaches no-flow boundaries in multidirections and only a little effect reaches the narrow channel, the pressure derivative curve will show an approximate unit slope straight line.

It is the difference between this situation and a closed reservoir that reflects these flow characteristics not only in a pressure drawdown curve but also in a pressure build-up curve with nearly the same curve shape.

If the connecting channel is very long, this approximate unit slope straight-line segment on the pressure derivative curve will last for a long duration.

d. Segment of flow in the area far away from the wellbore

If there is another fissure zone at the other side of the connecting channel, as described in Figure 5.134, the fissure zone becomes a new gas drainage area of the well. Its pressure and pressure derivative curves will separate gradually, displaying a pressure variation of leveling down due to the effect of the new gas drainage area.

Field examples: Wells BS-8 and BS-7

Geologically, the reservoir of the QMQ gas field is a typical complex fissured band. The main production zone of the field is the Majiagou formation in Ordovician. Fissures, pores, and caves are well developed inside the formation; large-scale highly angular fissure groups on well walls are seen from logging imaging of some wells, for example, Well BS-7. When a high flow rate was obtained in Wells BS-8 and BS-7, this band became the major exploration area. However, further analysis of well test data

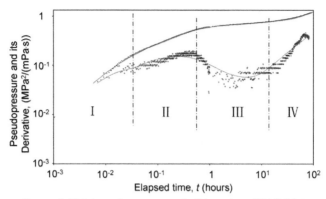

Figure 5.136 Log–log pressure buildup plot of Well BS-8.

showed that the accumulation rule of the reservoir is much less optimistic than that expected from the initial gas test results. Well test data indicated that the reservoir was composed mainly of local complex fissure systems. Although big fissure systems were encountered during the drilling of some wells, the development of fissures became poorer obviously in the exterior zones away from these wells. It is those dynamic characteristics that remind people to conduct further research so as to make a production plan in a proper and rational manner.

a. Pressure buildup curves of well BS-8

Pressure buildup tests were performed in Well BS-8 many times, and data acquired in several former tests could not reflect formation conditions faithfully due to the hump influence caused by retrograde condensation of gas. The pressure buildup curves obtained from the successful final test are shown in Figure 5.136.

Figure 5.136 shows that the curves in middle and late times are similar to those of model graph 5.135 except that an additional linear flow segment exists in the early curves. The curves can be divided into four segments as follow.

(i) Afterflow segment (I). This segment is very short.

(ii) Early linear flow segment (II). Acid fracturing stimulation during well completion made the natural fissures connect to each other and form fractures connecting to the bottom hole, thus forming the "uniform flow" shown in Figure 5.73. The flow may be either linear or bilinear. The early flow segment of Well BS-8 showed just these characteristics.

(iii) Radial flow in near-wellbore area (III). This segment lasts about 20 hours, leading to the pressure derivative curve of an approximate horizontal straight line, which is the radial flow regime. This segment indicates the flow process of the main gas drainage area near the bottom hole, as shown as zone A in Figure 5.134; it is this area and the high conductivity fractures formed by acid fracturing that maintain production with a high flow rate and the rate maintenance of Well BS-8.

(iv) Transition segment (IV). This segment shows a boundary effect to the gas drainage area with high permeability. The radius of the area (zone A in Figure 5.134) is estimated to be less than 300 meters from the time when the pressure derivative curve bends upward. After that, gas flow occurs in the zone with a low kh value (zone C in Figure 5.134).

The pressure derivative curve bends downward at the end. If test data reflect the formation conditions faithfully, there could be some other connecting gas drainage areas (e.g., zone B in Figure 5.134) in addition to the connecting channel.

Parameters obtained from primary analysis by well test interpretation software are as follow:

- Permeability of main gas drainage area $k = 20.47$ mD
- Diameter of the main gas drainage area is about 600 meters

An interesting phenomenon is that a very good match is obtained in the pressure history match verification of the pressure drawdown test, which was performed 1 year before with the interpretation results, proving that the interpretation results are surely reliable.

b. Pressure buildup curves of well BS-7

Several intervals of Well BS-7 have been conducted by the pressure buildup test. Similarly, test data acquisition of this well was influenced by the hump effect of retrograde condensation and abnormal high bottom-hole temperature. Results of a representative test are shown in Figure 5.137.

- The segment of early time. The existence of a linear flow segment indicates that the big fissures connected to wellbore were formed by acid fracturing, which played a crucial role in gas flow and high production of the gas well.
- The segment of radial flow in near-wellbore fissure zone. Up to 100 hours since shutting-in, the pressure derivative curve locates around a certain horizontal straight line even though there is a little fluctuation; 10 hours after shutting-in, the pressure derivative curve declines slightly, which indicates that there is a neighbor area with better developed fissures.

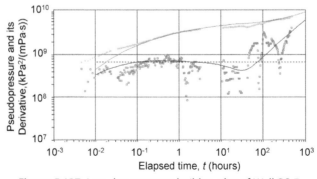

Figure 5.137 Log–log pressure buildup plot of Well BS-7.

- The segment of boundary response. The pressure derivative curve in this segment, 100 hours after shutting-in, bends upwards obviously, indicating that the property of the peripheral reservoir is poorer. From the beginning of transition the distance to the well is estimated as about 240 meters, and therefore the main gas drainage area is very limited.

Parameters obtained from primary analysis with well test interpretation software are:
Permeability of main gas drainage area $k = 2.1$ mD
Radius of gas drainage area $r \approx 240$ meters

A pressure history match verification of the interpretation results is shown in Figure 5.138; a very good match of theoretical model and measured pressure data is clearly observed.

2. Wells located in zone B (of Figure 5.134)

If a well is located in zone B of Figure 134, its flow characteristic is shown in Figure 5.139.

It is different from Case 1—the well is located in smaller fissured zone B and thus the gas flow in the near-wellbore fissure zone can only last a short time. However, such formations still have the rate-maintenance ability if connected to other well-developed fissure zones.

Field examples:

a. Well S44

Well S44 is located in the northwestern margin of the middle district of the JB gas field and is one of the earliest evaluation wells. Mawu-1—Mawu-4 formations were perforated, and the perforation thickness was 7 meters. Its well test curves obtained during the production test period attracted great attention due to their special shapes. The well seems to be in an area like zone B of Figure 5.134. Its pressure buildup curves can be seen in Figure 5.140.

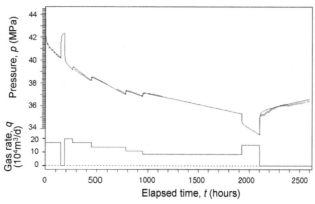

Figure 5.138 Pressure history match for well test interpretation results of Well BS-7.

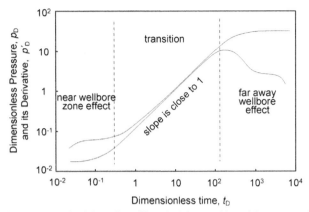

Figure 5.139 Log–log pressure buildup plot of beaded fissured band (Case 2: well located in zone B of Figure 5.134).

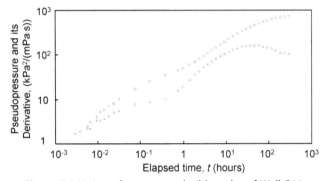

Figure 5.140 Log–log pressure buildup plot of Well S44.

A circular composite formation model was applied first. Certainly, the assumption of the circular composite formation is apparently quite farfetched for such a complicated geological situation. It is believed that more thorough understandings can be obtained if such test data are analyzed with numerical well test software combined with geological analysis and verification of production history.

b. Well BS-7 (interval of 4265–4309 meters)

Several intervals of Well BS-7 were perforated and tested. Because buried hill formations are composed of fissure bands vertically without obvious interval zonation, the relationship of those intervals is still unknown. If analysis is only carried out on the curve shape, the dynamic characteristics seem to be of several different mutual connected permeable zones, as shown in Figure 5.141.

The plot shows much resemblance to the curves of Well BS-7 and those in Figure 5.139. Although characteristic curves of some composite formations are also

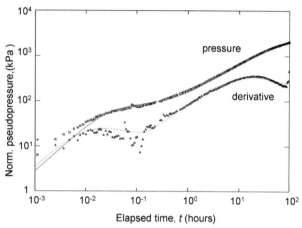

Figure 5.141 Log–log pressure buildup plot of Well BS-7 (4265–4309 meters).

similar to these curves in Figure 5.141, data analysis of this well should take a related model into consideration to tally with real conditions, as the formation is a fissured formation of combined or systematic structures.

3. Wells located in zone C (of Figure 5.134)

If a well is located in zone C of Figure 5.134, for example, the high conductivity channel created by acid fracturing, its well test model curves are as shown in Figure 5.142.

Figure 5.142 looks like flow curves with a long afterflow at first glance. As discovered by analysis, the wellbore storage coefficient will exceed 10 m³/MPa if the long approximate unit-slope straight line is regarded as an afterflow segment. However, such a high wellbore storage coefficient can never be possible in any wellbore conditions.

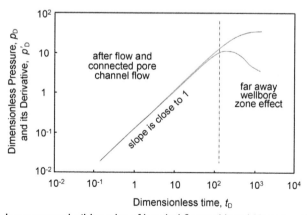

Figure 5.142 Log–log pressure buildup plot of beaded fissured band (Case 3: well located in zone C of Figure 5.134).

The reason for the aforementioned phenomenon is that this well locates in poor reservoirs when initially completed and contributes commercial flow only after acidizing or acid fracturing treatments. In addition, an extremely narrow zone with high conductivity and connecting with the borehole is created around the well by the acidizing operation, which is just like an enlarged wellbore formed by combining the extremely narrow zone and the wellbore itself. This extremely narrow zone with high permeability provides a path to flow the gas to the well even though the permeability of the reservoir around the well is very low. For this reason, an extremely high wellbore storage coefficient is obtained in the well test interpretation.

Field example: Well BS-6

Well BS-6 also penetrated the fissured reservoir in Ordovician. Geologically, the fissure development in this well is worse than that in Well BS-7 and those of some other wells. A well test was performed after acid fracturing. The well flowed for 2 days with an average gas flow rate of $20 \times 10^4 \, \text{m}^3/\text{day}$. The flowing pressure during the deliverability test and extend test declined continuously. The pressure buildup curves are shown in Figure 5.143.

It can be seen from Figure 5.143 that the pressure derivative curve is nearly a unit-slope straight line within 100 hours, which is the characteristic of an afterflow segment. But the wellbore storage coefficient will also exceed $10 \, \text{m}^3/\text{MPa}$, far beyond its normal range if the long approximate unit-slope straight line is regarded as an afterflow segment. Therefore, it is judged as the flow in the connecting channel. Due to the lack of radial flow, interpretation is not offered. A further thorough analysis can only be done by numerical well test software.

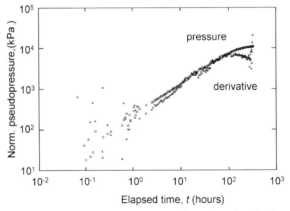

Figure 5.143 Log–log pressure buildup plot of Well BS-6.

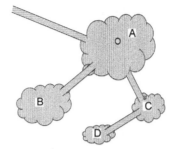

Figure 5.144 Schematic structure map of buried hill-type complex fissured zone.

5.7.3 Complex Fissured Zone

In practice, the real reservoirs in a fissured zone of buried hill–type formation may be much more complex than the "beaded band structures" mentioned previously. The possible pattern of this kind of structure is given schematically in Figure 5.144.

In a word, such a complex geological structural configuration may bring about several influences.

1. In view of hydrocarbon exploration, most favorable target zones are difficult to predict accurately by conventional reservoir lateral prediction methods, thus resulting in a low drilling success rate.
2. Great uncertainties are hidden in reserves estimation of such kinds of oil and/or gas fields. Very ridiculous results may be obtained in reserve estimation if imitating the calculation methods for sandstone reservoirs with static parameters.
3. Due to the heterogeneity and discreteness of the storage space of oil and/or gas, it is difficult to provide exact reservoir parameters by conventional logging analysis methods. Moreover, numerical simulation with these logging analysis results can hardly conclude sensible understanding when tallying the actual situation.
4. The dynamic analysis method is the only approach to reveal complicated reservoir conditions. With this method, even though target zones for new wells to be drilled cannot be given, abundant information about the tested zones and the reservoir structure where the tested wells locate is contained in the well test curves and from which sensible evaluation tallying the actual situation can be obtained. The typical shapes of the well test curves contain risk information in the exploration and development of such oil and/or gas fields actually. If great attention is paid, unnecessary economic loss can be avoided.

5.8 Characteristic Graph and Field Examples of Condensate Gas Wells

5.8.1 Geological Background and Focused Problems

Condensate gas fields are widespread in China, among which is the KKY condensate gas field in the southern margin of the Tarim Basin. It was discovered quite early; then such condensate gas fields as the BQ and QMQ gas fields in BH Bay Area, W-23 gas field in the ZY field, and many fields in the KC area of the Tarim Basin, for example, YH, TZLK, YKL, YTK, YD, YML, and JLK, were discovered in succession; some fields among them have been put into production.

Geologically, there is no difference between a condensate gas field and other gas fields. Condensate gas fields have been discovered in sandstone, carbonate, and other formations with various different boundary conditions. Only in the view of a reservoir structure, the formation models and their model graphs mentioned earlier are all applicable to condensate gas fields.

However, different from common dry gas zones, condensate oil may be generated in the wellbore or formations around the wellbore after condensate gas wells are producing, which causes many new problems for the acquisition and analysis of well test data. These problems are discussed later.

Phase Diagram of Condensate Gas

The phase diagram of condensate gas clearly indicates the phase transformation of the gas sample under different pressures and temperatures. Taking the YH gas field as an example, its measured phase relationship is shown in Figure 5.145.

The phase diagram reflects the conversion relationship of gas and liquid in the formation when the temperature and/or pressure changes. Figure 5.146 shows some important influence points in the phase diagram.

1. When the temperature is high, fluids are completely in a gaseous state or in a gas state zone, whereas when the temperature is low, fluids are completely in a liquid state or in a liquid state zone. The mixed phase region is between the two state zones, and the content of condensate oil in this mixed phase region varies within a range of 0—100% along with the temperature and/or pressure changes.

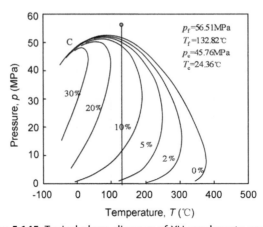

Figure 5.145 Typical phase diagram of YH condensate gas field.

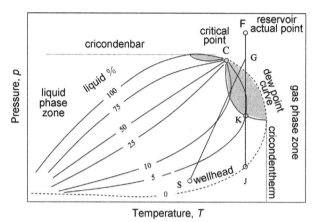

Figure 5.146 Typical phase diagram of a condensate gas.

2. The upper margin of the mixed phase region is the cricondenbar, where there is no gas phase at all if the pressure is higher than this cricondenbar. The cricondentherm is on the right side of the mixed phase region where there is no liquid phase (condensate oil) at all when the temperature is higher than this cricondentherm.

3. Phase change behavior in formation

   The real pressure and temperature point in formation is F: the pressure is p_F and the temperature is T_F. It can be seen that no condensate oil is precipitated at this point, and the gas is totally in a pure gaseous state under the original formation conditions.

   When the gas well is producing, pressure in the formation begins to decline, and phase behavior will change along the vertical line F—J. After the pressure reaches the dew point curve, the precipitation of condensate oil commences, and condensate oil accounts for 5% when the pressure reaches point K. While the content of condensate oil begins to decrease during the following pressure decline period, the gas will completely change into a gaseous state when the pressure reaches point J. This particular phenomenon is called "retrograde condensate phenomenon."

4. Phase change behavior in wellbore

   As for condensate gas wells, another important factor influencing data acquisition and the shape of test curves is the phase change behavior in the wellbore. Assuming that point G is the flowing pressure point when the well is producing and point S is the wellhead pressure point, then the pressure and temperature will vary along the line G—S from the bottom hole to the wellhead. It can be seen that the temperature decreases gradually from the bottom hole to the wellhead, which causes more condensate oil to precipitate and the retrograde condensate phenomenon to become more serious in the phase change process.

"Hump" Occurrence Due to Phase Change Behavior Inside the Wellbore During Shut-in Periods

As the wellbore pressure changes in an inverse manner, that is, it increases gradually during the pressure buildup process in the pressure shut-in test, the corresponding pressure varies along the direction of S→G→F in Figure 5.146. It can be seen that when the pressure approaches point G through the condensation region along the line S–G, the pressure keeps growing. Meanwhile, the precipitated condensate oil is decreasing, namely the condensate oil is continuously turning into natural gas. This gasification process caused by the retrograde condensate effect will make the fluid volume expand uninterruptedly, and the increasing pressure is accelerated by the expansion process. This abnormal retrograde condensation phenomenon is one of the main causes for "hump" occurrence.

Figure 5.147 shows a measured pressure buildup curve with a hump acquired in Well BS-8 in the QMQ gas field. The Ordovician buried hill reservoirs were penetrated in the well. Condensate gas was produced from the well, and the original condensate oil content was about 290 g/m^3. The test interval was 4246.0–4324.0 meters, and the pressure gauges were set at 3985 meters, 300 meters above the middepth of the gas zones (4285 meters).

It can be seen that
- Flowing pressure just before shutting in is 38.146 MPa
- Pressure increases to 39.008 MPa when the well has been shut in for 1 hour
- Pressure reaches its peak and then, 12.2 hours later, declines to 38.353 MPa, forming an obvious hump. The relative height difference of the hump is about 0.655 MPa.

This abnormal phenomenon occurring in the pressure buildup curve made data unable to be used normally for analysis.

Two causes leading to a hump occurring are concluded as follow:

1. Wellbore pressure increases abnormally caused by retrograde condensation. That the wellbore pressure increases abnormally is inevitable for some condensate gas wells,

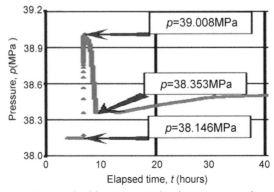

Figure 5.147 Pressure buildup curve with a hump acquired in Well BS-8.

which may not exist in some other gas wells. Factors forming a hump include formation permeability and casing program.

2. Pressure shift due to liquid load in wellbore. The mixture of condensate oil and natural gas has entered the wellbore flow toward the wellhead and then the condensate oil sinks to the bottom hole from the upper part of the wellbore during the oil–gas separation process after shutting in and thus liquid loading in the bottom hole forms. The level of the accumulated liquid keeps rising to offset part of the bottom-hole pressure, resulting in a continuous decline of the measured pressure at the point pressure gauge is set, as described in Figure 5.148.

a. Situation at beginning of shutting in

If the pressure gauge is set at the place H higher than the middepth of the gas zones, the pressure at measured points is

$$p_1 = p_{WS} - \Delta p_1 = p_{WS} - HG_{Dh} = p_{WS} - Hp_g g, \qquad (5.80)$$

where

p_{ws} = bottom-hole shut-in pressure, MPa

Δp_1 = initial difference between bottom-hole pressure and pressure at the measured point, MPa

H = distance between measured point and bottom hole, m

G_{Dh} = pressure gradient of well fluid at the beginning of shutting in, MPa/m

$$G_{Dh} = \rho_g g$$

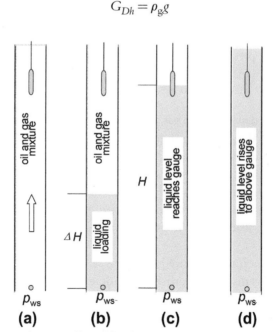

Figure 5.148 Interference of liquid level rising to measured bottom-hole pressure.

ρ_g = density of well fluid composed mainly of gas, kg/m^3

g = acceleration of gravity, 9.80665 m/s^2

b. Situation in the process of shutting in

After shutting in for a while, a liquid column with a height of ΔH is generated due to the liquid loading. The density of liquid ρ_o is much higher than that of gas ρ_g, that is, $\rho_o \gg \rho_g$, and the reading of pressure gauge will be

$$p_2 = p_{WS} - \Delta p_2 = p_{WS} - \Delta H \rho_0 g - (H - \Delta H)\rho_g g$$

$$= p_{WS} - H\rho_g g - \Delta H(\rho_0 - \rho_g)g. \tag{5.81}$$

Apparently, an additional pressure difference Δp_N occurs between bottom-hole pressure and pressure at the measured point, which is called "shift pressure":

$$\Delta p_N = \Delta H(\rho_0 - \rho_g)g. \tag{5.82}$$

This equation shows that shift pressure Δp_N is increasing as the liquid level is rising, that is, the height of liquid column ΔH is increasing.

c. Situation when the liquid level reaches the pressure gauges

When the liquid level reaches the pressure gauges, measured pressure p_3 can be expressed as

$$p_3 = p_{WS} - H \cdot \rho_0 g, \tag{5.83}$$

where ρ_o is density of the liquid.

Due to $\rho_o \gg \rho_g$, shift pressure due to liquid column level rising can be expressed as

$$\Delta p_N = H(\rho_0 - \rho_g)g. \tag{5.84}$$

For example, if the density of the gas and liquid mixture at the beginning of shutting in is ρ_g = 200 kg/m^3, the density of liquid column is ρ_o = 700 kg/m^3, and H = 300 meters, Δp_N = 1.47 MPa is calculated.

It shows that quite a serious pressure shift exists due to rising of the liquid level.

d. Situation when the liquid level is above the pressure gauges

Once the liquid level rises above the pressure gauges, and the liquid column is a pure condensate oil column, that is, ρ_o value is a constant, and thus the shift pressure will also be a constant.

It is conceivable that if the pressure gauges are set at the middepth of the gas zones, the shift pressure Δp_N discussed earlier will not exist. The best way to overcome a pressure shift due to liquid loading is to set the pressure gauges at the middepth of the gas zones. If there is liquid loading in the bottom hole before the test, setting the pressure gauges below the liquid level can avoid this pressure shift effect, even though it is impossible to measure the pressure in the middepth of pay zones directly and exactly.

The aforementioned analysis results imply that as for the problems of acquiring pressure data in the condensate gas wells, the first one, that is, the retrograde condensate effect, is difficult to avoid sometimes and the influence of the second one, that is, liquid-loading effect, is often more difficult. These problems must be resolved with all approaches possible, as data measured otherwise cannot reflect the formation conditions at all, even providing false information.

In order to verify the effect of accumulated liquid loading during testing of Well BS-8, the bottom–hole pressure gradient tests were conducted specifically at different times. The results are shown in Figure 5.149.

This well was shut in on May 23. Three pressure gradient tests shown in Figure 5.149 were performed on May 28, May 31, and June 5, respectively. Test results show the existence of the condensate oil column and water column. Both levels rise differently, as shown in Figure 5.150.

If the pressure gauge is set at or above 3800 meters, all formation information that could be reflected by pressure buildup curves will be covered completely by the shift pressure effect due to rising of those liquid levels.

In fact, a pressure buildup test begun on May 23 was divided into three stages. The pressure gauge was set at 4285 meters (middepth of the gas zones), 3800 meters (485 meters higher than the middepth of gas zones), and 4050 meters (235 meters higher than the middepth of the gas zones), respectively. The measured pressure values can be seen in Figure 5.151.

The following stages can be seen from Figure 5.151.
- In the first stage, the pressure gauge was set in the bottom hole and good pressure buildup data were obtained.
- In the second stage, the pressure gauge was lifted and set at 3800 meters above the liquid level as shown in Figure 5.150; the acquired pressure buildup curve declined and could not be applied at all.

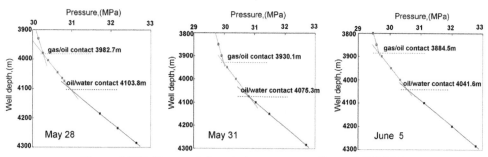

Figure 5.149 Measured downhole pressure gradient of Well BS-8.

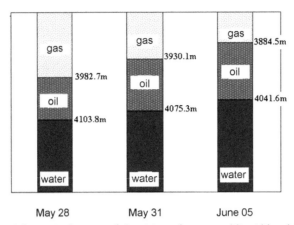

Figure 5.150 Schematic diagram of the rising of measured liquid level in Well BS-8.

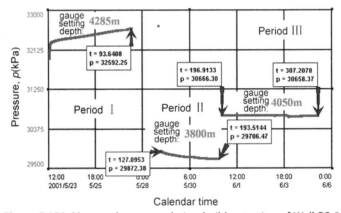

Figure 5.151 Measured pressure during buildup testing of Well BS-8.

- In the third stage, the pressure gauge was lowered to 4050 meters below the water level, and readings showed that the bottom-hole pressure had reached an approximate steady value in the late period of buildup.

Conversion of Gas Flow Rate During Well Test in Condensate Gas Wells

A gas flow rate is needed in the well test interpretation of condensate gas wells. This flow rate is usually composed of two parts, that is,

1. natural gas produced out to the surface
2. condensate oil produced out to the surface

During the early production stage of gas wells, both these parts are in the gaseous state in the formation. In this case the volume of the produced condensate oil must be converted into the subsurface gas volume. The gas flow rate should be the sum of the natural gas produced out to the surface and the gas converted from the produced condensate oil. The conversion method can be found elsewhere [73].

Well Test Analysis Problems Bought About by Accumulation Belt of Condensate Oil in Formation

During the producing process of condensate gas wells, a pressure drop funnel forms around the bottom hole. The pressure at the position, the distance between which the well is $r = r_N$, is equal to the dew point pressure, and the condensate oil will precipitate in the section between the bottom hole to this position, as shown in Figure 5.152.

Once the condensate oil zone is formed, flow in the formation will change seriously. The mobility $(k/\mu)_N$ in the condensate oil section will decrease significantly because oil viscosity μ_o is much higher than gas viscosity μ_g. In addition, a two-phase flow existing there makes the permeability decrease significantly. As a result, two different flow zones form accordingly, as shown in Figure 5.153.

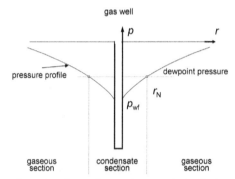

Figure 5.152 Condensate oil section formed near bottom hole during production of condensate gas well.

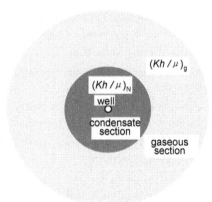

Figure 5.153 Mobility distribution during production of a condensate gas well.

The pressure buildup curves in this case will represent composite formation characteristics. Also, some differences exist between pressure drawdown and pressure buildup curves.

Two Ideas for Solving Problems in Condensate Gas Well Tests

1. Different countermeasures at different time intervals

The aforementioned analysis shows that the well test analysis of the condensate gas wells can be divided into two stages.

(1) Stage when the formation pressure is higher than the dew point pressure. In this case, natural gas flow in the formation is absolutely the same as common dry gas flow, and the analysis methods discussed earlier are completely applicable. The only difference is that a variable wellbore storage effect exists in acquired pressure data. Whereas the "hump" effect can be avoided to a great extent and the dynamic analysis research can be accomplished as long as suitable test methods are chosen, well test design is well done, and thus the pressure shift caused by accumulated liquid loading is minimized.

(2) Stage when the formation pressure is lower than the dew point pressure. This stage will last until the late stage of the production of the condensate gas wells. It can be classified into three situations, including

    a. Only a very small area near the bottom hole where condensation exists. In this case, condensate oil precipitation in the bottom hole can be treated as an additional skin factor. There is little difference in well test analysis between condensate gas wells and common gas wells.

    b. Local condensate oil section exists. There is a local condensate oil section near the borehole and composite formation characteristics will appear. In a pure gas zone where r is greater than r_N, it can be treated as a common dry gas zone; in a two-phase flow zone where r is smaller than r_N, it should be treated as a two-phase flow.

    c. Formation pressure is lower than critical condensation pressure in the whole reservoir. As for condensate gas fields developed by natural depletion, this stage is dominant in their production life of the fields. The whole gas reservoir is a two-phase zone in this stage and the flow in it should be treated as two-phase flow in the well test analysis.

2. Two ideas for solving this problemWell test analysis problems of condensate gas wells used to be very popular topics in well test research in the 1990s, for which plenty of professional papers dealing with these problems were published. Two ideas are concluded basically based on them, including:

(1) Precise solution obtained from the two-phase flow theory

It is impossible to achieve analytical solutions to this problem due to the complexity of the two-phase (or even three-phase) flow. As an alternative, the numerical solution is considered feasible. Surely, the compositional model in numerical simulation software and numerical well test methods developed recently lays a foundation for the methods of well test analysis in condensate gas reservoirs.

There are three problems influencing the research of this topic.

a. Difficulty in solution to this topic. Compositional model should be used to solve this problem; saturations of gas and condensate oil, relative permeability, and flow state should be determined based on the pressure distribution at different node points. In addition, the conventional method of numerical simulation should be changed to fit the requirement of well test analysis in the selection of time interval and grid division of formation so as to adjust itself to the data acquisition of the transient well test. It makes simple data analysis of a single well much more complicated and difficult.

b. Difficulty in data acquisition. Corresponding to the solution process, the acquisition of transient pressure test data should be more rigorous, more precise, and last longer; in addition, the requirement of working systems of a test well and adjacent wells is more rigorous as well. It is surely difficult to do in a condensate gas well under the normal production condition.

c. Urgency of field need. When a condensate gas field is put into development, problems to be solved with the transient well test method await the answers of the parties concerned, particularly the management personnel onsite. Conducting research on these problems with great effort is worthwhile in case of such urgent needs. However, there is a lack of impetus if it is only for theoretical discussion, and there are great difficulties in acquiring measured data onsite for verification of the correctness of theoretical models.

Therefore, a series of difficulties exist in a thorough study on a well test of condensate gas wells.

(2) Solutions aiming at practical requirements in the field

What are the practical requirements of well test and analysis when a condensate gas field is put into production? Simply speaking, the requirements can be classified into the following aspects:

a. To acquire and analyze bottom-hole flowing pressure p_{wf} and formation pressure p_R of production wells; this is indispensable to any condensate gas field developed with natural depletion or cyclic gas injection methods

b. To analyze additional flow resistance and the skin factor and their effects on gas well production after the retrograde condensate around the bottom hole has happened

c. To make clear the existence of a condensate oil section around the bottom hole of the production well, its size, and the mode of outward development if it exists

d. If a condensate oil section exists around the bottom hole of the production well, to make clear the decrease of flow coefficient of the condensate oil section and the flow coefficient ratio of the condensate oil section to the pure gas section

e. To ascertain the gas injection pressure of the gas injection wells, formation pressure around the injection section, and connection relationship between the injection wells and the production wells, etc.

Answers to these urgent problems in the field can be offered by conventional modern well test interpretation methods and well test interpretation software. They are introduced briefly later.

5.8.2 Model Graphs and Field Examples of Transient Test in Condensate Gas Well
Hump Caused by Wellbore Phase Change Behavior
As illustrated previously, two main factors influencing wellbore phase change behavior are as follow:

- Hump caused by retrograde condensation of condensate gas
- Hump and pressure shift caused by accumulated liquid loading inside wellbore

1. Hump caused by retrograde condensate process

The following factors must be considered when analyzing whether the hump phenomenon is caused by the phase change behavior induced by the retrograde condensate:

(1) A hump is very likely to occur when the amplitude of the bottom-hole pressure variation is large during well flowing or shutting-in periods, that is, the phase change behavior is induced easily if the producing pressure differential of the gas well is large and thus the amplitude of the subsequent shut-in pressure increasing is large.

(2) A hump may be induced easily by the wellhead shut-in operation. If the well is shut in by downhole shut-off valves, humps will not occur generally because there is no space in the wellbore available for phase changes.

(3) A hump is very likely to appear in low permeability formations. If formation permeability is low, a great production pressure difference is needed for production; when the pressure increases abnormally, the formation cannot make it balanced.

Humps on pressure buildup curves formed by phase change behavior are shown in Figure 5.154.

The left side of Figure 5.154a shows the normal situation that the slope of the pressure curve in the afterflow segment offsets 1 during abnormal pressure increasing, which causes the pressure derivative curve to tilt up simultaneously and exceeding the

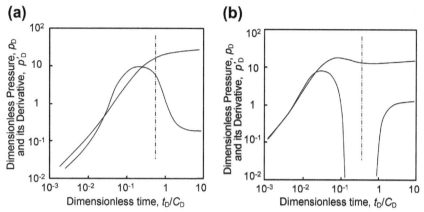

Figure 5.154 Schematic diagram of humps caused by retrograde condensate. (Left) Pressure increases abnormally. (Right) Occurrence of a funnel in a derivative curve.

pressure curve. The curves will come back to normal usually after the retrograde condensate process terminates.

The right side of Figure 5.154b shows that the hump effect is stronger than that shown in the left side of Figure 5.154. When the retrograde condensate occurs, the wellbore pressure exceeds the normal value, reaches its peak, and forms a typical "hump." Then when the retrograde condensate process finishes, the wellbore pressure becomes lower than the previous values and then gradually returns to its normal values. The pressure derivative curve is broken and looks like a funnel during occurrence of the hump.

If the data acquisition duration is short and only the preceding half segment of the funnel on the pressure derivative curve is acquired, the operators may be puzzled and cannot determine which reflections it is. If the test lasts longer to successfully acquire the later half segment of the funnel, the characteristics of the hump and its cause will be very clear.

2. Field examples

Figure 5.155 is a field example from the BQ condensate gas field. The test in this field example lasted nearly 1000 hours, after successfully passed the section of hump effect, acquired the data of the radial flow and boundary reflect segments, and all necessary information about the reservoirs was obtained.

Figure 5.156 is a field example in the YH condensate gas field, which only reflects the preceding half-segment of the funnel of the hump.

The test duration of this example in the YH condensate gas field is too short. It can be seen from comparison with the typical graph in Figure 5.154 and measured data in Well BS 74-1 in Figure 5.155. This example has only the preceding half-segment of the funnel of the hump and cannot be used for the interpretation and calculation of formation parameters.

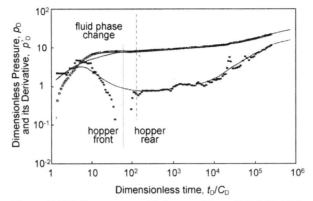

Figure 5.155 Test curves with hump effect of Well BS 74-1.

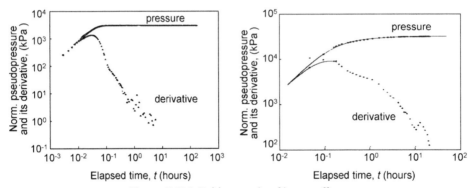

Figure 5.156 Field example of hump effect.

Normal Test Data Acquired before Condensation Happens in Formation

In condensate gas fields, when the pressure has not yet decreased to the dew point, test data are just the same as those measured in a dry well. Figure 5.155 is an example of a normal gas zone except for having a hump segment. Because Well BS 74-1 was tested in the early stage of exploration, there was no condensation phenomena happening at that time, and well test data could be completely interpreted with methods for common dry gas zones.

Figure 5.157 shows the test curves of Well YH 5 and Well YH 6 in the Yaha condensate gas field, both of which are of homogeneous formation reflections. The analysis method for them is the same as for dry gas zones.

Analysis of curve (a) in Figure 5.157 with well test interpretation software gives the parameters of Well YH 5 as follow:
- Flow coefficient $kh/\mu = 47.3$ mD·m /mPa·s
- Formation permeability $k = 2.93$ mD
- Skin factor $S = 8.0$

(a)

(b)

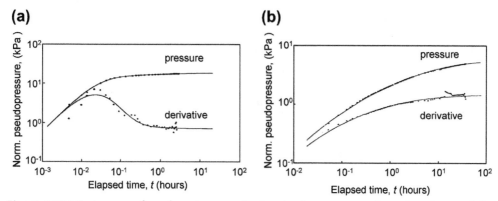

Figure 5.157 Test curves of condensate gas wells showing homogeneous formation characteristics: (a) Well YH 5 and (b) Well YH 6.

Analysis of the curves (b) in Figure 5.157 gives the parameters of Well YH 6 as follows:

- Flow coefficient $kh/\mu = 109.2$ mD·m/mPa·s
- Formation permeability $k = 5.93$ mD
- Skin factor $S = -5.49$.

Test Data Obtained from Well in Near Bottom-Hole Area with a Local Retrograde Condensation Zone

When formation pressure near the bottom hole is lower than the cricondenbar, there is a condensate oil zone in the formation, and thus there is a two-phase zone there as shown in Figure 5.153.

As the flowing coefficient in the two-phase zone is much lower than that in the pure gas zone, the composite formation, and the flow resistance in the outer inner zone of which is less than that in inner outer one, could be reflected in the characteristic graph of the pressure buildup curve, as shown in Figure 5.158.

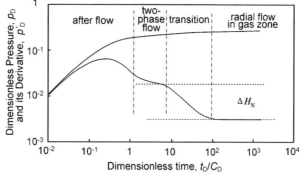

Figure 5.158 Model graph for a well with a condensate oil zone near its bottom hole.

The following parameters can be obtained from analysis of the aforementioned curves:

- Formation permeability k and flow coefficient kh/μ of pure gas zone
- Flow coefficient ratio of gas zone and two-phase zone

$$M_N = \frac{(kh/\mu)_N}{(kh/\mu)_g} = 10^{\Delta H_{ND}} \qquad (5.85)$$

- Expand radius of two-phase zone, r_N
- Skin factor of two-phase zone S_N and total apparent skin factor S_a

Field example: Well YH-3

Well YH-3 is an early appraisal well of the YH gas field. The JDK formation in Tertiary was penetrated. The flowing pressure was $p_{wf} = 44.27$ MPa during production, and the dew point pressure (p_d) was 51.06 MPa from the phase diagram, 6.79 MPa higher than the flowing pressure. It indicates that the condensate oil zone obviously formed in the formation. The formation pressure of this well was 55.79 MPa. The producing pressure differential Δp was as high as 28.24 MPa, as the additional producing pressure differential caused by the two-phase flow in the condensate oil zone was very high.

The pressure build-up curves of this well are shown in Figure 5.159.

Parameters obtained from analysis with well test interpretation software include:

- Flow coefficient of pure gas zone $(kh/\mu)_g = 152.1$ mD·m/(mPa·s)
- Formation permeability of pure gas zone $k = 8.9$ mD
- Flow coefficient ratio of gas zone to two-phase zone $M_N = 3.38$
- Expand radius of two-phase zone ≈ 50 meters
- Apparent skin factor of the well $S_a = 18.8$
- Skin factor of two-phase zone $S = 0.4$

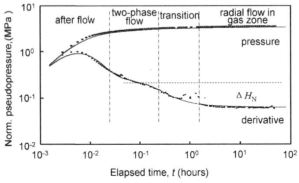

Figure 5.159 Log–log pressure buildup plot influenced by the two-phase flow effect of Well YH-3.

The production performance of the well can be concluded as follows: the formation permeability is medium and the flow coefficient is as high as 150 mD·m/(mPa·s). Meanwhile, the comprehensive flow coefficient of the oil and gas mixed area decreases to less than one-third of that of the pure gas zone. This causes obvious degradation of well production and the dramatic increase up to 19 or so of the skin factor. Moreover, the main influencing factor is still the effect of the condensation zone.

These analysis values are sufficient to make a basic evaluation on the production performance and the variation of the gas well for field management.

5.9 Characteristic Graph of Horizontal Wells (Model Graph M-16) and Field Examples

5.9.1 Geological and Engineering Background

Drilling horizontal wells is a main approach in the development of gas fields with thin layer reservoirs, fluvial channel sandstones, and other gas reservoirs with special configurations declared by many famous petroleum companies in the world. The drilling success rate aimed at special target zones increases continuously with the development of new horizontal well drilling techniques. However, well test data acquisition and analysis for horizontal wells are not good enough, for which the reasons can be concluded as follows.

1. Generally, a pressure gauge is set near the bottom of the straight section, that is, somewhere above the kick-off point, in the wellbore during data acquisition in the horizontal well. This position is far away from the target test interval in both vertical distance and linear distance in the wellbore. Moreover, the composition of well fluids in the horizontal section of the well is very complex, accumulated loading water and liquid there are very difficult to discharge out, and some gas dead corners are created there due to rising and falling of the horizontal section of the well. Therefore, it is very difficult for pressure data acquired during the test to reflect formation conditions correctly.

2. The reservoir is usually rising and falling underground, and because controlling and standardizing the borehole trajectory of the well are very difficult, the drilled horizontal borehole sometimes penetrates the reservoir intermittently or repeatedly, whose analytical model is therefore very hard to be set up.

3. Many parameters influence well test curves. To determine the effects of these parameters comprehensively, a test duration as long as about 8—10 logarithm cycles is demanded even for some simple formation models. Obviously, it is nearly impossible in field practice.

Due to these reasons, typical field examples are hardly seen, although many horizontal wells have been drilled in many fields. But it is believed that more well test results of horizontal wells will be used in performance analysis along with the improvement of drilling techniques and updating of testing methods.

If conditions on the horizontal well and the formation are met, the typical characteristic curves of this horizontal well have been offered by some researchers, including:

- The formation is infinite, horizontal, and homogeneous sandstone with uniform thickness, horizontal permeability k_X and k_Y, and vertical permeability k_Z
- After drilling into the formation, the horizontal section of the well is horizontal, the length of the horizontal section of the well is L_e, and the distance from the wellbore to the bottom of gas zone Z_W is a constant.

The position of the well in the formation is shown in Figure 5.160.

5.9.2 Typical Well Test Model Graph

1. Model graph with typical conventional conditions. These conditions include:

- The horizontal section of the well is sufficiently long, for example, the length of horizontal section $L_e = 300$ meters
- The penetrated gas zone is sufficiently thick, for example, the thickness of penetrated gas zone $h = 20$ meters
- The horizontal section of the well is approximately in the middle of the formation
- The horizontal section of the well has not been damaged seriously

The pressure buildup curves under these conditions are shown in Figure 5.161, and the related flow pattern is shown in Figure 2.22 in Chapter 2.

Figure 5.161 shows the following characteristic segments as listed here:

a. Afterflow segment

b. Vertical radial flow segment. Vertical radial flow will appear in the thick formations if a horizontal well was drilled through it (see flow pattern graph, Figure 2.22). Meanwhile, if the formation is very thin or the afterflow effect of the well is very serious, this segment may disappear or be covered.

c. Linear flow segment. This is an important characteristic segment in the well test curves of a horizontal well. For a well with a long horizontal section, the flowing

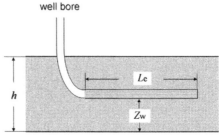

Figure 5.160 Position of a horizontal well in a formation.

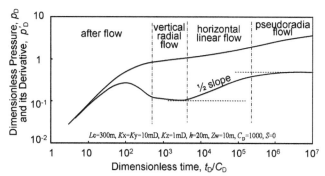

Figure 5.161 Typical flow characteristics of a pressure buildup curve in a horizontal well.

segment will be quite obvious. The pressure derivative curve becomes a rising straight line with a half-unit slope.

d. Pseudo radial flow segment. The pressure derivative curve of this segment is a horizontal straight line. The segment appears only when the reservoir is quite large.

2. Curves of a horizontal well in a very thin reservoir: Similar to those of fractured wells

If the reservoir is very thin, the vertical radial flow segment disappears and the linear flow segment of the horizontal well appears instantly after shutting in, as shown in Figure 5.162a, which is similar to the graphic characteristics of common fractured wells. If the horizontal section of the well was damaged during well drilling and/or completion, the curve will be similar to those of fractured wells with fracture skin characteristics, as shown in Figure 5.162b.

3. Curves of horizontal wells with a serious wellbore storage effect: Similar to those of a completely penetrating well

When a gas well is tested with the wellhead shut-in method, the wellbore storage coefficient is generally very high, and the effect of wellbore storage often makes the pressure buildup curves of the horizontal well similar to those of a completely

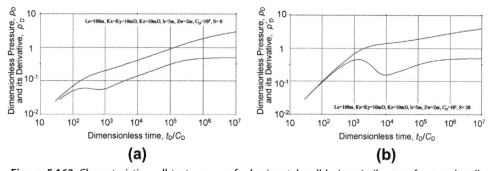

Figure 5.162 Characteristic well test curves of a horizontal well being similar to a fractured well.

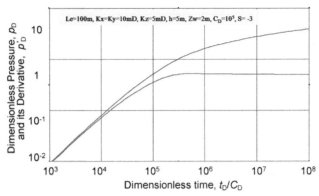

Figure 5.163 Characteristic well test curves of a horizontal well being similar to a completely penetrating well.

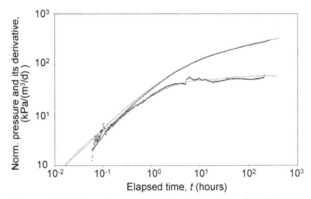

Figure 5.164 Log–log pressure buildup plot of Well T111.

penetrating well, especially for those horizontal wells with a short effective horizontal section, as shown in Figure 5.163.

Figure 5.163 shows that the characteristic segment of a horizontal well, the linear flow segment, does not appear at all.

Field examples

A horizontal oil well, Well T111, is temporarily taken as an example [45], as the author has no typical pressure buildup curves of horizontal gas wells available at the moment.

Well T111 penetrated thesandstone reservoir in Silurian. The length of the horizontal section of the well is 360 meters, the length of the effective interval is 270 meters, and the thickness of the oil zone is 0.9 meter. The pressure buildup curves of the well are shown in Figure 5.164 [45].

There are no characteristics of the horizontal well at all in Figure 5.164: with neither vertical radial flow segment nor half-unit slope linear flow segment of horizontal section, and the curves are just like typical curves of a completely penetrating well.

The interpretation results obtained from the interpretation with the well test interpretation software are as follow:

- Total skin factor $S = -6.67$
- Horizontal permeability $k_x = k_y = 194$ mD
- Vertical permeability $k_z = 0.008$ mD
- Wellbore storage coefficient $C = 0.7$ m^3/MPa.

6 SUMMARY

The content of this chapter is the core of this book. As mentioned in Chapter 1 of the introduction, the main purpose of the gas field dynamic study is to establish a dynamic model of gas wells and gas reservoirs. This dynamic model should take geological research results as its background, describe dynamic behavior with mathematic equations, vividly shed light on its dynamic characteristics in the form of pressure and pressure derivative characteristic curves, and finally make a description of the structure, distribution, and parameters of the reservoir.

Three dynamic characteristic graphs are introduced in this chapter, including pressure history graph, semilog and log—log plots of pressure, and log—log plot of the pressure derivative. These dynamic characteristic graphs describe the dynamic characteristics of gas reservoirs and gas wells from different aspects.

The pressure semilog analysis developed in the 1950s initiated the transient test analysis method, which has been widely applied as a conventional method until now. However, the semilog analysis method has its shortcoming. That is, it can only give such basic parameters such as reservoir permeability and well completion skin factor, while it cannot describe comprehensively the dynamic model characteristics of the reservoir. As a result, it is very limited in application.

Log—log plot and analysis of pressure and pressure derivatives developed in the 1980s pushed the transient well test to a new and higher level and formed the modern well test. A pressure derivative plot contains rich information about the reservoir structure and hydrocarbon flow regimes in the reservoir. If and only if the derivative plot is interpreted and analyzed carefully can an exact description be pinpointed for the reservoir and dynamic characteristics of the formation.

In this chapter, more than 10 types of reservoir commonly met in exploration and development of a oil or gas field were enumerated, which were introduced one by one with comparisons on the characteristics of their log—log plots, and verified by gas field examples as many as possible. The generation, application, and parameter calculation of these plots are all completed with the aid of well test interpretation software. Well test interpretation software is an effective, competent, and indispensable means; however, it is simply a means. To evaluate a gas well and gas reservoir rationally and remarkably on the basis of geological study results of the gas field, combined with completion process of the gas well, is the responsibility of the oil reservoir engineer and also the demonstration of his ability.

Characteristics reflected on the transient well test log—log plot are characteristics of the well and the reservoir in the short term. Due to the limited test time, information about boundaries far away from the well or information regarding the deep reservoirs cannot be fully revealed; it is just like that the whole life of a person cannot be evaluated comprehensively by only one thing he did at a specific time. Therefore, evaluations for the results of well test interpretation must be verified and improved in the pressure history of the gas well.

As for research on the dynamic performance of a gas well and gas reservoir, its ultimate purpose is to make a prediction for the future trend of the gas field base on the correct dynamic model rather than a correct evaluation. The model verified by previous pressure history verification can be used to predict the trend of field performance for quite a long time; the longer the verification time, the longer the duration of reliable prediction. It is undoubtedly very important for production planning.

CHAPTER 6

Interference Test and Pulse Test

Contents

Dynamic Well Testing in Petroleum Exploration and Development   © 2013 Petroleum Industry Press. Published by Elsevier Inc.
ISBN 978-0-12-397161-6, http://dx.doi.org/10.1016/B978-0-12-397161-6.00006-1   All rights reserved.   **411**

A multiple–well test normally means the interference test and the pulse test. In such a test at least two wells are required to form a well pair to conduct testing; hence the test is named the "multiple-well test." The method of the interference test was first applied in hydrogeology research and was called "water pumping test" or "hydrogeology exploration." During the test, while water was pumped out from one well, the water levels of the adjacent wells were observed to identify the connectivity among these wells. As the petroleum industry was developed, the interference test was introduced to studies in oil and gas fields, test means and data analysis methods were improved further, high accuracy and high resolution downhole pressure gauges were used to measure and record pressure data, and well test interpretation software was developed; all of these have played an important role and led to formation of the interference test method used in oil and gas fields.

In the mid–1960s, the pulse test was developed on the basis of the interference test. Essentially, the pulse test is a kind of interference test; however, during the test, an active well is required to change its flow rates several times so that the interference pressure fluctuates in the observation well. This method was initially considered as a completely new substitute for the interference test and was patented. After many years of field application, however, reservoir engineers, getting to know the method further, realized that it is not even as effective as the interference test with respect to obtaining the performance parameters of the dynamic reservoir model; furthermore, it takes more time in field operation, and because it requires the active well to be opened and shut in repeatedly, it brings about quite a lot of troubles. From a scientific and practical point of view, the pulse test is even not as good as the conventional interference test.

One important feature of the multiple-well test is that results obtained reflect not only reservoir information of the area around the testing wells, but that of a certain area controlled by the testing well pair, including actual interwell connectivity. As a result, its effect of understanding the reservoir is far superior to the normal single-well transient test.

Previous field practice and experiences have demonstrated that it is not easy to conduct successfully an interference test in exploration or the early development appraisal phases of oil or gas fields, especially gas fields. Because gas, as the medium of pressure transmission, is very compressible, the pressure change signal is very difficult to transmit from the active well to the observation well, which means that a large proportion of the tests cannot give any conclusion, especially if there are errors or omits in the test design.

Overall, the interference test, in particular the interference test in gas fields, is a project requiring prudent consideration. It is discussed in detail in the following sections of this chapter.

1 APPLICATION AND DEVELOPMENT HISTORY OF MULTIPLE-WELL TEST

1.1 Application of Multiple-Well Test

Multiple-well tests, either in gas fields or oil fields, serve the following purposes.

1.1.1 To Identify Formation Connectivity between Wells

As illustrated in Figure 6.1a, in some layered sandstone formations, the major pay zones of two wells penetrated the same horizon through are often interconnected. In some other sandstone formations with lithological boundaries, although the sand layers of the two wells correspond with each other, the effective pay zones may not correspond with each other, leading to nonconnectivity of two wells underground, as shown in Figure 6.1b.

If a gas field to be developed is a condensate one and is to be recovered by cyclic gas injection, and Well A is designed as a producer while Well B as an injector, the formation is as shown in Figure 6.1(b), then the development performance will be very poor. Therefore, it is very important to make the connectivity between wells clear by the interference test.

For water flooding oil fields, it is also very important to make clear the relations of interwell connectivity.

1.1.2 To Confirm the Sealing of Faults

Faults are well developed in eastern China, especially in the oil fields around BH Bay. Some oil and gas fields are cut into many small blocks by faults—the area of each block is

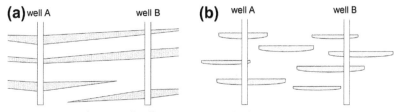

Figure 6.1 Comparison of interwell effective permeable formation. (a) Communication between effective zones. (b) Effective zones isolated by lithologic boundaries.

less than 1 km^2—so the connectivity relation between wells is very complicated. Due to their small down throw, small faults of class 2 and class 3 on the structure map may not isolate the zones as barriers to the oil or gas flow. For example, it has been observed that some oil producers in the ST area of the SL oil field received the influence of water injector wells on the other side of the fault. To make clear whether a fault is sealed or not, the interference test is no doubt a very effective method.

1.1.3 To Estimate Interwell Connectivity Parameters

Permeability and other formation parameters obtained from logging or coring analysis report are parameters at the points where the wells locate, and formation parameters obtained from single-well transient test are parameters of the area around the well. However, the interference test provides comprehensive average parameters of the wide area between wells, as shown in Figure 6.2.

Parameters obtained from multiple-well tests are as follow:
1. Flow coefficient of the area between testing wells: kh/μ
2. Storability parameter of the area between testing wells: $\phi\, h\, C_t$
3. Diffusivity of the area between testing wells: $\eta = \dfrac{k}{\mu \phi C_t}$
4. Permeability of the area between testing wells: k
5. Reserves per unit volume of the area between testing wells: $\phi\, h$

All of these parameters cannot be obtained by other methods.

1.1.4 To Identify the Vertical Connectivity of Reservoir

Some foreign researchers [46] presented research on the vertical permeability of the tight layer between two high permeability layers with the interference test and the pulse test, established its interpretation mode, and fulfilled its analysis, as illustrated in Figure 6.3.

As seen from Figure 6.3, layer 1 and layer 2 are isolated by a packer and are not communicated in the wellbore. Layer 1 is produced through the casing to generate pressure disturbance; pressure gauges are run into the perforated interval of layer 2 to monitor the interference pressure transmitted through the tight zone from layer 1.

This method was used in the Bigoray Nisku B reservoir in Alberta, Canada, to measure the vertical permeability for providing parameters in designing of the miscible flood secondary recovery plan [47].

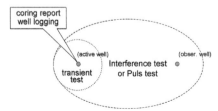

Figure 6.2 Parameters obtained from different methods are meaningful in different areas.

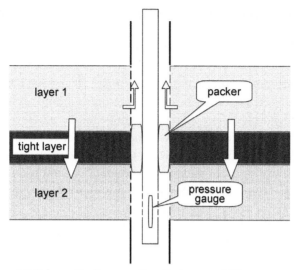

Figure 6.3 Schematic diagram of vertical permeability measurement.

1.1.5 To Study Formation Anisotropy

Interference tests and pulse tests involving 28 well pairs were conducted in the KL buried hill oil fields of the SL oil field. Comprehensive analysis on test results shows that the formation permeability within the area is highly directive (Figure 6.4).

As shown in Figure 6.4, permeability in the direction of the main fracture is as high as 1500 mD, but that in the direction vertical to the fault is less than 1 mD. This fact has influenced seriously on the producer spacing and water flood development [43, 44]. The analysis of actual production performance also confirms the aforementioned conclusion.

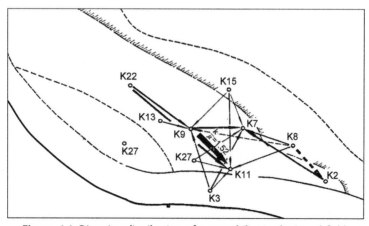

Figure 6.4 Directive distribution of permeability in the KL oil field.

1.1.6 To Study the Reservoir Areal Distribution and to Confirm the Results of Reserves Estimation

Interference test data were used effectively in reserve estimation in the JB gas field of China.

During the development preparation stage of the JB gas field, the impact of low permeability boundaries was noted from the pressure buildup curves. This raised the anxiety that the Ordovician system reservoir might be locally developed and divided into many small blocks. The interference test was the most effective method to eliminate this anxiety. The three wells of L5, L1, and SC1 in the middle of the central gas field were selected to form two well pairs for the interference test. The test lasted 10 months. It was proven that the wells were interconnected, but the reservoir was quite heterogeneous within the area, which is 4 km long from east to west.

This research sufficiently supported the result of reserves estimation and facilitated the determination of development of this gas field finally.

The situation of interference tests in the JB gas field is discussed in detail later in this book (refer to section 3.1 of chapter 3).

Each of the aforementioned research studies enhances our understanding of the dynamic models of oil or gas fields from different aspects. The dynamic models established on the basis of these research studies will reflect the reservoir conditions more realistically.

1.2 Historical Development of Multiple-Well Test

1.2.1 Multiple-Well Test Development Abroad

The development and application of multiple-well test date back to the 1930s. At that time, it was used as a hydrogeologic research method and was named "hydrogeologic exploration" or "pumping test," as it was called in the field. During the test, while water was pumped out from one water well to cause the water level to drop, the water levels of one or more adjacent wells of it were observed, and the connectivity between wells was identified from the observed variation of water levels of the adjacent wells. Even now the method is still used in some coal mines.

In 1935, Theis published his research result and presented the transient pressure solution for the interpretation of these data [49].

Jacob [50] named this test method the "interference well test" in his paper and applied the "type curve" in 1941. This type curve used in the interference test is the dimensionless pressure log-log normal plot described in Chapter 5, is the first type curve ever put forward and applied, and is still widely used up to now. King Hubbert [51] discussed this research and Muskat [52] also introduced it in his famous book *The Physical Principles of Petroleum Production*.

In the 1950s, this test method was widely used in oil and gas field research studies in the United States and the former Soviet Union. It is also during this period that the

Soviet Union designed and manufactured the ДГМ model piston-type downhole differential pressure gauge especially for the data acquisition of interference tests in oil fields. Some feature point methods for parameter calculation, such as the maximum value method and the initial disturbing point method, were also published during this period.

In 1964 Matthies [53] published his research results of the interference test in the Walfcamp oil field.

Between the 1970s and mid-1980s, research papers on the interference test were published constantly. In 1973, Earlougher and Ramey [55] presented the influence of various no-flow boundaries on the shape of the interference test curve and created the corresponding type curves. Ogbe and Tongpenyai [56, 57] studied the impact of active well wellbore storage and skin effect on the interference test curves. When fractures exist in the bottom hole, the interference test can help determine the fracture orientation, which was studied by Mousli and colleagues [58].

Much progress was made in the 1980s in the study on double porosity formations. With the advent of single-well test pressure and derivative type curves for double porosity formations, Deruyck and colleagues [59] created the type curves of the interference test in double porosity medium in 1982. Interpretation of these type curves, not only permeability k_f and other parameters of the fracture system, but also interwell storability ratio ω, interporosity flow coefficient λ, and other parameters, can be calculated.

The results of other research studies demonstrate that as the distance between the observation well and the active well increases, the value of the interference pressure becomes less than expected. Parameter r_D is introduced to correct the type curve; at the same time, with the increase of interwell distance, the characteristics of the double porosity medium diminished gradually and the test curve becomes similar to that of homogeneous formation [60, 61].

In 1966 Johnson and collegues presented the pulse test method [62], and later Kamal and created the interpretation type curve [63], which was once an indispensable interpretation tool for the pulse test. The parameters used in the type curve analysis were pulse amplitude and lag time obtained from cross plotting of the test data plot. This method is essentially a feature point method. In 1968, McKinley and co-workers [54] introduced some application examples of the pulse test in an oil field, which aroused great interest.

Between the end of the 1970s and the early 1980s, some researchers carried out studies whether the pulse test could be used in double porosity medium formation or fractured wells. It was shown that the pulse test could only analyze the approximate orientation and permeability of the fractures qualitatively, but could not work out parameters such as ω and λ [64, 65].

The pulse test in a layered reservoir is a complex issue, the data of which are very difficult to be interpreted. Prats [66] did some research on this, and Chropra [67] studied application of the pulse test for reservoir evaluation. After analyzing the pulse test in a pilot experimental area of the San Andres reservoir, he found that it was far from enough to use this method alone for layered heterogeneous formation and that the pulse test must be combined with the analysis of geological and petrophysical and single-well transient test analyses to obtain a comprehensive evaluation.

As can be seen from the development of multiple-well test research studies abroad, the method has long been a project of great interest to reservoir engineers. However, up until now it is just an experimental project, is not included in the productive test series, and needs to be improved and consummated continuously in practice.

1.2.2 Development of Multiple-Well Test in China

The field practice and theory study of the multiple-well test started quite early in China and remarkable achievements have been made in many oil fields.

As early as the 1950s, preliminary experiments of multiple-well tests were conducted in the YM oil field. In the 1960s, the DQ oil field purchased ДГМ model differential pressure gauges from the Soviet Union and made some field experiments.

From the late 1960s to the mid-1980s, to meet the needs of injector and producer design of water flood development of complicated fault block oil fields in the SL oil field, large-scaled interference test researches were conducted in the ST, DX, BN, and KL oil fields. Test data were acquired successfully from more than 100 well pairs and some of the research results had been published [68].

In 1980, full-scaled comprehensive interference test and pulse test researches were performed in a complicated buried hill reservoir, the KL oil field. Twenty-eight well pairs were tested one after another, and a full picture of the interwell areas of those well pairs, including permeability distribution, fracture development orientation, distribution of the storability parameter $\phi h C_t$, water flood effectiveness, and effecting factors, was obtained [43, 44]. The pulse test was performed successfully in the field for the first time in China, which provided valuable experience for later studies.

In the 1970s, interference test and pulse test achievements on quite a few well pairs were also achieved and applied in the oil field development research in ZS, WC, and XJK areas of the JH oil field. In the complicated fault formation of the DG oil field, extensive field experiments were also performed and much progress was made.

Between the late 1960s and early 1970s, HuiNong Zhuang and colleagues successfully designed and manufactured their own Model CY-733 glass piston-type differential pressure gauge and were rewarded the Chinese National Invention Award. Between the 1970s and the 1980s, almost all the early field interference tests and pulse test data were acquired by this instrument in the SL oil field, JH oil field, and DG oil field [68].

In 1986 the type curves for interpretation of the interference test in double porosity reservoirs were developed to solve the interpretation problem of interference tests in fissured double porosity reservoirs encountered in the oil fields [35]. This type curve was created at the same time as the type curves described elsewhere [59]. The latter matched the fracture flow section with the flow section of the total system and the transition section separately, while Zhuang Hui-Nong and Zhu Yadong [35] combined the three into one complete curve that looked exactly like the field actual data plot and was more characteristic of the double porosity medium.

In the 1990s, interference tests played an important role in the development preparation stage of the JB gas field. From September 1993 to August 1994, interference tests were conducted for more than 10 months between Well L–5 (active well) and Well SC-1 and Well L-1 (observation wells) in the central area of the JB gas field. Good results were obtained, which, combined with analysis of geological data, proved that the reservoir of the JB gas field in the Ordovician system was a continuously distributed but very heterogeneous formation. This description of the reservoir model provided powerful support for the reserves estimation and laid an important basis for development planning. This field practice and theoretical analysis of interference tests also demonstrated China's capability and level of the dynamic research on gas fields [69].

1.3 How to Perform and Analyze the Interference Test and Pulse Test

The multiple-well test has its unique advantages, but also faces special problems in field operation. In some new areas, definitive conclusions cannot be obtained even if considerable data have been acquired. The reason is that after the observation well has been monitored continuously for a long period, sometimes dozens of hours or even dozens of days, exact interference pressure signals generated by the active well cannot be received. Sometimes the observed downhole pressure in the observation well keeps fluctuating with various noises so that the test has to be halted with no results obtained.

Why does this happen? The reason might be that the transmission characteristics of interference pressure and the actual field conditions are not clearly known beforehand or that it is impossible to effectively identify the formation characteristics of a reservoir with an interference test without a dialectic thinking process.

1.3.1 Factors Affecting Interference Pressure Acquisition
The Value of the Interference Pressure is in a Very Small Order
When Well A is opened for flow, a pressure drop funnel is formed at the bottom hole. As illustrated in Figure 6.5, the funnel will extend outward as the well is flowing continuously.

As the pressure drop funnel extends outward, the pressure variation value decreases rapidly. At Well B, with a distance of r from Well A, this value becomes very small. Its specific value and the calculation method are discussed in detail later in this book. But within dozens of hours, this pressure variation value is often less than 0.01 MPa, or even less than 0.001 MPa.

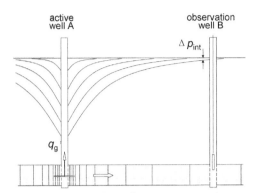

Figure 6.5 Transmission of interference pressure between wells.

As a result, it becomes very critical for the success of the interference test whether the downhole pressure gauge set in Well B can record such a small interference pressure variation.

Background Bottom-Hole Pressure of the Observation well should be Sufficiently Steady
It is surely difficult to hear clearly what others are speaking in a noisy public place. Similarly, there are some certain requirements for the background pressure in observation wells when an interference test between wells is performed.

1. There should not be much fluctuation and noise in the background pressure. Fluctuation here means the pressure going up and down per hour, the change amplitude varies from 0.01 to 0.1 MPa, and the noise means the pressure changes per minute or per second with amplitude between 0.001 and 0.01 MPa. The following factors can cause fluctuation and noises.
   (a) When the observation well is kept flowing, the downhole pressure may fluctuate and be very noisy. These fluctuation and noises are caused by the effect of turbulence flow in the borehole and pulses generated by decollement of the mixture of oil and gas in the borehole. The production variation itself also generates huge fluctuation of the bottom–hole flowing pressure.

   The author conducted a survey and summarized more than 100 well pairs in the field in which interference tests between wells were conducted successfully; none of them were conducted while the observation well was flowing.
   (b) The influence of operation in adjacent wells.

   Sometimes some wells near the observation well are under certain operations such as acid fracturing, well killing, injection, or shutting in or flowing etc. during the interference test. These operations may affect the bottom-hole pressure of the observation well.

(c) Cross flow effects of the observation well itself.

    For multilayer producers or a vertically heterogeneous thick reservoir, the contribution from each layer is different during production. This results in differential pressure in different layers within the formation, and when the well is shut in, cross flow will occur. Figure 6.6 clearly displays this severe fluctuation and noise.

    In addition to cross flow, Well KL203 was also subject to the impact of a phase change in the wellbore caused by water–gas separation and precipitation of the water. Fluctuation lasted for almost the whole period after the well is shut in.

(d) Pressure variation caused by the impact of tide.

(e) Pressure perturbance caused by other unknown factors.

    Some abrupt bottom-hole pressure changes, either a sudden increase or a decrease a step, are related to borehole conditions, for example, leakage in the wellhead or accidental opening of the vent valve. But there are some fluctuations, the causes of which cannot be found out eventually.

    If there are many unstable fluctuating factors in the background pressure, obviously it is difficult to conduct the interference test.

2. The background pressure should be in regular variation without an excessive upward or downward tendency. When an observation well is shut in to measure the interference pressure, its previous production history will affect the current pressure tendency.

(a) The pressure tends to increase if the observation well was shut in before long.

    The bottom-hole pressure of the observation well shut in before long, p_{ws}, is buildup pressure. It rises rapidly at the beginning after shutting in, and as time elapses, its increase becomes slower gradually and somewhat logarithmically.

(d) The pressure of the observation well in a developing field tends to decrease if it has been shut in for a long time.

    The pressure of this observation well having been shut in for a long time usually tends to decrease due to the depletion of formation pressure caused by the production of other wells within the oil and gas region.

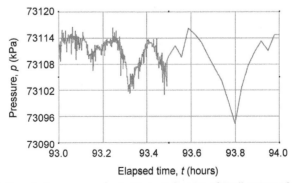

Figure 6.6 Bottom-hole pressure fluctuation and noise of Well KL203 after shutting in.

Before conducting the interference test, pressure gauges should be run in the observation well(s) to record the variation of the background pressure so that "pure interference pressure" can be distinguished.

The Test Duration is not Long Enough so that the Accumulated Interference Pressure is not Enough at All for Identification and Analysis

There is no abrupt "forward position" in the transmission of interference pressure from active Well A to observation Well B; instead, the interference pressure increases gradually from a very small value as the test time elapses. So this saying is not right that it takes 100 hours for the interference pressure to be transmitted from Well A to Well B.

If too optimistic geological parameters are used when making the interference test simulation design, expected interference pressure cannot be obtained during the designed test duration. In this case the test plan has to be modified and the test duration has to be extended, such a situation occurs in the field frequently.

Active Well and Observation Well are not Connected at All

If the active well and the observation well are not connected at all, it is certainly impossible to obtain the expected interference pressure no matter how long the test lasts.

If no interference pressure is obtained during the test, the operator will ask:

1. Are the active well and the observation well really not connected to each other?
2. Is there something wrong with the test design so that the expected interference pressure data cannot be acquired within the designed test duration?
3. Is it because we have not fully understood the background pressure pattern and consequently failed to identify the interference pressure, which already reached the observation well?

All in all, even having obtained interference pressure data normally acquired consecutively in several hundred or even thousands of hours' test duration, any questions expecting to be answered may not be answered. This is the most difficult issue faced when conducting the interference test and the main impedance of application of the interference test method.

1.3.2 Dialectic Consideration for Performing Multiple-Well Test Research in a Region

Operators should Avoid being in a Dilemma

As pointed out at the beginning of this chapter, the multiple-well test is often an experimental research project conducted for solving a specific problem within a region. How to deal with this kind of project is a seemingly simple but rather difficult problem for operators and test designers.

1. If definitive interference pressure is acquired in the observation and the connection parameters in the area between the testing wells are calculated, the test is no doubt very successful; however, such results can only be obtained when the formation conditions are very favorable, the test design is perfect, and operation of the test is arranged properly.

2. Interference pressure response cannot be observed if the test is conducted on both sides of a fault with big downthrows confirmed geologically, the physical properties of the testing wells are totally different, or the testing wells are very closely located. In these cases, the test can be regarded as successful and judgment of nonconnection between the testing wells can be concluded.

3. In cases of when the relationship between testing wells is not uncertain, especially when the connectivity between testing wells is expected but cannot be surely confirmed due to the failure in acquisition of the interference pressure response after a long test time, operators are put in a dilemma:

(a) There is not enough evidence for concluding immediately that the testing wells are not connected; this conclusion of "not connecting" is what the management personals of the field are not willing to accept.

(b) It is difficult to estimate how long the test will last if the test is kept carrying on, and to do so will increase the operation cost and delay the field production arrangement; it is also what operators are not willing to accept.

The Test Research should be Grouped and Classified According to Different Situations

1. The dialectics between connectivity and nonconnectivity

While conducting multiple-well test research in oil fields, designers usually do not consider well pairs with a clear relationship as study candidates, assuming that there is nothing worth studying. Instead, well pairs with a questionable relationship or with poor formation conditions and low production are most likely to be study candidates. Actually, this is a misunderstanding. Only after we have got a clear understanding of the characteristics of connective wells within a region can we determine its opposite characteristics—those of nonconnectivity.

2. Well pair classification

It is suggested that wells within a region be divided into three types while planning a multiple-well test in it.

- The first type: advantageous well pairs. Well pairs that are definitively connected each other and with good test conditions geologically. Good test conditions here mean that the distance between the testing wells is short, formation permeability is high, layers of the testing wells have clear correspondence, bottom-hole pressures of the observation wells are stable and easily measured, and the active well has high production and good flowing and shutting-in conditions.

- The second type: Negative well pairs. Well pairs that are confirmedly not connected with each other geologically, for example, there is a fault with a big downthrow between them or their formation pressure and/or fluid properties are totally different.

- The third type: Questionable well pairs. The connectivity relationship between them is not clear.

3. Arrange the test sequence and design the test plan properly

    (1) The test plan should include a certain number of well pairs in each of the three types mentioned earlier—especially the first type, that is, advantageous well pairs.

    (2) The test sequence: the well pairs of the first type should be prior tested. From the test results of these well pairs, which are connected each other, the connectivity regular trends and the actual formation parameters of the formation can be made clear.

    (3) It is possible that the expected interference pressure variation trends still cannot be known from the test results of selected favorable advantageous well pairs. In this case, serious modification of the whole test plan is necessary by extending the test duration and/or intensifying the activation magnitude of the active well or even selecting more favorable advantageous well pairs to be tested for complement.

        Testing questionable well pairs when no test results have been obtained from the advantageous well pairs often leads to failure of the whole test plan.

    (4) When the expected interference pressure variation trend has been obtained successfully from the advantageous well pairs, tests on the negative well pairs and questionable well pairs can be conducted further.

        The aforementioned planned test sequence demonstrates the dialectic consider process of this dynamic analysis project of the multiple-well test in the understanding of reservoirs. It is also the author's personal experience with many years of field practice.

2 PRINCIPLE OF INTERFERENCE TEST AND PULSE TEST

2.1 Interference Test

2.1.1 Test Methods

An interference test can be run with two or more wells, but the basic unit is still "the well pair" formed by two wells. In this well pair, one is called an "active well," which changes its working system or its flow rate during the test, either from flowing with rate q to shutting in or from shutting in to flowing with rate q so as to create a "disturbance" or "stimulation" to the formation pressure. The other one is called an "observation well," which is shut in and in a static state during the test; in which high-precision and high-resolution downhole pressure gauges are run to record the changes of interference pressure transmitted from the active well, as shown in Figure 6.7.

As very often in field operations, several wells are involved in the interference test, as illustrated in Figure 6.8.

No matter how many wells are involved in the test, one basic principle must be followed: at the same time interval, it is allowed that many observation wells are

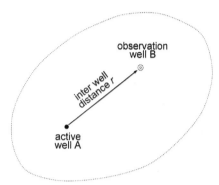

Figure 6.7 Well pair for interference test.

measuring the interference pressure response but there should only be one active well to change its flow state to generate a disturbance signal. Otherwise the following data analysis will be a mess.

1. If there is an interference pressure response, it is impossible to ascertain from which well the response comes.
2. While calculating the formation parameters, it is impossible to decide which active well's production rate should be used.
3. If formation parameters have been worked out, it is impossible to ascertain in which area they are.

This simple principle also applies to interference sources involved accidentally in the test. These include interference sources from the adjacent wells, such as hydraulic fracturing operation, well-flowing operation, injection, or well shut in due to mechanical failure.

Suppose there are four active wells—A1, A2, A3, and A4—and four observation wells—B1, B2, B3, and B4—in the test. The time sequence of stimulation, that is, well flowing and shutting in of active wells and running in of pressure gauges in observation wells, is as illustrated in Figure 6.9.

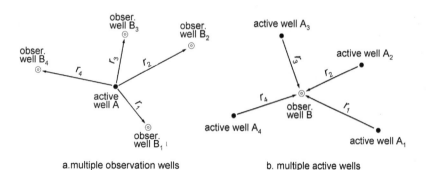

a. multiple observation wells   b. multiple active wells

Figure 6.8 Multiple-well interference tests.

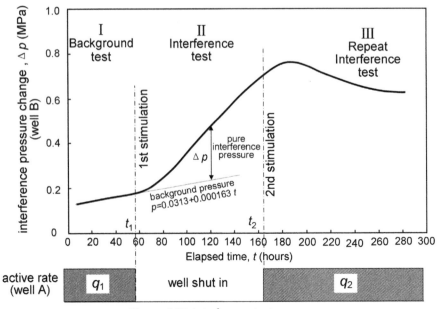

well type	operation schedule
active well A_1	
active well A_2	
active well A_3	
active well A_4	
obser. well B_1	
obser. well B_2	
obser. well B_3	
obser. well B_1	

━━━ active well flowing   ☐☐☐ obser. well monitering

Figure 6.9 Operation schedule of multiple-well interference tests.

The stimulation times of active wells should not overlap. Furthermore, due to the fact that there is a certain "time lag" of the interference pressure response, there should be certain intervals between the stimulation times, the length of which depends on the values of the formation parameters (permeability k, distance between wells r, etc.) and is determined in well test design.

An "interference pressure" transmitted from an active well is recorded in the observation well during interference test, as shown in Figure 6.10.

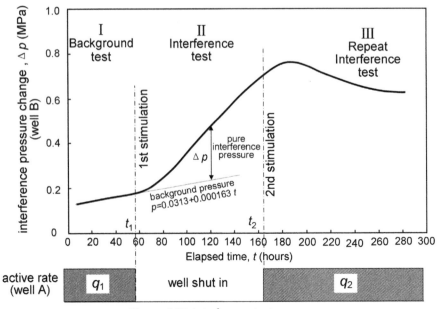

Figure 6.10 Interference test curve.

As seen from Figure 6.10, the test is divided into three stages:

1. Background pressure test stage (I). Even if no influence pressure is transmitted from the active well, the bottom-hole pressure of the observation well is still changing: remaining stable, increasing or decreasing with a role, or with some kinds of fluctuation and noises. Consequently, before recording the interference pressure, the background pressure of the observation well B should be monitored consecutively for a period of time, moreover, this period should be long enough and arranged in advanced.

   The purposes of monitoring the background pressure are as follow.

   (1) To identify whether observation well B meets the requirements of conducting the interference test. If any one of the following conditions exists, Well B is not suitable to be an observation well:

   - There is a fluctuation of 0.01 to 0.1 MPa and the reason for it cannot be identified
   - There are noises—the frequency of which is measured by seconds or minutes
   - There is an abrupt pressure jump up or pressure drop down that exceeds 1 MPa daily
   - The pressure occasionally jumps up or down a step with unknown reasons, etc.

   Just as described by the late academician, well test expert Tong Xianzhang, the bottom hole of any oil or gas well in a developing field is never quiet. Therefore, before monitoring the interference pressure in an observation well, the background pressure must be measured to exclude any observation wells that do not meet the requirements or to improve the monitoring methods to make the well suitable for being an observation well.

   (2) To find out the change role of the background pressure. As shown in Figure 6.10, the background pressure can be expressed by an analytic expression. Whether the pressure of the observation well has been influenced by interference pressure can be judged by the deviation of the measured pressure from the background pressure. In addition, a "pure interference pressure value" can be separated from the difference of the measured pressure and the value on the extrapolated trend line of the background pressure.

2. Interference test stage (II). This is the main stage of the interference test.

   (1) Pure interference pressure value Δp can be worked out from data in this stage. Draw a log–log plot of pure interference pressure Δp vs time t, and the formation parameters can be calculated by type curve matching.

   Pure interference pressure Δp is the pure pressure change on the basis of the background pressure caused by the stimulation effect of the active well.

   Δt is pure interference time. $\Delta t = t_1$ when the active, that is, when its flow rate is changed. At that time, personnel of the test team are usually working at well B and cannot get accurate time t_1 when well A located far away changes its working system so the value of t_1 has to be read from the operation log. This sometimes results in considerable errors in interpretation.

(2) Data of stage II are also the main stage of determining whether the testing wells interfere with each other.

In Figure 6.10, the interference of Well A has influenced Well B: Not long after active well A was shut in did the pressure of observation well B deviate from the background pressure and exhibit an increasing tendency. The shutting in of active well A should cause increased formation pressure, which is consistent with the deviation tendency of Well B.

The field conditions are complicated. Some abnormalities have been encountered during many years of field practice of the interference test, for example, before active well A changes its working system, that is, when $t < t_1$, the pressure of observation well B has already started changing, or the shutting in of Well A corresponds with a pressure decreasing in Well B—obviously both of these are very illogical. Sometimes the pressure of Well B increases and then decreases abruptly. All of these pressure changes of Well B are false appearances and should be eliminated from the interference pressure responses.

3. Repeated interference pressure test stage (III). For questionable interference test results, the repeated test, that is, stage III, is certainly the best way to eliminate any doubt.

At this stage, the active well usually resumes its original working system or flow rate, and the test is completed simply by extending the monitoring time of the observation well.

Formation parameters can also be analyzed by data in the repeated test stage; the pressure history of this stage can be used to verify the analysis results.

2.1.2 Parameter Factors Affecting Interference Pressure Response Value
Interference Pressure Response Estimation

No matter in oil fields or in gas fields, before conducting an interference test, estimations about when the interference pressure response can be observed and how much the interference pressure response is should be made. For example, the parameters of the test well pair in a gas field are as follow:

- Formation permeability $k = 2$ mD
- Formation thickness $h = 5$ meters
- Natural gas viscosity in place $\mu_g = 0.03$ mPa·s
- Distance between the active well and the observation well $r = 1000$ meters
- stimulation gas flow rate of the active well $q_g = 5 \times 10^4$ m^3/day

These parameters are input into the well test analysis software, and the pure interference pressure response can be calculated as shown in Figure 6.11.

As seen from Figure 6.11, the increase of the pure interference pressure response can be roughly divided into three stages:

1. Initial stage. Within the initial approximate 250 hours, the interference pressure response is close to zero, which means that no interference pressure response is observed for about 10 days.

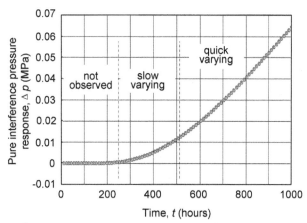

Figure 6.11 Pure interference pressure response estimation ($k = 2$ mD, $h = 5$ m, $\mu_g = 0.03$ mPa·s, $r = 1000$ meters, $q_g = 5 \times 10^4$ m³/day).

2. **Slowly increasing stage.** Within the next 250 hours, the total increase of the interference pressure response is about 0.01 MPa or 0.1 atm. At this stage the pressure rises slowly.
3. **Fast increasing stage.** During the coming period, the interference pressure response increases much faster—at a rate of 0.01 MPa per 100 hours.

   Although these three stages exist in each kind of formation, the specific increasing value and the duration of each stage are different, and sometimes the differences may be very significant. The influence of some formation parameters on the interference pressure response is discussed in the following sections.

Influence of Formation Fluid Mobility k/μ

Figure 6.12 shows interference pressure responses under the condition of different fluid mobilities. The input parameters are as follow: formation thickness $h = 5$ meters, distance between test wells $r = 1000$ meters, gas flow rate of the active well $q = 5 \times 10^4$ m³/day, and the values of k/μ are between 30 and 3000 mD/(mPa·s).

   As shown in Figure 6.12, there are some features in transmission of the interference pressure:
1. There is no abrupt change front during transmission of the interference pressure regardless of the value of k/μ. The interference pressure, starting from point zero, increases slowly at first and then faster gradually in the later stage of the test. Generally speaking, however, the accumulated increase of the interference pressure is still very small. For example, in Figure 6.12 the total accumulation of interference pressure is less than 0.1 MPa in nearly 1000 hours.
2. There must be a defined value that pressure gauges can measure clearly when we speak of "the time when the interference pressure is received." For example, if it is

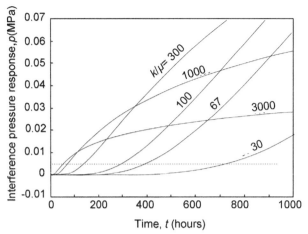

Figure 6.12 The influence of mobility k/μ on interference pressure response (k/μ_g = 30–3000 mD/(mPa·s), h = 5 meters, r = 1000 meters, q_g = 5 × 10^4 m^3/day).

defined as 0.005 MPa (the dotted line in Figure 6.12), it takes about 50 hours for formation with a mobility of k/μ = 3000 mD/(mPa·s)(in this case, i.e., k = 90 mD), while it takes about 700 hours for formation with a mobility k/μ = 30 mD/(mPa·s)(i.e., k = 0.9 mD).

3. In interference tests, the higher the mobility k/μ, the sooner the interference pressure reaches the observation well. But the absolute accumulated value of the interference pressure is very limited. It is because when k/μ is high, the pressure drop funnel of the active well itself is very small, and hence the funnel front is smaller.

4. In interference tests, the lower the mobility k/μ, the later the interference pressure reaches the observation well. But the absolute accumulated value of the interference pressure in the later stage may be a little higher. For example, when k/μ >300 mD/(mPa·s), the value of the accumulated pure interference pressure may reach 0.1 MPa or higher.

Generally speaking, just as demonstrated from the aforementioned analysis, it is rather difficult to observe the interference pressure response between gas wells, and therefore:

(1) High-precision pressure gauges are needed for continuous pressure measurement.

(2) The test needs a rather long time. Especially for low permeability formation (e.g., k < 1 mD), it requires dozens of days to observe a pressure change of 0.01–0.02 MPa—this is really not easy!

(3) The bottom-hole background pressure must be very stable. There should be no fluctuation or noises greater than the value of the aforementioned interference pressure.

It is a little bit easier for oil or water wells, especially for water wells, because the compressibility and viscosity of water are very small, which make it easier to create a pressure interference between wells, which is why the multiple-well test was first developed in hydrogeologic study.

Influence of Distance between Wells r on Interference Pressure Transmission

Figure 6.13 shows the transmission of interference pressure with different distances between wells. The other formation parameters of Figure 6.13 are $k = 3$ mD, $h = 5$ meters, $\mu_g = 0.03$ mPa·s, and $q_g = 5 \times 10^4$ m^3/day.

As can be seen from Figure 6.13, as the distance between testing wells increases, the time required for the same amount of interference pressure to arrive at the observation is consequently delayed:

1. When $r = 500$ meters, the transmission time is about 180 hours for 0.05 MPa of interference pressure
2. When the distance doubles, that is, $r = 1000$ meters, the transmission time for the same amount of interference pressure 0.05 MPa is 720 hours, three times longer than the former case.

It is seen that the transmission time is delayed at the rate of the distance square (r^2) so when selecting a test well pair, avoid conducting the test with the wells too far away from each other, as otherwise useful test data acquisition will be very difficult.

Influence of Stimulation Amount per Unit Thickness q_g/h on Interference Pressure Response

We can imagine that the bigger the stimulation amount of the active well is, the higher the interference pressure response observed at the observation well. Simulated computation proves that they are approximately in direct proportion to each other as shown in Figure 6.14.

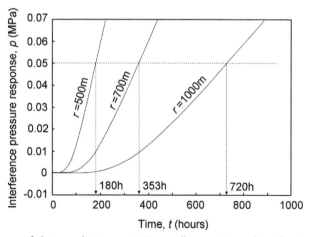

Figure 6.13 Influence of distance between testing wells r on time when the interference pressure reaches the observation well ($k = 3$ mD, $h = 5$ meters, $\mu_g = 0.03$ mPa·s, $r = 500-1000$ meters, $q_g = 5 \times 10^4$ m^3/day).

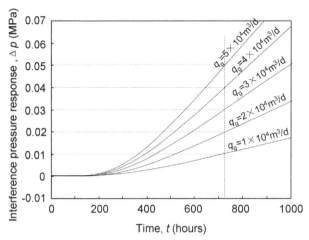

Figure 6.14 Influence of stimulation amount on interference pressure.

As can be seen from Figure 6.14, with the same values of distance between the test wells, formation permeability k and formation thickness h, when the stimulation amount increases from 1 to 5×10^4 m^3/day, the interference pressure response at the same moment increases from 0.01 to 0.05 MPa.

When planning the interference test with a well pair, the well with a lower production rate should be closed and adopted as the observation well, and the well with a higher production rate selected as the active well. In this way, better test results are obtained.

2.1.3 Type Curve Interpretation Method for Interference Test Data
Interpretation Type Curves for Interference Test in Homogeneous Formations
The interference test interpretation type curve for homogeneous formation is the earliest log—log type curve. To plot this type curve, the following equations are used.

$$\Delta p_D = -E_i\left(-\frac{1}{4(t_D/r_D^2)}\right),\tag{6.1}$$

where Δp_D is dimensionless pressure,

$$\Delta p_D = \frac{2.714 \times 10^{-5} kh T_{SC}}{q_g T_f p_{SC}} \cdot \Delta\psi\tag{6.2}$$

$\dfrac{t_D}{r_D^2}$ is dimensionless time,

$$\frac{t_D}{r_D^2} = \frac{3.6 \times 10^{-3} k\Delta t}{\phi \mu_g C_t r^2}.\tag{6.3}$$

Figure 6.15 is the type curve plotted with Equations (6.1).

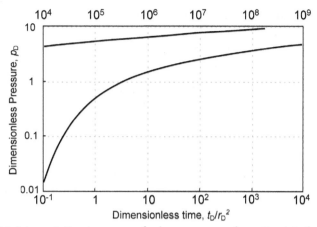

Figure 6.15 Interpretation type curve for homogeneous formation interference test.

Separation of Pure Interference Pressure

Three jobs must be done while interpreting test data that definitely contain interference pressure response as illustrated in Figure 6.10.

1. Work out the analytical expression of the background pressure variation.

Take the case illustrated in Figure 6.10 as an example, where the background pressure is displayed as an upward straight line, and it can be expressed as $p_{bp} = 0.0313 + 0.000163\ t$, or in common condition as

$$p_{bp} = a + b\,t \tag{6.4}$$

If there is no influence of pressure disturbance created by active well A, the pressure of observation well B should continue varying follow this role.

2. The separation of pure interference pressure Δp.

As seen from Figure 6.10 when $t > t_1$, the measured pressure obviously deviates from the role expressed by Equation (6.4). The pressure difference or the measured pure interference pressure Δp at time t can be calculated by

$$\Delta p = p_{mp} - \left(a + bt\right). \tag{6.5}$$

The corresponding time should also be converted into "interference time Δt" calculated from the stimulation of active well A.

The expression of Δt is

$$\Delta t = t - t_1 \tag{6.6}$$

Draw measured pure interference pressure Δp and interference time Δt. If the coordinate scales of this measured pure interference pressure plot are consistent with those of the type curve, type curve match analysis can be done and the parameters worked out.

Type Curve Match Analysis and Parameter Calculation

The methods of type curve match and parameter calculation have been discussed in detail elsewhere and are not repeated here (refer to section 5.4.1 of chapter 5).

The following two groups of parameters can be obtained from the interference test type curve match:

Flow coefficient

$$\frac{kh}{\mu_g} = \frac{3.684 \times 10^4 q_g T_f p_{SC}}{\mu_g T_{SC}} \cdot \frac{[\Delta p_D]_M}{\Delta \psi_M} \tag{6.7}$$

Storability parameter

$$\phi h C_t = \frac{3.6 \times 10^{-3} kh}{\mu_g r^2} \cdot \frac{\Delta t_M}{[t_D/r_D^2]_M} \tag{6.8}$$

$[\Delta p_D]_M$ and $\Delta \psi_M$ are pressure coordinates of the match point, that is, readings of the match point on the type curve plot and on the actual measured data plot, respectively. Just the same, $[t_D/r_D^2]_M$ and Δt_M are time coordinates of the match point on the type curve plot and the actual measured data plot, respectively.

With the obtained two groups of parameters, diffusivity coefficient $\eta = k/(\mu \phi C_t)$, permeability k, and reserves per unit volume ϕh can be calculated.

As mentioned earlier, during the 1970s and 1980s, many engineers repeatedly completed such match analysis manually, but nobody does it now, as a computer with well test interpretation software can conveniently provide the results of match analysis.

Pressure History Match Verification of Interpretation Results

The same as cases of a single well transient test, a theoretical model can be established with the interpretation results obtained from interference tests; and match vilification of the theoretical model with measured pressure data can validate the reliability of the interpretation results. This is also accomplished with the aid of well test interpretation software. Figure 6.16 is a schematic diagram.

If the tendencies of pressure variation of the theoretical model and the measured pressure are different, interpretation results should be corrected or modified.

There were usually no such verification process and corresponding plots in earlier interference test reports; when interpretation was done manually, such verification was impossible, but this must be done for analysis with well test interpretation software.

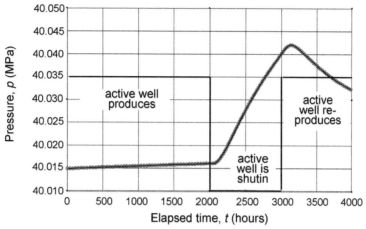

Figure 6.16 Pressure history match verification of interference test interpretation results.

Other Interpretation Type Curves for Interference Test Analysis

1. Pressure and its derivative type curves for homogeneous formation.

   Just like single well transient test analysis, interference test analysis also uses pressure and its derivative type curves as shown in Figure 6.17. This type curve further improves the precision of match analysis.

2. Interference test type curves for homogeneous formations with a boundary effect.

   Earlongher and Ramey [55] presented type curves for a test well pair in a rectangular reservoir with no-flow boundaries. The type curves are as illustrated in Figure 6.18.

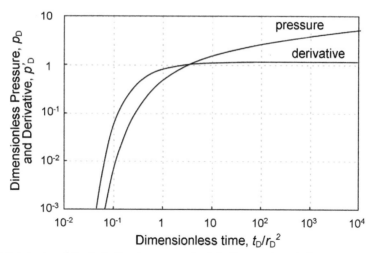

Figure 6.17 Pressure and its derivative type curve for interference tests in homogeneous formation.

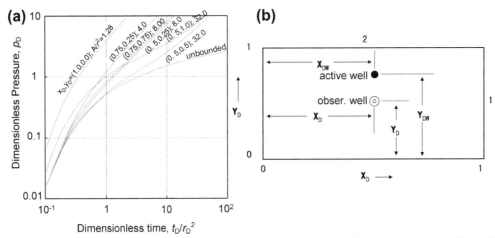

Figure 6.18 Interference test type curve for formation with boundary effect. (a) Type curves. (b) Well locations.

3. Interference test type curves for double porosity formations.

Figure 6.19 shows interference test type curves for double porosity formation developed by Deruych and colleagues [59]. The type curves include homogeneous flow curves and transition flow curves. The jointed application of this composite type curve was introduced in Figure 5.43 in Chapter 5.

4. Interference test type curves for double porosity formation that combines homogeneous flow curves and transition flow curves.

Zhuang Huinong and Zhu Yadong [35] developed interference test type curves that combined homogeneous flow curves and transition flow curves; these type curves were introduced in Figure 5.44.

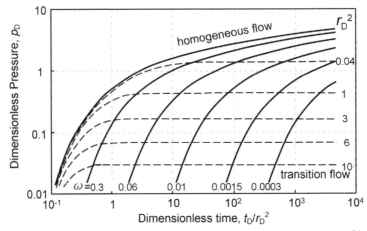

Figure 6.19 Interference test type curves for double porosity formation (consist of both homogeneous flow curves and transition flow curves).

5. Pressure derivative type curves for double porosity formations.

Similar to single well log–log type curves, interference test type curves for double porosity formation consist of pressure derivative type curves as well. This was presented in Figure 5.45 in Chapter 5 and is not repeated here.

2.1.4 Characteristic Point Interpretation Method for Interference Test

Interpretation methods for the interference test introduced earlier are all the type curve analysis method. Their remarkable characteristic is focusing on the establishment of a dynamic reservoir model, the pressure performance of the dynamic reservoir model, that is, the type curve, is compared with the actual performance of the reservoir, that is, actual pressure data. If they completely consist with each other, the model characteristics of the reservoir are confirmed and then are verified by pressure history verification.

This concept was not established yet in the early development stage of the interference test. Many interpretation methods developed at that time can be basically classified as the "characteristic point method," which laid its base on certain special points on the interference pressure curve, for example, maximum value point and initial pressure disturbance point, and the formation parameters were calculated with the locations of these characteristic points.

Shortcomings of characteristic point methods are as follow.

1. Some contingencies exist that may cause errors. The interference pressure itself is very small, often at the edge of accuracy control of the pressure gauges, so some jumps of measuring points may cause these characteristic points to drift, making them very difficult to identify. In addition, the selection of these characteristic points is often related to the plotting method passing through these test points.

2. The characteristic point methods are only applicable for homogeneous formations. All characteristic point methods published so far are only applicable for homogeneous formations; none of them is suitable for analysis and parameter calculation in double porosity formation or formation with boundary or other complicated formations.

3. They are difficult to include in the analysis sequence of well test interpretation software. Up until now, there is no large-scaled and commonly used well test interpretation software in which any of these methods have been integrated.

These methods are introduced here briefly for they once played a role in interference test analysis. For example, the maximum value point method and its application and related equation are introduced as the following.

If stimulation of the active well in an interference test follows the three stages of the first flowing, shutting in, and second flowing, there will be a maximum value point m on the interference pressure response curve at time t_m as shown in Figure 6.20.

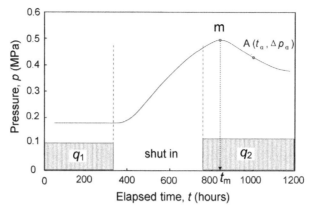

Figure 6.20 Parameter calculation by maximum value point method.

Diffusivity coefficient η can be calculated with the following equation:

$$\eta = \frac{k}{\mu \phi C_t} = \frac{r^2 t_1}{4 t_m (t_m - t_1) \ln \dfrac{q_2 t_m}{q_1 (t_m - t_1)}}. \tag{6.9}$$

Select point A in the interference pressure section on the interference pressure response curve, its pressure value is Δp_a and time is t_a, the flow coefficient kh/μ can be calculated with Δp_a and t_a:

$$\frac{kh}{\mu} = \frac{q_2 E_i \left[\dfrac{-r^2}{4\eta(t_a - t_1)} \right] - q_1 E_i \left[\dfrac{-r^2}{4\eta t_a} \right]}{1.068 \Delta p_a}. \tag{6.10}$$

In addition to the maximum value point method, there are some other characteristic point methods for parameter calculation, such as

- Integration method
- Differential method
- Initial pressure disturbance point method
- Secant method

They are not introduced here.

2.2 Pulse Test

2.2.1 Pulse Test Method

The pulse test is essentially a kind of interference test. What is different from interference test is that, during the test period, the active well (sometimes called the pulse well as well in the pulse test case) changes its working systems repeatedly, usually from flowing to shutting in and from shutting in to flowing again and again alternately; each working system lasts the same period of time. After the well has changed its working system more than three times, the pulse pressure response can be observed at the observation well as shown in Figure 6.21.

In Figure 6.21, the active well (the pulse well) had been flowing steadily for 2000 hours before the test started; the well was shut in at the 2000th hour and reopened at the 3000th hour. After several cycles of opening and shutting in, pulses of pressure change were obviously observed in the observation well.

The following can be obtained from the change pulse in Figure 6.21:

1. The formation is interconnected between the pulse well (active well) and the observation well.
2. The value of the pulse pressure is very small. Even the stimulation of 1000 hours of shut in following 1000 hours of flowing, the amplitude of the pulse pressure response is lower than 0.02 MPa.
3. Formation permeability and other parameters can be calculated from the test curve of Figure 6.21.

2.2.2 Kamal's Analysis Method for Pulse Test [63]

Since Johnson presented the pulse test method in 1966, many researchers have tried many methods to analyze test data. In 1975, Kamal presented his analysis method, which was included in its special series of technology books and recommended by the Society of Petroleum Engineers, and is introduced in this book. Kamal's method was specially created for homogeneous formations and was essentially a kind of characteristic point method. As described previously in the analysis of characteristic point methods for interference tests, the Kamal method has its own limitation in data analysis, bringing some inconvenience in operation and errors in interpretation.

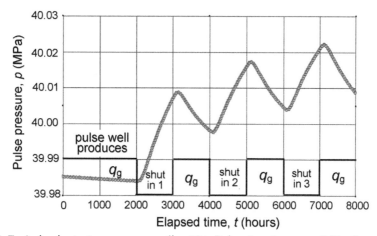

Figure 6.21 Typical pulse test pressure curve ($k = 30$ mD, $h = 5$ meters, $\mu_g = 0.03$ mPa·s, $r = 2000$ m, $q_g = 5 \times 10^4$ m³/day).

With the common application of well test interpretation software, pulse test data can also be interpreted by the type curve match method used for interference test analysis; this is discussed later.

Some Definitions

1. Pulse serial number.

Figure 6.22 gives the pulse serial numbers Kamal defined in the pulse test. The reason for numbering is that the different curve among Kamal's type curves is only applicable to its corresponding pulse.

Pressure pulse here means:

*   Two wave troughs with a wave peak in between is a pulse
*   Two wave peaks with a wave trough in between is also a pulse

Thus, in Figure 6.22, the first shut in forms pulse No. 0, the first flow forms pulse No.1, and the others are on the analogy of this. Consequently, odd pulses and even pulses are further classified,

*   Even pulse—pulse No. 0, pulse No. 2, pulse No. 4 …
*   Odd pulse—pulse No.1, pulse No.3, pulse No. 5 …

A different analysis type curve is applicable to a pulse with a different number.

2. Some characteristic quantities of pressure pulse.

Figure 6.23 gives some main characteristic quantities of a pressure pulse.

Draw the common tangent of the two wave troughs and then draw a tangent paralleling to the common tangent at the wave peak. Then draw a vertical line through the tangential point at the wave peak. The following characteristic quantities can then be obtained:

a. Δt_C = pulse period

Pulse period is the whole duration of the stimulation process mentioned earlier, that is, the whole duration of the pressure pulse—one shut in and one flow of the active well.

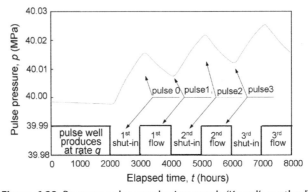

Figure 6.22 Pressure pulse numbering graph (Kamal's method).

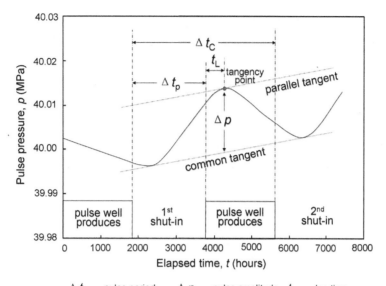

Figure 6.23 Definitions of symbols in pulse test graph analysis (Kamal method).

b. Δt_p = shut-in period

Shut-in stimulation period Δt_p is duration of the shut-in section of the pulse period as shown in Figure 6.23.

c. Δp = pulse amplitude

Draw a vertical line downwards from the tangential point until intersecting the common tangent; the height from the tangential point to the intersection with the common tangent is the pressure pulse amplitude—pulse amplitude for short.

d. t_L = lag time

Lag time t_L is the time difference between the end of the shutting in that generates the pressure pulse and the time when the pressure pulse reaches its peak.

e. Δp_D = dimensionless pulse amplitude

For oil wells,

$$\Delta p_D = \frac{0.54287 kh}{\mu B q} \cdot \Delta p \tag{6.11}$$

For gas wells,

$$\Delta p_D = \frac{2.714 \times 10^{-5} kh T_{SC}}{q_g T p_{SC}} \cdot \Delta \psi \tag{6.12}$$

f. $[t_L]_D$ = dimensionless lag time

$$[t_L]_D = \frac{3.6 \times 10^{-3} k t_L}{\mu \phi C_t r_w^2} \tag{6.13}$$

g. F' = pulse time ratio

$$F' = \frac{\Delta t_p}{\Delta t_C} \tag{6.14}$$

Eight Type Curves

In his paper, Kamal presented eight type curves, which are categorized into two classes.

1. Class 1: Type curves for calculation of flow coefficient kh/μ

The typical pattern of this class of type curves is shown in Figure 6.24. As can be seen from Figure 6.24, the vertical coordinate of this type curve is the dimensionless pulse amplitude $\Delta p_D[t_L/\Delta t_C]^2$ and the horizontal coordinate is the ratio of lag time to pulse period $t_L/\Delta t_C$.

There are four type curves in this class, namely:

- Type curve for the first even pulse (i.e., pulse No. 0)
- Type curve for the first odd pulse (i.e., pulse No. 1)
- Type curve for the other even pulses (i.e., pulse No. 2, No. 4......)
- Type curve for the other odd pulses (i.e., pulse No. 3, No. 5......)

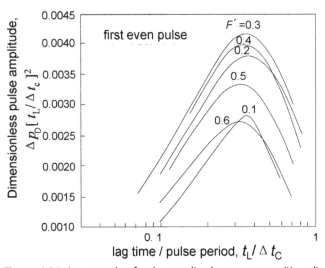

Figure 6.24 An example of pulse amplitude type curves (Kamal).

2. Class 2: Type curve for the calculation of storability $\phi h C_t$

The typical pattern of this class of type curves is shown in Figure 6.25. As can be seen from Figure 6.25, the vertical coordinate of this type curve is the dimensionless delay time $(t_L)_D/r_D^2$ and the horizontal coordinate is still $t_L/\Delta t_C$.

There are four type curves in this class, namely:

- Type curve for the first even pulse (i.e., pulse No. 0)
- Type curve for the first odd pulse (i.e., pulse No. 1)
- Type curve for the other even pulses (i.e., pulse No. 2, pulse No. 4......)
- Type curve for the other odd pulses (i.e., pulse No. 3, pulse No. 5......)

Parameter Calculation

After having obtained a qualified pulse test curve, parameters are calculated as follow.

1. Select the pulse to be analyzed, identify its serial number and its even or odd category, and then select the suitable type curve for analysis.

2. Follow the way illustrated in Figure 6.23 and draw the common tangent of the pulse troughs (or pulse peaks) before and after the pulse peak (or pulse trough) to be analyzed. Draw the tangential line parallel to the common tangent at the pulse peak (or trough).

Sometimes it is difficult to draw the tangential line directly due to scattering and jumping of the test data points. Computer plotting software can be used to make the match curve of the test data points and then select a proper tangential point.

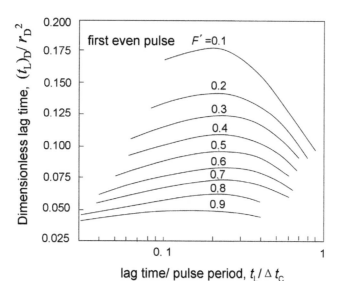

Figure 6.25 An example of lag time type curves (Kamal).

3. Determine Δp and t_L. Once the tangential point is selected, lag time t_L and pulse amplitude Δp can be determined by graphic locating method, and then lag time ratio $t_L/\Delta t_C$ can also be obtained.

4. Look up the type curves to determine the dimensionless pulse amplitude and dimensionless lag time and calculate the parameters.

   • Calculation of flow coefficient kh/μ:

   Look up type curves (Figure 6.24) with obtained $t_L/\Delta t_C$ to get the dimensionless pulse amplitude $[\Delta p_D(t_L/\Delta t_C)^2]_{type-curve\ looking-up}$. Calculate the flow coefficient with the following equations:

   For oil wells,

   $$\frac{kh}{\mu} = \frac{1.842q[\Delta p_D(t_L/\Delta t_C)^2]_{type-curve`looking-up}}{\Delta p[t_L/\Delta t_C]^2} \tag{6.15}$$

   For gas wells,

   $$\frac{kh}{\mu_g} = \frac{3.684 \times 10^4 q_g T p_{SC}}{T_{SC}} \cdot \frac{[\Delta p_D(t_L/\Delta t_C)^2]_{type-curve`looking-up}}{\Delta\psi[t_L/\Delta t_C]^2} \tag{6.16}$$

   When looking up the type curves, attention should be paid that there are many curves in each figure, each curve has a value of F' as its parameter, and the type curve with the same F' value should be selected. For example, in the case of Figure 6.23, the duration of shutting in and flowing are the same, so $t_p = (\Delta t_C)/2$, and $F' = 0.5$, so the type curve of $F' = 0.5$ should be selected.

   Equations (6.15) and (6.16) are both in CSU units.

   The calculation of storability parameter $\phi h C_t$:

   Similarly, look up the type curves in Figure 6.25 with the obtained $t_L/\Delta t_C$ to get the dimensionless lag time $[(t_L)_D/r_D^2]$. When looking up the type curves, use the value $F' = \Delta t_p/\Delta t_C$ to determine the curve to be selected.

   Storability parameter $\phi h C_t$ is calculated with Equation (6.17):

   $$\phi h C_t = 3.6 \times 10^{-3} \frac{kh}{\mu} \cdot \frac{t_L}{r^2[(t_L)_D/r_D^2]_{type-curve`looking up}} \tag{6.17}$$

After calculating the flow coefficient (kh/μ) and the storability parameter $(\phi h C_t)$, permeability k and diffusivity η can also be calculated.

The aforementioned analyses are usually done manually. As the used interpretation method is a kind of the characteristic point method, the operational pattern is not consistent with the type curve match method commonly used now. There seems to be no well test software containing Kamal's analysis method so far.

2.2.3 Pulse Test Analysis by Conventional Interference Test Type Curve Methods

As mentioned repeatedly earlier, the pulse test is essentially a kind of interference test, but the active well changes its working systems or flow rates repeatedly to make several stimulations so as to make repeated interference pressures consequently. Thus, for test data with a clear pressure pulse response, if only the first one or two interference pressure changes are captured, expected results can be obtained with the interference test type curve methods.

1. If the background pressure variation has been monitored before the pulse test, all parameters can be worked out by analyzing the front of the initial pressure pulse during the pulse test.

   As demonstrated in Figures 6.21, 6.22, or 6.23, because the bottom–hole background pressure was acquired at the early stage of the pulse test, the analysis became very simple. All required parameters can be obtained by analysis with conventional interference test data interpretation methods.

2. If the background pressure was not acquired during the pulse test, the front of the initial pulse can be regarded as the background pressure and the back edge of the initial pulse as the interference pressure for analysis. In this way the value of pure interference pressure variation Δp_1 is obtained for calculating all the formation parameters, as illustrated in Figure 6.26.

Pressure History Verification

In the previous chapters describing establishment of a dynamic model from the reservoir description, the importance of pressure history verification in well tests has been emphasized repeatedly. This is equally important in pulse test data analysis.

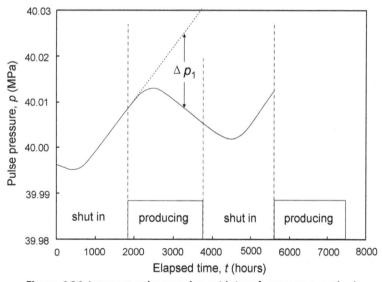

Figure 6.26 Interpret pulse test data with interference test method.

Formation parameters, no matter how they are obtained through the pulse tests, must be put into the dynamic model to verify the performance of the theoretic model against the actual measured pressure. If they are consistent with each other, the analysis results are reliable; otherwise a correction or modification needs to be made.

Three different interpretation results are displayed in Figure 6.27: Figure 6.27a shows that interpreted permeability is too high, Figure 6.27b shows interpreted permeability is too low, and Figure 6.27c shows interpreted permeability is appropriate and consistent with that of the formation. The correctness of the parameters is confirmed by the pressure history verification.

It should be especially noted that Kamal's method is not applicable if the formation is adouble medium or has boundaries. However, "interference test type curve match analysis methods for pulse test" are applicable. As mentioned earlier repeatedly, the type curves of interference test analysis for homogeneous formation, for double porosity medium, for formation with boundary effect, and type curves of them with well storage and skin effects have been developed so interpretation can be done for any cases in those reservoirs. What is more, most of these type curves have been included in the conventional well test interpretation software and can be used readily at any time.

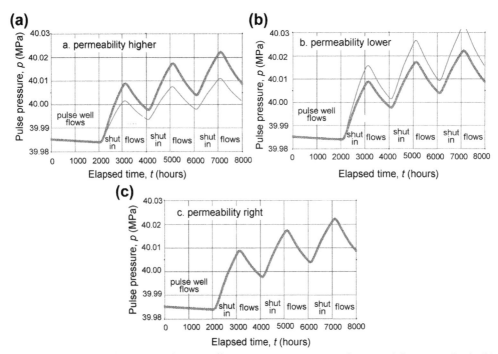

Figure 6.27 Pressure history verification effectiveness. (a) Interpreted permeability is too high. (b) Interpreted permeability is too low. (c) Interpreted permeability is appropriate.

2.3 Multiple-Well Test Design

2.3.1 Principle of Multiple-Well Test Design

As discussed earlier, due to the peculiarity of its test methods, the success ratio of the multiple-well test is quite low. Well test design should be made properly beforehand to increase the success ratio. The well test design consists of the following three aspects.

1. According to the geologic features of the oil or gas field and the problems expected to be solved, divide the target well pairs into three classes, that is, advantageous well pairs, negative well pairs, and questionable well pairs. In field operation, the sequence of "advantageous well pairs first, negative and questionable well pairs next" should be adopted to study the reservoir characteristics dialectically and gradually. Adjustment should also be made as per the actual situation to ensure satisfactory results.
2. Based on the specific formation parameters, make several simulated designs and select the most reliable one.
3. Make a specific and detailed field implementation plan. These are discussed in detail as follows.

2.3.2 Multiple-Well Test Simulated Design

It will be illustrated by an example.

Interference Test

Both the active well and the observation well are gas wells, the distance between them is $r = 1000$ meters.

1. Parameters of active well A:
   - Formation permeability $k = 4$ mD
   - Formation thickness $h = 6$ meters
   - Formation porosity $\phi = 10\%$
   - Gas viscosity in place $\mu_g = 0.03$ mPa·s
   - Formation pressure $p_R = 40$ MPa
   - Gas production rate $q_g = 5 \times 10^4$ m³/day
2. Parameters of observation well B:
   - Formation permeability $k = 2$ mD
   - Formation thickness $h = 4$ meters
   - Formation porosity $\phi = 10\%$
   - Gas viscosity in place $\mu_g = 0.03$ mPa·s
   - Formation pressure $p_R = 40$ MPa
3. Average parameters

   For the simulation design, the average parameters of the two wells can be taken temporarily: $k = 3$ mD, $h = 5$ meters, $\phi = 10\%$, $\mu_g = 0.03$ mPa·s, $p_R = 40$ MPa, stimulation production rate $q_g = 5 \times 10^4$ m³/day, and distance between wells $r = 1000$ meters.

With these parameters, the interference pressure variation can be calculated by well test interpretation software; the calculation result is shown in Figure 6.28.

As can be seen from Figure 6.28:

Within 200 hours after active well A is shut in, almost no interference pressure variation can be seen at the observation well.

When the active well has been shut in for 350 hours, the accumulated interference pressure can be 0.01 MPa ≈ 0.1 atm.

When the active well has been shut in for 720 hours, the accumulated interference pressure can be 0.05 MPa ≈ 0.5 atm. If 0.05 MPa is considered as the lower limit value of the interference pressure for the interference test parameter analysis, the interference test should at least last 30 days or 1 month.

If we want to reduce the test duration but still get the same test results, we can increase the stimulation amount, for example, increasing to $q_g = 10 \times 10^4 \, \mathrm{m^3/day}$, and the simulation result in this situation is shown in Figure 6.29.

As seen from Figure 6.29, the time needed to observe 0.01 MPa of interference pressure is reduced to 300 hours and that to observe 0.05 MPa of interference pressure is reduced to 520 hours.

It can also be seen from the simulation that even if the formations are well connected, it is impossible to obtain obvious interference pressure and the test results in a very short time such as 3–5 days. If there is serious variation in the formations between the active well and the observation well, some layers are interconnected while others are not, only part of the stimulation amount of the active well works, the test results will be reduced greatly. So these uncertainties or "risk factors" should be taken into account when preparing the simulation design of the interference test.

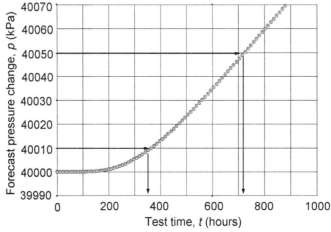

Figure 6.28 Simulated interference pressure variation ($q_g = 5 \times 10^4 \, \mathrm{m^3/day}$).

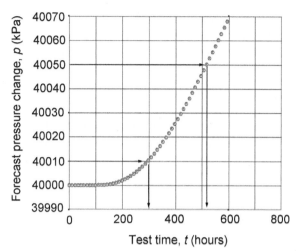

Figure 6.29 Simulated interference pressure variation ($q_g = 10 \times 10^4$ m³/day).

Pulse Test

For the same formation parameters, that is, $k = 3$ mD, $h = 5$ meters, $\phi = 10\%$, $\mu_g = 0.03$ mPa·s and $p_R = 40$ MPa, if we want to design a pulse test consisting of at least three pressure pulses, that is, pulse test result graphs No. 1, No. 2 and No. 3 are needed, the active well must be shut in three times and flowed twice. In the design $\Delta t_p = 1000$ hours and $\Delta t_C = 2000$ hours are taken, as shown in Figure 6.30.

As can be seen from Figure 6.30, with the aforementioned parameters, a good pulse test curve was obtained when the pulse period Δt_C is 2000 hours. The designed total test time is very long, however. Excluding the background pressure test period, the pulse

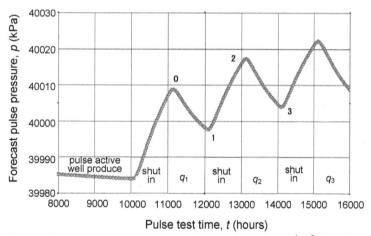

Figure 6.30 Simulated pulse pressure change of pulse test ($q_g = 5 \times 10^4$ m³/day, $\Delta t_p = 1000$ hours, $\Delta t_C = 2000$ hours).

interference test itself still required 8000 hours (equivalent to 1 year), which is unaffordable in the field. The main approach to shorten the test time is to reduce the pulse period Δt_C, for instance, shorten pulse period to 240 hours, 0.5 is taken as the value of F', that is, $\Delta t_p = 120$ hours, the simulation results are as follow and shown in Figure 6.31.

Although the total pulse test duration is still as long as 720 hours, that is, 1 month, the interference pressure curve is no longer good for analysis. Especially when the accuracy and resolution of the pressure gauges are reduced slightly, the measured data curve can only be approximately a straight line going upward at a certain slope.

If the field operation can afford a total test duration exceeding 1 month, further simulation designs can be arranged, for example, $\Delta t_C = 480$ hours and $\Delta t_p = 240$ hours, and the simulated results are as in Figure 6.32.

It can be seen from Figure 6.32 that when $\Delta t_C = 480$ hours, a pulse test curve can be obtained that can be used for analysis, but the total test time is extended to 50 days. If this is acceptable in the field, the test can be conducted as per the schedule.

2.3.3 Make Multiple-Well Test Field Implementation Plan
After the simulation design, the framework arrangement has been determined about the multiple-well test feasibility, the duration of each test stage, and the selection of test well pairs, a detailed field implementation plan or program for its practical execution is needed. The implementation plan should be in a written report and include the following:
1. Basic geologic condition of the well test area
   - Structure map of the test area and location of the testing wells
   - Well location map indicating the active and observation wells involved in the multiple-well test

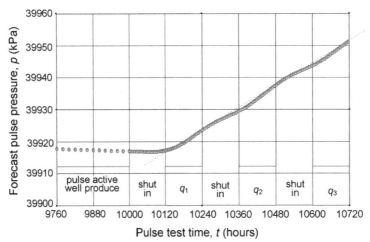

Figure 6.31 Simulated pulse pressure change of pulse test ($q_g = 5 \times 10^4$ m³/day, $\Delta t_p = 120$ hours, $\Delta t_C = 240$ hours).

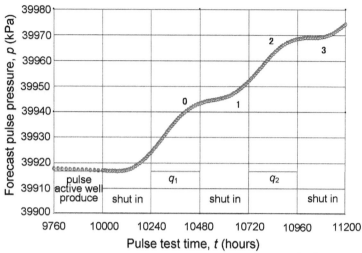

Figure 6.32 Simulated pulse pressure change of pulse test ($q_g = 5 \times 10^4$ m³/day, $\Delta t_p = 240$ hours, $\Delta t_C = 480$ hours).

- Summary of logging interpretation results of the tested layers
- Interconnection profile map of the tested layers between the testing wells
- Drilling and completion basic data of each testing well
- Summary of well test results of the testing wells
- Physical property of fluid

2. Purpose of the multiple-well test
   Describe the problems to be solved by the test in detail.

3. Test methods to be applied
   - Interference test well pairs (well names of each well pair)
   - Pulse test well pairs (well names of each well pair)
   - Active wells that need to additionally conduct single well pressure buildup test (well names.)

4. The selection and sequence of test well pair.
   The sequence can be made as per Figure 6.9.

5. Simulated interference or pulse pressure curves of each testing well pair
   As per the methods described in this chapter, plot the simulated interference or pulse pressure curves of each well pair and determine test duration, stimulation production rate and specific time intervals in accordance with the information provided by the simulation results.

6. Work out the test operation procedures
   It includes well names, well category, specific time intervals of the observation well to monitor the background pressure, times when pressure gauges are run into the well and pulled out of observation, time when the active well changes its working

system, that is, changes its flow rate, and how high the flow rates are in each flow period.

7. Departments in charge of operation, safety, environment protection, and other issues

3 FIELD EXAMPLES OF MULTIPLE-WELL TEST IN OIL AND GAS FIELD RESEARCH

Multiple-well tests have been conducted in China for more than 40 years, and there are many successful field examples of its application—some of which are presented here. The purpose is to illustrate what problems have been solved by multiple-well tests. It is believed that this will be helpful for future research on oil and gas fields.

During petroleum development in most areas except Sichuan in China, efforts mostly focused on oil field development in the past. Hence most of the multiple-well test achievements were about oil wells. With recent large-scaled gas field development in the middle and western region of China, great changes have taken place. The interference test study in the JB gas field and the SLG gas field has just demonstrated this progress. The reservoir research methods are similar for both oil wells and gas wells. However, it is more difficult to perform the multiple-well test in gas fields.

3.1 Interference Test Research in JB Gas Field [69]

In 1988, industrial gas flow was obtained while testing Well SC1, a discovery well in the Ordovician system. After that, quite a lot of detailed prospecting wells were completed and impressive industrial gas flow was commonly acquired, leading to the discovery of the JB gas field. But it was found from acquired transient test data that many pressure derivative curves of these wells went up at the late stage. While evaluating the reservoir structure of the Ordovician system, experts expressed their concern about whether effective reservoirs are distributed continuously and integrally. As the JB gas field was assumed as an uncompartmentalized gas field for reserve estimation, whether the reservoir was distributed continuously or not would directly impact the reliability of the reserve estimation of the whole gas field. Of course the effective way to answer this question directly was to conduct an interference test study in the gas field.

The interference test of the JB gas field was performed in the area near the discovery well—Well SC1 of the mid area of the central region from September 21, 1993 to August 1, 1994. Well L5 was selected to be the active well; both Well L1 and Well SC1 are the observation wells—they formed two well pairs to conduct the tests simultaneously. The test lasted 10 month and 10 days, during which around 200,000 data points were recorded by high-precision electronic pressure gauges. The analysis and research of test data were done with well test interpretation software, combining regional geologic, seismic, and logging research results. The analysis results made clear that:

- Within a range of 4 km from the east to the west, the main pay zones of the three wells, Ma-Wu$_1^3$, were interconnected, which meant that formation in the Ordovician system was distributed continuously.
- The formation was seriously heterogeneous. The formation permeability of the area near Well SC1 west to Well L5 located in the middle was obviously better than that of the area near Well L1 east to Well L5.
- The formation near the testing well groups exhibited the characteristics of a double porosity system with a large value of storativity ratio ω, which meant that the fissure system still played an important role in natural gas accumulation and flow.
- The single well transient test conducted simultaneously at Well L5 and other wells also validated the aforementioned conclusion.

These interference test results supported the reserves estimation from the point of the reservoir dynamic model and gave the green light to the examination and verification of the JB gas field reserves and also paved the way to development of the gas field and the subsequent transmission of its gas to Beijing.

Compared with many interference test results obtained both at home and abroad, the interference tests in the JB gas field are unprecedented in several aspects:

- The test object was low permeability gas formation whose k value was only around 1.5 mD and the distance between the testing wells was as far as 1800 meters, which were very rare both at home and abroad.
- The tests lasted more than 10 months, which was also rare both at home and abroad.
- Throughout the tests, high-precision electronic pressure gauges were used and about 200,000 data points were recorded. It was quite remarkable to acquire such an amount of pressure data as well.
- The tests were conducted during the gas field exploration and early development appraisal stage. At this stage because the distance between the testing wells was quite large and production facilities were absent, it was very difficult to conduct those tests under such subsurface and surface conditions. However, all these difficulties were overcome and the tests were completed successfully.
- The test results influenced the development progress of a big uncompartmentalized gas field. Its role and effectiveness set an excellent example in the research of the gas field interference test.

3.1.1 Geological Conditions of JB Gas Field

The well groups of the interference test were located in the central—southern area of the central region of the JB gas field, near the discovery well, Well SC-1, as shown in Figure 6.33.

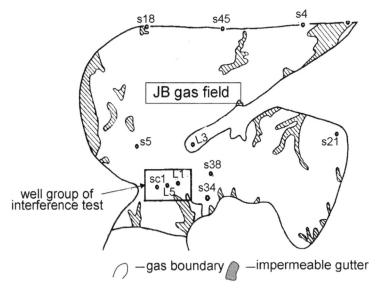

Figure 6.33 Location of the well group for interference test.

It can be seen from the structure map that the well group was located on the northern side of the Well S21–Well S34–Well S48 uplift in the southern area of the central region. Each well of the testing well group perforated formation Ma-Wu$_{1-2}$ of the Ordovician system. The test well group was representative in this region regarding both structural location and pay zones.

In the well group, Well L5 was the active well and was flowing for the deliverability test and production test and then shut in to generate pressure disturbance in the formation. At the same time, high-precision electronic pressure gauges were run in Well L1 and Well SC1 for continuous observation. The location of the testing wells is shown in Figure 6.34.

The fine solid line in Figure 6.34 is the denudation line of formation Ma-Wu$_1^1$. It can be seen that the major pay zones of the well group are complete. The corresponding formation profile of the three wells is shown in Figure 6.35.

It can be seen from Figure 6.35 that Ma-Wu$_1^3$, the major pay zones of the test well group, correspond with each other. So it could be predicted that it was very likely to obtain an interference pressure response in the test.

According to parameter statistics from geologic analysis, the total connected thickness of the connected layers of Ma-Wu$_1^2$, Ma-Wu$_1^3$, and Ma-Wu$_2^2$ was 7 meters for the well group of Well L5–Well L1; and the total connected thickness of the connected layers of Ma-Wu$_1^1$, Ma-Wu$_1^3$ and Ma-Wu$_2^2$ was 6.6 meters for well group Well L5–Well SC1. However, it was also found through zonal gas tests that while Well L5 was commingled producing, not every layer contributed—only layer Ma-Wu$_1^3$ was producing gases, so the

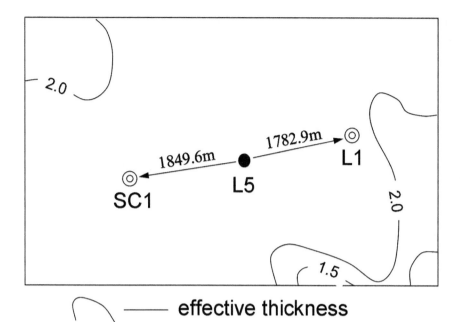

Figure 6.34 Relationship of wells involved in interference test.

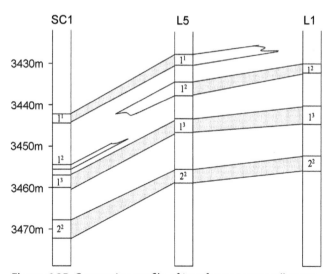

Figure 6.35 Connection profile of interference test well group.

actual effective connected thickness for the well group of Well L5–Well L1 was only 2.8 meters and that for the well group of Well L5–Well SC1 was 3.0 meters.

3.1.2 Well Test Design and Operation

A thorough design was made beforehand for the interference test well group of Well CL5, Well L1, and Well SC1.

1. Well group arrangement

   Two well groups were formed by the three wells. Well L1 and Well SC1 were both observation wells to monitor jointly the interference pressure response caused by the shut in and flowing of the active well, Well L5.

   Shutting in for a long period of time, bottom-hole pressures in both Well L1 and Well SC1 had been stable. High-precision electronic pressure gauges were run in the holes to monitor the bottom-hole pressure.

   Well L5 was opened for the deliverability test and then for the production test; the flow test lasted 178 days and then the well was shut in for 50 days, thus creating a pressure increase disturbance and a pressure drop disturbance.

   The operation items and schedule of the whole test are shown in Figure 6.36.

2. Simulation of interference test

   The following parameters provided by geologic research were selected for simulation:
   - Permeability $k = 1.734$ mD (average of Well L1 and Well SC1)
   - Effective thickness $h = 6.6$ meters (the average connected value of the three wells)
   - Porosity $\phi = 0.0466$ (average value)
   - Storativity ratio $\omega = 0.285$ (measured value of Well L1)
   - Interporosity flow coefficient $\lambda = 0.44 \times 10^{-7}$ (measured value of Well L1)
   - Gas relative density $\gamma_g = 0.5817$ (Well L5)
   - Pseudo critical pressure $p_{pc} = 4.758$ MPa (Well L5)
   - Pseudo critical temperature $T_{pc} = 192.1$K (Well L5)
   - Distance between testing wells $r = 1800$ meters

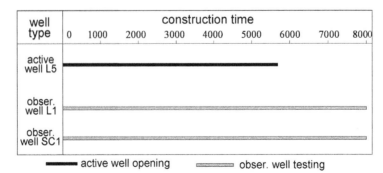

Figure 6.36 Schedule of interference test in JB gas field.

- Formation pressure $p_R = 31$ MPa
- Formation temperature $T = 105\,°C$

The simulation result is shown in Figure 6.37.

It can be seen from Figure 6.37 that if 0.1 MPa is taken as the upper limit of the interference pressure acquisition, the relationship between the stimulation of production rate and its acquisition time is as shown in Table 6.1.

Finally, based on gas consumption in the field, the well flowed at a rate of $4 \times 10^4\,m^3/day$, and so the first stage of the test lasted about 200 days (6–7 months). During actual implementation, it was extended to 5800 hours (about 8 months). The field practice proved that the initial design fitted the actual situation and ensured the success of the interference test.

3.1.3 Test Results

The measured interference test results of the well group of Well L5–Well SC1 and the well group of Well L5–Well L1 are shown in Figures 6.38 and 6.39, respectively.

As demonstrated in Figures 6.38 and 6.39, after Well L5, the active well, was opened for about 1000 hours, the bottom-hole pressure of the observation wells, Well L1 and Well L5, decreased obviously, and their decreasing slopes were:

Well L1: $\Delta p/\Delta t = 0.0020$ MPa/day

Well SC1: $\Delta p/\Delta t = 0.00088$ MPa/day

Well SC1 and Well L1 were completed in June, 1990, and January, 1992, respectively, and never flowed after that. The nearby gas wells such as Well S5 (completed in January,1992), Well S36 (completed in January 1991), Well S34 (completed in May 1992), and Well S38 (completed in April 1992) were all completed more than 1 year before and never flowed since then. So it was reasonable to say that the pressure decrease

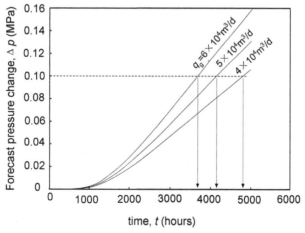

Figure 6.37 Forecast of interference pressure change.

Table 6.1 Relationship between Production Rate and Acquisition Time

Stimulation of production rate, 10^4 m³/day	Acquisition time, hour
6	3550 (about 148 days)
5	4200 (about 175 days)
4	4800 (about 200 days)

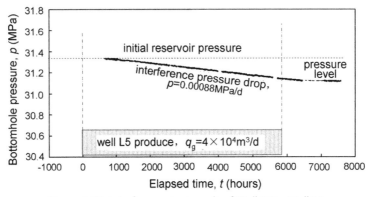

Figure 6.38 Interference test result of Well SC1−Well L5.

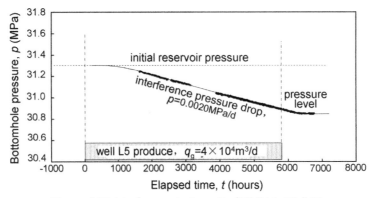

Figure 6.39 Interference test result of Well L1−Well L5.

of Well L1 and Well SC1 was caused by flowing of the active well, Well L5. The time when the pressure started to decrease was approximately 1000 hours after Well L5 was opened for flow, which is consistent with the simulation results shown in Figure 6.37.

The active well, Well L5, was shut in on May 21, 1994, and thus imposed another reverse disturbance on the formation pressure. About 1000 hours later, the bottom–hole pressures of Well L1 and Well SC1 resumed equilibrium again. This verified the interference test results once again. All these proved clearly that these three wells were interconnected.

3.1.4 Parameter Calculation

Interference test data of well groups of Well L5—Well L1 and Well L5—Well SC1 were interpreted by the methods introduced in Section 6.2 of this chapter with the help of well test interpretation software (Saphir or FAST). Double porosity formation pressure and its derivative type curves were used; the interpretation results were as follow.

The pure interference pressure plot of the two testing wells was drawn first as shown in Figure 6.40. Although few data in some intervals in the middle part of the measured pressure were missed, the variation rule of the pure interference pressure was still very clear. Then interpretation was made on the pure interference pressure with interference test log—log type curves in double porosity formation, and a match map was obtained as shown in Figure 6.41.

The interpretation results listed in Table 6.2 were obtained from the aforementioned analysis.

1. The formation permeability of the testing well group is quite low and the permeability on the side Well SC1 locates is about twice of that on the side Well L1 locates. This means that the formation is obviously heterogeneous.

2. The value of reserve parameter $\phi\,h$ is 0.1—0.2 meter, which means that the effective reservoir accumulation volume is $10-20 \times 10^4$ m^3/km^2, that is, the effective pore volume per km^2 of gas field is about $10-20 \times 10^4$ m^3; suppose the formation pressure is 31 MPa, converted to surface standard volume, the effective pore volume per km^2

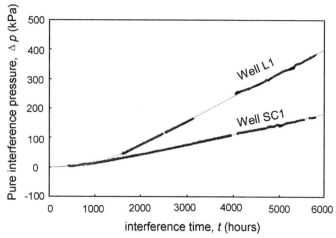

Figure 6.40 Pure interference pressure of Well L1 and Well SC1.

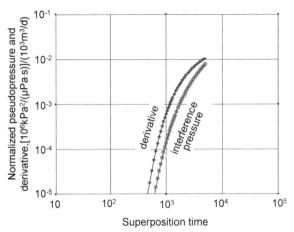

Figure 6.41 Interference test type curve matching of well pair Well L1—Well L5.

of gas field should be $3000-6000 \times 10^4$ m³. This implies that the interference test can be used to verify the dynamic reserves in place.

3. The storativity ratio ω is 0.28, which means that nearly one-third of the natural gas was stored in fissure system. So fissure system is both flow passage and storage space in the JB gas field.

Table 6.2 Interpretation Results of Interference Test of Well SC1—Well L5 and Well L1—Well L5

Well pair	Well SC1—Well L5		Well L1—Well L5	
Items	Correlated connected layer	Effective pay zone	Correlated connected layer	Effective pay zone
Flow coefficient, kh/μ mD·m/ (mPa·s)	411.95	518.14	181.99	242.32
Permeability, k mD	1.3008	3.5993	0.5514	1.8356
Reserve parameter, $\phi\, h$ m	0.2073	0.2502	0.0925	0.1232
Storability parameter, $\phi\, hC_t$ m/MPa	5.1744×10^{-3}	6.2455×10^{-3}	2.2587×10^{-3}	3.0073×10^{-3}
Connected porosity, ϕ %	3.14	8.31	1.32	4.40
Diffusivity, η m²/h	2.8661×10^2	2.9866×10^2	2.9006×10^2	2.9008×10^2
Storativity ratio, ω (dimensionless)	0.28		0.28	
Interporosity flow factor, λ (dimensionless)	0.32×10^{-7}		1×10^{-7}	

The single-well pressure buildup test had been conducted in each of the three wells and all of them exhibited the characteristics of a double porosity medium. The log–log curve of Well L1 was introduced in Chapter 5. The formation parameters of the three wells obtained from the transient test are listed in Table 6.3. It can be seen that the figures in Table 6.3 are very close to the interference test results.

3.2 SLG Gas Field Interference Test Research

Since its discovery at the beginning of this century, the SLG gas field has caught the attention of the world for the amount of its reported reserves. However, after a thorough geologic study, it was found that its main pay zone, the Permian system formation, was plain subfacies thin sandstone of the fluvial facies deposition. The main storage space of the gas was the coarse sandstone deposited in the high-energy channel of the composite body of the river channel. The reservoir was cut by impermeable lithological boundaries, and the single tiny sand bodies were isolated from each other, leading to a low rate of effective reservoir distribution. The reservoir was also cut by impermeable restraining barriers vertically, and the chances of superimposition and communication of the sandstones were rare. All these led to very limited available reserves controlled by a single well.

Reasonable well pattern and spacing became key factors in the development of this gas field so an interference test design was made and implemented. The interference test well groups were selected at the area around Well S6 with the smallest well spacing after twice infill drilling had been conducted. This was also the region that had been studied the most thoroughly in both geology and dynamic performance.

Interference tests were performed successfully in nearly 1 year's effort in data acquisition and analysis from 2007 to 2008. First, a complete interference pressure curve of the well pair, Well S6-j3 and Well S38-16-2, was obtained, which confirmed that

Table 6.3 Interpretation Results of Pressure Buildup Test in Well L5, Well L1, and Well SC1

Well name Items	Well L5	Well SC1	Well L1
Flow coefficient kh/μ, mD·m/(mPa·s)	628	538	345
Permeability k, mD	1.47 (whole well) 4.06 (effective zone)	1.41 (whole well)	1.0 (whole well) 2.23 (effective zone)
Storativity ratio ω, dimensionless	0.28	0.2	0.285
Interporosity flow factor λ, dimensionless	0.75×10^{-8}	1.04×10^{-6}	1.40×10^{-8}
Skin factor S, dimensionless	-4.61	-4.01	-0.98
Wellbore storage coefficient C, m^3/MPa	1.798	2.53	3.012

these two wells were interconnected. This was the first time that interwell pressure interference was observed among more than 1000 wells in this area during 8 years, which conformed that the formation was connected between the wells about 400 meters apart, and innovatively determined the critical value of well spacing for interconnected formations in the Permian system.

At the same time, interwell connection parameters such as flow coefficient kh/μ, formation capacity kh, permeability k, and storativity factor $\phi\, hC$ were calculated with the obtained interference pressure curve combined with the interference test log—log type curve; all of these basic parameters were provided for further adjustment of field development.

3.2.1 Overall Geological Conditions of Well Group of Interference Test

The interference test well group of Well S6 is located around Well S6 on the east side of the central region of the SLG gas field. The infill drilling operation in this region was carried out twice in 2003 and 2007, respectively. The well spacing was about 400 meters from east to west and about 600 meters from south to north; the well location and spacing are shown in Figure 6.42.

Strata correlation from the east to the west sides of Well S6 showing the connectivity of the gas layers is shown in Figure 6.43.

As shown in Figure 6.43, the Strata correlation shows that the major pay zones of Well S6 extend poorly in the east—west direction and that most layers pinch out before reaching its nearest offset wells. In addition, some thin effective layers may be distributed continuously, but due to the fact that effective sandstones of fluvial facies deposition exist mostly in the form of a lenticular body, the connectivity correlation of the thin layers in Figure 6.43 follows no regular rules geologically so interference tests must be used to confirm the actual interwell connectivity.

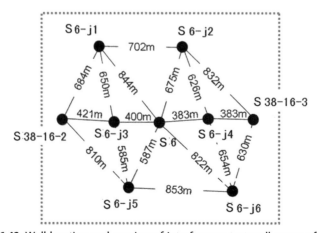

Figure 6.42 Well location and spacing of interference test well group of Well S6.

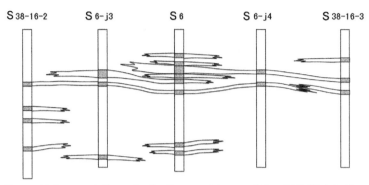

Figure 6.43 Connectivity correlation of effective sandstones of the Well S6 interference test well group.

3.2.2 Interference Test Well Group Design and Implementation

As emphasized repeatedly at the beginning of this chapter, the particularity of the interference test makes the success rate of its field implementation quite low—very often the operation was caught in a dilemma: a definitive conclusion on the reservoir connectivity could not be made, even the validity of the test data themselves might be doubtful. So before the design of the interference test of the well group of Well S6, detailed analysis on the nine wells in the group was conducted and then, based on the analysis, the wells were divided into the advantageous group, negative group and questionable group; the key interference test well pairs were further selected for simulation and the test design.

Dynamic Monitoring and Analysis Results of Well S6

Well S6 was the first gas well put into production in this region. Since it was put into production test in 2001, Well S6 had been monitored for a long period.

1. Modified isochronal tests performed in 2001 from which the deliverability equation was established and AOFP estimated.
2. Initial pressure build up tests performed after the deliverability test in 2001 from which a preliminary dynamic model was established for Well S6 with the numerical well test method and parameters near the wellbore of Well S6 calculated.
3. The dynamic model of Well S6 was improved based on the analysis of production test data in 2002 and 2003. It was confirmed that Well S6 was located in a limited closed rectangular block 800 meters long and 320 meters wide, with an area about 0.26 km^2 and a controlled gas reserve of $0.29 \times 10^8 \text{ m}^3$.
4. Follow-up dynamic monitoring research lasting until 2005 further verified the validity of the dynamic model and confirmed that the area of the controlled block of Well S6 was only 0.22 km^2 and the controlled reserves about $0.26 \times 10^8 \text{ m}^3$.

These research results are discussed in detail in Section 8.4 of Chapter 8.

Static Pressure Analysis when the Second Group of Infilling Wells was put into Production
A second group of infilling wells around Well S6, that is, Well S6-j1, Well S6-j2, Well S6-j3, Well S6-j4, Well S6-j5, and Well S6-j6, was completed in 2007. The well locations of the original wells and these infilling ones are shown in Figure 6.42.

After these wells were completed and put into production, static pressure tests were conducted on all nine wells of the group at about the same time, the results of which are summarized in Table 6.4.

1. The initial static pressures of infilling Well S6-j3 and Well S6-j4, located on the east and west sides of Well S6, respectively, were far lower than the hydrostatic column pressure. This indicated that they had been influenced by the production of early producers Well S6, Well S38-16-2, and Well S38-16-3.

2. Further analysis discovered that the bottom–hole pressure of the infill well, Well S6-j3, was 11.10 MPa, close to that of Well S38-16-2 located on the western side (its bottom–hole pressure was 9.35 MPa), which indicated that these two wells may be interconnected. However, the bottom–hole pressures of Well S6-j3 and Well S6 located on the eastern side (its bottom–hole pressure was 4.99 MPa) were very different, which indicated that they were very unlikely interconnected.

3. Different from Well S6-j3, the initial bottom–hole pressure of Well S6-j4, which had never produced remarkably, decreased to 21.82 MPa, but was still much higher than that of nearby well S38-16-3 located on its eastern side (13.78 MPa), even higher than that of nearby Well S6 located on its western side (4.99 MPa). This implied that even though they were somewhat related, their interconnectivity was very limited.

4. Depart from the well array in the east–west direction where Well S6 located, Well S6-j1, and Well S6 -j2 located away about 600 meters on the northern side, and Well S6-j5 and Well S6-j6 located away about 600 meters on the southern side; all these wells kept their initial formation pressure of above 30 MPa. This indicated that these

Table 6.4 Summary of Static Pressure Test of All Wells in Well S6 Group

Pressure type	Well name	Test date	Buildup duration, days	Middepth of formation, m	Static pressure at middepth, MPa
Initial	S6-j1	2007–10-10	26	3347.80	30.72
formation	S6-j2	2008-3-12	92	3336.50	30.46
pressure of	S6-j3	2008-4-24	187	3356.00	11.10
infill wells	S6-j4	2008-3-8	121	3329.00	21.82
	S6-j5	2008-3-8	96	3347.75	30.36
	S6-j6	2008-5-15	45	3346.50	30.68
Current	S6	2008-3-6	73	3323.90	4.99
formation	S38-16-2	2008-3-4	71	3355.50	9.35
pressure of original wells	S38-16-3	2008-3-2	69	3326.50	13.78

wells had not been influenced by the long-term production of the original wells, that is, Well S6, Well S38-16-2, and Well S38-16-3, and that they were not interconnected with these original wells.

These analyses on static pressure data provided important information for the interference test study on the infilling S6 well group and determination of the well group for the interference test.

Determination of Interference Test Well Group

According to the classification principle described in Section 6.1, the potential well pairs of the interference test for the S6 infilling well group were classified as follow.

1. The advantageous well group: well pair S6-j3—S38-16-2

    The three reasons of this classification are:

    a. Similar bottom-hole static pressures implied that the effective pay zones were likely to be interconnected

    b. Well S6-j3 had the field test conditions to be an observation well. Since it was completed, fractured, and tested in October 2007, the well had been shut in for pressure monitoring

    c. Offset well S38-16-2 could be used as an active well to generate disturbances on formation pressure by well producing and shutting in

    As a result, well pair S6-j3 and S38-16-2 was selected as the key interference test well group. Before the test, the interference pressure was predicted and the interference test design was made.

2. Negative well group

    Observation wells of the negative well group included Well S6-j1, Well S6-j2, Well S6-j5, and Well S6-j6. The initial pressures obtained from monitoring showed that they did not have any relationship with any surrounding wells that had produced gas. Therefore, although they could be used to continuously monitor the pressure change throughout the whole interference test, considering the fact that many years of continuous gas production of the early producers (Wells S6, S38-16-2, and S38-16-3) had not been able to cause the bottom-hole pressure of these wells to decrease, it seemed impossible to observe any interference pressure response within 1 or 2 months' interference test.

3. Questionable well group

    The well pair of S6-j4 and S38-16-3 was classified as the questionable well group. Because these two wells, and perhaps Well S6 on the western side as well, displayed both the possibility of interconnection and the difference in dynamic performance, they could be listed as a potential well pair of the interference test to identify the interconnectivity, and their pressure response could be used for comparing and analyzing the test results of the advantageous well pair of S6-j3 and S38-16-2.

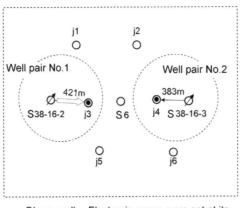

⊙ Obser. well ---Electronic gages were set at its
    wellhead to monitor interference pressure variation

∅ Active well --- Flowed and shut-in for creating
    pressure disturbance

○ Auxiliary observation well---Electronic gages were
    set at its wellhead to monitor pressure variation

Figure 6.44 Locations of interference test well group of S6 infill well region.

Simulation and Design of Interference Test

1. Target well group of simulation and design

The interference test target well group consisted of the observation well, Well 6-j3, and the active well, Well S38-16-2. In Figure 6.44 they are listed as well pair No.1.

2. Simulation and design of interference test

According to the averaged formation parameters of Well S6-j3 and Well S38-16-2, the following parameters were selected for simulation and design:

- Zone: No. 8 sand formation in the Permian system
- Permeability $k = 0.5$ mD
- Effective thickness $h = 10$ meters
- Distance between testing wells $r = 421$ meters
- Natural gas viscosity in place $\mu_g = 0.015$ mPa·s
- Deviation factor of natural gas $Z = 0.913$
- Formation porosity $\phi = 10\%$
- Gas saturation $S_g = 60\%$
- Formation temperature $T_f = 378$K
- Gas flow rate of active well $q = 0.4 \times 10^4$ m^3/day

With the well test design module of FAST well test interpretation software, the simulated interference pressure change curve of the target well group was obtained as shown in Figure 6.45.

It can be seen from Figure 6.45 that the test needed nearly 7 months. As the amount of the interference pressure change was very small, high-precision electronic pressure gauges were required to run into observation well S6-j3 to monitor its

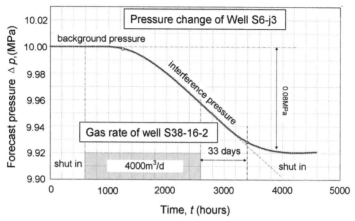

Figure 6.45 Simulation and design of interference test for well pair S6-j3 and S38-16-2.

pressure continuously. The active well, S8–16–2, was required to change its flow rates during the test, first from shutting in to opening and then from opening to shutting in, to generate formation pressure disturbance.

3. Operation schedule

The field operation schedule was made according to the simulation and design of the interference test, as illustrated in Table 6.5.

In the field, it took nearly 1 year to complete the interference tests of infilling well group S6 successfully.

Table 6.5 Operation Schedule of Interference Test of Infilling Well Group S6

Well name \ Time	Feb.	Mar.	Apr.	May	Jun.	July	Aug.
S6-j3							
S38-16-2							
S6-j4							
S38-16-3							
S6							
S6-j1							
S6-j2							
S6-j5							
S6-j6							

Obser. well, electronic gages were set at its wellhead to monitor
Active well, flowed and shut-in for creating pressure disturbance
Auxiliary obser. well, electronic gages were set at its wellhead to monitor

3.2.3 Interpretation of Interference Test Data

Results of Interference Tests

The interference pressure change curve acquired from Well S6-j3 is shown in Figure 6.46. It can be seen from Figure 6.46 that Well S6-j3 was remarkably influenced by the opening and shutting in of Well S38-16-2:

1. During the initial shut in of Well S38-16-2, the pressure of Well S6-j3 came to be stable. The pressure at the later stage was about 8.591 MPa.
2. About 20 days after Well S38-16-2 was opened for production, the pressure of Well S6-j3 decreased; this decrease lasted nearly 100 days with a rate of about 0.002 MPa per day; and the pressure decline was close to a straight line and could be expressed as

$$p = 8.568 - 8.888 \times 10^{-5}(t - 3121.3).$$

3. After Well S38-16-2 was shut in, the pressure decline of Well S6-j3 began slowly, and finally the pressure tended to be stable but dropped to 8.317 MPa, about 0.274 MPa lower than that before Well S38-16-2 was opened to create a disturbance.

From the aforementioned interference pressure change tendency, the conclusion could be made clearly and correctly that the perforated layers of the interference test well pair S6-j3 and S38-16-2 were interconnected. This was the first time that an interwell connection was identified since the discovery of the SLG gas field at the beginning of the 21st century and among more than 1000 exploration and production wells. Undoubtedly this was of course innovatively an important and instructive guide for the development of the SLG gas field and the selection of reasonable well spacing.

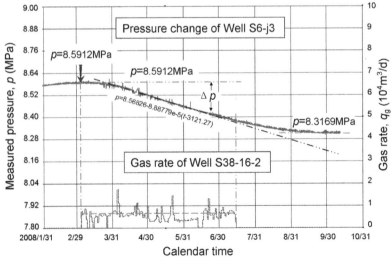

Figure 6.46 Relationship between interference pressure changes in Well S6-j3 and gas flow rate of Well S38-16-2.

Interpretation of Interference Test Data

The conventional type curve analysis method was used to interpret the aforementioned interference test data. The pressure measurement points of Well S6-j3 were subject to the influence of the ambient temperature and they fluctuated or "jumped" to some extent. So while the pure interference pressure was separated from the measured pressure, the measured points were replaced by the approximate expression of the straight line of the interference pressure change. Pure interference pressure Δp was calculated and plotted on the log–log paper as illustrated in Figure 6.47, and the parameters were obtained by type curve matching.

The coordinates of match point M were $\Delta(p^2)_M = 1$ and $\Delta t_M = 1000$ on the measured interference pressure plot and $p_{DM} = 0.07$ and $(t_D/r_D^2)_M = 0.15$ on the interference pressure type curve, as shown in Figure 6.47.

The connectivity flow coefficient calculation equation for the interference test

$$\frac{kh}{\mu} = \frac{q_g Z T_f}{7.8523 \times 10^{-2}} \cdot \frac{(p_D)_M}{\Delta(p^2)_M} = 12.74 q_g Z T_f \cdot \frac{(p_D)_M}{\Delta(p^2)_M}$$

was used. The physical property parameters of the well group were selected as follow: $Z = 0.913$, $T_f = 378K$, $q_g = 0.59 \times 10^4$ m^3/day, and the following interwell connectivity parameters were obtained:

- Connectivity flow coefficient: $\dfrac{kh}{\mu} = 181.58 \dfrac{\text{mD·m}}{\text{mPa·s}}$
- Connectivity formation capacity: $kh = 2.724$ mD·m
- Connectivity formation permeability: $k = 0.27$ mD

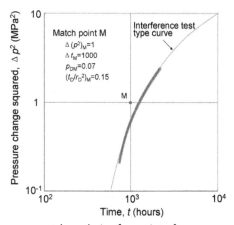

Figure 6.47 Type curve match analysis of pure interference pressure of Well S6-j3.

The equation of storativity was

$$\phi h C_t = \frac{3.6 \times 10^{-3} kh}{\mu_g r_g^2} \cdot \frac{\Delta t_M}{(t_D/r_D^2)_M}.$$

and the connectivity storativity was calculated as $\phi\, h C_t = 2.459 \times 10^{-2}$ m/MPa.

Pressure Monitoring Results of Other Observation Wells

Pressure change curves in other observation wells involved in the interference test were also acquired.

1. The interference pressure monitoring result of Well S6-j4 and S38-16-3

   The pressure monitored in Well S6-j4 decreased slowly and linearly with a rate of about 0.00374 MPa/day (1.36 MPa/year) from the beginning to the end during the test, as shown in Figure 6.48.

   It can be seen from Figure 6.48 that, on the one hand, Well S6-j4 was indeed connected with the offset wells, which led to the pressure decreasing. On the other hand, during the test lasting 10 months, the repeated alternating of flowing and shutting in of Well S38-16-3 and Well S6-j2 had not resulted in any obvious pressure changes in Well S6-j4, which meant that the connection was seriously impeded.

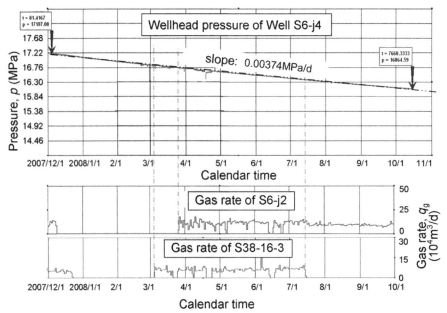

Figure 6.48 Relationship between pressure variations of Well S6-j4 and gas flow rate of offset wells (Well S38-16-3 and Well S6-j2).

2. The monitoring results of other auxiliary observation wells

Just as predicted in the interference test design of the S6 infilling well group, no influence of any nearby early producers was observed in wells S6-j1, S6-j2, S6-j5, and S6-j6, indicating that they were not interconnected.

3.2.4 To Identify Rational Well Spacing in SLG Gas Field by Interference Test Results

1. The basis of identifying rational well spacing

There were four categories of information:

(a) Geological research results on the reservoir. These are mainly research results on the reservoir sedimentary facies, which had been described briefly at the start of this section and could be summarized as follows: the main gas bearing reservoirs were separated into many isolated gas bearing sand bodies by lithologic boundaries on the plane; each of them has a very limited area only. It was the existence of these lithologic boundaries that caused the effective layers controlled by each gas well isolated from each other and there was no communication among wells. When the distance between wells was too long, sand bodies may have existed that were not controlled by any wells, which could reduce the ultimate recovery of the gas field greatly.

(b) Initial pressure data of infilling wells. Just as described in Table 6.4, the initial pressures of Well S6-j1, Well j2, Well j5, and Well j6 among the nine wells of the infilling well group, measured after completion, remained the same level as the original pressure. From this, it could be predicted that these wells were not connected with the other wells (S6, S38-16-2, and S38-16-3) of the group that were put into production earlier. Each of them was located in an isolated gas bearing sand body. That is to say, within 600 meters of Well S6 from the south to the north, wells were not interconnected.

(c) Long-term dynamic description research of Well S6. Just as what is described thoroughly in Chapter 8 and mentioned briefly in the interference test design, Well S6 was a key well located in the center of the infilling well group. After as long as a 5-year dynamic tracing study, it was confirmed that this well controlled a 750 × 320-meter closed rectangular block. Due to long-term gas production, static pressure near the bottom hole dropped to 4.99 MPa currently, which was far lower than that of the surrounding wells.

(d) The interference test study of infilling well group S6. This was the most direct and convincing evidence used to identify the rational well spacing. After a long-term field test and careful analysis, it was verified that Well S6-j3 and Well S38-16-2 were interconnected and that Well S6-j4 and Well S38-16-3 seemed to be interconnected but the interconnection was impeded greatly. The other four wells, that is, wells S6-j1, S6-j2, S6-j5, and S6-j6, were not connected with the original wells of the group.

2. Some understandings on rational well spacing in SLG gas field
Based on the four studies just mentioned, the gas bearing regions possibly controlled by each well of the S6 infilling well group and the interwell relationship of those wells were made clear, as illustrated in Figure 6.49.

The area controlled by each single well described in Figure 6.49 could be regarded as a miniature of the whole SLG gas field.

It can be seen that the effective gas bearing area control by a single well in the SHZ gas reservoir of the SLG gas field was no more than 0.2–0.25 km^2; and at least four to five wells should be drilled per square kilometer to achieve effective development. In the well pattern arrangement, the critical distance between wells in the direction of east–west was about 300–400 meters, whereas that in the direction of south–north was no more than 600 meters; otherwise the ultimate gas recovery might be decreased.

3.3 Gas Well Interference Test Study in Fault Block Y8 of SL Oil Field

The interference test conducted in fault block gas reservoir Y8 began in March 1969. It was the first block where the interference test was conducted successfully in the SL oil field, the first block where the interference test was conducted successfully in a gas field, and also one of the blocks in China where one of the earliest interference tests was conducted successfully. It must also be pointed out that the success of this test attributed partly to the accidental blowout of active well Y8-12.

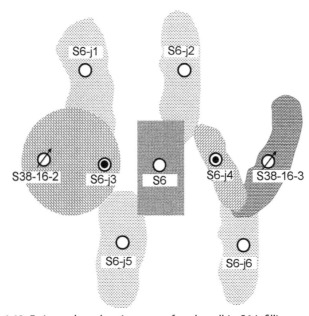

Figure 6.49 Estimated gas bearing area of each well in S6 infilling well group.

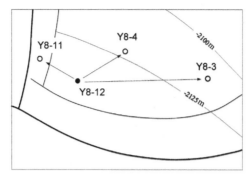

Figure 6.50 Structure well locations of Y8 fault block.

This test was planned to identify the interwell connectivity of interval No.1 of the SHJ formation of the Tertiary system in fault block Y8 and to determine the seal of the fault between Well Y8-12 and Well Y8-11. It had been about 3 years since the Interference Test Research Group of SL Oilfield Research Institute was established, nothing had been obtained from the field tests, and no interference pressure change in the field was acquired.

Before the test formally started, the valve of the Christmas tree of Well Y8-12 ruptured during the hot water flushing operation and a blowout occurred. This no doubt caused a large-scale "formation pressure perturbance." With this accident, the research group rushed to run in the differential pressure gauges in the offset wells and recorded the interference pressure caused by Well Y8-12 successfully. This was the first successful interference test!

The structure and well locations of fault block Y8 are shown in Figure 6.50. The geologic parameters of the four wells involved in the test are summarized in Table 6.6. The obtained interference test curves are shown in Figures 6.51 and 6.52.

Table 6.6 Geologic Parameters of Interference Test well Group of Fault Block Y8

	Active well			Observation well			Distance between both wells, m	Remarks
No.	Well name	Horizon	Perforated thickness, m	Well name	Horizon	Perforated thickness, m		
1	Y8-12	S_1^{1-2}	3.1	Y8-3	S_{1zhen}	2.2	960	Both were gas wells in the same block
2	Y8-12	S_1^{1-2}	3.1	Y8-4	S_11-2	11.3	360	Both were gas wells in the same block
3	Y8-12	S_1^{1-2}	3.1	Y8-11	S_{1zhen}	2.6	190	Both were gas wells, located on different side of the fault

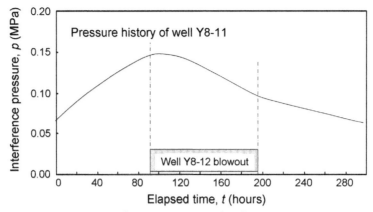

Figure 6.51 Interference test curve of Wells Y8-12 and Y8-11.

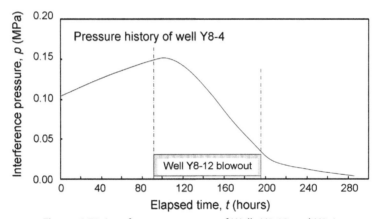

Figure 6.52 Interference test curve of Wells Y8-12 and Y8-4.

It can be seen from Figures 6.51 and 6.52 that no matter where the observation wells are, whether in the same fault block with the active well or the observation well and the active well were in the different side of the fault, remarkable interference pressure responses were received. The results given in Table 6.7 were obtained by interpretation of the curves.

Table 6.7 Summary of Interference Test Results of Fault Block Y8

Well pair Item	Y8-12—Y8-4	Y8-12—Y8-11
Flow coefficient kh/μ, mD·m/(mPa·s)	56,500	30,400
Permeability k, mD	156.9	217.1
Diffusivity η, m^2/h	1393	1105
Storability parameter $\phi\, h\, C_t$, m/MPa	0.225	0.153
Reserve parameter $\phi\, h$, m	6.75	4.56

3.4 Test Research on Connectivity between Injector and Producer in Fault Block

3.4.1 Research of Connectivity between Injector and Producer in ST Block 3, SL Oil Field

There are complicated fault systems in eastern China, especially in the oil fields around BH Bay. Those faults affect the effectiveness of water flood development seriously. The locations and status of the fault provided by seismic analysis were often too vague, and some faults may be not sealed. Fortunately, accurate judgment can be made by the interference test.

ST Block 3 in the SL oil field is one of the regions developed by early water flood. Because the water flood effectiveness among injectors 3-4-15, 3-2-15, and producer 3-3-15 was always questionable, interference tests were conducted there. The well group location is shown in Figure 6.53.

Among the three wells involved in the tests, Well 3-4-15 and Well 3-2-15 were active wells and Well 3-3-15 was the observation well. Well 3-4-15 and Well 3-2-15 injected alternatively to generate formation pressure disturbance; observation well 3-3-15 was shut in to maintain stable bottom-hole pressure and to monitor the interference pressure changes. The geologic condition of the three wells is summarized in Table 6.8.

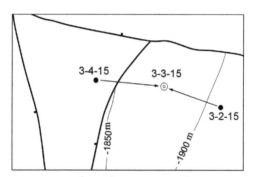

Figure 6.53 Location of interference test well group in ST Block 3.

Table 6.8 Geologic Parameters of Well groups 3-4-15, 3-2-15, and 3-3-15

	Active well			Observation well			Distance	
No.	Well name	Horizon	Effective thickness, m	Well name	Horizon	Effective thickness,m	between both wells, m	Remarks
1	3-4-15	S_21-S_26	22.5	3-3-15	S_21-S_22	10.4	480	Separated by fault
2	3-2-15	S_21-S_23	18.0	3-3-15	S_21-S_22	10.4	450	In the same fault block

Table 6.8 shows that the horizons of the three wells correlate with each other. However, as can be seen from Figure 6.53, Well 3-4-15 and Well 3-3-15 were separated by a fault. Normally the injection at injector 3-4-15 should not have any influence on producer 3-3-15, while the injection at injector 3-2-15 should influence producer 3-3-15, as they were in the same fault block. However, the interference test results came as a surprise, as shown in Figures 6.54 and 6.55.

It can be seen from Figure 6.54 that a few hours after Well 3-4-15 started water injection with a rate of 200 m³/day from its shutting-in status, the bottom-hole pressure of producer 3-3-15 increased rapidly. After injection was stopped at Well

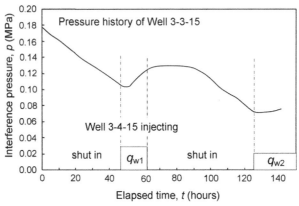

Figure 6.54 Interference test curve of Well 3-4-15 and Well 3-3-15.

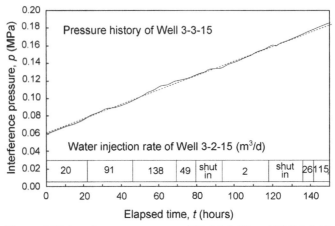

Figure 6.55 Interference test curve of Well 3-2-15 and Well 3-3-15.

3-4-15, the bottom-hole pressure of producer 3-3-15 decreased rapidly with the same rate before injection in Well 3-4-15. Later when the injection was conducted at a rate of 88 m³/day in Well 3-4-15, the pressure of Well 3-3-15 increased again. This demonstrated that the fault did not function as a barrier between the injector and the producer.

On the contrary, a different result was observed from Figure 6.55. This interference test began 2 days after completion of the interference test illustrated in Figure 6.54. The injection at Well 3-4-15 caused the bottom-hole pressure of Well 3-3-15 to have an increased tendency. Afterward, injector Well 3-2-15 increased its injection rate from 20 to 138 m³/day and then lowered the rate and stopped injection; later it resumed the injection and once raised the rate up to 115 m³/day. However, because the repeated changes of the injection rate did not make any impact on the pressure of producer Well 3-3-15, it was identified that the injection in Well 3-2-15 was useless for the pressure maintenance of Well 3-3-15.

The results of these two tests verified each other and the obtained conclusions were unquestionable.

Parameters of the formation between Well 3-4-15 and 3-3-15 calculated based on the interference test curve are as follow:

- Flow coefficient $kh / \mu = 1715$ mD·m/(mPa·s)
- Effective interconnectivity permeability $k = 32$ mD
- Storability parameter $\phi \, hC_t = 8.3286 \times 10^{-4}$ m/MPa
- Diffusivity $\eta = 7218$ m³/h

3.4.2 Research on Isolation of the Fault in Well Y18 Area of SL Oil Field

Well Y18 area locates in the HK area of the SL oil field. The producing zone is a limestone reservoir of SHJ formation. After acidizing treatment, the wells were tested and high flow rates were obtained, but the production rates declined rapidly. A series of interference tests were arranged to make clear the relationship of interwell connectivity and isolation of the faults; the structure map and the location of the wells are shown in Figure 6.56.

There were four wells involved in five tests conducted one after another. The geological parameters of this well group are summarized in Table 6.9.

It can be seen from Table 6.9 that the horizons of the four wells involved in the test correlated with each other. But from the structure map it can be seen that two well pairs were in the same fault block while the other three well pairs were separated by faults. So by such a group of tests, the formation connectivity, connectivity parameters, and the isolation status of the fault, and hence the reservoir, can all be further understood or made clear.

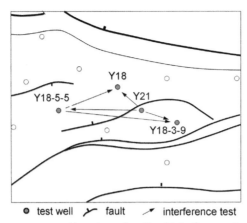

  ● test well   ⤙ fault   ⟋ interference test

Figure 6.56 Structure map and location of wells of interference test in Well Y18 area.

All interference tests of the well pairs in Table 6.9 achieved good results—among them are the two pairs shown in Figures 6.57 and 6.58, respectively.

The interwell connectivity parameters obtained from interpretation of the interference test curves are summarized in Table 6.10.

The following could be concluded from the interpretation results:

- All the well pairs showed good interwell connectivity so the fault between the tested wells did not isolate the flow between them obviously.

Table 6.9 Geological Parameters of Interference Test of Y18 Well Group

| | Active well | | | Observation well | | | Distance | |
No.	Well name	Horizon	Perforated thickness, m	Well name	Horizon	Perforated thickness, m	between both wells, m	Remarks
1	Y18-5-5	S_16	10	Y18	S_16-8	12.8	1150	In the same fault block
2	Y18-5-5	S_16	10	Y18-3-9	S_15-6	15.4	2200	Isolated by fault
3	Y21	S_11-6	9.2	Y 18-5-5	S_14	0.6	1540	Isolated by fault
4	Y21	S_11-6	9.2	Y 18-3-9	S_11-2	2.4	700	In the same fault block
5	Y21	S_11-6	9.2	Y 18	S_16-8	12.8	640	Isolated by fault

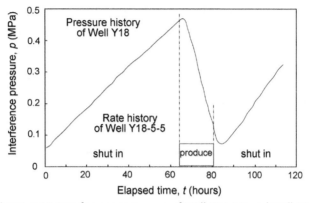

Figure 6.57 Interference test curve of Well Y18-5-5 and Well Y18.

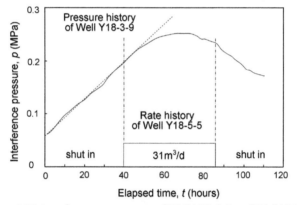

Figure 6.58 Interference test curve of Well Y18-5-5 and Well Y18-3-9.

Table 6.10 Summary of Interference Test Interpretation Results of Well Y18 Area

Well pair Item	Y18-5-5 -Y18	Y18-5-5 -Y 18-3-9	Y21 -Y18-5-5	Y21 -Y18-3-9	Y21 -Y18
Flow coefficient kh/μ, mD·m/ (mPa·s)	85	120	19.5	19.1	17.0
Permeability k, mD	22.3	28.8	11.9	9.9	4.6
Diffusivity η, m²/h	84,600	41,040	42,840	14,330	8100
Storability parameter $\phi h C_t$, m/MPa	3.570×10^{-6}	1.039×10^{-5}	1.617×10^{-6}	4.736×10^{-6}	7.457×10^{-6}
Reserve parameter ϕh, m	2.38×10^{-3}	6.927×10^{-3}	1.078×10^{-3}	3.157×10^{-3}	4.971×10^{-3}

- The permeability of the area between the tested wells was on the high side of medium, which was the main reason causing the wells to have a high oil production rate.

- The reserve parameter $\phi\,h$ values of the area between the tested wells were all very small, the average was $\phi\,h = 3.7 \times 10^{-3}\ \mathrm{m}^3$, which was equivalent to 3700 m³ of subsurface pore volume per square meter. This was the main reason causing the flow rate of the wells to decline rapidly when producing.

3.4.3 Efficiency Analysis of Injection in Fault Block B96

The oil-bearing area in Fault Block B96 of BN oil region, SL oil field, looked like a triangle screened by faults on the two sides and by water on the third side. The perforated zones of the three wells were the same zone of the second interval of SHJ formation in the Tertiary system, and distances from injector Well B106 to the two producers Well B96 and Well B49 were about the same. Their locations were as shown in Figure 6.59.

The tests were conducted on two well pairs; the corresponding geologic conditions are summarized in Table 6.11.

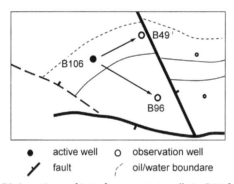

Figure 6.59 Locations of interference test wells in B96 fault block.

Table 6.11 Geologic Parameters of Interference Test Wells in B96 Fault Block

| | Active well | | | Observation well | | | Distance | |
No.	Well name	Horizon	Perforated thickness, m	Well name	Horizon	Perforated thickness, m	between both wells, m	Remarks
1	B106	S_2	6.9	B 96	S_2	7.6	650	In the same fault block
2	B106	S_2	6.9	B 49	S_2	14.7	500	In the same fault block

As illustrated in Figure 6.59 and Table 6.11, two tests were arranged: the active well was B106 in both tests and the observation wells were B49 and B96, respectively. The test results are plotted in Figures 6.60 and 6.61, respectively.

As demonstrated in Figure 6.60, when Well B106 injected water normally, the bottom-hole pressure of Well B49 kept on increasing with a slope of about 0.024 MPa/day. After injector well B106 was shut in, the pressure of Well B49 tended to become stable 30 hours later. During 150–250 hours, the pressure was completely stable.

Pressure monitoring was suspended for 20 hours due to pulling out the gauges of the hole, and then pressure gauges were rerun to resume monitoring the interference pressure change. At the beginning, the pressure of B49 was still stable. However, 20 hours after water injection resumed in B106 at a rate of 120 m^3/day, the bottom-hole

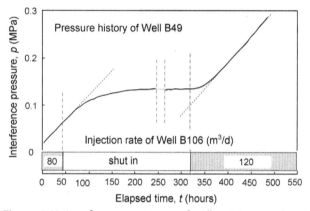

Figure 6.60 Interference test curve of well pair B106 and B49.

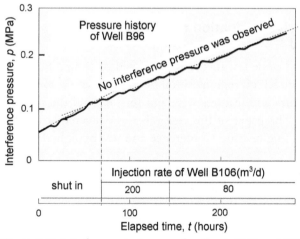

Figure 6.61 Interference test curve of well pair B106 and B96.

Table 6.12 Summary of Results of Interference Test of Well Pair B106 and B49

Well pair Item	B106—B49 (the first stage)	B106—B49 (the second stage)	B106—B96
Flow coefficient kh/μ, mD·m /(mPa·s)	430	435	/
Permeability k, mD	39.8	40.3	/
Diffusivity η, m²/h	802	652	/
Storability parameter $\phi\, hC_t$, m/MPa	4.465×10^{-4}	5.557×10^{-4}	/
Reserve parameter $\phi\, h$, m	2.98	3.70	/

pressure of B49 rose again at a rate of 0.02 MPa/day. This clearly demonstrated that the two·wells were interconnected and that Well B49 was influenced by the water injection in B106.

Analysis of the interference test curve gave the following results (see Table 6.12). As seen from Figure 6.60 and Table 6.12, the permeability of the area between Well B106 and Well B49 was 40 mD and the effectiveness of water injection was good. Reserve parameter $\phi\, h$ of the area between the two well was around 3 meters, which meant that the pore volume of the reservoir between the two wells was about 30×10^4 m³ per square kilometer.

However, there was no sign of water injection effectiveness between B106 and B96 (Figure 6.61). The pressure of Well B96 kept on increasing. The reason was suspected that this well was shut in before the test and the formation pressure of the area controlled by the well had been rising constantly. The pressure of Well B96 did not rise faster due to the influence of water injection at B106, it rose more slowly instead. This in fact reflected the characteristics that the pressure built up more slowly at a later stage. Further analysis needed to be conducted with geological data.

3.5 Comprehensive Evaluation of Multiple-Well Tests in KL Palaeo-Burial Hill Oil Field

Interference test and pulse test methods were used to comprehensively evaluate a fractured reservoir in the KL oil region, SL oil field, from which complete new ideas about the reservoir structure and features were put forward and the corresponding dynamic model established. The paper of this research was delivered at the First Beijing International Petroleum Engineering Conference and was published in JPT [44].

3.5.1 Overall Geological Condition of KL Oil Region

The KL oil field was a small reservoir in the GD oil region in the JY depression. Burial hill limestone fissured formations in the Ordovician system were penetrated. On the southwest of the reservoir was the KG2 fault. The formation dipped northeastward and formed a faulted nose structure. The area of the oil region was about 4 km², half of

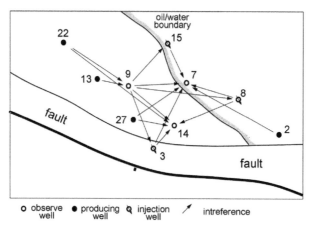

Figure 6.62 Location of interference test wells in KL oil field.

which was in the waterway of the Yellow River. When interference tests were conducted, there were eight producers and three injectors. The well locations are shown in Figure 6.62.

After 3 years of development, the daily crude oil production dropped from 500 to 60 m^3. Two or 3 months after water injection, the producers all experienced water breakthrough, and the water cut rose to 50–90%, but the ratio of total oil produced to OOIP was only 3.6%. At that moment, recovery methods needed to be adjusted urgently so a group of interference and pulse tests were designed to study the connectivity between the wells, permeability in different directions, reserve parameter distribution in the area between the wells, and other parameters of the limestone fissured formation so as to provide references for the next development adjustment.

3.5.2 Test Arrangement and Achieved Results
The whole test lasted 40 days and was divided into six periods. The observation wells consisted of KG7, KG9, and KG14 and the other five wells were taken alternatively as the active wells. Twenty-eight tests, including interference tests and pulse tests, were performed. The schedule was as illustrated in Figure 6.63.

It can be seen from Figure 6.63 that

1. In period I, Well KG7, Well KG9, and Well KG14 were the observation wells and injection Well KG8 was the active well. Formation pressure perturbation was caused by stopping injection and then resuming again.
2. In period II, the observation wells were the same as period I, but the active well was changed to Well KG3. The stimulation method adopted was the same as period I. During this period, the injection in Well KG8 and Well KG15 had been suspended temporarily due to pump station failure.
3. In period III the active well was changed back to Well KG15.

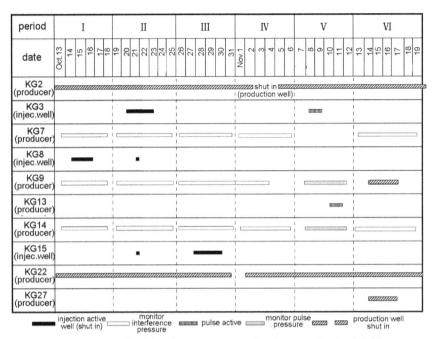

Figure 6.63 Schedule of interference and pulse tests in KL oil field.

4. In period IV, the test was conducted alternately. Well KG9 continued monitoring since period III. The active wells were changed to Well KG2 and Well KG22. Formation pressure perturbance was caused by shutting in and producing the oil producer.

5. The pulse test was performed in period V. The water injection rate was pulsed to cause pulse perturbance in Well KG3 and Well KG13, and observation was made in Well KG9 and Well KG14.

6. The test purpose of period VI was to identify the relationship among the original observation wells KG7, KG9, and KG14. Wells KG7 and KG14 continued monitoring the interference pressure, but Well KG9 was changed to an active well to generate formation pressure perturbance by shutting in—flowing—shutting in. During the implementation, Well KG27 was opened accidentally for flowing due to misunderstanding of the test plan by the production crew. The simultaneous perturbance brought some difficulties for data analysis, but also presented some interesting phenomena on the interference pressure curve shape.

Test Results of Period I

The test results of period I were as illustrated in Figure 6.64. It can be seen from Figure 6.64 that:

1. When the test began, the pressure of Well KG7 went up at a rate of 0.055 MPa/day, which was apparently influenced by water injection of the offset wells. Two hours

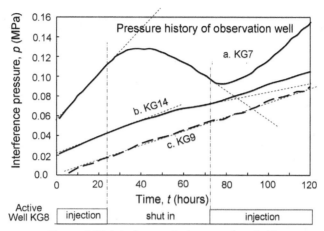

Figure 6.64 Interference test curve of period I.

after Well KG8 stopped injection, Well KG7 responded and its rising pressure smoothed out and then declined. However, when Well KG8 resumed injection, the pressure of Well KG7 started to rise again.

2. In sharp contrast to this was curve b in Figure 6.64. During the test, the pressure of Well KG9 increased constantly at a rate of 0.018 MPa/day. The injection suspension in Well KG8 did not obviously impact its pressure change tendency. These two wells should be interconnected, but the distance between them was too far (1050 meters) and the permeability of the area between them was too poor to acquire the interference pressure response.

3. For Well KG14 and Well KG8 with a little shorter distance between, an interference pressure response was observed but was very weak. When Well KG8 changed to flowing from shutting in, the pressure rising slope of Well KG14 changed slightly from 0.02 to 0.013 MPa/day.

As a result, injection well KG8 mainly had an effect on Well KG7, but its influence on Well KG9 and Well KG14 was very weak.

The connectivity parameters of these three well groups are calculated and listed in Table 6.13.

Test Results of Period II

The test results of period II are illustrated in Figure 6.65.

This group of curves was quite different from those of period I. The injection in Well KG3 had a remarkable influence on Well KG9 and Well KG14, but was weak on Well KG7.

1. The pressure of Well KG7 increased at a slope of 0.055 MPa/day initially. The injection suspension of Well KG3 just had a slight influence on the rising slope,

Table 6.13 Summary of Multiple-Well Test Interpretation Results of KENLI Palaeo-Burial Hill Oil Field

Items Well group	Distance between wells spacing, m	Stimulation quantity, m³/day	Flow coefficient kh/μ, mD·m /(mPa·s)	Fluidity k/μ, mD/(mPa·s)	Diffusivity η, m²/h	Storability parameter $\phi\, hC_v$ m/MPa	Reserve parameter $\phi\, h$, m
7-2	1028	170.2	3243	345.5	1.09×10^{-3}	8.26×10^{-4}	0.454
7-3	760	170	3392	164.5	0.519×10^{-3}	2.135×10^{-3}	0.998
7-8	537	248	3345	133.1	4.203×10^{-4}	2.211×10^{-3}	1.215
7-9	174	174	8139	688.8	2.174×10^{-3}	1.040×10^{-3}	0.572
7-15	453	140	735	39.5	1.247×10^{-4}	1.637×10^{-3}	0.900
7-27	610	50	489	48.4	1.528×10^{-4}	0.889×10^{-3}	0.489
9-3	710	170	2278	169.7	5.356×10^{-4}	1.182×10^{-3}	0.650
9-3 (pulse test)	710	175	2056	146.9	4.636×10^{-4}	1.232×10^{-3}	0.677
3-9 (in 1978)	710	52	2023	128.0	4.039×10^{-4}	1.391×10^{-3}	0.765
9-13 (in 1978)	326	199	2305	97.9	3.092×10^{-4}	2.071×10^{-3}	1.139
9-15	579	151	1510	51.3	1.619×10^{-4}	2.590×10^{-3}	1.426
9-22	765	116	1675	373.1	1.177×10^{-3}	0.395×10^{-3}	0.217
14-3	313	170	2401	33.5	1.05×10^{-4}	6.30×10^{-3}	3.464
14-8	677	225	20053	121.8	3.844×10^{-4}	14.49×10^{-3}	7.968
14-9	603	174	3472	4519.8	1.427×10^{-2}	0.68×10^{-4}	0.0372
14-15	874	151	2685	183.1	5.581×10^{-4}	1.336×10^{-3}	0.735
14-22	1360	118	2105	1385.4	4.373×10^{-3}	0.134×10^{-3}	0.073
14-27	400	50	706	51.1	1.614×10^{-4}	1.215×10^{-3}	0.669

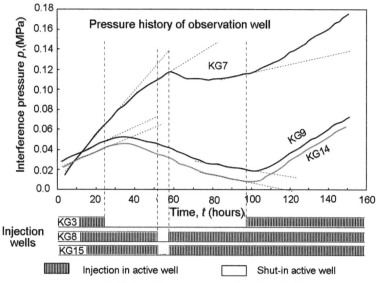

Figure 6.65 Interference test results of period II.

lowering it to 0.033 MPa/day. The power failure at the pump station during the test resulted in 3.37 hours of suspension of injection in Well KG8 and Well KG15 and a sudden pressure drop in Well KG7, thus forming a turn.

2. The injection suspension in Well KG3 caused the pressure of both Well KG9 and Well KG14 to drop and then rise again; furthermore, the shapes and amounts of variation of both wells were very similar, as if of just monitored in the same well.

Test Results of Periods III and IV

As shown in Figure 6.63, the test of periods III and IV were performed alternately.

1. The pressure of Well KG9 was monitored by downhole gauges continuously. After the test of period III was finished, the gauges continued to monitor the pressure response of the test of period IV for 4 more days.

2. In Well KG7 and Well KG14, the pressure gauges were pulled out of the hole after the test of period III finished and were run back into the hole on the next day for monitoring the pressure response of the test of period IV.

During the test of period IV, producer Well KG2 and Well KG22 were shut in one after another to generate as pressure perturbation, but Well KG22 was not shut in as per the original plan, which resulted in overlap with the shut in of Well KG2.

The test results of periods III and IV were as illustrated in Figure 6.66.

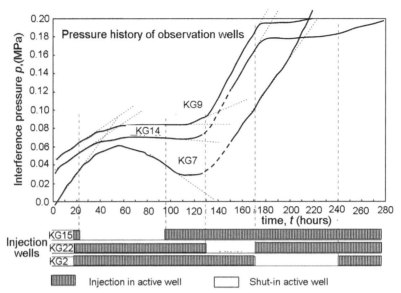

Figure 6.66 Interference test results of periods III and IV.

It can be seen from Figure 6.66:

1. The interference pressure response in Well KG9 was monitored continuously throughout periods III and IV, and the received influence was very obvious:
   - The pressure of Well KG9 was apparently influenced by the water injection of Well KG15. The injection suspension in Well KG15 made the originally rising formation pressure stable and constant, but when Well KG15 resumed injection, the formation pressure of Well KG9 turned to rise again.
   - The shut in of Well KG22 had a great impact on Well KG9. The originally stable bottom–hole pressure turned to rise at a rate of 0.06 MPa/day due to the shut in of Well KG22. When Well KG22 produced again, the pressure of Well KG9 turned stable again.
   - As the influence of Well KG2 overlapped and in the opposite direction with that of Well KG22, it was quite difficult to analyze its influence. But from the fact that the pressure of Well KG9 turned to rise at the end, it seemed that Well KG9 had been influenced by Well KG2, but this influence was much less than that from KG22.
2. The interference pressure response in Well KG14. Well KG14 was tested separately during periods III and IV, with a 20-hour break in between. Figure 6.66 combined the test results of both periods. It can be seen that:
   - The shape and amount of the interference pressure variation in Well KG14 during period III was almost the same as Well KG9.

- The interference pressure curve in period IV followed the tendency of that in period III and its shape was still very similar to that of Well KG9. As the later period extended comparatively longer, the obvious influenced effect of the delayed shut in of Well KG2 had been observed.

3. The interference pressure response in Well KG7. Different from the two wells just given, Well KG7 was very sensitive to the influence of the stimulation of Well KG15, but not so sensitive to the influence of the shut in of Well KG22.

It should be pointed out especially that the overlapped influence of the flowing and shut in of Well KG22 and Well KG2 seemed to reach a certain balance in Well KG7. It was after 30 hours that the influence of Well KG2 was greater and caused the interference pressure ti increase. Later, when Well KG2 was flowed individually, the interference pressure in Well KG7 began to decline within 10 hours (not shown in Figure 6.66).

Test Results of Period V

Very good pulse test data of four well pairs were acquired during this period. These were the pulse test data first acquired in China and were very significant. The obtained pulse test curves are shown in Figures 6.67 and 6.68.

Parameters were calculated with the Kamal type curve match method, and some of the results are listed in Table 6.13.

Test Results of Period VI

This last period of the interference test was provided to learn the relationship of the three observation wells. As described earlier, the pressures of Well KG9 and Well KG14 behaved in a very similar way, which predicted that these two wells must be well interconnected. But how high was the permeability in the area between them? This needed to be determined by testing. Figure 6.69 shows the test results.

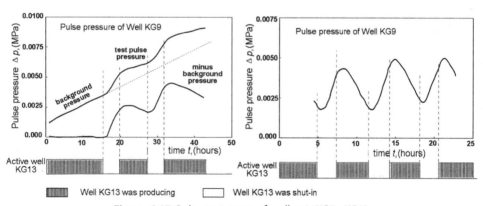

Figure 6.67 Pulse test curve of well pair KG9–KG13.

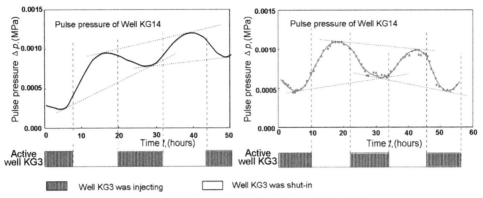

Figure 6.68 Pulse test curve of well pair KG14–KG3.

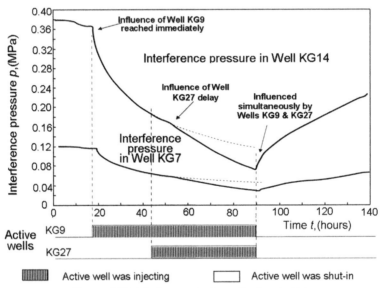

Figure 6.69 Interference test result of period VI.

It can be seen from Figure 6.69 that
1. The flowing and shutting in of Well KG9 instantly caused an abrupt pressure drop or increase in Well KG14; the flowing and shutting in of Well KG27 also influenced Well KG14.
2. The flowing or shutting in of Well KG9 influenced Well KG7 a little bit, but the influence was much smaller and delayed a lot. This demonstrated that the permeability in different directions was different.

All aforementioned interference test and pulse test data were analyzed with the type curve match method, and the results are listed in Table 6.13.

3.5.3 Analyzing the Characteristics of Formation Dynamic Model with Multiple-Well Test Results

Through the comprehensive multiple-well test analysis mentioned earlier, further understanding about the KL palaeo-burial hill reservoir model was achieved:

1. The formation between wells (between producer and injection well, between producer and producer) within the reservoir was generally interconnected.
2. The connection permeability between wells varies greatly and is different in different directions; two high mobility belts existed:
   - The high mobility belt along with Well KG22—KG9—KG14 direction
   - The high mobility belt along with Well KG9—KG7—KG2 direction

These two high mobility belts aligned roughly with the direction of the main fractures that dominated the formation of the reservoir.

Because the fractures in the burial hill limestone reservoir dominate the value of permeability, there is a direct causality between the directional distribution of permeability and the tension fractures developed during the faulting process.

Distribution of the mobility in the aforementioned high mobility belt is summarized in Table 6.14.

Based on the mobility categories, the mobility k/μ distribution map was plotted, as shown in Figure 6.70.

3. The reserve parameter was very low in the high mobility belt direction

Table 6.15 lists the calculated values of reserve parameter $\phi\,h$ of each well group in the high permeability direction. It can be seen that in the area along the direction of wells KG22—KG9—KG14, $\phi\,h < 0.6$ meter, even as low as 0.037 meter, which meant that the available reserve was only $10-20 \times 10^4$ tons per square kilometer,

Table 6.14 Categories of Well test Groups as per Mobility

High mobility category [k/μ >300 mD/(mPa·s)]		Low mobility category [k/μ <300 mD/(mPa·s)]	
Well group	k/μ value mD/(mPa·s)	Well group	k/μ value mD/(mPa·s)
7–2	345.5	9–8	<10
9–22	373.1	14–3	33.5
7–9	688.8	7–15	39.5
14–22	1385.4	7–27	48.4
14–9	4519.8	14–27	51.1
		9–13	97.9
		14–8	121.8
		3–9	128
		7–8	133.1
		7–3	164.5
		9–3	168.6
		14–15	183.1

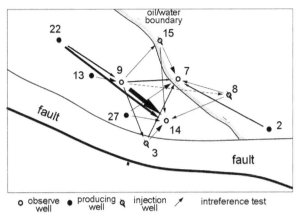

o observe   • producing   ⵁ injection   ✗   intreference test
    well   well   well

Figure 6.70 Distribution of mobility k/μ of KL palaeo-burial hill oil field.

and that in the area between wells KG9 and KG14 was less than 40,000 tons per square kilometer. With such small available reserves, it was no wonder that the production rate declined so rapidly there. The reserve parameters of test well groups are listed in Table 6.15.

4. Changes of water saturation

The interference test was performed successfully between Well KG9 and Well KG3 when these two wells were first brought into production. Two years later, the formations were flooded, the interference test was conducted again, and a very good test curve was obtained. The log–log curve matches of these two tests are shown in Figure 6.71. It can be seen that the curve moved remarkably leftward after water

Table 6.15 Categories of Reserve Parameter of Each Test Well Group

Low reserves parameter category ($\phi h < 0.6$ m)		High reserves parameter category ($\phi h > 0.6$ m)	
Well group	**ϕh value (m)**	**Well group**	**ϕh value (m)**
14–9	0.0372	9–3	0.650
14–22	0.073	14–27	0.669
9–22	0.217	14–15	0.735
7–2	0.454	3–9	0.765
7–27	0.489	7–15	0.900
7–9	0.572	7–3	0.998
		9–13	1.139
		7–8	1.139
		9–15	1.426
		14–3	3.464
		14–8	7.968

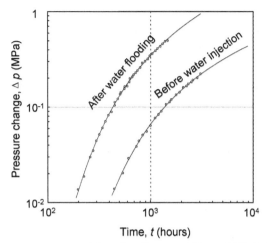

Figure 6.71 Interference test curve matching of well groups KG3 and KG9 before and after flooding.

flooding. The interpretation results of the tests before and after flooding are summarized in Table 6.16.

Substitute the diffusivity value in Table 6.16 into the following equation,

$$\frac{\eta_{after}}{\eta_{before}} = \left(\frac{k}{\mu\phi C_t}\right)_{after} \Big/ \left(\frac{k}{\mu\phi C_t}\right)_{before} = \frac{(k/\mu)_{after}}{(k/\mu)_{before}} \cdot \frac{(C_t)_{before}}{(C_t)_{after}} = 1.326 \cdot \frac{(C_t)_{before}}{(C_t)_{after}}$$

$$= 1.126$$

(6.18)

is obtained, and then

$$\frac{(C_t)_{after}}{(C_t)_{before}} = 1.178.$$

(6.19)

Substituting the expression of C_t into Equation (6.19), the following expression of oil saturation S_o was obtained:

$$S_{o\,after} = \frac{\left[C_o S_{o\,before} + C_w\left(1 - S_{obefore}\right)\right]\dfrac{C_{t\,after}}{C_{t\,before}} - C_w - C_f}{C_o - C_w}$$

(6.20)

Table 6.16 Comparison of Interference Test Interpretation Results of Well Pair KG3 and KG9 before and after Water Flooding

Test time	Flow coefficient kh/μ, mD·m/(mPa·s)	diffusivity η, m²/h
1978.6	2023	4.039×10^{-4}
1980.10	2278	5.356×10^{-4}

The compressibility coefficients of oil (C_o), water (C_w), and formation (C_f) from PVT analysis of the sample in that region are as follow: initial oil saturation $S_{o\ before}=$ 0.72, so it was calculated that $S_{o\ after}=0.5$, which meant the oil saturation after flooding was reduced by 22%. A large amount of the water-flooded fracture system resulted in cross flow; a high pressure was maintained, which made it more difficult for the oil in matrix and fissures to be recovered, and led to the oil wells fully flooded and shut down.

These were the understandings achieved on the KL palaeo-burial hill oil field through multiple-well tests.

4 SUMMARY

This chapter introduced use, development history, engineering design methods, data acquisition and analysis methods, and many typical examples of the interference test and pulse test in Chinese oil and gas fields.

The interference test was the first transient test method applied in the field. In hydrogeologic research it was called "hydrogeologic exploration." The pulse test is essentially a kind of interference test; the only difference is that it requires the active well to change its working status or flow rates more than three times so that the measured interference pressure can form a shape like a "wave," or a pulse. Based on current knowledge, the interference test has wider applications than the pulse test: it can provide the characteristics of oil and gas reservoir descriptions in more aspects, it can save test time, and it is more convenient for field operations.

The multiple-well test can be used to identify interwell connectivity and to calculate interwell connectivity parameters: connectivity flow coefficient kh/μ, connectivity storability parameter $\phi\ hC_t$, and the interwell reserve parameter $\phi\ h$ per unit of area (e.g., per square meter). For a description of the oil and gas reservoir, these parameters are irreplaceable and cannot be obtained from any other methods. However, there are special difficulties in the field implementation of multiple-well tests. Therefore, making a proper well test design as per the dialectic thinking process is the key to success.

The multiple-well test, stepping forward side by side with the Chinese petroleum industry, has 40 years of development history in China. At the early stage in the water flood development of the faulted oil fields in eastern China, interference tests were used to study the fault sealing property, interwell connectivity, the effectiveness of water flood development, and so on. These research results achieved were crucial to the development of the oil and/or gas fields. This book makes a brief introduction on this for readers' reference. The comprehensive analyses and evaluation on multiple-well tests in KL limestone oil fields in the Ordovician system presented in this book showed considerable profoundness in the completeness of test data, the stringency of data analysis, research on oil field permeability development patterns, and understanding of water flood effectiveness. The related reports were delivered at the Beijing International Petroleum Engineering Conference and later published in JPT.

The interference test study on the JB gas field can be regarded as a good model of the field application of multiple-well tests. The test conditions were very poor: the permeability of the formation was very low (1.5 mD) and the distances between wells were very long (1800 meters); they brought about many difficulties in data acquisition. However, 10 months were scheduled by the field management, electronic pressure gauges were used to monitor the interference pressure changes from the beginning to the end, and convincing results were achieved eventually. The following understandings on the reservoir were obtained by data interpretation: The Ordovician system distributes in a wide area continuously; the reservoir varied greatly horizontally and the formation was severely heterogeneous; the formation behaved with the characteristics of a double porosity medium system; and the fissures were dominant in natural gas storage and flow. These analysis results supported the static reserves estimation from the point of a reservoir dynamic model, gave the green light to the reserves examination of the JB gas field, and contributed to smooth development of the gas field and the project of gas transmission to Beijing.

The interference test study performed in the S6 infilling well area of the SLG gas field was another excellent example of multiple-well tests in a gas field. It was the first time since discovery of the SLG gas field at the beginning of the 21st century that an interwell connectivity relationship was identified and confirmed among over 1000 producers, which provided practical and effective references for the solution to difficult problems encountered during gas field development, such as rational well spacing and block areas controlled by a single well. It also demonstrated the unique role played by interference tests in gas field development research.

Coalbed Methane Well Test Analysis

Contents

Dynamic Well Testing in Petroleum Exploration and Development   © 2013 Petroleum Industry Press. Published by Elsevier Inc.
ISBN 978-0-12-397161-6, http://dx.doi.org/10.1016/B978-0-12-397161-6.00007-3

During the process of carbonification of the coalbed, a large amount of methane is generated. Generally, the methane migrates to the sandstone reservoirs adjacent to the coalbeds and is usually called "coal-derived gas." However, there is still a proportion of methane remaining in the coalbeds, either adsorbed on the surface of micropores of the coal matrix or existing in the coalbed cleats or cleavages of the coalbed as free gas. This portion of methane remaining in coalbeds is called "coalbed methane."

Coalbed methane was discovered as early as in coal mining and is regarded as mash gas. The overflow of mash gas is often the main factor causing a coal mine explosion. Mash gas is actually a very important kind of resource. It is clean and efficient energy if it can be produced by drill holes with a small diameter and transmitted to users before the coal mines are opened. Such holes are coalbed methane wells.

The production of coalbed methane depends on the storage of coalbed methane in the coal reservoirs, which is studied mainly by the performance of coalbed methane wells besides by coring. Therefore, the coalbed methane well test is an important means in understanding coalbed methane reservoirs.

1 COALBED METHANE WELL TEST

1.1 Function of Coalbed Methane Well Test in Coalbed Methane Reservoir Investigation

According to foreign literatures and domestic experiences achieved in recent years, coalbed information acquired by the well test includes the following.

1.1.1 Effective Permeability of Fissures or Cleats in Coalbed

Although there are many approaches to measuring the permeability of coalbed, the well test is the only effective way.

The permeability of common rocks can be measured by core analysis in a laboratory. Theoretically, this approach is also applicable to coal cores. Nonetheless, it is very difficult to maintain the original fissures in coal cores, as they are rather fragile. As a result, this approach is unable to provide the true permeability of coalbed cleats.

To get the permeability by well logging, it is required to make logging-type curves based on the results of core analysis. As the reliability of the coring is questionable, the way of permeability determination by logging is limited.

It should be specially pointed out that, no matter by well logging or coring, the obtained understanding is only about the drill hole under static conditions. However, it is quite different by the well test: the permeability obtained represents the comprehensive value in a region the fluids flow through under dynamic conditions. Therefore, the permeability determined by the well test is the most representative value.

1.1.2 Average Reservoir Pressure

Coalbed pressure is a key parameter in coalbed methane production, and the original pressure marks the original absorption conditions of coalbed methane. The variation of the coalbed pressure indicates the change in desorption conditions, whereas the pressure measurement and calculation can only be done by relying on well testing.

1.1.3 Damage and Improvement of Coalbeds

Coalbeds are damaged more or less in the process of drilling and completion, and coalbed damage increases the resistance in production. The skin factor provided by well test analysis can characterize the degree of coalbed damage quantitatively. If the coalbed methane wells are stimulated, the stimulation effect can also be identified by the S value.

1.1.4 Evaluation of Fracturing Effects

Fracturing in a coalbed will creates big fractures in the down hole, and the created fractures can be propped by adding sand during the operation to form effective flow channels. The length of the propped fracture x_f and the flow conductivity within the propped fracture F_{CD} can be calculated by well test data analysis, and then the fracturing effects can be evaluated.

1.1.5 Identification of Coalbed Connectivity and Calculation of Connectivity Parameters

Due to geologic conditions, coalbeds are not developed continuously and uniformly in every direction, and the development of fissures, which enable methane flows, is especially heterogeneous. As a result, the interwell connectivity is influenced dramatically. An interference test between wells can be used not only to identify the connectivity of coalbeds, but also to calculate the interwell connection permeability k and the connected storativity $\phi h C_t$ by the quantitative analysis of interference test curves. As the C_t value of a coalbed is one to two orders of magnitude greater than that of common sand reservoirs and there is almost no other way to get it except theoretical method and well testing, the interference test is the only effective way to determine the compressibility coefficient C_t of the coalbed.

1.1.6 Determination of Pore Volume of Coalbed

The storativity coefficient $\phi h C_t$ can be determined by interference testing. Then the value of pore volume can be determined by excluding the influence of C_t.

1.1.7 Analysis of the Development Direction of Fissures

The development of fissures in a coalbed is often influenced by geostatic stress and other factors and so is different in different directions; this status can be made clear by conducting the interference test in several well groups arranged in different directions and then calculating the permeability differences in different directions.

1.1.8 Detection of Flow Boundaries in Coalbed

If impermeable boundaries are caused by geologic factors in the coalbed, the feature and patterns of boundaries and the distance between the well and the boundaries can be determined by well testing.

1.2 Differences between Coalbed Methane Well Test and Common Oil or Gas Well Test

Because the coalbed methane well test is conducted in completely different reservoirs, there are differences in many aspects, such as the well test method, the theoretical flow model, and the data analysis method.

1.2.1 Although it is a Gas Well Test in the Coalbed, Fluid seen During Coalbed Methane Well Testing is often Water

When a coalbed methane reservoir is opened initially, there is often no free gas (but free gas exists in some coalbed methane reservoirs in the United States), as the reservoir pressure is usually higher than the critical desorption pressure. Injection/falloff well testing is often used to measure the reservoir permeability in this situation. Therefore, the tested wells are often water wells, and the tested reservoirs are fully filled with water. Accordingly, test data are analyzed with the water well test analysis method.

1.2.2 Although the Coalbed seems to be Double Structured, Well Test Data in a Coalbed do not Show Flow Characteristics of the Double Porosity Medium like in Oil or Gas Reservoirs

In a geological structure, the coalbed contains two different media of a coal matrix and cleats. However, the coal matrix is an impermeable medium. When producing with water drainage and pressure reduction in a coalbed reservoir, the flow consists of three stages of flow in fissures: transition flow and a two-phase flow of water and gas in the total system; what is different is that this process may last months, years, or even decades. In contrast, well testing only lasts several days in a common oil or gas well, and therefore only a single-phase flow or two-phase flow in a homogeneous reservoir appears, which has been proved by both theoretical analysis and field practices. Even though some farfetched interpretation of test data with double porosity medium models exists, it has been discovered that all of them are actually variable wellbore storage effects caused by this kind of special test technique.

1.2.3 Purpose and Analysis Methods of a Coalbed Methane Well Test are Different and Depend on Production Stages

1. The purpose of the coalbed methane well test is mainly to find high-yield zones in the exploration period.

   During the early period when coalbed methane wells are perforated, the coalbed fissures are fully filled with water, but this water cannot flow into the coal matrix, and the methane does not flow in the fissures before it is desorbed as pressure drops. Therefore, flow during the testing is actually "the single phase water flow uniform medium," which has been proved by much injection/falloff well test data obtained in coalbed methane wells in China.

   Based on well test research during this period, fissures with high flow conductivity in coalbed methane production zones can be discovered and verified. This high flow conductivity is the main characteristic of high production zones and is thus the major study subject of exploration wells and production appraisal wells in coalbed methane zones. This period is also the predominant one for coalbed methane well test and well test analysis.

2. It is two-phase flow during well testing in the exploitation period of coalbed methane wells

   After water is drained from the coalbed methane wells and the pressure is reduced, methane begins to be desorbed and flow into the fissures. During this period, fluid in the coalbed is a mixture of gas and water, and the water in the fissures is still unable to enter the coal matrix. After desorbed from the coal matrix flowing into the fissures, the gas is almost unable to flow back into the coal matrix. Therefore, what happens and is tested in pressure buildup testing is two-phase flow in homogeneous formations.

The well test for two-phase flow is still a theoretical study subject. Furthermore, for coalbed methane zones producing under low pressure, there is neither the necessity nor the possibility to conduct large-scaled well test research. This has been illuminated in some Chinese and foreign literatures.

1.2.4 Some Differences in Well Testing Analysis Caused by Different Physical Properties of the Coalbed

- The differential equation describing the flow process is different
- Desorption of methane adds a desorption compressibility C_d onto the reservoir compressibility, which makes it 10 to100 times higher
- The permeability of coalbeds is usually very low
- The reservoir pressure is very low in coalbed methane production
- Some very simple testing methods, such as water tank test and slug test, are often used

2 FLOW MECHANISM AND WELL TESTING MODELS IN A COALBED

2.1 Structural Characteristics of a Coalbed and Flow of Coalbed Methane

2.1.1 Structure of Coalbed and Reserve of Methane

A typical coalbed unit is shown in Figure 7.1.

The definition of a coalbed unit is consistent with that of a flow unit defined in Chapter 2. As seen from Figure 7.1, the coal matrix is cut by face cleats and end cleats. Here, cleats are like fissures referred to in oil or gas reservoirs. In coalbeds, face cleats are generally the fissures in the main development direction, whereas end cleats are fissures in the direction perpendicular to the face cleats. Their development depends on coal rank, coalbed type, structure type, earth stress, mineral compositions in coalbed, and so on. What is more, their development is often enhanced by the later tectonic movements.

The size of the coal matrix between cleats is about 2−20 mm. The methane is adsorbed on the surface of the micropores on the end planes. Because the area of the

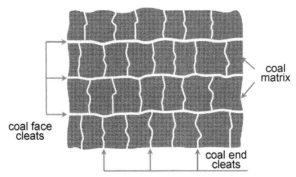

Figure 7.1 Diagram of the physical structure of a coalbed unit.

micropores is very large, the volume of methane adsorbed by the coal matrix is as much as two to seven times that of gas adsorbed by the same volume of other reservoirs.

When coalbed pressure drops, the methane adsorbed on the surface planes begins to be desorbed into free gas, flows to the coalbed methane wells along the cleats (fissures), and is produced.

2.1.2 Flow Process in Coalbed Methane Production
The Three Flow Stages in the Coalbed Methane Production Period
Flow in a coalbed can be divided into three stages from the commencement of water drainage after perforation to the end of gas production:
1. Stages of single-phase flow of water in fissures
2. Stages of transition flow in local areas when methane is desorbed
3. Stages of two-phase flow of water and gas when methane is desorbed in the whole coalbed

Essential differences exist between coalbed and common double porosity medium reservoirs:
1. Unlike in common reservoirs where the gas is compressed and stored in pores of the matrix, coalbed methane is desorbed and reserved in the coal matrix. Hence it does not participate in the flow in the fissures during the early period.
2. When methane is desorbed from the coal matrix, it follows the Fick's diffusion law rather than Darcy's law followed by oil and gas in common reservoirs.
3. When the flow in fissures (cleats) is transited to the flow in the total system, the fluid composition changes. During the stages of flow in fissures, the flowing fluid is mainly water, whereas in the stages of flow in the total system, the flowing fluid consists of both gas and water or is dominated by the gas phase.

Long-Term Methane Desorption and Production and Short-Term Methane Well Testing
The three flows just described cannot occur successively during one well testing process. The reason is that they happen in different stages of the coalbed methane production process, which may last for months, years, or even decades. The foregoing flow state is replaced by the successive one when it has finished and would not exist again. The process of well testing only lasts a few days and can only be in one flow state:
1. When injection/falloff testing is operated in a coalbed methane well at its early stage, what flows in the fissures is only water. No flow occurs in the coal matrix. Therefore, the flow is single-phase water flow in a homogeneous medium of fissures.
2. When the coalbed pressure drops significantly, the desorbed methane gas is mixed with the water in the cleats and then flows together to the bore holes. During the well testing process, gas and water two-phase flow happens. Because the flow takes place only in cleats, no flow happens in the coal matrix and the coal matrix acts only as a "source" generating methane gas. Therefore, the flow model is two-phase

(gas and water) flow with the gas source in homogeneous reservoir. This is discussed in more detail in the section describing the theoretical well test model under desorption conditions.

3. In the initial stage of water drainage, it is possible to form a desorption area in some near-wellbore locations in coalbeds with low cleat permeability; there is a water area outside the desorption area. As a result, the flow model is a composition flow model.

2.2 Seven Typical Dynamic Models of Coalbed Methane Well Test

Well test models applicable to a coalbed methane well are different in different production stages. There are seven typical models summarized based on the following theoretical analysis:

1. Model of water single-phase flow in homogeneous fissures or in cleats. This model is applicable to injection/falloff well testing conducted when coalbed methane wells are just opened or perforated.

2. Model of water single-phase linear or bilinear flow. This model is applicable to injection/falloff testing conducted after a large-scale fracturing operation and completion.

3. Model of gas and water two-phase flow in homogeneous fissures or cleats. This model is applicable to coalbeds where there are large amounts of primary free gas.

4. Model of water and gas two-phase flow with sources. This model is applicable in the stage of production by desorption when pressure of the coalbed methane formation decreases.

5. Model of gas single-phase flow with sources. This model is applicable under the conditions of that the methane is desorbed and the water saturation in the coalbed is very low or the distribution of water saturation is stable and close to irreducible water saturation.

6. Model for composition formations. This model is applicable under the conditions of that the near-wellbore zone is the methane desorption area because of a serious pressure drop and the zone far away from the wellbore is still single-phase water area.

7. Model of gas and water two-phase linear or bilinear flow. This model is applicable to fractured wells when the methane is desorbed or there is a large amount of free gas in the reservoir in the early period.

Among these seven models, the first and the second ones can be used to analysis satisfactorily with the available well test interpretation software; the fifth and the sixth ones can be used to interpretation with the approximate method: modify the compressibility in the desorption area by adding C_d, interpret data, and calculate the parameters with the available well test interpretation software. This is enough for meeting the needs of field production. As is known, this approach is also used in other countries.

There are no very good solutions by conventional well test analysis methods for the third, fourth, and seventh flow models, as two-phase flow is involved in them. However, the numerical well test software developed in recent years can provide approximate solutions to them.

2.3 Characteristics of Water Single-Phase Flow in Homogeneous Fissures and Well Test Data Interpretation Methods

The water single-phase flow in homogeneous fissures is the flow state when the coalbed methane wells are just opened. No matter if the drill stem test (DST), the injection/ falloff test, or the water tank injection test is being used, the flow states are all single-phase flow of water in homogeneous fissures.

Single-phase flow in a homogenous reservoir is commonly met in conventional oil and gas well testing, as well as in hydrogeology testing in coal mines; thorough investigation and studies have been done, and there has been a great deal of discussion about its specific analytical methods. Therefore, it will not be introduced separately any more here. Nonetheless, differences resulted from the testing method are discussed in later sections about injection/falloff testing.

2.4 Single-Phase Flow Under the Condition of Methane Desorption and Well Test Analysis Method

2.4.1 Coalbed Conditions

After methane gas is desorbed, there are two phases of water and gas in coalbeds. If only a small proportion of water exists or if the water is distributed uniformly and stably, its influence on the gas phase is mainly demonstrated in the permeability of methane gas, that is, the relative permeability of methane gas under certain water saturations. In such cases, the transient flow of methane gas under desorption conditions can be discussed alone to study the methane flow in coalbed. The following assumptions are made for this discussion:

1. There is only single-phase methane gas in coalbeds, and the water phase just influences the relative permeability of the gas.
2. Fissures or cleats are distributed uniformly in coalbeds.
3. The desorption of methane gas is completed instantaneously or the pressure at the moment of desorption is equilibrated.
4. The duration of desorption may vary depending on the distribution of fissure density. If the desorption cannot be finished instantaneously, a desorption duration adjusting coefficient τ is set, and $\tau = 1$ means instantaneous desorption.
5. The content of the desorbed gas follows Langmuir's isothermal absorption law.
6. There is some damage near the wellbore, and the damage is expressed with skin factor S, that is, a pressure difference caused by skin effect Δp_s exists, which includes the influence of inertial flow.

2.4.2 Flow Equation

Under the assumptions just given, the flow equation of methane gas can be expressed as

$$\frac{\partial(\rho\phi)}{\partial t} - q_r - \frac{1}{r}\cdot\frac{\partial}{\partial r}\left(r\rho\cdot\frac{k}{\mu}\cdot\frac{\partial p}{\partial r}\right) = 0, \tag{7.1}$$

where ρ is gas density; p is pressure; ϕ is porosity of fissures; r is distance to the well; t is time; and q_r is quantity of desorbed gas, expressed here based on the diffusion law:

$$q_r = \left(-\frac{\partial M_d}{\partial p}\right)_P \cdot \left(\frac{\partial p}{\partial t}\right)_r; \tag{7.2}$$

M_d is mass of desorbed gas in per unit volume of coal; and

$$M_d = \rho_g^0 \cdot V_d,$$

where ρ_g^0 is gas density under the standard state; V_d is volume of desorbed gas under the standard state; and

$$V_d = V_L\left(\frac{p_i}{p_i + p_L} - \frac{p}{p + p_L}\right), \tag{7.3}$$

where V_L is Langmuir volume; p_L is Langmuir pressure; and p_i is initial pressure.

Equation (7.2) can be rewritten as

$$\left(\frac{\partial M_d}{\partial p}\right)_P = \rho_g^0 \cdot \frac{\partial V_d}{\partial p}$$

$$= \rho_g^0 \frac{\partial}{\partial p}\left(\frac{V_L p_i}{p_i + p_L} - \frac{V_L p}{p + p_L}\right)$$

$$= \rho_g^0\left(-\frac{V_L p_L}{(p + p_L)^2}\right). \tag{7.4}$$

Then substitute it into Equation (7.1), to which Equation (7.5) is obtained:

$$\frac{\partial(\rho\phi)}{\partial t} - \frac{1}{r}\cdot\frac{\partial}{\partial r}\left(r\rho\frac{K}{\mu}\cdot\frac{\partial p}{\partial r}\right) + \frac{\rho_g^0 V_L p_L}{(p + p_L)^2 \tau}\cdot\frac{\partial p}{\partial t} = 0. \tag{7.5}$$

In this equation, τ is the adjusting coefficient of coalbed methane desorption.

Coalbed methane gas containing mainly methane follows the real gas law:

$$\rho = \frac{pM}{ZRT}, \tag{7.6}$$

where M is mole mass of gas, its value is equal to the relative molecule mass; Z is Z factor of real gas; T is temperature; and R is universal gas constant.

In thermodynamics the isothermal compressibility of C is expressed as

$$C = \left[\frac{1}{p} - \frac{1}{Z}\cdot\frac{\partial Z}{\partial p}\right]_{T}. \tag{7.7}$$

The pseudo pressure $\psi(p)$ is

$$\psi(p) = 2\int \frac{p\cdot dp}{\mu Z}.$$

By substituting the aforementioned equations into Equation (7.5), Equation (7.8) is obtained:

$$\frac{1}{r}\cdot\frac{\partial}{\partial r}\left(r\frac{\partial\psi(p)}{\partial r}\right) = \frac{1}{k}\left[\frac{\rho_g^0 RTV_{\mathrm{L}}p_{\mathrm{L}}\mu Z}{Mp(p+p_{\mathrm{L}})^2\tau} + \phi\mu C\right]\frac{\partial\psi(p)}{\partial t}. \tag{7.8}$$

Express the coefficient of $\dfrac{\partial\psi(p)}{\partial t}$ as

$$ICD = \frac{\rho_g^0 RTV_{\mathrm{L}}p_{\mathrm{L}}\mu Z}{kMp(p+p_{\mathrm{L}})^2\tau} + \frac{\phi\mu C}{k}. \tag{7.9}$$

ICD refers to the inverse coefficient of diffusivity of coalbed methane reservoirs. Compared with the inverse coefficient of diffusivity of conventional reservoirs, $IHD = \phi\mu C/k$, a pressure depending term $\dfrac{\rho_g^0 RTV_{\mathrm{L}}p_{\mathrm{L}}\mu Z}{kMp(p+p_{\mathrm{L}})^2\tau}$ is added, which turns Equation (7.8) into a nonlinear equation and makes it more difficult to be resolved.

2.4.3 Analyzing Coalbed Methane Well Test Data by Conventional Well Test Interpretation Software

The mathematic equation on which conventional gas well test software is based is given as

$$\frac{1}{r}\cdot\frac{\partial}{\partial r}\left(r\frac{\partial\psi(p)}{\partial r}\right) = \frac{\phi\mu C}{k}\cdot\frac{\partial\psi(p)}{\partial t}. \tag{7.10}$$

Compared with Equation (7.8), the desorption depending term C_d is added to the coefficient of the term of the differential of pseudo pressure against time; and so, under the coalbed methane conditions, compressibility C is replaced by C':

$$C' = C + C_d. \tag{7.11}$$

Research results show that the value of C_d is one to two orders of magnitude greater than that of C and changes with pressure. To simplify the interpretation, an approximation is taken that the pressure in most parts of the area where coalbed methane is desorbed is considered close or equal to the average reservoir pressure $\bar{p}$; as a result, C_d is approximated as a constant, namely:

$$C_d = \frac{\rho_g^0 RTZV_L p_L}{\phi M \bar{p}(\bar{p} + p_L)^2 \tau} = \text{constant}. \tag{7.12}$$

Table 7.1 lists the results calculated based on data of the San Juan Basin by the Gas Research Institute (GRI).

When desorption compressibility is added, the total compressibility increases significantly, and its value is about

$$C_t = 9.74 \times 10^{-1} \text{MPa}^{-1}.$$

With this C_t value, the pressure–time relationship can be found by the conventional method of solving equations, and conventional well test interpretation software can also be used to interpret coalbed methane well test data under the desorption condition.

As for the calculation of desorption compressibility C_d, Readers can refer to Saulsberry's paper [70], in which the formula for calculating C_d is provided.

The formula for calculating C_d under the imperial unit system is

$$C_d = \frac{B_g \rho_c V_L (1 - \alpha - W_c) p_L}{32.0368 (p_L + \bar{p})^2 \phi}. \tag{7.13}$$

Table 7.1 Components of Compressibility of Coalbeds in San Juan Basin

Component		Value (MPa^{-1})	Percentage
C_d	Desorption compressibility	0.913	93.7
C_f	Compressibility of coalbed	0.0478	4.9
$S_g\, C_g$	Compressibility of methane	0.0128	1.3
$S_w\, C_w$	Compressibility of coalbed water	0.000384	0.04

Converted into the legal unit system of China, the aforementioned formula is expressed as

$$C_d = \frac{1.102 B_g \rho_c V_L (1 - \alpha - W_c) p_L}{\phi (p_L + \bar{p})^2}, \tag{7.14}$$

where B_g is volume factor of methane gas, m^3/m^3 (in standard conditions):

$$B_g = 1.949 \times 10^{-4} \frac{ZT}{\bar{p}}; \tag{7.15}$$

Z is Z factor; T is reservoir temperature, °R; $\bar{p}$ is average reservoir pressure, MPa; ρ_c is density of coal, g/cm^3; α is ash content, decimal; W_c is humidity, decimal; p_L is Langmuir pressure, MPa; V_L is Langmuir volume, m^3/ton; and ϕ is porosity, decimal.

These formulas are cited from *A Guide to Coal Bed Methane Reservoir Engineering* prepared by GRI.

2.4.4 Characteristics of Well Test Curves and Well Test Analysis When Desorption Happens Only in Part of a Region

In the initial period of water drainage, a desorbed area exists in the formation near the bottom hole, and the size of this desorbed area depends on the critical desorption pressure.

Suppose the radius of the desorbed area $r_d = 50$ m and the fluid flowing within the desorbed area is mainly methane gas; when $r > r_d$, all the fluid flowing in the coalbed is water and it is assumed that the coalbed is infinite, as shown in Figure 7.2.

The parameters used in characteristic analysis are listed in Table 7.2.

Figure 7.2 Desorption happens only in the formation near the bottom hole.

Table 7.2 Parameters for Falloff Curve Analysis in Partial Desorption

Parameter	Desorbed Area (Internal Area)	Undesorbed Area (External Area)
Permeability k, mD	1	1
Coalbed thickness h, m	10	10
Porosity ϕ	0.03	0.03
Skin factor S	0	/
Wellbore radius r_w, m	0.09	/
Wellbore storage coefficient C, m^3/MPa	0.0148	/
Radius of desorbed area r_d, m	50	/
Total compressibility C_t, MPa^{-1}	0.974	0.0482
Fluid viscosity μ, mPa·s	0.014	0.5
Production rate of methane gas q, m^3/day	1000	/

Typical characteristic curves obtained by well test interpretation software based on the aforementioned parameters are shown in Figure 7.3.

It can be seen from Figure 7.3:

1. The pressure and pressure derivative curves show the characteristics of a homogeneous reservoir during the early period. About after 1 hour, the pressure falloff accelerates suddenly. This is because the methane gas flow is limited by the water region outside the desorbed one.

2. During the late period of the typical curve of pressure falloff, the pressure derivative curve rises to a rather high level, indicating that flow has already influenced the water region. As a result, the characteristic of a composite formation of gas and water regions is shown.

In field practice, it is very difficult to measure such a complete characteristics test curve because it takes a very long time. In general, the characteristics of restricted flow caused by

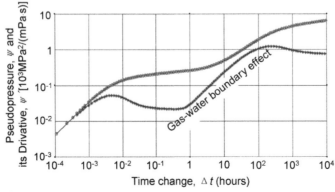

Figure 7.3 Match of pressure falloff curve of partial desorbed coalbed methane layer.

restriction of the gas—water boundary can appear within 100 hours, which makes the pressure derivative curve go upward. Well test data reflecting such circumstances can still be interpreted with the composite model in well test interpretation software.

3 INJECTION/FALLOFF WELL TEST METHOD FOR COALBED METHANE WELLS

3.1 Equipment and Technology for Injection/Falloff Well Testing

Because most coalbed methane wells are shallow and their depths are about 1000 meters, the pressure of the perforated coalbed is often lower than the hydrostatic pressure. In the initial well completion stage, fluid in the coalbed fissures is water only or water containing a very small amount of free methane gas only; the completion conditions are simpler than those of conventional oil or gas wells. However, the well head should be sealed and equipped with a valve in order to produce methane gas and perform injection/falloff well testing.

3.1.1 Test String

A multiflow evaluator or other DST test strings with a surface water injection pump can be used, but this will inevitably increase the test cost.

Now there are downhole test strings specifically for the coalbed methane well test, which are equipped with the basic parts, including packers and valves for flowing and shutting in the well. Such strings can satisfy the needs of the test in shallow water wells with low pressure.

If a well test is conducted with the wellhead shut in, it can be done only with tubing equipped with a surface valve for flowing and shutting in, as well as water injecting pumps or elevated water storage tanks.

3.1.2 Measuring Instruments

1. Bottom-hole pressure gauges. Bottom-hole electronic memory pressure gauges are usually used. Mechanic pressure gauges with high precision can also be used.
2. Wellhead flowmeter. A wellhead flowmeter is used to measure and record the change of the water injection rate with time for well test analysis. An electronic data acquisition system that can acquire all kinds of data automatically is commonly used abroad.
3. Wellhead pressure gauges. Self-recording electronic pressure gauges are commonly used. Acquired pressure data are used to control the injection pressure of the pump and monitor the point when the pressure is dropped to 0 after stopping of the injection used to determine the "abnormal changing point of pressure data," which is the beginning of the variable wellbore storage effect and is very important for next test data analysis.

3.1.3 Water Injection Pump

Water injection is needed for the injection/falloff test. The injection pump should have a high output pressure but low discharge rate to meet the requirements of the well test in a low permeability coalbed.

For reservoirs with very good water adsorption ability, a water injection pump can be replaced by an elevated water storage tank to inject water into the well.

3.1.4 Testing Process

1. Connect the testing instrument and tools, run them down to the depth of the tested coalbed, and set the packer to minimize the volume of the wellbore connected with the coalbed so as to reduce the effect of wellbore storage.
2. Inject water into the coalbed reservoir from the wellhead with a water injection pump. Record the changes of the injection rate and pressure with time.
3. Shut in the well, that is, stop injection to measure the bottom–hole pressure falloff curve.
4. Pull the pressure gauges out of the hole and analyze test data.

3.2 Design of Injection/Falloff Well Test

3.2.1 Selection of Shut-In Method

A bottom-hole shut in is preferred to reduce the wellbore storage effect. Particularly, the bottom-hole shut in can make the pressure to avoid reaching the "abnormal changing point of pressure data" during shutting in for those coalbed methane wells whose pressure is lower than the hydrostatic pressure. This is of great importance, as if pressure reaches the "abnormal changing point of pressure data" it may be useless for analysis or application.

3.2.2 Calculation of Injection Pressure

The formula for calculating the injection pressure p_{inj} is

$$p_{inj} = \sigma_{min} - 0.0098\rho_w D_c, \tag{7.16}$$

where p_{inj} is injection pressure of the water injection pump, MPa; σ_{min} is minimum in situ stress, MPa; ρ_w is density of water, g/cm^3; and D_c is depth of the coalbed, m.

For example, if $D_c = 1000$ meters, $\rho_w = 1$ g/cm^3, and $\sigma_{min} = 15.5$ MPa,

$$p_{inj} = 15.5 \text{ MPa} - (0.0098 \times 1 \times 1000)\text{MPa} = 5.7 \text{ MPa}.$$

In the water injection operation, the actual injection pressure can be a little higher than the designed value. The reasons of doing so are that the radius of influence can be increased in effective duration and so more representative data can be acquired on the one hand and that the higher pressure may generate microfissures in the area near the bottom hole so as to improve the water absorption conditions there and reduce skin effect on the other hand.

3.2.3 Calculation of Water Injection Rate

Equation (7.17) is for calculating the water injection rate, and the unit used in it is the statutory units of China:

$$q_{\text{inj}} = \frac{0.471kh(p_{\text{inj}} + 0.0098\rho_w D_c - \bar{p})}{\mu B\left(\lg\dfrac{8.085 \times 10^{-3}kt}{\mu\phi C_t r_w^2} + 0.87S\right)}. \tag{7.17}$$

The values of coalbed permeability k, skin factor S, and the average reservoir pressure $\bar{p}$ are all unknown before testing; those of the neighboring wells can be used for test design. In fact, because the injection rate can be determined as soon as the injection pressure is known, the calculated q_{inj} is only a reference injection rate before testing.

3.2.4 Determination of Water Injection Volume

Total water injection volume Q_{sj} determines the radius of the advancing front of injected water R_j. Usually a R_j value is set in advance, and water injection volume Q_{sj} can be calculated with

$$Q_{sj} = \pi R_j^2 h\phi. \tag{7.18}$$

3.2.5 Determination of Influence Radius and Injection Duration

According to the assumptions concerning conventional sandstone reservoirs and carbonate reservoirs in Chapter 2, the reservoirs are deemed an elastic medium. When the injected water enters the reservoirs, a pressure buildup funnel is generated, the radius of which r_i is calculated with Equation (2.8).

To find out the reservoir parameters within a given influence range, the value of influence radius r can be set in advance and then correspondence time t_j can be calculated conversely with Equation (7.19):

$$t_j = \frac{\mu\phi C_t r^2}{0.0144k}. \tag{7.19}$$

This time t_j can also be used as the basis to determine the injection duration.

3.2.6 Effect of Coalbed Elastioplasticity

A coalbed medium is not a conventional sandstone or limestone reservoir. Its compositions are mainly compounds of organic carbons, and it has elastoplasticity. When it is pressurized by water injection, a part of the energy is adsorbed and plastic deformation happens, and the so-called "pressure equilibrium point" is reached. Particularly when the amount of injected water is rather small, such a pressure balance point is reached very easily. Real field data shown in Figure 7.4 are all under the condition that the amount of water injected is very small and that an abnormal change occurs due to the pressure approaching the pressure equilibrium point.

3.3 Data Examination and Analysis Methods of Injection/Falloff Well Testing

Data examination and analysis methods of well testing data were introduced in detail in Chapter 5, and the meaning of which is that an abnormal change of well test data, if any, is identified and evaluated based on analysis of the operation during the testing process. Then these abnormal changes are eliminated based on the evaluation so that data truly reflect the tested reservoir.

Abnormal changes occur very commonly in the operation process of the injection/falloff test of coalbed methane wells because of its special techniques in test operations.

3.3.1 Variable Wellbore Storage Effect in Injection/Falloff Test Process

Figure 7.5 shows water level variations in the wellbore during the injection/falloff test process.

1. Water level when injection is just finished (see a in Fig. 7.5)

At this moment, the water level is at the wellhead, and the wellhead pressure $p_H > 0$.

Because the whole wellbore is filled with water, the pressure of the water column on the bottom hole is p_w, which can be calculated as

$$p_w = 0.0098 \rho_w D_c, \tag{7.20}$$

where ρ_w is density of water, g/cm^3; and D_c is depth of coalbed (i.e., the bottom hole), m.

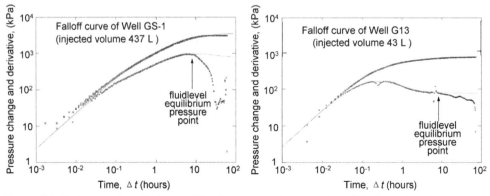

Figure 7.4 Real examples of abnormality of pressure changes when falloff reaches the pressure equilibrium point.

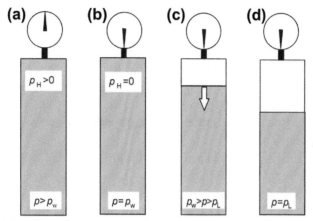

Figure 7.5 Water level changes in a wellbore during the injection/falloff test process.

Because wellhead pressure $p_H > 0$, the bottom-hole pressure $p = p_w + p_H > p_w$. At this moment, the wellbore storage effect is caused by fluid compression, and wellbore storage coefficient C_1 can be calculated as

$$C_1 = V_w \cdot C_w, \qquad (7.21)$$

where C_1 is the wellbore storage coefficient caused by fluid compression, m^3/MPa; V_w is volume of fluid column in wellbore, m^3; and C_w is compressibility of water, MPa^{-1}.

For example, in a 1000-meter-deep well, its coalbed is separated by a bottom-hole packer, and only the volume of $2\frac{1}{2}''$ tubing is considered, then $V_w = 3.02\ m^3$. If the compressibility of the water is calculated with the average tubing pressure, 5.5 MPa, $C_w = 4.5 \times 10^{-4}\ MPa^{-1}$. As a result, $C_1 = 1.4 \times 10^{-3}\ m^3/MPa$ is obtained from Equation (7.21).

2. When wellhead pressure is reduced to 0, bottom-hole pressure $p = p_w$. At this moment, "the wellhead pressure equilibrium point" is reached (see b in Fig. 7.5).

From this point, the fluid level draws down gradually, and the process of variable fluid level falloff is thus formed.

3. Variable fluid level falloff (see c in Fig. 7.5)

In the process of variable fluid level falloff, the mechanism of the wellbore storage effect is completely different. During this process, wellbore storage coefficient C_2 is calculated with Equation (7.22):

$$C_2 = 101.97\ V_u/\rho_w, \qquad (7.22)$$

where C_2 is the wellbore storage coefficient with variable fluid level, m^3/MPa; V_u is volume of wellbore of one unit of length, m^3/m; and ρ_w is density of the fluid, g/cm^3.

For instance, for 2½" tubing, if $V_u = 3.02 \times 10^{-3}$ m^3/m and $\rho_w = 1$ g/cm^3, $C_2 = 0.307$ m^3/MPa.

Compared with wellbore storage coefficient C_1 calculated in case a, it can be found that $C_2/C_1 = 219$, which means that the wellbore storage effect suddenly increases 200 times.

The sudden increase in the wellbore storage coefficient makes the wellbore storage effect having disappeared emerging abruptly, and the test curve showing an "abnormal change."

4. Fluid level in the wellbore decreases continually, the point where $p = p_L$ will be reached finally (see d in Fig.7.5).

The initial value of p_L is the bottom–hole pressure at the beginning. Once the fluid level reduces to its equilibrium point, the bottom–hole pressure will accordingly reduce to its limit pressure p_L, which is called the "fluid level pressure equilibrium point."

When the fluid level equilibrium point is reached, the pressure derivative curve will go down suddenly. Such a phenomenon happened in both of the two wells shown in Figure 7.4.

According to the theory of graph analysis in Chapter 5, the abnormal going down of the pressure derivative curve generally shows the existence of a constant pressure boundary. However, in a coalbed filled fully with water, the so-called "constant pressure boundary" does not have any geologic basis. In fact, this is because the injection amount is too small, the coalbed plasticity has adsorbed the influence of the injected water, and so made the pressure balanced.

3.3.2 Inspection of Abnormal Changes of Test Curves Resulted from Wellhead Pressure Equilibrium Point p_w and Fluid Level Equilibrium Point p_L by Real Field Test Data

Example 1: Data examination and analysis of the injection/falloff test of Well WS-1

Well WS-1 is an appraisal well for coalbed methane exploration. After the well is completed, an injection-falloff test is conducted. To improve the bottom–hole condition, fracturing is then implemented and the injection/falloff test is done again after fracturing.

1. Data examination and analysis of injection/falloff test after fracturing

Well WS-1 is tested with Baker's downhole tools, but the well is shut in at the wellhead. Figure 7.6 shows the test curve.

The locations of wellhead pressure equilibrium point p_w and fluid level pressure equilibrium point p_L can be seen clearly in Figure 7.6.

Point A is the pressure of the fluid column in the casing recorded when the test instrument is run into the well and when water is poured in from the water tank. Point B is the pressured recorded when the test instrument is run into the bottom hole.

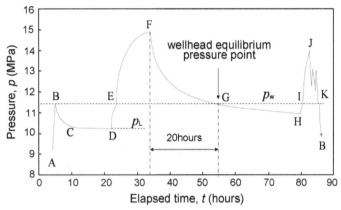

Figure 7.6 Pressure curve of injection/falloff test of Well WS-1 after fracturing.

Because the water in the tubing is squeezed out continually when the test instrument is being run down in the well and the fluid level is kept at the wellhead, the pressure recorded at point B is wellhead equilibrium pressure p_W. Also, because the water in the wellbore is absorbed into the reservoirs, which causes the fluid level to lower, the pressure decreases from point B to point D, which is a lower balance level. At this moment, bottom–hole pressure is fluid level equilibrium pressure p_L.

Because the tubing is filled with water before water injection, the fluid level returns to the wellhead and the pressure rises to point E, which is equal to the pressure at point B or p_W. E—F is the water injection process, and F—H is the falloff process. After the time corresponding to point H, the in situ stress is measured; before measuring, water is poured into the tubing once more until the pressure reaches point I, which equals the pressure at points B and E, that is, equal to wellhead equilibrium pressure p_W. Similarly, the pressure returns back to point K, and equals p_W again after earth stress testing.

Draw a horizontal straight line through points B, E, I, and K, which intersects with the falloff curve at point G. It can be determined that the pressure at point G is equal to p_W. The time from point G to point F is $\Delta t \approx 20$ hours. It can be seen from the log–log plot in Figure 7.7 that an abnormal change of the pressure derivative curve occurs just at this point corresponding to point G.

2. Data examination and analysis of pressure abnormal change point of Well WS-1 before fracturing

The test of Well WS-1 before fracturing is a typical example affected by p_L. As can be seen from the injection/falloff test history curve in Figure 7.8, the pressure reaches stable value p_L and maintains this level just as the same before water injection from the time t_Q corresponding to the point Q until that corresponding to point R, that is, the end of the falloff test. Thus, when $t > t_Q$, in the log–log plot pressure derivative curve does certainly go down and appears the abnormal change as shown in Figure 7.9.

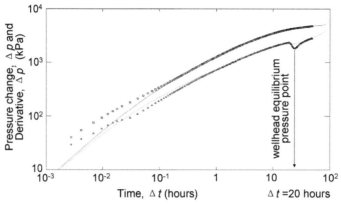

Figure 7.7 Diagram of pressure abnormal variation point in the testing curve of Well WS-1.

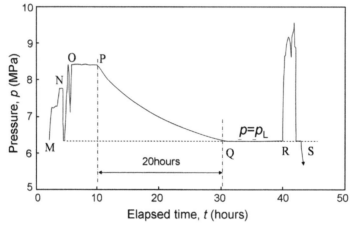

Figure 7.8 Pressure history of injection/falloff test of Well WS-1 before fracturing.

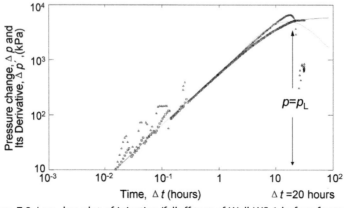

Figure 7.9 Log–log plot of injection/falloff test of Well WS-1 before fracturing.

Example 2: Data examination and analysis of injection/falloff test of Well WS-2

It can be seen from the injection/falloff test pressure history plot (Figure 7.10) that during the frequent well flowing and shutting in earth stress testing T-U, the pressure returns to the value that corresponds to wellhead pressure $p_H = 0$ every time. This is just the point of the wellhead pressure equilibrium point; at this point the bottom-hole pressure $p = p_w$.

Draw the horizontal straight line of $p = p_w$, which intersects with the pressure falloff curve at point R. Thus, the time difference between point R and well shut-in point Q is obtained at approximately $\Delta t \approx 10$ hours. And this time is just the pressure abnormal change point of the pressure derivative curve on the log–log plot (Figure 7.11).

The pressure derivative curve goes up suddenly on the log–log plot from the time corresponding to point R, and its slope is 1, just the characteristics of the flow blocking boundary (Figure 7.11). However, it is known from evaluation and analysis that such characteristics of the pressure derivative curve are not reflections of the formation variation at all. If it is interpreted as the reflection of flow blocking boundary incorrectly, the real formation conditions cannot be revealed. In fact, similar mistakes are found in some published papers.

3.3.3 A Few Comments on Data Examination and Analysis

It can be concluded from the aforementioned data examination and analysis of the injection/falloff test in coalbed methane wells that

1. Wellhead pressure equilibrium point p_w and fluid level equilibrium pressure point p_L are the major factors causing a pressure abnormal change in an injection/falloff test in coalbed methane wells. Abnormal pressure data cannot be used for well test interpretation in a coalbed.

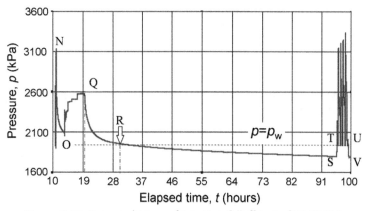

Figure 7.10 Pressure history of injection/falloff test of Well WS-2.

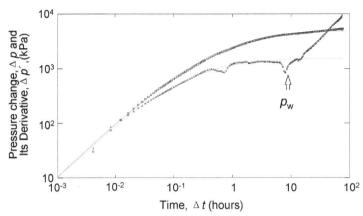

Figure 7.11 Log–log plot with abnormal pressure change of injection/falloff test of Well WS-2.

2. Data examination and analysis are effective ways to determine the values of p_w and p_L and to judge the quality of pressure data. An interpreter should find out every turning point on the acquired pressure curves and the corresponding time by comparing the operating steps during the test process recorded in test operation reports, find out the time when the pressure reaches p_w and p_L, and find out the factors causing the pressure abnormal changes. Also, during the test the wellhead pressure must be monitored carefully and continuously for acquiring the time when wellhead pressure drops to 0, that is, the time the pressure reaches p_w.

3. Downhole shut in is the best way to eliminate the influence of the value of p_w. When the well is shut in at downhole, the wellbore storage coefficient C is very small and the phenomenon of variable fluid level recovery will not occur so that the pressure will not reach p_w.

4. Other accidental factors in operation, such as leakage of any one of the testing tools, repeated well flowing and shutting in caused by loose valves, operation interference of neighboring wells, or pressure gauge failure, can all result in an abnormal change of pressure data.

4 ANALYSIS AND INTERPRETATION OF INJECTION/FALLOFF TEST DATA OF COALBED METHANE WELLS

4.1 Interpretation Methods

4.1.1 Model Types

As introduced in Chapter 2, data of the injection/falloff test in coalbed methane wells belong to the first or second models in the seven well test models, and data of both these models can be analyzed effectively with the conventional well test interpretation software used for conventional oil, water, or gas wells. In addition, typical well test models and the related graph analysis principles are also applicable.

Because most of the tested coalbed wells are exploration wells or development appraisal wells, the test duration is short, and the influence range is small, most of the tested coalbeds show the characteristics of homogeneous reservoirs or homogeneous fractured reservoirs. These two well test models are listed as M-1 and M-4 in Chapter 5.

4.1.2 Data Interpretation Procedure

Based on experiences obtained in field data analysis, analysis and interpretation of injection/falloff test data should follow the procedures given here.

1. Data examination, analysis, and preprocessing of data

    The examination and analysis methods are introduced in Section 3 of this chapter. Delete abnormal data caused by the faults of testing technology confirmed by examination and analysis and then process acquired pressure data by interpretation software.

2. Type curve match analysis

3. Verification of pressure history match

    Different from a conventional oil or gas well test, during verification of a pressure history match in some stimulated wells, the influence of the fracturing and the fluid drainage after fracturing on the reservoir pressure, as well as the influence of the injection during the testing operation, should be taken into consideration.

    A real field example is shown here to illustrate the significance of such an influence in well test analysis.

4.2 Real Field Example

4.2.1 Well Ex 1—A Coalbed Methane Well Completed with Fracturing

Figure 7.12 shows the flow rate history during fracturing and injecting in Well Ex 1. It can be seen from Figure 7.12 that

- During the injection/falloff test, 3.97 m^3 of fresh water is injected and the reduced rate is 7.92 m^3/day, which is a very small volume
- Before the injection/falloff test, fracturing treatment was conducted, 689.1 m^3 of fluid was injected, and the reduced rate was 8000 m^3/day
- After fracturing, 172.72 m^3 of fluid is flown back, and the reduced rate was 50 m^3/day

Clearly, if flow rate changes during the injection test are only taken into consideration in well test interpretation, a large error would occur in the interpreted results. The result of such an interpretation that only considers the flow rate history in the injection test is shown in Figure 7.13.

Figure 7.13 shows that a large deviation exists between theoretical and measured pressure curves in the injection period, although the log—log pressure curve

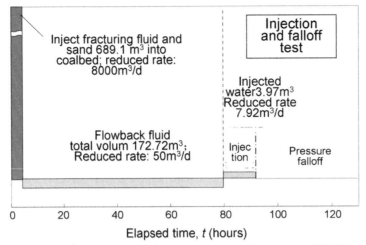

Figure 7.12 Flow rate history during fracturing and injection of Well Ex 1.

match can roughly meet the requirement of analysis. Therefore, the result is not reliable.

The results of interpretation in which the whole flow rate history during the fracturing and injection indicated in Figure 7.12 is taken into consideration are shown in Figure 7.14.

Clearly, the interpretation results are reliable, which are verified and proved by both very good type curve matching and very good pressure history matching. The graph type is the same as the typical model graph M-4 discussed in Chapter 5.

The final calculated parameters are as follow:
- Permeability of the coalbed $k = 4.495$ mD
- Skin factor $S = -5.84$
- Half-length of fracture $x_f = 42.8$ meters

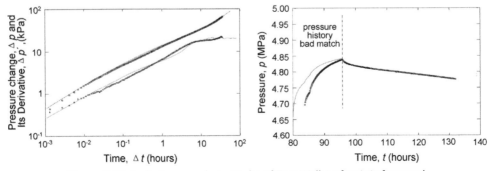

Figure 7.13 Preliminary analysis results of Test Well 1 after it is fractured.

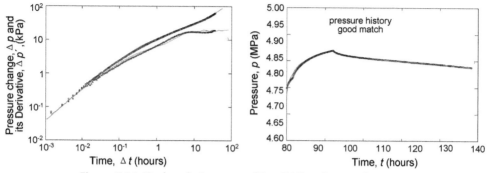

Figure 7.14 Final analysis curves of Test Well 1 after it is fractured.

- Reservoir pressure $p = 4.554$ MPa
- Wellbore storage coefficient $C = 0.007$ m^3/MPa
- Influence radius $r_i = 121$ meters

4.2.2 Well Ex 2—A Perforated Completion Coalbed Methane Well

1. Figure 7.15 illustrates the testing operation. This well is tested after perforated completion. There are five coalbeds of Shanxi formation in the well. The total thickness of the five beds is 9.6 meters. Data are well acquired.
2. Analysis by interpretation software. As judged from the log–log plot, coalbeds are a homogeneous medium, consistent with the M-1 model graph in Chapter 5.

The log–log plots are matched quite well in the type curve analysis by the well test interpretation software, as shown in Figure 7.16. The obtained excellent pressure history match verification is shown in Figure 7.17.

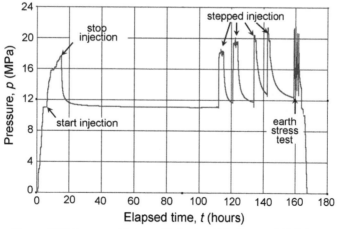

Figure 7.15 Pressure history of injection/falloff test of Well Ex 2.

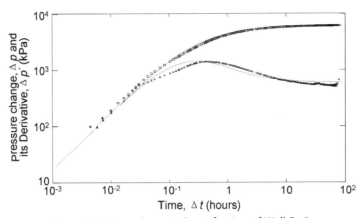

Figure 7.16 Log—log match verification of Well Ex 2.

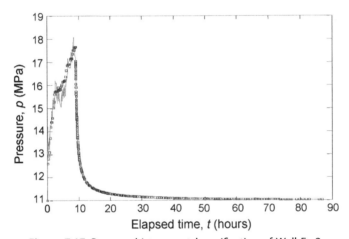

Figure 7.17 Pressure history match verification of Well Ex 2.

From Figure 7.17 it can be observed that the model curve and measured data are matched very well. The interpreted parameters are as follow:

- Coalbed pressure $p_R = 11.006$ MPa
- Permeability of the coalbed $k = 0.078$ mD
- Skin factor $S = -0.2$
- Influence radius $r_i = 44$ meters
- Wellbore storage coefficient $C = 0.002$ m^3/MPa

5 SUMMARY

This chapter introduced the special problems and their solutions in coalbed methane well test analysis. Unlike the common natural gas layers, methane gas in a coalbed is not reserved in pores or fissures in the form of compressed gas, but is adsorbed on the surface of the micropores of the coalbed matrix. Hence, most coalbed methane wells do not produce gas when they are initially perforated; what is produced from the coal cleats (fissures) is basically water. This causes some special problems in the coalbed methane well test: even though named as gas wells, the well test is mostly a water well test during the exploration and development appraisal stages; although coalbeds have double structures, their test data do not demonstrate such double medium characteristics as those of common fissured reservoirs sometimes do.

The production process of coalbed methane wells, from water swabbing for reducing pressure to methane gas flowing out by desorption in local areas and then methane gas being put into production by desorption in a wide range, lasts several stages and a very long time spanning several years—over a decade and even decades. Each of the stages is not repeatable or reversible. However, the well test process only lasts a few days and is only a very short period in one of its production stages. Therefore, it is impossible to reproduce all stages of the production process in a well test. What can be done in a well test is to observe and analyze the reservoir characteristics under certain production conditions only.

Therefore, the coalbed methane well test is divided roughly into seven typical models based on different coalbed methane production stages and actual field demands. Among these seven models, two can be interpreted by well test interpretation software used for interpretation of the common oil and gas well test: one is dominated by water flow at the initial stage and the other is dominated by methane flow in the late period. The well test analysis of gas and water two-phase flow can be conducted by numerical well test software.

The analysis of the injection/falloff well test is the key task in the coalbed methane well test, and is the main subject discussed in this chapter as well. Many test data acquired by wellhead shut in are distorted seriously when pressure decreases to wellhead equilibrium pressure p_w or reaches fluid level equilibrium pressure p_L. If the inherent reasons cannot be found, combined with what happened in the operation process, a completely wrong interpretation will be made. Such cases are not rare in the past. The "data examination and analysis method" introduced in this chapter, targeting injection/falloff well test data, can eliminate such mistakes. If the relevant factors are taken into consideration in the test design, such abnormal changes may be avoided.

This chapter gave analysis and interpretation of some typical field examples.

CHAPTER 8

Production Test in Gas Field and Dynamic Description of Gas Reservoir

Contents

Dynamic Well Testing in Petroleum Exploration and Development   © 2013 Petroleum Industry Press. Published by Elsevier Inc.
ISBN 978-0-12-397161-6, http://dx.doi.org/10.1016/B978-0-12-397161-6.00008-5   All rights reserved.   **527**

Methods of thoroughly studying a gas well, gas zone, and gas reservoir through research on the dynamic characteristics of gas wells have been illustrated from many aspects in the previous chapters, they are:

1. Comprehensive introduction of dynamic research and new idea of dynamic description research of a gas reservoir in different stages of gas field development with the well test method (Chapter 1).

2. Measure the original reservoir pressure by means of a gas well test so as to ascertain the uncompartmentalization conditions of natural gas reserving in the reservoirs, acquisition methods of dynamic reservoir formation pressure during gas production, and research methods of pressure distribution in areas of the gas field (Chapter 4).

3. Ascertain the initial deliverability, dynamic deliverability, relevant procedures, and analysis approaches about the deliverability test and the geological and completional factors affecting gas well deliverability. In particular, the stable point LIT deliverability equation for vertical wells and horizontal wells is highlighted (Chapter 3).

4. Calculate the parameters of near-wellbore formation (i.e., permeability k; double porosity parameters ω and λ) and well completion quality parameters (i.e., skin factor S; fracture parameters F_{CD}, x_f, and S_f), and primarily establish the dynamic reservoir model for near-wellbore formation (Chapters 5 and 7).

5. Identify connectivity among different wells or different zones in a reservoir through interference test and pulse test, and calculate connectivity parameters of formations between tested wells (Chapter 6).

By these tests and analyses a relatively thorough understanding of the wells and the reservoirs the wells locate in different aspects has been achieved, but neither overall and comprehensive understanding nor a comprehensive description of gas reservoir has been obtained, and many key questions such as how gas reservoirs the wells locate and their outer boundaries distribute in planar and vertically; how much their recoverable dynamic reserves and their stable gas production are; how long their stable gas production periods are; and how much gas can be produced during their stable gas production periods cannot be answered accurately.

How to answer these questions? The only effective way is to perform a production test in representative gas wells drilled in the gas reservoirs and to establish a "perfect dynamic models" of the reservoirs controlled by the wells and the whole gas reservoirs by analysis of acquired dynamic performance data during the production test. This method can supplementary modify the description of the internal reservoir conditions combining static geological data and can provide a dynamic prediction function by the dynamic model expressed by partial differential equations virtually. In other words, it is able to display not only the static physical properties of the reservoirs in the interpretation results, but also a series of crucial gas field production indices acquired from gas wells in production, such as how bottom-hole pressure and formation pressure decline, how pressure distribution in the formations varies with time, and which parts of the reservoir where the reserves may be remained subsurface and fail to be produced the during natural flow period.

This chapter introduces dynamic description methods for gas reservoirs and successful field examples.

1 PRODUCTION TEST IN SPECIAL LITHOLOGIC GAS FIELDS IN CHINA

A gas field production test means that some gas wells in a gas field that have not fully put into development are selected and tested gas production first in the preparatory stage of field development for learning problems probably met in the field after putting formally into development and finding out the methods to solve those problems from monitored detailed production performance data. Because the production test of a well lasts much longer than the initial well test, the influence range of production of the well, and so the influence range of the caused pressure variation have extended widely in the formation, and so much of the information about formation, such as heterogeneity far away from the wellbore, no-flow boundary distribution, and size of fissured zone , is reflected in pressure variation data, and therefore much reservoir information is consequently provided.

1.1 Special Lithologic Gas Field in China

Many of the gas fields discovered and put into development or not formally put into development but a lot of human and financial resources spent on their development in preceding studies in recent years in China are acknowledged as special lithologic gas fields. It is true that there is no final conclusion yet on geological conditions for the classification of special lithologic gas fields in academic circles, but it is also true that many very special problems difficult to solve with conventional development theories are discovered in field practice in these gas fields.

As for development program design methods illustrated in common textbooks, they basically aim at homogeneous sandstone formations and consider that the reservoirs extend homogeneously to a broad lateral gas-bearing area, a line well pattern or lattice well pattern can be adopted, controllable area and reserves of each single well are determined in terms of well spacing density, different development indices can be calculated with general numerical simulation software, and the gas flow rates of each single well and the whole gas field can be determined in accordance with prescribed values of related industrial standards and that is all.

However, field practices are quite different. The following are some typical examples.

1. QMQ gas field

More than 10 wells were completed during the early development of this field, and their deliverability varies from one well to another. Among them is Well BS-7, which has the highest deliverability, its daily gas output in the initial period was as high as several million cubic meters per day, and its original oil in place(OGDP) within its controlled geological area was more than 10 billion cubic meters; however, its flowing pressure and wellhead flow rate declined quickly during the production test period and even could not maintain stationary flow in the back-pressure test. Based on a thorough study of the well geological environment, this gas field penetrated the buried hill carbonate reservoirs, natural gas storage space in it is a complex system of fissure — solution cavity, the OGDP obtained from static reserves calculation methods cannot reflect the real field conditions at all, and such a gas field should be undoubtedly classified into a special lithologic gas field.

2. SLG gas field

The Neopaleozoic gas fields in the Ordose Basin are fluvial sedimentary lithologic gas fields, and the SLG gas field is a representative of one of them. During the evaluation of field reserves, a reservoir prediction with geophysical methods was applied, the boundaries of sandstone reservoirs were delineated roughly, gross width and thickness of the reservoirs were determined, several hundred billion cubic meters of OGDP was estimated by the volumetric method, uniform well spacing with a line drive well pattern was adopted, and very good calculated development indices were finally obtained with conventional numerical simulation methods.

However, it was discovered after a sedimentary facies study that the sandstone reservoirs seemed to be widespread are not as simple as they looked; in fact, the effective reservoirs controlling natural gas storage are mainly thin sandstone of fluvial plain subfacies, or the sandstones of residual sand body on narrow and long meandering river bed, or scattered point bars; and they are characterized by small gas reserves stored in each effective single sandstone, being cut by impermeable mudstones, different geological ages, and none of vertically mutual connectivity. These were proved later by production tests in many exploration wells. It was confirmed by dynamic reservoir description of the wells that the area of the block controlled by each well is small—its dynamic reserves was less than 20 million cubic meters. Apparently fields like these are of special lithologic gas fields.

3. Volcanic gas field

Extensive distribution of a gas reservoir in volcanic rocks was found in China in the past several years. Thanks to fissure development in eruptive facies volcanic rock mass, a considerable gas flow rate was obtained from some exploration wells in the gas well test; these types of reservoirs are defined as important targets in petroleum exploration and development.

However, the structure of eruptive facies volcanic reservoirs is very complicated; there are vertically distributed channel facies vocalic crater rock mass, which is like a funnel, accumulation of explosive facies volcanic rock splashdown and debris flow and rhyolite formed from coagulation of effusive facies magma, etc. Gas fields in this kind of reservoir are undoubtedly acknowledged special lithologic gas fields.

In addition to these commonly accepted typical types, some other gas fields are considered as special lithologic gas fields due to their special geological structures; among them are oolitic shoal limestone gas fields in eastern Sichuan, fissured gas fields in the Lower Paleozoic being cut by delves in the Ordose Basin, faulted block sandstone fields in eastern China, and offshore shoal and sandbar gas fields on the mud diapir structure belt in Yingehai Depression.

1.2 Production Test: An Effective Way to Solve Problems in Development of Special Lithologic Gas Reservoirs

1.2.1 Problems in Development of Special Lithologic Gas Fields

A series of difficult problems have been encountered during the development of the aforementioned gas fields with special geological structures.

1. Difficulty in determining initial deliverability of gas wells.

Usually the gas deliverability of single wells over the whole field can be known as only a few selected appraisal wells are needed to conduct a normal deliverability test in a gas field in a homogeneous sandstone formation. Design of the development program depends mainly on numerical simulation to calculate a series of development indices, and the used geological model is established from the contour diagrams of permeability k and effective thickness h based on logging data.

However, the aforementioned method for gas fields in homogeneous sandstone formations is completely invalid for special lithologic gas fields. The gas deliverability of production wells in an area enclosed by effective boundaries delineated by geological research and drilled according to the design by the aforementioned methods varies greatly, a large majority of those wells have no industrial deliverability, and it is hard to determine the industrial deliverability of other gas producing wells with conventional methods.

2. Stability of gas well deliverability is unpredictable.

Many gas wells in special lithologic reservoirs are proved to have an industrial production rate, some even with an absolute open flow potential (AOFP) as high as several million cubic meters by the well test, but their productivity declines very rapidly; the wellhead pressure of some wells declines so much that pipe transmission is unable to maintain only in one or two years or even in several months, and normal industrial production of them has to be ended.

3. The cumulative production of a single well is very low.

The productivity of some gas wells in special lithologic reservoirs declines very rapidly and the production life of them is extremely short, resulting in extremely low cumulative production. Take gas wells in the SLG gas field for instance, the cumulative production of single wells averages less than 10 million cubic meters. This means many new wells have to be drilled uninterruptedly during the gas development process, which decreases the benefit of gas development dramatically.

1.2.2 Protection Test is the Only Effective Approach of Gas Reservoir Research
New Connotation of Gas Field Production Test

With regard to various special lithologic gas fields, the problems mentioned earlier cannot be solved effectively at all under the present technical conditions relying only on the static geological research methods, the reservoir prediction based on geophysical prospecting, and the estimation of formation parameter distribution obtained from logging and geological research. What is the effective research approach to study such gas reservoirs in the preparatory stage of field development? The only answer to this question is the production test.

The gas field production test is not new. The method was introduced in some monographs on field development in the mid-20th century and was also stipulated as a measure in the oil exploration and development industry standards of China (e.g., Well Test and Production Test Standard, SY/T 5981, SY/T 6171). Here it comes up for discussion again due to the new connotation contained in it.

1. Pertinent production test design.

The previous production test contained only simple contents, centralizing on the requirements for the production test procedure from the aspect of operational process, and involving too less even no consideration in data acquisition and analysis for gas reservoir study. It once happened in fields that, although several hundred million cubic meters of natural gas had been produced from some production test wells, bottom-hole pressure, even wellhead pressure, has never been measured, so that no useful information for a reservoir study can be provided.

It is required that a perfect design must be made prior to the production test for new research on the gas dynamic reservoir description; this design puts forward the planned gas flow rate during the production test, proposes overall requirements for data acquisition centering at bottom-hole pressure, and simulates and matches the whole process pressure history with well test design software to provide complete basic data for the follow-up dynamic analysis.

2. Rigorous and continuous data acquisition.

Data required to be acquired during the production test include:

(1) Precise flow rate metering: it involves accurate recording of the gas flow rate during the deliverability test and also accurate and uninterrupted recording of the gas flow rate throughout the whole extended production test period. A detailed daily gas flow rate can be recorded continuously with the aid of electronic metering instruments; with common metering instruments, exact values of an average daily gas flow rate should be provided at least.

(2) Precise bottom-hole pressure records: an electronic pressure gauge is used to record bottom-hole pressure data continuously during deliverability and pressure buildup tests; whereas for production test stages, the point measurement method can be considered to record historical variations of flowing pressure. In an offshore gas field of China, good quality pressure history data have been acquired and recorded uninterruptedly with permanent electronic pressure gauges installed at the bottom-hole of horizontal wells in recent years so that very good results have been obtained from research of dynamic gas reservoir description.

(3) Sample analysis: sampling and analysis of natural gas produced from the well should be conducted regularly to identify physical components and their changes with time; water sampling and analysis should also be conducted to make clear the water cut and the change of water properties.

Dynamic Gas Reservoir Description Centering on Gas Well Deliverability

Some new developments of gas reservoir studies, namely the new idea for dynamic description of gas wells and gas reservoirs centering on the evaluation of gas well deliverability, were particularly introduced in Chapter 1. Data needed for doing such studies are flow rate and pressure acquired during the production test. Due to the fact that the production test lasts a long time, a great deal of natural gas is produced from the reservoir, and overall scanning of effecting reservoir controlled by the gas well is accomplished during percolation of the natural gas and by which much information about the reservoir characteristics is brought so that it is possible to obtain the deliverability and its stability of the gas well, many other parameters such as the permeability of the area near the wellbore, completion parameters of gas wells, and boundary parameters and producing reserves of gas blocks by thorough analysis of these information, and all of these enable engineers to have an overall understandings of a reservoir, to establish mathematic equations and dynamic models describing this reservoir, and subsequently to verify the reliability of the analysis results and predict the dynamic trend of the reservoir in the future.

1.3 Procedure of Production Test in Gas Wells

According to gas field development experiences in China in recent years and by referring to foreign literature and information, selected gas wells for the production test should be typical and representative in the reservoir. That is, results of the production test in those selected wells can surely answer the key questions that will most likely be encountered in the follow-up formal development of the whole gas field. The procedure of the production wells is basically shown in Figure 8.1.

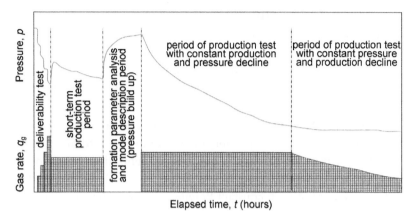

Figure 8.1 Procedure of production test.

1.3.1 Arrangement for Production Rate of Gas Test Wells in Different Periods

Figure 8.1 shows a general period arrangement for the production test. This arrangement is made to meet the needs of a reservoir dynamic study and to adapt to the rule of gas deliverability change of gas wells. The arrangement at different periods of production test is illustrated as follows.

1. Deliverability test period. Normative methods are usually chosen at this period to verify gas deliverability during the initial production period of gas wells or to establish the stable point LIT deliverability equation with the selected stable production point in the initial flowing hours. The gas flow rate in different production test periods will be further estimated and arranged on the basis of deliverability analysis results.

2. Short-term production test period. Stable production with a constant rate for about a month is arranged for learning primarily about the stability of gas well deliverability. As for wells conducted on a modified isochronal test, a short-term production test sometimes follows an extended test. Due to a long producing duration, relatively stable flow rate, and wide influence range, this period will prepare very good conditions for data acquisition and dynamic model establishment in the next step of the pressure buildup test.

3. Pressure buildup test period. This is the main period when the source of data acquired for reservoir dynamic model preliminary establishing, that is, a dynamic model of the gas well can be initially constructed based on the interpretation results of reservoir parameters, completion quality parameters, heterogeneity of the formation near-wellbore, and boundary distribution extracted from acquired data in this period.

4. Period of production test with constant flow rate. This is the most important period of the whole production test. The duration of this period depends on actual gas reservoir conditions. Some semideep gas wells can keep their bottom-hole flowing pressure very stable during producing with a controlled constant rate; their flowing

pressure declines less than 1 MPa with every 10 million m^3 of natural gas produced. The controlled reserves of such a well must be more than 1 billion m^3; the reservoir these wells locate can be considered as homogeneous formations when designing its development program. In some production test wells, flowing pressure declines very rapidly, they almost cease the flow or their wellhead pressure has been lower than export pressure after only 10 million m^3 of natural gas was produced, and controlled area and dynamic reserves of one well can be basically estimated accordingly. Also, the condition of some other wells is possibly between the two extreme ones mentioned previously, the formation boundaries cannot be recognized in a short duration, and the production test should last a rather long time, even as long as more than a year.

5. Period of production test with constant flowing pressure. The well enters the period of production test with constant pressure and production decline when its wellhead pressure declines to less than export pressure and the gas flow rate has to be decreased gradually to keep on production testing.

Sometimes some wells will have the production test extended to a blow–down period depending on practical conditions in the field.

1.3.2 A Typical Pressure History of Gas Production Test Wells in Field
A real production test well is given in Figure 8.2 to show its test procedure and pressure history.

Figure 8.2 shows that the whole production test lasts 3 years and can be divided into six sections (I—VI) illustrated as follows.

Modified Isochronal Test Period
The deliverability test was performed from the end of March to May in 2001, during which sections Iand II are transient and extended deliverability test periods of the

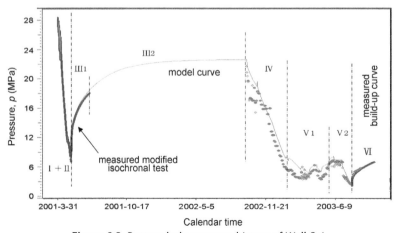

Figure 8.2 Bottom-hole pressure history of Well S-4.

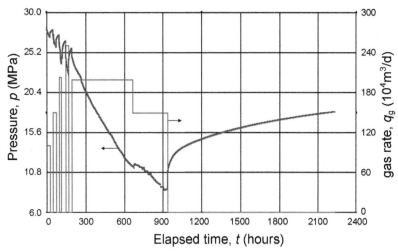

Figure 8.3 Gas flow rate and pressure history during deliverability test of Well S-4.

modified isochronal test. In this period, high-precision downhole pressure gauges are applied to continuously record bottom-hole pressure, simultaneously detailed variation of gas flow rate is measured, and both are shown in Figure 8.3.

Pressure Buildup Test Period

Section III in Figure 8.2 is the pressure buildup test period, which lasts 14 months due to surface construction for production test, in which III1 is measured bottom-hole pressure variation with high precision electronic pressure gauges; III2 is the succeeding section of the measured pressure buildup—bottom-hole pressure failed to be measured in this period. After section III2 ends, the well was put into production in September 2002 and wellhead tubing and casing pressures have both been monitored since then.

Long-term Production Test Period

A long-term production test of Well S-4 lasted for a year from September 2002 to September 2003. Daily gas, water and oil flow rates, and wellhead tubing and casing pressures were monitored continuously during the test, and detailed dynamic variation data are shown in Figure 8.4.

Figure 8.4 shows clearly that the long-term production test of Well S-4 can be divided into three periods.

1. Period of production test under constant flow rate (section IV). During this period, the gas flow rate was basically 6×10^4 m^3/day, and wellhead tubing and casing pressures declined sharply from 17 to 6 MPa and reached the lower export pressure limit; this period ended then.

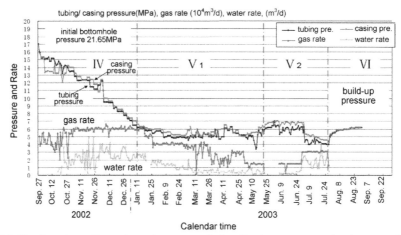

Figure 8.4 Monitored gas flow rate and pressure during the long-term production test of Well S-4.

2. Period of production test under constant flowing pressure (section V). Since the wellhead pressure had decreased to the lower limit of export pressure at the end of the period of the production test with constant production, the following production test at a lower gas flow rate had to be chosen to maintain the lowest export pressure and keep the gas well natural flowing. This period consisted of two sections of V1 and V2: the former was a pressure buildup test begun at the end of May, the well was shut in for the gas flow rate had decreased down to less than $1.5 \times 10^4 \, \mathrm{m^3}$/day; and the latter was producing the well again when the shut-in wellhead pressure had risen a little, the gas flow rate was elevated to $3 \times 10^4 \, \mathrm{m^3}$/day; but the tubing pressure declined rapidly to 4 MPa and the well had to be shut down.

3. Final pressure buildup period (section VI). Thanks to plenty of dynamic data acquired from the six sections, explicit understandings had been obtained for such well conditions as reservoir structure, boundary distribution, and controlled dynamic reserves of the well. Detailed discussions about the dynamic reservoir description with these data are given later.

Actually, with the long-term production test finished in 2003, the well kept extended producing in an open flow state intermittently until 2006 and then the well was adjusted to intermittent producing, as shown in Figure 8.5.

The production test process of Well S-4 shown in Figure 8.5 is the whole process of a gas well from its putting into production to depletion. The monitoring of flow rates and pressures, including bottom-hole pressure and wellhead pressure, was carried out throughout the test, based on which a complete dynamic model of the reservoir controlled by the well can be established, and the validity of the model can be traced and verified continuously and the future dynamic trend predicted continuously. In fact, it has been done by combining with application of the numerical well test.

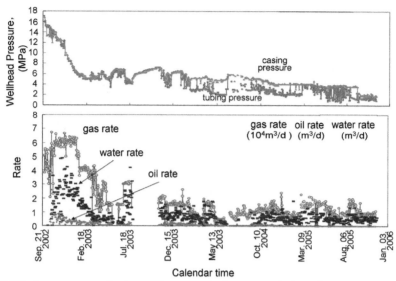

Figure 8.5 Whole process monitoring of flow rates and pressure during a long-term production test in Well S-4.

1.4 Dynamic Reservoir Description Based on Production Test Data of Gas Wells

A new method of dynamic gas reservoir description was illustrated in detail in Chapter 1. Figure 1.4 showed the processing flow chart of the method. Field examples of dynamic analysis in this book are all conducted based on this train of thought. This train of thought has been formed and improved gradually and continuously in recent years, earlier studies and operations onsite were quite simple, and it became clearer and clearer as the study went on that each step of its flow chart was substantiated gradually and became the guide of present studies consequently.

The dynamic description, combined with the gas well production test, commences from completion of the first exploration well and runs throughout the whole exploration and development process, particularly in the development preparatory stage; all research works are basically focused on this dynamic description.

Dynamic analysis is closely integrated with geological study. If there is no such connection and geological conditions of the research targets have not been made clear, the interpretation result will be confusing because it is not unique. However, dynamic study methods are irreplaceable in some aspects. They enable the discovery and thorough understanding of what cannot be observed in the geological research, and the quantitative analysis to be made for each study target, so as to complement each other to reach comprehensive understanding of the gas reservoir.

The well test study has been developed to a higher level at present. Due to the development of modern well test theories, continuous updating of test and analysis methods, and application of high precision electronic pressure gauge and well test interpretation software, it can absolutely play a key role in the dynamic description of gas reservoirs. In several key gas fields discovered and developed in China in recent years, the dynamic description method has been applied to some extent and many valuable experiences and lessons have accumulated, which have proven invaluable. Also, it is believed that if we can learn something from which not only gas field development can be accelerated, but also the development benefits can be enhanced.

When discussing the analysis methods in the former chapters, many field examples of well test analysis and application during exploration and development were introduced. These examples will be used again during comprehensively introducing further the research procedure, data acquisition requirement, obtained understandings, and the roles the study results paid in this chapter.

2 DYNAMIC GAS RESERVOIR DESCRIPTION IN DEVELOPMENT PREPARATORY STAGE OF JB GAS FIELD

2.1 Geological Conditions of JB Gas Field

The marine limestone of the Ordovician Majiagou Formation was penetrated during drilling exploration of the JB gas field. The reservoir is a low-dip monocline that is high in the northeast and low in the southwest and in which there are a series of low-amplitude nose-like structures developed in the northeast to southwest direction. There are erosive delves formed by tides in the marginal zones of the structure, and the formation of Mawu-$_{1-2}$, the major pay zone, is cut by 10 main delves, which caused its absence and formed gas flow barriers. In addition, dome and basin structures develop in some areas of the field. Generally, although the reservoir is large and distributed in the area completely, some no-flow boundaries with huge lateral changes and deep penetration into the gas reservoir exist, which brings great uncertainty into future field development. Moreover, related geological research results show that the pay zones are thin, the permeability and porosity are both extremely low, reservoir parameters among the well are greatly different, and no gas is produced at all during well testing in some exploration wells (e.g., Well S-23) in some parts of the gas zones, which shows that geological research results unable to resolve the problems possibly meet in the prediction of various indices in future field development.

2.2 Focuses of the Problems

As the large uncompartmentalized gas field first discovered in China, the following questions must be answered before developing:

1. Whether the reservoirs are distributed continuously and whether the reserves calculated with the static volumetric method are reliable

2. What are the exact deliverability and maximum deliverability of a gas well, and whether its production can be stable

3. How much are dynamic reserves controlled by different wells in different areas

4. How to divide the development unit, to set the parameter field for numerical simulation and well spacing and well pattern in the design of the development program

If these questions cannot be answered in time correctly, then the reserves cannot be examined and approved by the related national administration, and the initiation and implementation of the gas field development will be influenced consequently.

2.3 Dynamic Study at the Preparatory Stage of Gas Field Development

The research on field development started very early in the JB gas field. A comprehensive production test and dynamic analysis were conducted for 26 exploration wells and some development and appraisal wells in the main part of the field from 1991 to 1996, and their results answered practically the aforementioned key problems. The practice after the gas fields were put into production and commenced gas supply to Beijing proved that the initial understandings of the field were in time and correct and also has been highly appraised by postevaluation of a CNPC specialist team.

However, the JB gas field was developed as early as in the 1990s when the gas reservoir dynamic description was still at the exploratory stage; because the study work only dealt with the analysis and study of short-term production test data, established dynamic reservoir models were of the primary level and improvements needed to be done. Also, because the study did not combine with the pressure history during the subsequent production of these wells for a further tracing study of those models, research results could not be supplied about the shape, area, and dynamic reserves of the areas controlled by the wells or gave any prediction of the dynamic trend of the main gas producing wells in the JB gas field.

2.3.1 Short-term Production Test Integrating Deliverability and Pressure Buildup Tests

A modified isochronal deliverability test with an extended short-term production test was conducted in the JB gas field first in China. Reservoir features are concluded as follow on the basis of related tests in 26 early appraisal wells.

1. The initial AOFP of the wells was estimated by means of a modified isochronal deliverability test

2. The conditions of stable deliverability of the gas wells were primarily made clear through extended production tests following the transient deliverability test

3. A primary characteristics evaluation of the reservoir model in the area near the wellbore was done based on the pressure buildup test following the extended

production test; the obtained reservoir features included structural features, reservoir extension, boundary distribution and their features, reservoir parameters, and completion quality

4. Dynamic models of single wells obtained in the preliminary evaluation were verified by the pressure history match of single wells during the short-term production test and were primarily validated

As emphasized repeatedly in this book, the purpose of the well test is to describe gas reservoirs and gas wells dynamically. Based on the aforementioned studies (1–4), a dynamic model (theoretical model) of the well has been established by well test interpretation software and then verified primarily by a pressure history match in a short-term production test so that it can be used immediately to predict the short-term performance of the well under the conditions of different production rates. This is the dynamic prediction, including the prediction of stable production at different flow rates, of a single well.

The following field example is given for illustration.

Well S-62 is a gas well in the south of the JB gas field. The perforated interval of this well is $Mawu_1^1$–$Mawu\ _1^4$ containing all major pay zones. The pay zones are 13.6 meters thick and are composed of light gray fine or silt crystalline dolomite with developed fissures.

A three-stage test was adopted:

a. Four transient point modified isochronal test
b. Extend test, which is, in fact, a short-term production test
c. Final pressure buildup test

This arrangement was adopted for the dynamic test in all 26 wells. The obtained pressure history plot of the test is shown in Figure 8.6.

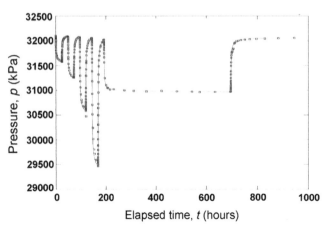

Figure 8.6 Observed pressure history of Well S-62.

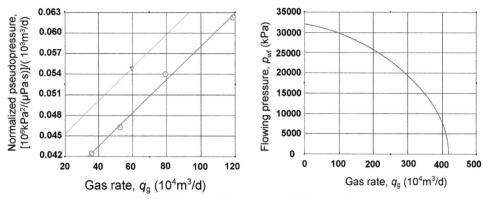

Figure 8.7 Deliverability curve and IPR of Well S-62.

Deliverability Analysis

The deliverability equation and AOFP (q_{AOF}) of this gas well can be obtained from data acquired in the first and second stages in Figure 8.6; the relationship between the gas flow rate and bottom-hole flowing pressure, that is, IPR, can be obtained as shown in Figure 8.7.

The AOFP calculated from Figure 8.7 is

$$q_{AOF} = 41.87 \times 10^4 m^3/day.$$

AOPFs of all 26 wells were analyzed with the same method, and the calculated AOFP values ranged from 8 to 65×10^4 m^3/day and averaged 28.86×10^4 m^3/day.

Deliverability data provided by the modified isochronal test became the main basis of production planning in gas field development.

Analysis of Deliverability Stability

As emphasized repeatedly in Chapter 3, the AOFP of a gas well is an instant gas deliverability index during the initial production period of the well. Gas well production stabilization is controlled by boundaries in the area near-wellbore and controllable dynamic reserves of the well. These indices can be determined comprehensively and concretely by means of a dynamic model study or made clear approximately from the performance of pressure history during the producing periods of the well.

It was discovered from the short-term production tests that there are three types of trends of pressure history of wells in the JB gas field, that is, stable type, basically stable type, and unstable type with rapid building up, as shown in Figure 8.8. Well S-62 is of the first type (a).

It shows that the deliverability stability of the gas wells in the JB gas field is quite good, which was proved by the subsequent practice of production.

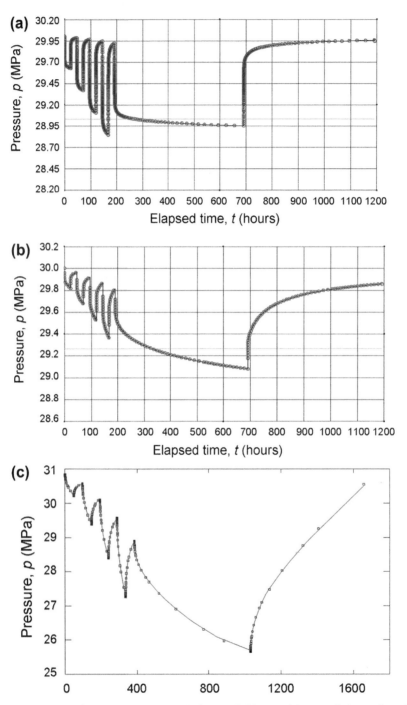

Figure 8.8 Three types of pressure history trend of JB gas field: (a) stable type, (b) basically stable type, and (c) unstable type with rapid buildup.

Features of Reservoir Model

The pressure buildup test conducted after the deliverability test provided the information for confirming the reservoir model.

As introduced in Chapter 5, a specified shape of a pressure log—log plot corresponds to a specified flow feature and is decided by the reservoir type. According to the analysis and interpretation of pressure log—log plots of the JB gas field by well test interpretation software, most wells present "composite formation" characteristics.

1. The reservoir where a number of high productivity wells locate like the composite formation in which the flow condition of the inner zone is better than that of the outer zone (model graph M -7) in Table 5.5—among these wells are Well S-5 (Figures 5.93 and 5.94), Well S-155 (Figures 5.95 and 5.96), Well S-81, Well S-52, Well S-84, Well S-100, and Well S-175.

2. The reservoir where some other wells locate like model graph M-8 in Table 5.5, representing composite formation in which the flow condition of the inner zone is worse than that of the outer zone—among these wells are Well S-13 (Figures 5.97 and 5.98) and Well S-62.

   To sum up, these two types of gas wells make up the majority of the 26 test wells. It means that major pay zones in the JB gas field are connected distributed and vary greatly, that is, are very heterogeneous.

3. The gas flow is limited by no-flow boundaries in the area around some gas wells near delves—among these wells are Well S-8 (Figures 5.104, 5.105, and 5.106) and Well G19-11.

4. Very few wells (e.g., Well S-6) show the characteristics of being in a constant-volume closed reservoir. The perforated interval of Well S-6 is Mawu-4, not a major pay zone. As realized by geological research, this pay zone is distributed locally. Both research results are absolutely consistent (Figure 5.109).

5. Group- and/or series-distributed fractures were encountered during drilling in the western marginal zones, and the trend of the pressure history of those wells is as shown as in Figure 8.8(c). Among those wells are Well S-71 and S-181; the pressure log—log plots of them are shown in Figures 5.129 and 5.130.

An explicit description was made primarily for the dynamic reservoir model of the JB gas field based on the analysis and study of short-term production test data and can be concluded as follows.

1. Laterally, the Ordovician major pay zones are distributed connectedly and with heterogeneous variation. If the gas wells were drilled in the area with a high kh value, they will have quite high deliverability, but kh values become slower in the outside areas of these wells, which bring a certain effect on production stabilization. Also, if the gas wells were drilled in the area with a low kh value, they will have low deliverability, but kh values very often become higher in the outside areas of these wells, which is very favorable for deliverability stabilization.

2. A delve plays a key role in deliverability limitation. Although wells located at the margin of a delve very often have quite high production, the no-flow boundaries nearby will affect the stability of the production. If the well is drilled deep into a delve or a buried pit, major pay zones will be absent, and it is possible that no gas flows at all from the well.

3. Small constant-volume gas blocks do exist, but are seldom seen in the gas zones.

The aforementioned model analysis laid the foundation for high and stabilized production of the JB gas field.

2.3.2 Interference Test Analysis

As an important method of determining the dynamic reservoir model, the interference test was introduced in detail and the JB gas field was a typical example of it, as discussed in Chapter 6.

Precious understandings for the gas reservoir description were obtained on the basis of the interference test of a well group made up of Well L-5, Well L-1, and Well SC-1 (observation wells), which lasted 10 months. It was confirmed that the major pay zones are well connecting in a large area of Ordovician limestone and seriously heterogeneous. The reservoir has somewhat double porosity medium features, but the ω value is quite large, indicating that natural gas storage and flow are both dominated by fissures (refer to Figures 6.38–6.41 and Table 6.2).

2.3.3 Dynamic Reserves Test in Well Block S-45

A dynamic reserves test was performed for a major block, the Well S-45 block of the JB gas field, in 2000. There is a gas-bearing area of 408 km^2, proved gas reserves of 256.76×10^8 m^3, and 28 production wells in this block. Identified by dynamic characteristics, the whole block is in the same pressure system.

The meaning of being in the same pressure system was illustrated explicitly in Chapter 4. A gas field being originally one pressure system may probably be composed of multiple gas reservoirs either superposed vertically or extended laterally. The initial formation pressures, by the regression analysis, are in a "natural gas static pressure gradient straight line" (refer to Figure 4.4). However, when gas wells in the field are producing, the formation pressures of the wells will generally decline; after producing for a period of time, the formation pressures of the wells will decrease a certain value, pressure data can be divided into several groups, and the initial pressure gradient line may "disperse" onto several different pressure gradient lines. Wells whose pressures are in the same group still are in the same static pressure gradient line and are also considered being in the same pressure system. The different groups are possibly isolated by a boundary or barrier or have only a very weak connection between each other, which is regarded as connected during gas migration but independent of each other in field development.

In the dynamic study of development of the JB gas field, such mutually relatively independent blocks were differentiated in accordance with the relations of pressure performance. Among them is Well S-45 block, the relatively independent one with the largest area and the richest reserves.

A dynamic reserves test of Well S-45 block started in April 2000. All 28 wells in the block were shut in simultaneously for a pressure buildup test, the pressure became stabilized 3 months later, and the pressures of different wells were much closed and averaged 29.5 MPa, 1.2 MPa less than the original pressure. The cumulative gas output was 5.0041×10^8 m^3 until then.

Results calculated from aforementioned data are as follow:

1. Gas production per unit of pressure drop of the block: 4.17×10^8 m^3/MPa
2. Average single-well gas production per unit of pressure drop: 0.149×10^8 m^3/MPa
3. Estimated average single-well cumulative gas production: 3.83×10^8 m^3
4. Estimated cumulative gas production of Well S-45 block: 107.2×10^8 m^3
5. Percentage of dynamic reserves in proven OGIP: 42%

It shows that even though part of the reserves is not yet controlled by drilled wells, the development effect of the block is fairly good. How to find out uncontrolled areas in the block is the next topic of dynamic study.

2.3.4 Static Pressure Gradient Analysis of JB Gas Field

During the stage of exploration and developing preparation of the JB gas field, a regression analysis of the well-measured initial static pressures of 44 gas wells was conducted, and the relationship between initial static pressure and depth was obtained:

$$p = 24.90 + 0.00193h.$$

This equation indicates pressure gradient value G_{Dh} is 0.00193 MPa/m, close to the pressure gradient of dry gas, and accordingly the field is identified as an uncompartmentalized gas field consisting of good connecting reservoirs.

All key questions just mentioned have been well answered by the results of the aforementioned dynamic evaluation and study done from 1991 to 1996, which paved the way for the JB gas field to pass the reserves examination and approval successfully and to be put into development smoothly. A postevaluation of reserves was approved in 2002, which was based on implementation of the development program of exploration wells in which the annual output was 1.2 billion cubic meters from 1997 to 1998 and the overall development program in which the annual output was 3 billion cubic meters in 2000. The postevaluation of reserves confirmed that the overall knowledge about the JB gas field was correct.

3 SHORT-TERM PRODUCTION TEST AND EVALUATION OF GAS RESERVOIR CHARACTERISTICS IN KL-2 GAS FIELD

The KL-2 gas field is a very large field with great potential and is one of the main gas resource areas for "west to east natural gas transmission project" in China as well. The research of such a gas field has been a focus for many departments concerned due to its great impact on smooth implementation of that macro project. In addition, new challenges are brought forward in the study of dynamic gas field description because of variations of well test data in this gas field. Various focal questions, which many departments concerned before the field was put into development, were eventually answered one by one on the basis of repeated and through analysis of short-term production test data obtained from several gas wells in the field.

Well test analysis of the KL-2 gas field gave rich information about the gas reservoir, the deliverability of gas wells, and evaluation of drilling and completion technology, and also made a comprehensive and complete description of the field.

3.1 Geological Condition

The KL-2 gas field is located in the middle northern margin of Kuche Depression and is a faulted anticline, 18 km long from east to west and 4 km wide from south to north. There are dozens of small faults inside the structure, and the fault throw ranges between 20 and 80 meters. The structural map of the field is shown in Figure 8.9.

The drilled gas wells penetrated the sandstone in the lower Tertiary and Cretaceous. Among them is the Bashijiqike Formation Group in Cretaceous, the major pay zone. The total length of the gas-bearing intervals is very long—it is about 500 meters in Well Kela-2 and among those is the 350-meter-long Bashijiqike Formation Group, and about 340 meters in Well Kela-201 and among those is the 273-meter-long Bashijiqike Formation Group. The interbeds within the intervals are generally thin, and the thicknesses of them are mostly less than 2 meters. Except those at the bottom of the Bashijiqike Formation Group, the interbeds have an extended length of 60—70 meters. Core analysis and logging interpretation results show that the porosity (ϕ) averages 12.6% and that the permeability (k) averages 49—53 mD with great variations in vertical distribution.

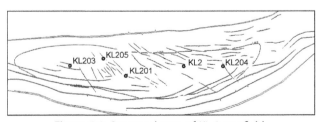

Figure 8.9 Structural map of KL-2 gas field.

It was confirmed from gas sample analysis that dry gas was produced from the field, and there is more than 97% of methane and very low sulfur in the content of the produced gas.

3.2 Procedure and Results of Well Test Analysis

The KL-2 gas field is a very deep and very thick field with abnormal high pressure; unprecedented difficulties and challenges were encountered in the study of this field with dynamic methods. However, thorough understandings of the internal conditions of the gas field were finally gained by gradually adjusting the testing and analysis methods and deepening the knowledge about the field step by step.

3.2.1 Test of Well KL-2

Understanding Gained From Testing Well KL-2

Well KL-2 was a discovery well that penetrated all gas-bearing intervals, which made a breakthrough in the exploration target. Fourteen intervals of this well were tested one by one, and the following understandings of dynamic characteristics of the reservoirs were obtained from those tests:

1. Almost all of the major tested intervals produced gas, confirming the existence of a gas pay zone with a huge thickness.
2. Every tested interval has a high deliverability, and the gas flow rates are as high as $500-700 \times 10^4 \, m^3/day$.
3. The reservoirs have very abnormal high formation pressures. Test data of each interval show that the pressure coefficient is over 2. It indicates that the gas layers have an extremely high potential energy.

Deficiency in the Testing of Well KL-2

However, the gas test failed to reveal completely and clearly the features of the reservoir. The main reason is that a method of zonal testing was adopted to test this gas reservoir with a huge thickness and a very good vertical connection. Mutual vertical interference of gas zones caused abnormal test curves and a "partial perforation" effect so that reservoir parameters cannot be calculated normally, and of which interval the deliverability (AOFP) obtained from the tests is cannot be recognized; therefore, the total deliverability of the well cannot be determined.

The tested intervals of Well KL-2 are listed in Figure 8.10.

As seen from Figure 8.10, for a tested gas well with a total length of intervals of 500 meters, the perforated interval is only 4 to 8 meters or even as thin as 1 to 2 meters in each; the test results are just like what is observed through a small aperture in a wall, it is very hard to get an overall picture. But as a gas well test of an exploration well, it is understandable to adopt such a conventional test method. Figure 8.11 shows typical pressure buildup curves obtained from the test.

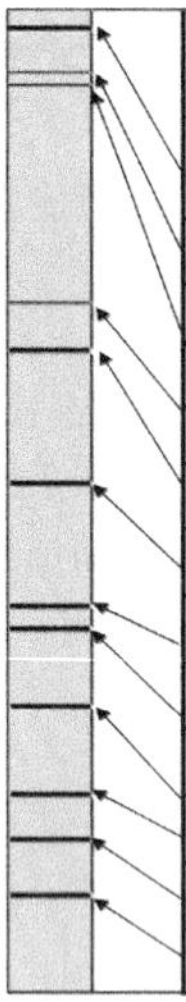

test No.	horizon	interval m-m	perforated thickness m	gas rate 10³m³/d	water rate m³/d
14	E	3567—3572	5	605.694	0
13	Cretaceous (K$_{bs}$)	3590—3591	1	664.329	0
12		3593.5—3595.5	2	652.037	0
11		3711—3713	2	553.503	0
10		3740—3750	10	717.145	0
9		3803—3809	6	682.065	0
8		3876—3884	8	0	0
5		3888—3895	7	237.755	0
6		3918—3925	7	0	0.406
4		3937—3941	4	0	33.6
3	Cretaceous (K$_b$)	3950—3955	5	0	0.13
2		3984—4071	8	2.500	2.61
1		4066—4071	5	0	3.6

Figure 8.10 Tested intervals of Well KL-2.

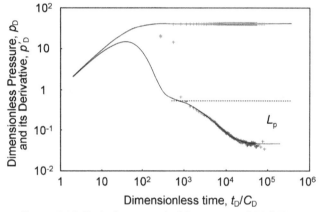

Figure 8.11 Typical pressure buildup curves of Well KL-2.

Figure 8.11 is a type curve of a well with partially perforated formation, very similar to model graph (M-6) in Table 5.5 in Chapter 5.

The following two points should be noted especially:

1. The perforated interval of this test is only 2 meters thick, and the test curves featured partially perforated clearly show that the above and below intervals are combined vertically together, verifying good vertical connectivity of gas zones, and which can be used further in semiquantitative analysis to calculate vertical formation permeability.

2. Any interpretation and analysis deviating from the actual formation conditions should be avoided.

There used to be the following analysis results during the processing above well test curves:

1. There is a constant-pressure boundary in the gas reservoir. This interpretation undoubtedly has no geological basis. In fact, the gas zone has no such so-called "constant-pressure boundary" actually.
2. The gas well was damaged. It is true that a very high skin factor may be calculated from analysis on the curve features, but this "skin effect" is caused mainly by "incomplete penetration" of the well rather than the quality problem of well completion.

3.2.2 Test of Well KL-201

Well KL-201 is an appraisal well. Before a preliminary development program of the KL-2 gas field was made, Well KL-2 had been completed and planned to be tested. A long interval commingled production of all pay zones was supposed to be chosen at that time, and so the total deliverability and the total storability coefficient of the well seemed to be the most needed parameters. It is really a pity that the zonal test method for stratified strata was adopted for Well KL-201, except for the longer testing time and thicker test intervals than Well KL-2 were designed.

The tested intervals of Well KL-201 are listed in Figure 8.12. The measured pressure buildup curves are shown in Figure 8.13.

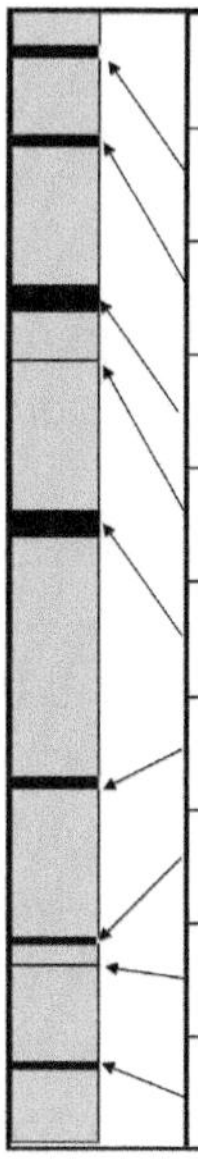

Test No.	horizon	interval m-m	perforated thickness m	gas rate $10^3 m^3/d$	skin factor
9	Paleogene (E)	3600—3607	7	92.158	
8		3630—3640	10	158.544	
7		3665—3695	30	211.060	300
6	Cretaceous (K$_{bs}$)	3712—3714	2	139.074	
5		3770—3795	25	376.681	163-454
4		3883—3892	9	306.730	
3		3926—3930	4	212.184	133
1		3936—3938	2	72.503	209
2	(K$_b$)	4016—4021	5	0	

Figure 8.12 Tested intervals of Well KL-201.

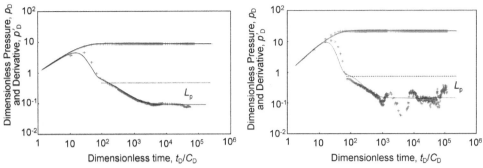

Figure 8.13 Pressure buildup curves of Well KL-201.

Some important features of the reservoir are reflected in test data:

1. Good vertical connectivity of major pay zones is further proved by test data, which are fully consistent with the geological research results. As noted for the wells in such reservoirs, all gas zones should commingling produce in development.

2. Because the test lasted longer than Well KL-2, the pressure derivative curve in the log–log plot had a longer radial flow horizontal straight line segment, which implied that faults in the area near Well Kela-201 do not pay obvious roles of hindering gas flow.

3.2.3 Test of Well KL-203

The next drilled well, Well KL-203, was tested by commingled testing, including deliverability test and pressure buildup test of the whole well after taking the lessons learned from previous practices.

Test Technology

A full opening annulus pressure respondent (APR) and perforating gun for combining the operation of tubing conveyed perforating (TCP) and testing were applied in the test. The structure of the test string is shown in Figure 8.14.

It can be seen from Figure 8.14 that two parts of testing string affected data acquisition and analysis.

1. Pressure gauges were installed too far away from the tested interval. Pressure gauges were installed at the upper and lower measuring points: the lower gauge was set at 3520.15 meters, about 178 meters above the top of the perforated interval of the gas zone (3698.5 m) and about 287 meters above the middepth of gas zone (3807.5 m) and the upper gauge was set at 3404.61 meters, about 294 meters above the top of the perforated interval of the gas zone and about 403 meters above the middepth of the gas zone. Many problems are brought about when the gauges are not installed at the middepth of the gas zone.

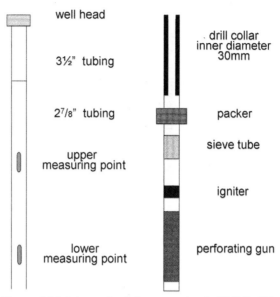

Figure 8.14 Schematic of the test string in Well KL-203.

2. A small diameter drill collar limited the deliverability of the well. There was a drill collar, of which the length was 106.5 meters and the inside diameter was 38.1 mm between the test interval and pressure gauges in the test string. The collar was used for protecting the downhole pressure gauges from damage caused by vibrating during perforation. Even only 100 meters long, this drill collar generated great frictional resistance in a high yield condition during the well test, limited the deliverability to pay its upper-most role, and caused distortion of the calculated parameters in data interpretation.

Arrangement of Production Proration and Acquisition of Pressure Data During the Test
The test procedure of the test is shown in Table 8.1.

The whole test included several steps: the former seven flows and shut ins were for getting the transient points, the eighth flow was the extended test of the modified isochronal test, and the eighth shut in was the first pressure buildup test, the ninth flow for sampling, and the ninth shut in for the second pressure buildup test.

The acquired pressure history is shown in Figure 8.15. The gas production rate and pressure records during the deliverability test are listed in Table 8.2.

The aforementioned records of pressure and flow rate clearly display that the gas wells in the gas field with high pressure and very thick gas zones like Kela-2 have very vigorous gas deliverability in commingled production.

However, it was discovered in the further analysis of test data that there were many hidden abnormal phenomena and some of them directly affected parameter calculation in data interpretation.

Table 8.1 Test Procedure of Well KL-203

Period	Time	Operations
April 5 Packer setting	18:10	Rotated forward eight cycles, pressurize 13 tons for packer setting
Initial flow	17:15	Perforated with a pressure of 34 MPa and surface open the well up, measure the gas rate with separator after a short blowout
April 7 Initial shut in	14:00	Surface shut in
Second flow	20:00	Surface open the well up with 6.35-mm choke
April 8 Second shut in	2:00	Surface shut in
Third flow	8:00	Surface open the well up with 8.73-mm choke
Third shut in	14:00	Surface shut in
Fourth flow	20:00	Surface open the well up with 10.3-mm choke
April 9 Fourth shut in	2:00	Surface shut in
Fifth flow	8:00	Surface open the well up with 7.14- and 9.53-mm chokes
Fifth shut in	14:00	Surface shut in
Sixth flow	20:00	Surface open the well up with 10.32- and 9.53-mm chokes
April 10 Sixth shut in	2:00	Surface shut in
Seventh flow	8:00	Surface open the well up with 15.88- and 14.29-mm chokes
Seventh shut in	12:00	Surface shut in
April 11 Eighth flow	10:00	Surface open the well up with 7.14- and 9.53-mm chokes
April 17 Eighth shut in April 20	20:00	Surface shut in
Ninth flow	15:00	Surface open the well up with 7.14- and 9.53-mm chokes
April 21 Ninth shut in	10:00	Downhole shut in
April 22 Well killing	8:00	Operated RD valve with annular pressure of 23 MPa, reverse circulation and kill the well
April 23 Packer unset	6:30	Lifted pipe string (640 KN↓590KN) and unset the packer

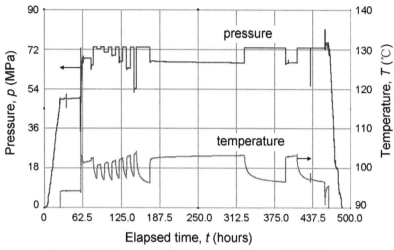

Figure 8.15 Pressure and temperature history of Well KL-203.

Abnormal Phenomena and Analysis of Pressure Data

Abnormal phenomena of pressure data provided rich dynamic information on gas well performance, especially on wellbore. On the one hand, abnormal phenomena hindered the normal analysis of reservoir parameters with pressure data; on the other hand, phase behavior changes in high productivity gas wells were disclosed for the first time because interesting phenomena were recorded by high precision electronic pressure gauges, which enabled reservoir engineers to take some measures to improve operation techniques of testing and thoroughly study dynamic reservoir features with data.

1. Abnormal pressure and temperature variations

    (1) Frequent fluctuation and "noise" of downhole pressure after shut in

    Frequent fluctuations of downhole shut-in pressure and more frequent "noises" superposed on them were recorded during each shut in of Well KL-203.

    a. Maximum amplitudes of the fluctuation and the noise are about 0.01 and 0.001 MPa, respectively.

    b. Frequencies of the fluctuation and the noise are about 8–10 and 70–80 times/hr, respectively.

    c. Pressure fluctuation and noise always existed from the initial flow to the ninth shut in

    d. Pressure fluctuations at the same measuring point recorded with two pressure gauges were completely consistent, indicating that the pressure gauges were working normally during testing, whereas the shapes of those fluctuations recorded at the upper and lower measuring points were a little bit different.

    e. When the acquisition interval of pressure points is small (say, several seconds), both pressure fluctuation and noise are observed simultaneously, whereas

Table 8.2 Production Rate and Pressure Records During Deliverability Test of Well KL-203

Choke, mm	Orifice plate, mm	Tubing/casing pressures, MPa	Flowing pressure, MPa	Flow rate, m³/day			Production pressure difference, MPa	Production duration, hours
				Oil	Gas	Water		
11.11	101.6	54.32/15.2	67.968		925,316		5.166	10
12.7	101.6	48.71/15.1	64.783	0.62	1,141,887	0.35	8.351	3.0
6.35	76.2	60.37/15.7	72.271	Trace	418,776	1.20	0.859	6.0
8.73	82.55	59.52/15.0	70.793	Trace	661,056	0.44	2.327	6.0
10.32	101.6	56.30/15.0	69.244	0.66	880,946	4.20	3.867	6.0
11.91(9.53±7.14)	88.9/101.6	51.74/15.0	66.607	2.75	1,179,450	4.26	6.497	6.0
14.05(14.29±10.32)	101.6/114.3	46.01/15.0	63.444	1.73	1,364,336	5.00	9.651	6.0
21.36(14.29±15.88)	88.9/114.3	26.70/16.0	54.056	1.64	2,010,035	6.96	19.032	4.0
11.91(9.53±7.14)	88.9/82.55	52.13/14.0	65.904	0.46	1,134,848	2.99	6.985	15.4
11.91(9.53±7.14)	88.9/82.55	52.20/14.5	66.025	0.18	1,133,458	3.51	6.775	10

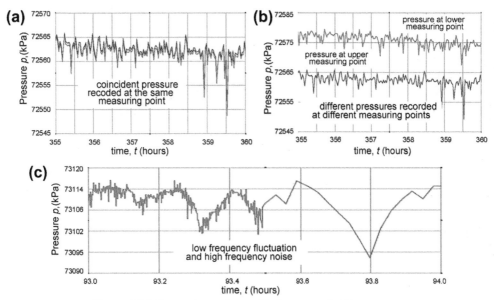

Figure 8.16 Comparison of magnified pressure buildup curves.

when the acquisition interval (in terms of second) is large (say, several minutes), only pressure fluctuation can be seen.

Some parts of it are shown in Figure 8.16.

f. Figure 8.16a shows pressure data recorded at the same time by two pressure gauges at the upper measuring point during the eighth shut-in period, it can be seen that the pressure fluctuations recorded are very consistent, which proved that the two pressure gauges were working in very good conditions.

g. Figure 8.16b shows pressures recorded simultaneously with the upper and lower pressure gauges during the eighth shut-in period, and the pressure curve shape seems different.

h. Figure 8.16c shows the superposition of low-frequency fluctuation and high-frequency noise during the second shut-in period.

(2) Pressure decreased at the late stage of pressure buildup

Figure 8.17 shows the eighth pressure buildup curve of Well KL-203. As seen from data recorded by pressure gauges at the lower measuring point in the well, the abnormal phenomenon appearing during the seventh, eighth, and ninth shutting in lasted rather long; it was also observed from data recorded by the pressure gauges at the upper measuring point in the well that the pressure declined during all shutting in, suggesting that this phenomenon is relevant to the depth at which the pressure gauges locate.

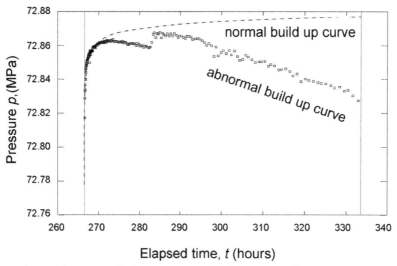

Figure 8.17 Abnormal pressure decline on pressure buildup curve (the eighth shut in by pressure gauges at the lower measuring point).

(3) Reverse peak value of abrupt drop appearing in drawdown curve at the very early stage of well opening

Corresponding to the pressure drop occurring in the pressure buildup curves, a reverse peak of the pressure drawdown curve occurred instantly when the well was just opened, and then the curves kept rising (Figure 8.18).

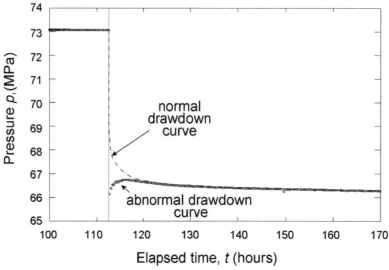

Figure 8.18 Early pressure drawdown curve with a reverse peak of Well KL-203.

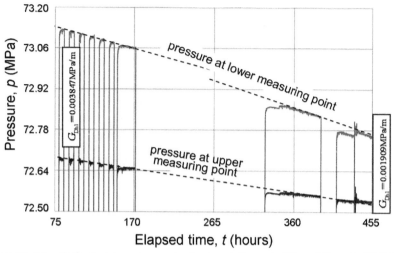

Figure 8.19 Downhole shut-in pressure gradient variations at different times in Well KL-203.

The phenomenon was observed during all flowing periods and in every test curves plotted with data acquired by the pressure gauges at the upper and lower measuring points. It implies that the influencing factor always exists in the whole test process of Well KL-203.

(4) The pressure gradient in the wellbore continuously varied throughout all shut-in periods.

Figure 8.19 shows the shut-in pressure variation recorded by the pressure gauges at the upper and lower measuring points. The distance between the two measuring points is 115.54 meters, and the static pressure gradient in the well-bore after shutting in can be calculated with pressure data recorded at different test points at the same moment.

The measured pressure curves acquired at the upper and lower measuring points are not parallel to each other (Figure 8.19). The converted pressure gradient is decreased from 0.003847 MPa/m in the initial shut in to 0.001909 MPa/m in the ninth shut in, and the relative difference between them both is more than 100%.

(5) Static pressure gradient in wellbore decreased gradually with time

The static pressure gradient decreased with time gradually both in different shut-in periods and during the same shut-in period. Taking the eighth shut in, the one with a long shut-in duration, for example, the pressure gradient in the wellbore was 0.002535 MPa/m at the beginning of the shut-in period and decreased to 0.002374 MPa/m at the end of the period; a big difference appeared between the pressure acquired at the two measuring points in the same shut-in period (Figure 8.20). Also, the same phenomenon appeared in the ninth shut-in period.

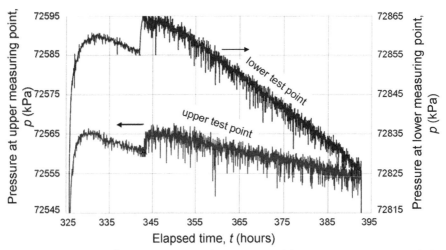

Figure 8.20 Comparison of pressure measured at upper and lower measuring points during the eighth shut in.

(6) Abrupt peaking of bottom-hole temperature occurring at the instant when the well was just shut in

Figure 8.21 shows an abrupt upward peaking of the bottom-hole temperature occurring at the instant when the well was just shut in. Such abrupt upward peaking of the bottom-hole temperature occurred every time when the well was just shut in during the whole test process. Furthermore, the higher the production rate and the lower the flowing pressure before shutting in, the larger the height of peaking.

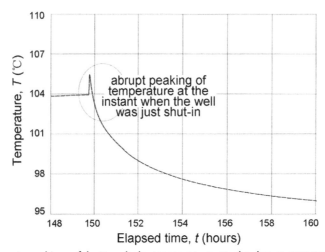

Figure 8.21 Abrupt peaking of bottom-hole temperature at the lower measuring point at the beginning of shut in.

2. Causes of abnormal pressure behavior during shut in

It is found from testing technical conditions and onsite test operation records that phase changes in the wellbore during flowing or shutting in are the main reasons causing the aforementioned phenomena; special examples are:

a. Fluids in the wellbore were mixture of gas and water when well was just shut in; after that the heavy fluid components fell down to the lower part of the wellbore due to gravitational separation effect and load water formed.

b. Bottom-hole load water was brought upward and out of the wellhead by the gas flow during flowing of the well.

c. Working fluids that leaked into the formation discharged continuously when the well was producing, causing the density of the well effluent to decrease gradually.

d. Cross-flow was induced by formation heterogeneity.

This kind of analysis process is "data examination and analysis" mentioned several times in this book. Curves with a standard shape introduced in textbooks are seldom seen among data acquired in fields, and more or less abnormal variations generated due to test technical factors or complicated conditions inside the formation are observed very often. To identify the factors causing these abnormal phenomena and to eliminate the influence of these factors are the major task of data examination and analysis and data pretreatment.

A detailed analysis for the abnormal phenomenon of acquired data of Well KL-203 is given here.

(1) Pressure fluctuation and noise are caused by gravitational separation and cross-flow between the thin layers after well shutting in

After well shutting in, the gas bringing liquid stopped flowing and the heavy fluid components in the mixture slipped and fell down to the bottom hole. Part of them combined to form liquid columns like plugs and fell downward; when their bottom reached the depth where the pressure gauges were set, the measured pressure readings increased consequently, whereas after it passed through the measuring point, the measured pressure readings were decreased consequently. Therefore, pressure fluctuations at the measuring point appeared (Figures 8.16 and 8.20).

In the ninth shut-in period, pressure fluctuation and noise still existed, despite that the effect of gravitational separation was eliminated to some extent due to downhole shut in. This suggests that noises mainly came from a deeper formation.

K_{1bS} Formation, the major pay zone of Well KL-203, can be divided into five intervals; the permeability of them varies greatly, as shown in Table 8.3.

A high permeability formation certainly contributes a great deal of gas in the production and is associated with a great voidage of reserves and a big

Table 8.3 Permeability of Different Intervals of K_{1bs} Formation

Horizon	Interval No.	Lithology	Permeability, mD	Porosity, %
K_{1bs}	I	Extremely fine sandstone	6.27	11.45
	II	Fine sandstone	48.04	13.55
	III	Fine sandstone	77.79	13.10
	IV	Fine sandstone	9.59	9.78
	V	Conglomerate with high shale content	Extremely low	Extremely low

pressure decline. The pressure difference among the intervals after shutting in would make gas flow vertically in the formations and the pressures would mostly be balanced via the wellbore. That is, when the well is shut in by switching off the shut-off valve, the sandface in the hole has not shut in yet and the cross-flow in the formation forms via wellbore, causing pressure fluctuation and noise.

(2) Estimation of the height of load water column in wellbore

Take the eighth shut in as an example. Fluid in the wellbore began to slip down to the bottom hole after shutting in. It means that while the density of the well effluent in the upper part of the wellbore was decreasing, load water in the lower part of the wellbore was being formed.

Supposing that the sectional areas of the upper and lower parts of the wellbore are equivalent, the height of the load water column in the wellbore can be calculated roughly as follows.

a. Fluid density near the measuring point in the initial period of shutting in is

$$\rho_1 = \frac{G_{Dh1}}{g} = 0.2587 \text{ g/cm}^3.$$

Converted density under average wellbore pressure is

$$\overline{\rho}_1 = \frac{69.35}{72.8} \cdot \rho_1 = 0.2464 \text{ g/cm}^3.$$

b. Fluid density at the measuring point in the late period of shutting in is $\rho_2 = 0.2422$ g/m^3 and the average fluid density is $\overline{\rho}_2 = 0.2308$ g/m^3.

c. Suppose the height of the load water column is ΔH, wellbore length is H, and section area is S, then

$$\overline{\rho}_1 \cdot H \cdot S = \overline{\rho}_2 (H - \Delta H) \cdot S + \rho_{\text{water}} \cdot \Delta H \cdot S.$$

Consequently,

$$\Delta H = 77.7 \text{ m},$$

that is, the height of the load water column in the wellbore is about 78 meters at the late period of the eighth shutting in.

(3) Shut-in pressure at the measuring point declined due to the load water in the lower part of the wellbore

At the late period of shutting in of Well KL-203, "reverse buildup" occurred and the pressure declined; this was mainly caused by a continuous increasing load water in the lower part of the wellbore. To verify this conclusion, the following calculation analysis was performed.

Pressure data acquired by the pressure gauge at the lower measuring point in the eighth shut-in period are selected for the analysis, as shown in Figure 8.22.

a. Pressure gauges at the upper and lower measuring points were installed at 3404.61 and 3520.15 meters, respectively, and the middepth of the gas zones was at 3807.5 meters, that is, distances from the upper and lower gauges to the middepth of gas zones were 402.89 and 287.35 meters, respectively.

b. Maximum pressure measured by the lower gauges at the time just after shutting in was 72.860 MPa.

c. Converted pressure gradient from the measured pressures at the upper and lower gauges at the time just after the shutting in was $G_{Dh0} = 0.002535$ MPa/m, and the average density of well effluent was $\rho_1 = 0.2464$ g/cm^3.

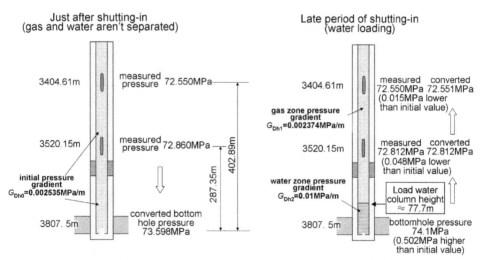

Figure 8.22 Pressure declines due to water loading after shut in.

d. Converted pressure at the middepth of gas zones at the time just after shutting in with the obtained pressure gradient was

$$p_{\text{bottom hole}} = p_{\text{measured}} + (3807.5 \text{ m} - 3520.15 \text{ m}) G_{\text{Dh0}} = 73.598 \text{ MPa}.$$

e. Suppose 66 hours later after well shutting in, the pressure in the middepth of the gas zones kept rising normally, the bottom-hole static pressure should be recovered to 74.1 MPa, 0.502 MPa higher than that at the time just after shutting in.

Considering that the height of the load water column in the lower wellbore is 77.7 meters, the pressure at the lower measuring point converted from the bottom-hole pressure should be

$$p_{\text{measured at lower point}} = p_{\text{bottom hole}} - 0.01 \text{ MPa/m} \times 77.7 \text{ m}$$
$$-0.002374 \times (287.35 \text{ m} - 77.7 \text{ m}) = 72.812 \text{ MPa}.$$

This converted pressure value was basically consistent with the measured pressure of the lower gauges and 0.048 MPa lower than that at the time just after shutting in (see the right side of Figure 8.22). It is proved that load water does cause a declining pressure at the measuring point.

f. The converted pressure also can be obtained with the same way for the upper gauges:

$$p_{\text{measured at upper point}} = p_{\text{bottom hole}} - 0.01 \text{ MPa/m} \times 77.7 \text{ m}$$
$$-0.002374 \text{ MPa/m} \times (402.89 \text{ m} - 77.7 \text{ m}) = 72.551 \text{ MPa}$$

The value is lowered by 0.015 MPa than that at the time just after shutting in. It shows that both pressures at the upper and lower measuring points declined to different degrees due to the load water, and the pressure shift measured by the lower gauges is more serious than that by the upper gauges, completely consistent to the phenomena in Figure 8.20. It is verified that the conclusion that reverse pressure buildup at the measuring points due to load water in wellbore is surely correct.

(4) Abnormal pressures during well flowing

Figure 8.23 shows the measured pressure variation in the early period of the eighth flow.

Load water at the bottom hole in the eighth shut-in period would be carried out by gas flowing in the ninth flow.

Load water has a balanced part of bottom-hole pressure and made pressure data lower during the early flowing period. The pressure will recover to higher normal values when the load water has nearly been discharged away from the wellbore.

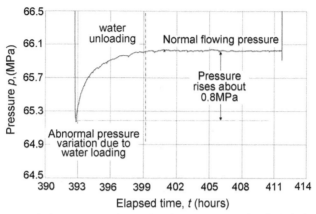

Figure 8.23 Pressure variation at measuring point due to water unloading during the well flowing period (ninth flow).

This analysis was in line with the measured pressure variation trend shown in Figure 8.23.

Pressure data measured by lower gauges dropped abruptly to about 65.2 MPa instantly when the well was opened and then recovered to about 66.0 MPa with a rising amplitude of around 0.8 MPa; if converted in terms of water column with a gradient of 0.01 MPa/m, the height of the blowout water column was 80 meters at least, which was very close to the calculated load water height in the eighth shut-in period.

(5) Cause analysis of pressure gradient decline in wellbore from the first to the ninth shut-in periods.

Measured pressure data at the ends of nine shut-in periods at both upper and lower measuring points are marked in Figure 8.24.

As displayed in Figure 8.19, the pressure gradient calculated with measured pressures was declining continuously, and pressure data displayed in Figure 8.24 are listed in Table 8.4.

The variation of shut-in static pressure gradient is caused by the density change of the well effluent. Table 8.4 shows that the density of the well effluent decreased from 0.3923 to 0.2036 g/cm^3 at the end of the ninth shut in. It implies that the liquid content of well effluent kept declining.

During the drilling and completion of Well KL-203, the loss of drilling fluid was 434.9 m^3, excluding liquid losses in other operations, for example, well washing, in completion procedures.

Only 27.71 m^3 of the water was recovered. Although both the temperature and the flow rate of this well were so high that part of the water flowing out from the wellhead could not be metered, this well seemed to still discharge continuously the

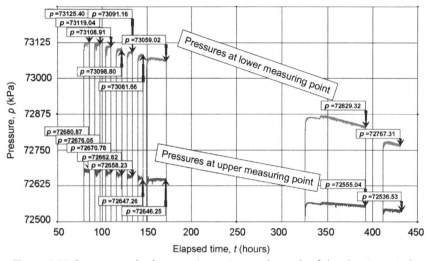

Figure 8.24 Pressures at both measuring points at the ends of the shut-in periods.

load water with a declining rate, which caused the density of the fluids in the wellbore to decrease. Also, a large amount of fluid was discharged in the eighth flow lasting 154 hours between the seventh and eighth shut-in periods, which made the pressure gradient vary greatly, and the density of the fluid in the wellbore decreased by 33.6%, from 0.3643 to 0.2420 g/cm^3.

(6) Analysis of instant and abrupt variation of temperature at the time when the well was just shut in

As shown in Figure 8.21, the recorded temperature increased abruptly and formed a shape point at the time when the well was just shut in. It is due to the "water hammer" formed by fluid inertia occurred when the well was shut in.

Table 8.4 Pressure Gradient and Fluid Density Variation at the End of Shutting ins

Time	Pressure at lower measuring point, MPa	Pressure at upper measuring point, MPa	Static pressure gradient, MPa/m	Fluid density, g/cm^3
First shut in	73.12540	72.68087	0.003847	0.3923
Second shut in	73.11904	72.67605	0.003834	0.3910
Third shut in	73.10891	72.67078	0.003792	0.3867
Fourth shut in	73.09880	72.66262	0.003775	0.3849
Fifth shut in	73.09116	72.65823	0.003747	0.3821
Sixth shut in	73.08166	72.65224	0.003717	0.3790
Seventh shut in	73.05902	72.64625	0.003573	0.3643
Eighth shut in	72.82932	72.55504	0.002373	0.2420
Ninth shut in	72.76731	72.53653	0.001997	0.2036

When pressure rises abruptly at the instant of well shutting in, the following formula is tenable for the gas in a constant volume in the wellbore:

$$pV = ZRT, \tag{8.1}$$

where

p = average wellbore pressure, MPa
V = closed wellbore volume, m³
T = average wellbore temperature, K
Z = deviation factor
R = universal gas constant

In this case the volume is a constant; it is approximately in a thermal insulation condition in a very short period; when the average pressure in the wellbore increased abruptly from 38 MPa of the flowing pressure to 68 MPa of the shut-in pressure (the seventh shut in), to equalize both sides of Equation (8.1), the well temperature rose correspondingly. After a moment, the temperature in the wellbore declined rapidly caused by the heat loss because the flowing temperature in the wellbore was higher than the static formation temperature and so the shape point of the temperature disappeared.

Causes of various abnormal phenomena are analyzed here, but the cause analysis is not the purpose—the ultimate purpose is to correct abnormal data caused by those factors. The most serious problem influencing pressure data analysis is the decline of pressure buildup curves in the late period of shutting in; this problem is specifically discussed later.

3. Converting the measured shut-in pressure to that at the middepth of gas zones

Strictly speaking, the analysis of pressure buildup curves should mean the analysis for sandface pressure at the middepth of gas zones. Due to the various causes mentioned earlier, especially the effect of load water and well temperature variation, some difficulties were encountered in converting the pressure at the measuring point to that at the middepth of gas zones in Well Kela-203.

According to aforementioned analysis, the reverse buildup of the shut-in pressure was mainly caused by fluids slipping into the lower part of the wellbore and forming the load water accordingly after the well shutting in. Suppose load water height ΔH increased in direct proportion with time:

$$\Delta H = A\Delta t, \tag{8.2}$$

where

A = water loading rate, constant, m/hr
ΔH = height of load water column, meters
Δt = shut-in time, hours

The pressure in the middepth of gas zones $p_{\text{bottom hole}}$ can be expressed as

$$p_{\text{bottom-hole}} = p_{\text{lower measuring point}} + G_{\text{Dh}}(287.35 - \Delta H) + 0.01\Delta H$$
$$= p_{\text{lower measuring point}} + 287.35 G_{\text{Dh}} + (0.01 - G_{\text{Dh}})\Delta H \tag{8.3}$$

or

$$p_{\text{bottom hole}} = p_{\text{upper measuring point}} + G_{\text{Dh}}(402.89 - \Delta H) + 0.01\ \Delta H$$
$$= p_{\text{upper measuring point}} + 402.89\ G_{\text{Dh}} + (0.01 - G_{\text{Dh}})\ \Delta H \tag{8.4}$$

Put Equation (8.2) into it, then

$$p_{\text{bottom hole}} = p_{\text{lower measuring point}} + 287.35 G_{\text{Dh}} + (0.01 - G_{\text{Dh}})A\ \Delta t \tag{8.5}$$

and for pressure at the upper measuring point:

$$p_{\text{bottom hole}} = p_{\text{upper measuring point}} + 402.89\ G_{\text{Dh}} + (0.01 - G_{\text{Dh}})A\ \Delta t \tag{8.6}$$

$A = 1.177$ m/hr as estimated by load water for the eighth shut in. Moreover, G_{Dh} also varies with time, and the linear equation for the eighth shut in is

$$G_{\text{Dh}} = 0.002535 - 2.439 \times 10^{-6}\ \Delta t.$$

Put it into Equations (8.5) and (8.6):

$$p_{\text{bottom hole}} = p_{\text{lower measuring point}} + 0.7284 + 8.085 \times 10^{-3}\Delta t$$
$$+ 2.871 \times 10^{-6}\Delta t^2 \tag{8.7}$$

$$p_{\text{bottom hole}} = p_{\text{upper measuring point}} + 1.0213 + 7.804 \times 10^{-3}\Delta t$$
$$+ 2.871 \times 10^{-6}\Delta t^2 \tag{8.8}$$

Convert the pressure at measuring points in the eighth shut in with Equation (8.7) or (8.8), and the converted bottom-hole pressure buildup curve is shown in Figure 8.25.

Figure 8.25 shows that the shape of the pressure buildup curve has gotten back to normal; the abnormal reverse buildup in Figure 8.17 has been excluded.

It should be pointed out that the aforementioned conversion consists of several steps, and an assumption that load water height ΔH increased in direct proportion with time was made but this assumption still needs to be further ascertained.

4. Prediction of pressure variation patterns at different measuring points at different depths

As described previously, due to slipping of fluids in the wellbore caused by gravity separation, the pressure gradient in the upper wellbore decreased and load water formed in the bottom hole so that the pressure distribution in the wellbore was distorted. That is, various pressure variations could be acquired at different measuring points, and these pressure variations might be very different from that acquired at middepth of the gas zones.

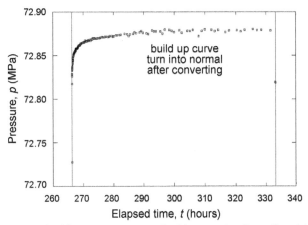

Figure 8.25 Pressure buildup curve converted with water loading effect (eighth shut in).

Which measuring point is optimal for pressure acquisition? Undoubtedly, the middepth of the gas zones is the best choice. If there is load water in the bottom hole, pressure gauges should be installed under the water level. This problem is demonstrated further later.

First let us assume that the height of the load water column rises in direct proportion with time, as given in Equation (8.2): $\Delta H = A\Delta t$, and that the average gradient in the wellbore varies linearly:

$$G_{\text{Dh}} = B - C\Delta t. \tag{8.9}$$

Consequently,

$$p_{\text{measured}} = p_{\text{bottom hole}} - Bh - [(0.01 - B)A - Ch]\Delta t - AC\Delta t^2,$$

where h is the distance between the measuring point and the middepth of the gas zones.

Take the eighth shut in of Well KL-203 as an example: $A = 1.177$ m/hr, $B = 0.002535$ MPa/m, and $C = 2.439 \times 10^{-6}$ MPa/m·hr, so

$$p_{\text{measured}} = p_{\text{bottom hole}} - 0.002535h$$
$$- (0.008786 - 2.439h)\Delta t - 2.870 \times 10^{-6}\Delta t^2. \tag{8.10}$$

Here is an example to illustrate different pressure variations at different measuring points. Suppose the middepth of the gas zones is 3700 meters and the pressure at this point is 74.00 MPa and remains unchanged, the measured pressure variations at different measuring points are predicted as follows. Different depths of selected measuring points are, respectively, 3700 meters (middepth of gas zones), 3680 meters (20 meters above the middepth), 3650 meters (50 meters above the

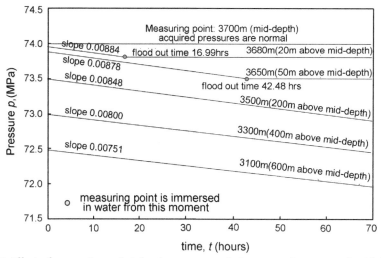

Figure 8.26 Effect of measuring point depth upon recorded pressure in a gas well with load water.

middepth), 3500 meters (200 meters above the middepth), 3300 meters (400 meters above the middepth), and 3100 meters (600 meters above the middepth); pressure variation curves at these measuring points are shown in Figure 8.26.

Figure 8.26 shows

(1) Constant bottom–hole pressure (74 MPa) was measured correctly by the gauge settled at the middepth of 3700 meters

(2) When the measuring point was at 3680 meters, that is, 20 meters above mid-depth, the pressure gauge was immersed at $\Delta t = 16.99$ hours, prior to which the pressure decreased at a slope of 0.00884 MPa/hr, but it became constant after which and 0.2 MPa lower than pressure at middepth

(3) When the measuring point was 3650 meters, that is, 50 meters above middepth, the gauge was immersed at $\Delta t = 42.48$ hours, prior to which the pressures decreased at a slope of 0.00878 MPa/hr and after which the pressure was constant but 0.5 MPa less than pressure at middepth

(4) When the measuring points were 3500, 3300, and 3100 meters, that is, 200, 400, and 600 meters above middepth, respectively, the gauge was never immersed by the load water, and the three measured pressure curves declined at constant but different slopes; the shallower the measuring point, the smaller the slope of the measured pressure: the slopes of the pressure curves measured at 3500, 3300, and 3100 meters were 0.00848, 0.00800, and 0.00848 MPa/hr, respectively. This has been proved by measured pressure data in Well KL-203.

The following conclusion can be obtained based on the aforementioned analysis: When testing gas wells producing gas with water like Well KL-203, the pressure gauges should be installed below the middepth of the gas zones; if there has been load water at the bottom hole prior to test, the pressure gauge should be set below the water level. Otherwise a distorted curve shape of pressure data caused by phase changes in the wellbore due to fluid slipping and load water forming may make correct data analysis, including parameter calculation, impossible.

Validation of Interpretation Results of Pressure Buildup Curves and the Dynamic Reservoir Model

1. Interpretation with the pressure data converted to bottom hole

An abnormal pressure buildup curve and its conversion were analyzed in detail earlier, Figure 8.25 shows the converted pressure buildup curve of the eighth shut in (i.e., plotted with converted data); the log–log plot drawn with well test interpretation software is shown in Figure 8.27.

Figure 8.27 shows that pressure data are characterized by typical homogeneous formation and in line with the model graph (M-1) discussed in Chapter 5:

(1) The horizontal straight line of radial flow on the pressure derivative curve extended until the test finished, indicating none of faults effects upon gas flow in gas zones.

(2) The gas zone permeability calculated in interpretation is very high: $k = 75.0$ mD. But it needs to be verified for its accuracy, for it was obtained in the interpretation with converted pressure buildup curve.

(3) Total formation coefficient of the reservoir is $kh = 13,990$ mD·m.

(4) Extremely high skin factor $S = 473.5$ is obtained from the interpretation.

The extremely high skin factor might be caused by a small diameter drill collar ($\phi = 30$ mm) in the test string. A 106-meter-long drill collar hindered the gas

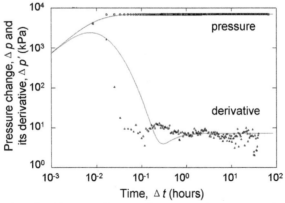

Figure 8.27 Log–log plot of Well KL-203 drawn with pressure data converted to bottom hole (eighth shut in).

Table 8.5 Parameters Calculated with Ninth Buildup at Lower Measuring Point

Parameter Interpreted period	Permeability, mD	Extrapolated pressure, MPa	Skin factor	Wellbore storage coefficient, m³/MPa
Initial shut in	74.86	73.134	565.62	0.1111
Second shut in	77.01	73.129	159.56	0.1987
Third shut in	79.97	73.123	283.55	0.1074
Fourth shut in	74.29	73.117	321.05	0.1132
Fifth shut in	76.52	73.124	427.35	0.2844
Sixth shut in	76.06	73.134	556.07	0.1551
Seventh shut in	77.00	73.126	735.18	0.1635
Eighth shut in	75.00	72.891	471.65	0.1624
Ninth shut in	76.00	72.801	463.98	0.0132

flow significantly, and the measuring point was above the collar, the flow resistance caused by which and reflected by the measured pressure was combined into the turbulent flow effect of the reservoir, so an incredible high skin factor was obtained in the interpretation.

(5) Wellbore storage coefficient is $C = 0.16$ m³/MPa; it is within the normal range under the wellhead shut-in condition.

2. Results of interpretation with pressure data acquired at measuring point

The shape of the pressure buildup curve is influenced by phase changes in the wellbore during the shut-in period, and the curve in the late stage of shut in is influenced especially seriously. Pressure data in the early stage are influenced as well but can still be used approximately for parameter calculation because the pressure builds up very quickly and the pressure increments are quite large at that time.

Parameters were calculated with measured pressure data of the lower measuring point during the ninth shut-in period, and the results are listed in Table 8.5.

Table 8.5 lists the following interpreted parameters:

(1) Permeability $k \approx 76.3$ mD (average value)

(2) Skin factor $S = 159-735$

(3) Wellbore storage coefficient $C \approx 0.16$ m³/MPa (for wellhead shut in) and $C = 0.0132$ m³/MPa (for downhole shut in)

(4) Extrapolated formation pressure $p_R = 73.134$ MPa (from initial shut in)

As can be seen from the interpretation results:

a. The formation permeability is quite high; permeability is as high as 76.3 mD, which is very favorable for achieving high gas well deliverability.

b. The value of the skin factor is extremely high but the formation is not damaged seriously.

The skin factor is extremely high but was due to hindering flow caused by a 106.53-meter-long drill collar and screen pipe in test string and turbulence

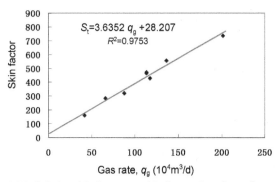

Figure 8.28 Relationship between skin factor S and gas flow rate q_g.

flow in formation. Figure 8.28 shows the relationship between skin factor and gas flow rate.

The regression of data gives a true skin factor value of 28.2 only, indicating that the formation damage caused by drilling, completion, and other operations is not serious. However, the non-Darcy flow coefficient D is very high:

$$D = 3.6352(10^4 \, \mathrm{m^3/day})^{-1}.$$

It suggests very strong effects of turbulence flow and drill collar flow choke.

c. The formation pressure decreases rapidly and abnormally

Table 8.5 shows the extrapolated formation pressure decreased 0.333 MPa from 73.134 MPa at the first shut in to 72.801 MPa at the ninth shut in, and only $0.1 \times 10^8 \, \mathrm{m^3}$ of gas was produced between these two shut ins; obviously the pressure declined too quickly.

However, this abnormal pressure decline may be caused by the static pressure gradient variation and the effect of load water at the bottom hole.

Table 8.6 lists the measured and extrapolated formation pressures obtained from pressure data recorded by gauges at the upper and lower measuring points.

Table 8.6 Comparison of Measured and Extrapolated Formation Pressures at Upper and Lower Measuring Points

Item Pressure measuring point	Extrapolated pressure of initial shut in, MPa	Extrapolated pressure of ninth shut in, MPa	Extrapolated pressure difference, MPa	Measured pressure in initial shut in, MPa	Measured pressure in ninth shut in, MPa	Measured pressure difference, MPa
Upper	72.700	72.565	0.135	72.681	72.537	0.144
Lower	73.134	72.801	0.333	73.125	72.767	0.358

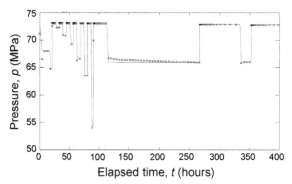

Figure 8.29 Pressure history match verification for well test interpretation results of Well KL-203.

Table 8.6 shows that the pressure drop at the upper measuring point is apparently less than that at the lower measuring point and that pressure analysis results vary with measuring points. It implies that to extrapolate formation pressure with pressure data acquired at a measuring point is inadequate. Also, this error is also due to a load water effect in the wellbore.

The pressure history match verification using the aforementioned interpretation results is shown in Figure 8.29.

d. Wellbore storage coefficient values (C) range normally

Although the C value does not reflect formation conditions, it can be used to verify whether any mistakes occur during interpretation. Values obtained from pressure data acquired at a lower measuring point in the former eight shut ins range from 0.11 to 0.28 m^3/MPa and average 0.16 m^3/MPa, which is consistent with wellhead shut-in conditions in high pressure gas wells. Also, a downhole shut in was adopted in the final shut in (the ninth shut in); the interpreted value decreased by an order and to 0.0132 m^3/MPa, which is also consistent with the technical conditions onsite. The feasibility and correctness of the interpretation operation with software are proved indirectly from this fact.

Deliverability Analysis

A modified isochronal test was performed in Well KL-203, and the former seven flows are for transient deliverability test points and the eighth flow for the stable deliverability point. Deliverability analysis with pressure data measured at both upper and lower measuring points was completed, while there is a big difference between the calculated deliverability and the actual one because the flow was hindered seriously by a 106.53-meter-long drill collar with an inner diameter of 38.1 mm located below the measuring point. To deal with this problem, the friction resistance during

Table 8.7 Deliverability test data of Well KL-203 (pressure data acquired at 3520.15 meters)

Period	Choke size, mm	Gas flow rate, 10^4 m³/day	Flow or shut-in duration, min	Bottom-hole pressure, MPa
Initial shut in		0	330	73.12540
Second flow	6.35	43.0761	390	72.24080
Second shut in		0	360	73.11904
Third flow	8.73	67.1984	360	70.79786
Third shut in		0	360	73.10891
Fourth flow	10.32	89.6922	360	69.25845
Fourth shut in		0	360	73.09880
Fifth flow	11.91	117.9400	360	66.60292
Fifth shut in		0	360	73.09116
Sixth flow	14.05	136.4300	360	63.44329
Sixth shut in		0	360	73.08166
Seventh flow	21.36	201.0000	240	54.04378
Seventh shut in		0	1320	73.05902
Eighth flow	11.91	113.4800	9240	65.90137
Eighth shut in		0	4020	72.82932

the flowing periods was calculated with the vertical pipe flow calculation formula, bottom-hole pressure was converted, and another deliverability analysis with converted pressure data was performed; the treatment process is discussed separately as follows.

1. Deliverability calculation with pressure data acquired at measuring point

Flow rates recorded on surface and pressure data acquired at lower measuring points of 3520.15 meters are listed in Table 8.7.

The laminar–inertial–turbulent equation is established with the pseudo pressure method:

$$\psi_R - \psi_{wf} = 0.002219q_g + 1.379 \times 10^{-5}q_g. \tag{8.11}$$

The AOFP calculated from Equation (8.11) is

$$q_{AOF} = 354.74 \times 10^4 \ m^3/day.$$

The exponential equation is

$$q_g = 0.236(\psi_R - \psi_{wf})^{0.504}. \tag{8.12}$$

The AOFP calculated from Equation (8.12) is

$$q_{AOF} = 342.35 \times 10^4 \ m^3/day.$$

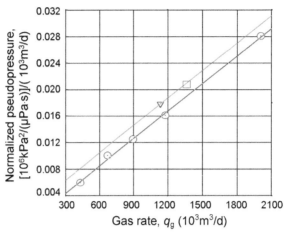

Figure 8.30 LIT deliverability plot.

Refer to Figures 8.30 and 8.31 for detailed information of the analysis.

Flow rates recorded on surface and pressure data acquired at the upper measuring point of 3404.61 meters are listed in Table 8.8.

The LIT equation is established with the pseudo pressure method:

$$\psi_R - \psi_{wf} = 0.002199 q_g + 1.481 \times 10^{-5} q_g. \qquad (8.13)$$

The AOFP calculated from Equation (8.13) is

$$q_{AOF} = 341.83 \times 10^4 \text{ m}^3/\text{day}$$

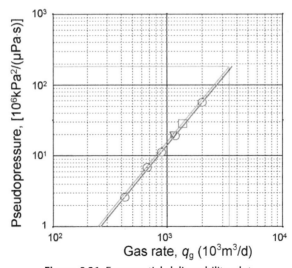

Figure 8.31 Exponential deliverability plot.

Table 8.8 Deliverability Test Data of Well KL-203 (Pressure Data Acquired at 3404.61 Meters)

Period	Choke size, mm	Gas flow rate, 10^4 m^3/day	Flow or shut-in duration, min	Bottom-hole pressure, MPa
Initial shut in		0	330	72.68087
Second flow	6.35	43.0761	390	71.75555
Second shut in		0	360	72.67605
Third flow	8.73	67.1984	360	70.21680
Third shut in		0	360	72.67078
Fourth flow	10.32	89.6922	360	68.54470
Fourth shut in		0	360	72.66262
Fifth flow	11.91	117.9400	360	65.72894
Fifth shut in		0	360	72.65823
Sixth flow	14.05	136.4300	360	62.39094
Sixth shut in		0	360	72.64726
Seventh flow	21.36	201.0000	240	52.33416
Seventh shut in		0	1320	72.64625
Eighth flow	11.91	113.4800	9240	65.19294
Eighth shut in		0	4020	72.55504

and the exponential deliverability equation is

$$q_g = 0.2421(\psi_R - \psi_{wf})^{0.501}. \tag{8.14}$$

The AOFP calculated from Equation (8.14) is

$$q_{AOF} = 328.74 \times 10^4 \text{ m}^3/\text{day}.$$

2. **Deliverability calculation with converted bottom-hole pressure data**

Limited by the string structure adopted in the test, the pressure gauge could not be installed at the midpoint of the gas zones. The deliverability calculated with pressure data acquired at the measuring point is a little bit too low and so measured pressure data need to be corrected. At present, some software for vertical pipe flow analysis are available; with any one of these software the static pressure and flowing pressure acquired at the measuring point can be converted to bottom-hole static pressure and flowing pressure. However, there is the difficulty of correctly selecting the well effluent density and wellbore friction coefficient encountered in the converting process. The primarily converted flowing pressures and static pressures are listed in Tables 8.9 and 8.10.

The LIT equation is obtained with pseudo pressure:

$$\psi_R - \psi_{wf} = 0.004555q_g + 2.00 \times 10^{-6}q_g. \tag{8.15}$$

Table 8.9 Converting Bottom-hole Static Pressure Prior to Well Producing With Measured Static Pressure Gradient Between Upper and Lower Measuring Points

Time	Pressure at lower measuring point, MPa	Pressure at upper measuring point, MPa	Static pressure gradient between both measuring points, MPa/m	Converted static pressure, MPa
End of initial shut in	73.12540	72.68087	0.003847	73.53522
End of second shut in	73.11904	72.67605	0.003834	73.52748
End of third shut in	73.10891	72.67078	0.003792	73.51287
End of fourth shut in	73.09880	72.66262	0.003775	73.50095
End of fifth shut in	73.09116	72.65823	0.003747	73.49032
End of sixth shut in	73.08166	72.65224	0.003717	73.47763

The AOFP calculated from Equation (8.15) is

$$q_{AOF} = 850.59 \times 10^4 \ m^3/day$$

and the exponential deliverability equation is

$$q_g = 0.1210(\psi_R - \psi_{wf})^{0.577}. \tag{8.16}$$

The AOFP calculated from Equation (8.16) is

$$q_{AOF} = 703.71 \times 10^4 \ m^3/day.$$

Table 8.10 Converting Bottom-hole Flowing Pressure with Vertical Pipe Flow Software

Time	Measured flowing pressure, MPa	Converted pressure, MPa Averaging method	Cullender's method
End of second flow	72.24080	73.37557	73.37556
End of third flow	70.79786	73.00956	73.00937
End of fourth flow	69.25845	72.92506	72.92425
End of fifth flow	66.60292	72.51286	72.50944
End of sixth flow	63.44329	71.43873	71.42966
End of seventh flow	54.15812	70.37366	70.30196
End of eighth flow	65.90137	70.87354	70.87087

3. AOFP calculated with stable point LIT deliverability equation and estimation of maximum deliverability

According to the calculation method of the "stable point LIT deliverability equation" elaborated in Chapter 3, during the pseudo steady state, the LIT deliverability equation is expressed as

$$p_R^2 - p_{wf}^2 = Aq_g + Bq_g^2, \tag{8.17}$$

where

$$A = \frac{29.67\overline{\mu}_g \overline{ZT}}{kh}\left[\log\frac{0.472r_e}{r_w} + \frac{S}{2.302}\right] \tag{8.18}$$

$$B = \frac{12.89\overline{\mu}_g \overline{ZT}}{kh} \cdot D. \tag{8.19}$$

The following parameters are selected on the basis of well test interpretation results of Well KL-203:

$r_e = 500$ meters

$r_w = 0.09$ meter

$T = 376$ K

$D = 2 \times 10^{-3}$ $(10^4\ \mathrm{m^3/day})^{-1}$ [calculated with Equation (2.14)]

$\mu_g = 0.025$ mPa·s

$p_R = 73.134$ MPa

$p_{wf} = 65.90137$ MPa (the eighth deliverability test point, refer to Table 8.7)

$q_g = 113.48 \times 10^4\ \mathrm{m^3/day}$ (the eighth deliverability test point, refer to Table 8.7)

$S = 471.65$ (the eighth deliverability test point, refer to Table 8.5)

The values of A and B are obtained by calculation with the aforementioned parameters:

$$A = 5.8096 \times 10^4 (kh)^{-1}$$
$$B = 0.242(kh)^{-1}$$

Put them into Equation (8.17):

$$p_R^2 - p_{wf}^2 = [5.8096 \times 10^4 q_g + 0.242q_g^2](kh)^{-1}. \tag{8.20}$$

Then put p_R, p_{wf}, and q_g of the eighth test point into Equation (8.20); the kh value is inversely calculated to be

$$kh = \frac{5.8096 \times 10^4 q_g + 0.242q_g^2}{p_R^2 - p_{wf}^2} = 6559\ \mathrm{mD \cdot m}.$$

Consequently, the deliverability equation is obtained:

$$p_{\mathrm{R}}^2 - p_{\mathrm{wf}}^2 = 8.8573q_{\mathrm{g}} + 3.6895 \times 10^{-5}q_{\mathrm{g}}^2. \tag{8.21}$$

In Equation (8.21), coefficient A is obtained under the condition of factor skin $S = 471.65$. If test conditions are improved, the hindering flow caused by the drill collar is eliminated and skin factor S will decrease to 28.2. Put this value into Equation (8.18) and coefficient A will decrease to

$$A = \frac{278.89}{kh}\left[3.419 + \frac{28.2}{2.302}\right]$$

$$= 0.6663.$$

Coefficient B remains unchanged, and the deliverability equation will be

$$p_{\mathrm{R}}^2 - p_{\mathrm{wf}}^2 = 0.6663q_{\mathrm{g}} + 3.6895 \times 10^{-5}q_{\mathrm{g}}^2. \tag{8.22}$$

The deliverability plot drawn from Equation (8.22) is shown in Figure 8.32, and Figure 8.33 shows the corresponding IPR curve.

It is known from the aforementioned predicted results that the AOFP of Well KL0-203 can probably increase to as high as 6020×10^4 m³/day with great potential benefit; it is far much higher than the analysis results of measured pressure data and converted pressure data. In view of the reservoir conditions of Well KL-203, the permeability of the reservoir is 60–70 mD, the effective thickness is more than 100 meters, the formation pressure is higher than 73 MPa, and the drilling and completion conditions are also good. That the three important factors influencing deliverability (high kh value, high reservoir pressure, and low skin factor) are perfect assures the realization of such deliverability indices. The predictions given earlier are promising to be proved in the future practices at that gas field.

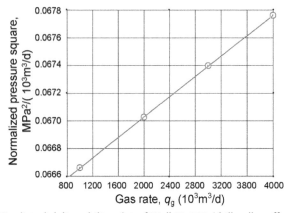

Figure 8.32 Predicted deliverability plot of Well KL-203 (drill collar effect eliminated).

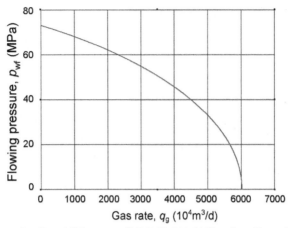

Figure 8.33 Predicted IPR curve of Well KL-203 (drill collar effect eliminated).

Understandings Gained in Dynamic Test of Well KL-203

Undoubtedly, understandings for the KL-2 gas field have been improved greatly due to the testing of Well KL-203.

1. From the pressure buildup curve of Well KL-203, no fault effects appeared when the influence radius reached 1400 meters; faults adjacent to the southeast of the well and other analogous small faults shown in the structure and well location map (Figure 8.9) do not play any roles in hindering the gas flow.
2. The deliverability of all pay zones in the well was obtained from the test. If the hindering flow caused by the drill collar is eliminated, the AOFP of Well KL-203 can reach $q_{AOF} = 6000 \times 10^4$ m^3/day. Such a high deliverability undoubtedly creates favorable conditions for high-efficiency development of the KL-2 gas field.
3. It is recognized after well testing that the value of the permeability-thickness product of the formations near Well KL-203 is extremely high and its reliable estimated value is about 7000 mD·m. This is one of the main conditions of the high yield of gas wells.
4. The total skin factor of all zones of Well KL-203 is 28.2, indicating that there is some, but not serious, damage and that stimulation treatment is necessary.

It is a pity that there are something undone during the testing of all pay zones of the well so that the understanding of the reservoir parameters was limited at the semiquantitative level. The reservoir model cannot be confirmed ultimately, even though many arcane and interesting phenomena in the wellbore caused by facies changes and unusual rules induced by accumulation and exchanging between layers of load water have been recognized or understood from millions of high-precision pressure data. These remaining problems were resolved eventually during the testing of Well KL-205.

3.2.4 Well Test Analysis of Well KL-205

Improvement of Test Technology

After learning from the lessons and experiences gained from Wells KL-2, KL-201, and KL-203, comprehensive improvement was made in the methods, procedures, and techniques in the test design for Well KL-205.

1. A special test design was chosen, including special test string structure, combining operation of TCP, and testing and all pay zones (of which the length is 163 meters) being perforated at the same time without dropping the perforating gun, and pressure gauges being installed at a depth of 3771 meters, 18 meters only above the top of the gas zones (3789 meters) to eliminate basically the influences of facies change of load water in the wellbore.

2. Back-pressure test methods were adopted for the deliverability test to avoid facies changes in the wellbore due to frequent well opening and shut-in operations.

3. During each flowing with various chokes and shutting in, the pressure gradient in the whole wellbore was tested to learn the gas and liquid distribution in the wellbore.

Well Test Analysis Results

Thanks to careful geological and technical designs, various abnormal phenomena that happened during the testing of Well KL-203 were eliminated completely and overall and explicit understandings of the dynamic characteristics of the KL-2 gas field were gained from the testing of Well KL-205.

1. Well KL-205 is a high productivity gas well with high permeability and an extremely high kh value. The permeability of the formation increases gradually from 5.6 mD in the area near the wellhole to about 40 mD in the outside area, and the kh value is as high as 5940 mD·m. Well KL-205 locates in the northeast of Well KL-203, and Well KL-203 is drilled in a high permeability area, the kh value calculated from deliverability test point is as high as 6000 mD·m, so such test results are logical.

2. Although some small faults distribute around Well KL-205, pressure buildup data analysis shows that they bring no hindering flow effect on well production.

3. After 1900×10^4 m^3 of natural gas had been produced from this well, the reservoir pressure did not decline, which further confirms good connectivity between Well KL-205 and the peripheral formation.

4. The AOFP of all pay zones was calculated to be 1320×10^4 m^3/day from back-pressure test data, which indicates that it is an extremely high deliverability gas well. The permeability of the area Well KL-205 locates is not the highest; if the well location is shifted somewhat toward the peripheral formation, its deliverability may be even higher.

5. The measured skin factor of Well KL–205 is about zero, decreasing gradually with producing and being scoured of the well.
6. The measured non–Darcy coefficient of the well is $D = 1.8 \times 10^{-2}\ (10^4\ \mathrm{m^3/day})^{-1}$. This is the first time of obtaining the exact value of the non–Darcy coefficient in this area. This value not only meets the needs for this parameter of development program design, but also provides a comparison reference for the calculated D value theoretically.

The following are specific situations of the test and the interpretation results.

Test Operation

Figure 8.34 shows the schematic diagram of the test string structure.

Test procedures are illustrated briefly as follow:

1. Run the assemble of the combining operation of TCP and test tools into the planned depth, set the packer by pressure, and then use nitrogen pressurization to knock the ball seat off so as to prepare pressure gauges to be set at the screen pipe.
2. Ignite and perforate, blowout on $\phi 7.94$-mm choke, and then replace the choke to control and measure the gas flow rate.
3. Run memory electronic pressure gauges in the hole in the screen pipe at a depth of 3771 meters to measure pressure continuously.

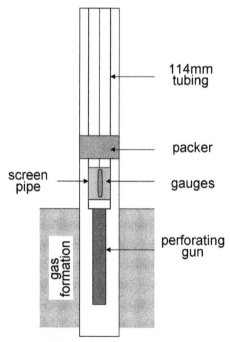

Figure 8.34 Schematic of test string in Well KL-205.

4. Open the well for the back-pressure test and extend production and then shut in the well for the buildup test.

Pressure Data Acquisition

Figure 8.35 is the measured pressure history graph of Well KL–205 after data correction.

It can be seen from Figure 8.35 that the acquired pressure buildup curve does not exhibit any abnormal phenomenon appearing in Well KL–203. After producing cumulatively $1900 \times 10^4 \text{ m}^3$ of gas, the measured shut-in static pressure decreased 0.01558 MPa from the original 74.50832 to 74.49274 MPa; no obvious variation occurred.

It can also be seen from Figure 8.35 that the well produced with 11.11-mm choke twice, the gas flow rates were approximately the same but the flowing pressures were very different: the former one was $p_{wf1} = 72.48162$ MPa and the later $p_{wf2} = 72.73945$ MPa, higher than the former one. Apparently, bottom-hole conditions have been improved dramatically by gas flow scouring.

Figure 8.36 shows the flowing pressure gradients under different chokes and static pressure gradients acquired in the test.

Figure 8.36 shows that the static pressure gradient in the wellbore was generally about 0.00273 MPa/m and was slightly decreasing with time, indicating that the liquid content of the well effluent is very small and remains unchanged basically, as shown in Table 8.11. These data are very important basis for the design of gas pipe string later on.

The gas flow rate and other data recorded at the surface during the deliverability test of Well KL–205 are listed in Table 8.12.

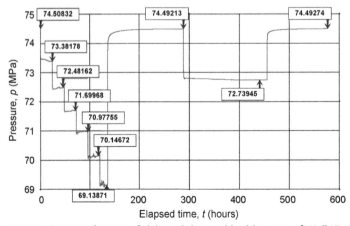

Figure 8.35 Pressure history of deliverability and buildup test of Well KL-205.

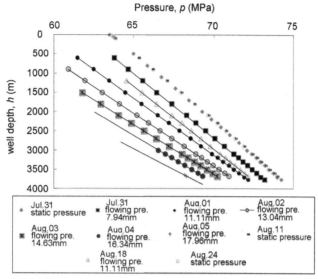

Figure 8.36 Flowing pressure and static pressure gradients of Well KL-205.

Deliverability Evaluation

Deliverability is evaluated with pseudo pressure and the LIT analysis method, and a related calculation can be seen in Table 8.13. However, the pseudo pressure in Table 8.13 is intermediate data calculated and used by software during the analysis process.

With the aforementioned pressure and flow rate data and well test software FAST, the LIT deliverability equation is established as follows:

$$\Delta\psi = 37.29q_{\mathrm{g}} + 0.0763q_{\mathrm{g}}.$$

Table 8.11 Flowing Pressure and Static Pressure Gradients of Well KL-205 (In Order of Test Time)

Flow rate, 10^4 m³/day	Diameter of choke, mm	Time	Gradient value, MPa/100 m	Remarks
0	0	July 31	0.273	Shut–in pressure
69.275	7.94	July 31	0.282	Flowing pressure
106.137	11.11	Aug. 1	0.316	Flowing pressure
148.573	13.04	Aug. 2	0.369	Flowing pressure
175.427	14.63	Aug. 3	0.384	Flowing pressure
206.713	16.34	Aug. 4	0.424	Flowing pressure
244.196	17.96	Aug. 5	0.465	Flowing pressure
0	0	Aug. 11	0.270	Shut–in pressure
107.586	11.11	Aug. 18	0.314	Flowing pressure
0	0	Aug. 24	0.267	Shut–in pressure

Table 8.12 Gas Flow Rate and Other Data Recorded at Surface During Deliverability Test of Well KL-205

Diameter of choke, mm	Tubing/casing pressures, MPa	Gas rate, m³/day	Average flowing temperature, °C	Converted flowing pressure at middepth of gas zones, MPa
7.94	61.95/2.54	692750	103.38	73.3818
11.11	59.64/0.00	1061368	103.34	72.4816
13.04	57.95/1.69	1485733	103.63	71.6997
14.63	56.41/3.27	1754265	103.85	70.9776
16.34	54.21/0.00	2067129	/	70.1467
17.96	51.69/0.10	2441955	104.15	69.1387
21.57	46.33/0.00	3004417	/	/

Table 8.13 Calculation for Deliverability Analysis of Well KL-205

Choke, mm	Gas flow rate q, 10^4 m³/day	Bottom-hole pressure p, MPa	Pseudo pressure ψ, MPa²/mPa·s	Pseudo pressure difference $\Delta\psi$, MPa²/mPa·s	$\Delta\psi/q$
0	0	74.5083	176357	/	/
7.94	69.2750	73.3818	173524	2833	40.895
11.11	106.1368	72.4816	171256	5101	48.061
13.04	148.5733	71.6997	169282	7075	47.620
14.63	175.4265	70.9776	167456	8901	50.739
16.34	206.2129	70.1467	165351	11006	53.243
17.96	244.1955	69.1387	162793	13564	55.546
11.11	108.0305	72.7395	171906	4451	41.201

The AOFP is calculated to be

$$q_{AOF} = 1295.2 \times 10^4 \text{ m}^3/\text{day},$$

where
q_g = gas flow rate, 10^4 m³/day
ψ = pseudo pressure, MPa²/ mPa·s
and the obtained deliverability plot is shown in Figure 8.37.

As shown in Figure 8.35, the flowing pressure increased obviously when flowing with ϕ 11.11-mm choke in the late period of the test, which indicates that the completion conditions improved or the skin effect has been reduced during the test and it is possible to increase the AOFP of the well accordingly. The improved deliverability test point is marked by symbol "∇" in Figure 8.38, and enhanced AOFP was calculated based on this point to be

$$q_{AOF} = 1319.3 \times 10^4 \text{ m}^3/\text{day}.$$

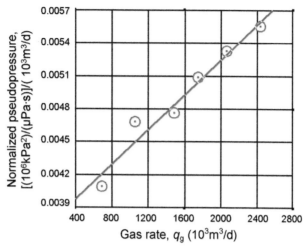

Figure 8.37 LIT deliverability plot with pseudo pressure of Well KL-205.

The AOFP was also calculated with pseudo pressure and the exponential method (the deliverability plot shown in Figure 8.39); the result is much higher than that calculated with the LIT method:

$$q_{AOF}(\text{exponential}) \; = \; 1998.4 \times 10^4 \; \text{m}^3/\text{day}.$$

Such a great difference in the results calculated with two different methods is caused by flow rates during the test and were less than one-fifth of the AOFP, which in fact caused obvious deviation of the extrapolated results of the two methods. Figure 8.40

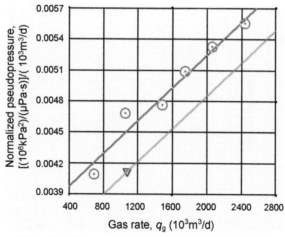

Figure 8.38 LIT deliverability plot with pseudo pressure of Well KL-205 (after improvement of bottom-hole conditions).

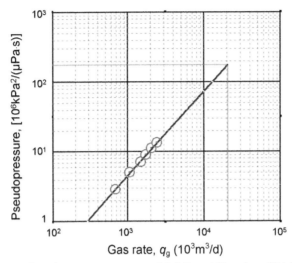

Figure 8.39 Pseudo pressure exponential deliverability plot of Well KL-205.

shows IPR curves plotted with different methods, which show the different trend of two IPR curves, and the calculated AOFP difference is reflected obviously in Figure 8.40.

In consideration of that the LIT method takes turbulent flow effect into account with production squared item (q_g^2) and so gives more correct result, an AOFP calculated with this method, about $1300 \times 10^4 \, \text{m}^3/\text{day}$, is recommended to be actual deliverability, or AOFP.

Reservoir Model Parameters Obtained From Pressure Buildup Analysis

Transient test analysis was performed with the pressure history shown in Figure 8.35. The analysis focused on final shut-in data and the log–log curve match was conducted

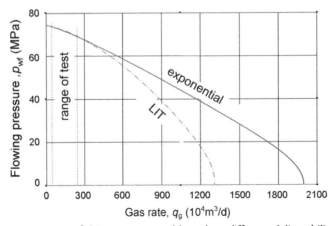

Figure 8.40 Comparison of IPR curves created based on different deliverability equations.

Table 8.14 Interpreted Wellbore Parameters of Well KL-205

Period	Gas flow rate, 10^4 m^3/day	Skin factor	Non-Darcy flow coefficient, (m^3/day)$^{-1}$	Wellbore storage coefficient, m^3/MPa
Initial flow	69.275	1.2	1.8×10^{-6}	0.108
Second flow	106.137	1.2	1.8×10^{-6}	0.108
Third flow	148.573	0.9	1.8×10^{-6}	0.108
Fourth flow	174.892	0.9	1.8×10^{-6}	0.108
Fifth flow	206.713	0.9	1.8×10^{-6}	0.108
Sixth flow	244.196	0.7	1.8×10^{-6}	0.108
Initial shut in	0	0.7	0	0.486
Extended flow	108.030	0	1.8×10^{-6}	0.108
Final shut in	0	0	0	0.108

repeatedly with the interpretation software of Fast, Work-Bench, and PAN; results interpreted with PAN were chose at last.

1. Interpretation results
   (1) Model: triple composite formation
       No reflection of no-flow boundary appeared in the test; and the outer boundary was infinite boundary.
   (2) Formation parameters
       $k_1 = 5.55$ mD
       $k_2 = 14.1$ mD
       $k_3 \approx 40$ mD
       $r_{1,2} = 40$ meters
       $r_{2,3} = 400$ meters
       $p_i = 74.516$ MPa

Skin factor and non-Darcy flow coefficient are listed in Table 8.14.
The plots of the aforementioned analysis results are shown in Figures 8.41, 8.42, and 8.43.

2. Interpretation results analysis
   (1) The reservoir the well penetrated becomes better gradually from near its bottom hole to far away from the bottom hole. The permeability of peripheral area of the well is about 40 mD, nearly the same as that of the adjacent Well KL-203.
   (2) No no-flow boundary effect appears during the test, even though there are faults in the south of the well as indicated in the structural map. It suggests that small faults in the blocks do not hinder gas flow and that it is not necessary to take them into account in gas field development.

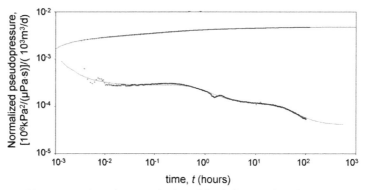

Figure 8.41 Log–log match plot of final shut in of Well KL-205.

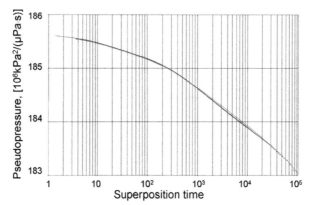

Figure 8.42 Semilog match plot of final shut in of Well KL-205.

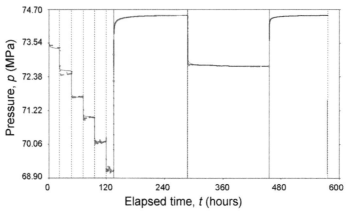

Figure 8.43 Pressure history match verification of Well KL-205.

Table 8.15 Deliverability Improvement of Well KL-205

Period	Choke, mm	Gas flow rate, 10^4 m³/day	Flowing pressure, MPa	Producing pressure differential, MPa	Skin factor
Second period of back-pressure test	11.11	106.17	72.4925	2.0155	1.2
Extended test	11.11	108.03	72.7440	1.7640	0

(3) The completion quality of this well is quite good. The true skin factor of the well is decreased gradually from 1.2 in the initial flowing period to zero later, indicating that the bottom-hole condition improved continuously with the lapse of production time due to the scouring effect of the gas flow.

(4) Another evidence of improvement of the bottom-hole condition is that Well KL-205 produced with ϕ 11.11-mm choke twice in the deliverability test, the first production period with this choke is the second period of back-pressure test with a gas flow rate of 106.14×10^4 m³/day; and the second period is the extended test with gas flow rate of 108.03×10^4 m³/day, slightly higher than the former, but the producing pressure differential is lower than that of the former (Table 8.15).

(5) The non-Darcy flow coefficient during the flowing periods of Well KL-205 is quite low, and its value calculated with variable skin factor processing method by well test interpretation software introduced in Chapter 3 is $D = 0.018 \times (10^4 \text{ m}^3/\text{day})^{-1}$, close to that of common gas wells. It implies that there was no serious turbulent flow in the bottom hole due to thick pay zones; even the flow rate of the well was very high, which was quite different from the extremely high non-Darcy flow coefficient $D = 3.6352 \times (10^4 \text{ m}^3/\text{day})^{-1}$ of Well KL-203 caused by its irrational casing program. These parameters can be used for the design of a development program.

(6) The wellbore storage coefficient is $C = 0.1$ m³/MPa, which is within the normal range of a high-pressure gas well, and provides other evidence of the correctness of well test interpretation results from the other side.

3.3 Gas Reservoir Description of KL-2 Gas Field

The analysis and study of the dynamic performance of the KL-2 gas field is a typical example for a solution to the key problems of field development via the short-term production test of exploration wells and development and appraisal wells. These single well evaluations make it possible to delineate the outline of the characteristics of the gas reservoir dynamically.

3.3.1 Dynamic Model of Gas Wells Established Primarily Through Short-term Production Test

Short-term production tests were conducted following the well test in a discovery well, Well KL-2, and three development and appraisal wells, Well KL-201, Well KL-203, and Well KL-205. The short-term production test consisted of both the deliverability test and the short-term extended test, and accorded with the arrangements for the production test well illustrated at the beginning of this chapter. At the very beginning, conventional layered well test methods were tried in Well KL-2 and Well KL-201 to make clear reservoir parameters of each interval in the vertical direction. However, it was discovered in field practice that neither layered parameters nor aggregative parameters of all layers in the well can be obtained with this method. Due to good vertical connectivity of the main pay zones, the method of the commingled well test, for example, all pay zones of the well tested together, was chosen for calculation of the deliverability and aggregative parameters of all pay zones of the well. Unfortunately, the commingled well test of Well KL-203 failed to obtain the aggregative parameters of all pay zones of the well caused by serious hindering of high-rate gas flowing in the string and acquiring of abnormal pressure data due to the its structural defects of test tools. For the later test in Well KL-205, the test string structure was improved and test methods and procedures were selected to meet the requirement of the field; representative and explicit knowledge about the deliverability of the wells, structure, and parameters of the reservoir of the KL-2 gas field were eventually obtained.

A preliminary description of the gas wells and the reservoir based on the aforementioned analysis of short-term production test data is concluded as follows.

1. All gas wells drilled on the structural axis of the anticline of KL-2 gas field are high productivity wells; their AOFP can be as high as more than 1000×10^4 m^3/day or even exceed $2000-3000 \times 10^4$ m^3/day in some regions and so the requirement of "higher yielding and fewer wells" in the development design can be met.

2. Formation flow capacity of the reservoir has been determined by dynamic analysis. The *kh* value is about 1000 mD·m in the area near the bottom hole of Well KL-205 and is as high as 6000 mD·m 400 meters away from the bottom hole; the *kh* value can also reach 6000–7000 mD·m in the area near Well KL-203. It is such a high formation flow capacity of the reservoir that creates the high deliverability condition for the KL-2 gas field.

3. The KL-2 gas field is an abnormal high-pressure, high-energy gas field; the pressure coefficients of the wells in the field calculated by test data are greater than 2, which is also one of the important factors of its high deliverability.

4. The pressure gradient analysis shows that all static formation pressures of different gas wells are on the same "natural gas static pressure gradient line," which means that the KL-2 gas field is an uncompartmentalized gas field.

5. Although there are more than 10 small faults in the gas-bearing area according to the structural map plotted based on geophysical prospecting, transient tests have confirmed that these faults do not hinder gas flow at all. Therefore, the whole field can be developed just like a complete and integrated gas reservoir.
6. The main pay zones of the KL-2 gas field distribute in the range over 200 meters vertically in the Cretaceous Bashijiqike Formation and are isolated by interbeds, but the dynamic behavior appearing during well tests is just like a thick formation, which means that the pay zones are connected to each other vertically and the interbeds do not isolate the pay zones at all.
7. It was confirmed from dynamic analysis that the skin factor of Well KL-205 completed finally is nearly zero and proved that the drilling and completion techniques had been improved constantly can fully satisfy the technical requirements for developing such gas zones with huge thickness and extremely high pressure.
8. According to the analysis of Well KL-205 test data, it was acknowledged that the non-Darcy flow coefficient of normal wells in this area is about $D = 0.018(10^4 \, \mathrm{m}^3/\mathrm{day})^{-1}$, which provides this basic parameter for the development program design.
9. Pressure gradient data under various different high deliverability conditions obtained from the dynamic data of well tests provide a reliable basis for the design of well completion techniques.

These analysis results provided knowledge of the gas wells and the reservoir from various different facets.

3.3.2 The Established Dynamic Model of Gas Wells and Reservoir is Not Yet Perfect

The dynamic model of gas wells and reservoir established based on the aforementioned descriptions is only primary and not perfect; it needs to be improved further, and the reasons are as follow:

1. The characteristics of peripheral boundaries of the gas wells and the reservoir have not been identified from the analysis results of the aforementioned well tests.
2. Controlled dynamic reservoirs of the whole gas reservoir, even of each gas well in it, have not been determined.
3. Subsequent deliverability decline and dynamic variation trend of the gas wells cannot be predicted reliably.
4. Most of the production test wells in the KL-2 gas field drilled earlier in the evaluation period have not been put into production as production wells; when the field was put into normal development, another 10 more production wells were drilled, and they are the basis for maintaining the deliverability of the whole field and their dynamic behavior still needs to be monitored carefully for further study.

Maybe the KL-2 gas field is just like "a vigorous and healthy young man" at present and "he" has not realized that it is necessary to go to hospital to have a physical examination. However, he should know that everyone will be old and feeble and inevitably attacked by diseases. If the changes cannot be learned about and grasped in time, the consequences are unpredictable and uncertain.

4 TRACING STUDY ON GAS RESERVOIR DYNAMIC DESCRIPTION OF SLG GAS FIELD

4.1 Overview of SLG Gas Field

1. The SLG gas field was estimated to be a large-scale gas area with several hundred billion cubic meters of OGIP. It was partly put into production in 2006 and about 1000 gas wells had been put into production, and the gas production rate over the whole field reached about 1000×10^4 m^3/day by the end of 2007.

2. From 2001 to 2007, the study group that the author participated in performed a tracing study on several dozen key wells in the SLG gas field and made the whole process tracing description and prediction for these gas production wells since they were put into production until depleted.

3. The relative perfect dynamic models were established for the key wells, for the modeling various reservoir parameters of the area near the wellbore and those of well completion, the configuration of the block controlled by the well, the distances of the nearby boundaries to the well, the dynamic reserves controlled by the well and the dynamic deliverability and dynamic formation pressures of the well have been measured or calculated well by well; in addition, the initial static pressure gradient of gas wells over the whole SLG area was analyzed and the conclusion of that SLG gas field was impossible to be a compartmentalized field was obtained.

4. The analytical well test model was established with the analytical well test interpretation method for each tested well; the numerical well test analysis model was established with the numerical well test method for each key well and each important well and thus the special problems of well test interpretation in the lithologic formation in this area that the analytical well test model could not describe were resolved; in addition, it was the first time interpreting well test data with the numerical well test method in China.

5. Combined with sedimentary facies analysis in geological study, some key indices such as configuration, effective reservoir area, controllable reserves of the controlled block of the single well, and some constructive suggestions on the next step of field development were brought forward clearly for the first time in the situation and under the condition without quantitative analysis results in static geological study; these understandings have been commonly acknowledged by different concerned parties.

This study result was published at the natural gas development annual conference of CNPC in 2005 and was compiled into a symposium series of the natural gas development annual conference. This study has comprehensively implemented the new thoughts of the gas reservoir dynamic description in Chapter 1 (refer to Figure 1.4). It proves sufficiently that the gas reservoir dynamic description methods are integrated closely with field practices and are scientific and vigorous as well.

4.2 Geological Situation of SLG Gas Field

4.2.1 Structural Characteristics and Production Horizons

The SLG gas field lies in the west of the Yishan Slope of the Ordos Basin and is composed of lithologic gas reservoirs. Its configuration is a wide and gentle west-inclined monocline. Although there were multiple low and flat nose uplifts in parallel on the monocline, natural gas distribution is controlled mainly by sand bodies and physical properties.

The major production horizons of the SLG gas field are sandstones of Shihezi formation in Permian, and some wells produce gas from the Shanxi Formation in Permian or Lower Paleozoic.

4.2.2 Sedimentary Microfacies of the Formation

The sedimentary facies of the Shihezi Formation in the SLG gas field is braided river course deltaic plain subfacies. The main specific microfacies include channel lag, river island, channel deposit, and natural barrier, and their model graph is as shown in Figure 8.44.

Although there are many types of microfacies in a braided river delta, gas producing zones are only limited in those coarse-grained facies belts in river midchannel bars in mainstream channels. As shown clearly in Figure 8.44, the effective gas pay zones were

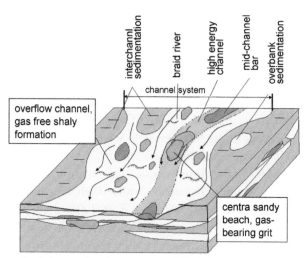

Figure 8.44 Schematic diagram of sedimentary model of Shihezi formation in the SLG gas field.

only encountered during the drilling of gas wells at the central part of high-energy channels in the river channel complex. The encounter ratio in drilling of the formation in the same sedimentary period is very low; the effective sand beds are very thin and very narrow; and the effective sand beds at different depths and formed in a different geologic age are absolutely separated from each other by impermeable barriers.

4.3 Dynamic Description Process of SLG Gas Field

Understanding of the SLG gas field underwent a hard process from its discovery in the exploration period to its initial reserves evaluation and to its being put into final development. At the very beginning, it was thought that the whole gas bearing area was of the high deliverability type according to AOFP values of individual wells calculated with simple methods and data obtained in initial periods of well tests. After that, the ranges of the sand layer were recognized by geophysical prospecting, and several hundred billion cubic meters of OOIP was obtained with the simple volumetric method, and the gas field was inferred to be a huge uncompartmentalized one.

However, the subsequent production test in some wells and primary analysis results of these production test wells dissented from the aforementioned inferences. Above all, the judgment of being an uncompartmentalized gas field was denied, as effective sand layers are in fact cut into very small broken blocks, which raises many inevitable difficult problems in the following field development. Tracing studies on production test well performance lasting for 7 years have made clear and confirmed the geological structure and deliverability decline rule of the reservoir, and a special development policy was formulated based on these studies. After many twists and turns, the rational development mode for this area was ultimately chosen and gas field development started.

4.3.1 Preliminary Knowledge Obtained in 2001 and 2002

From March to September 2001, several exploration wells, including Well S-4, Well S-5, Well S-6, Well S-10, and Well T-5, were put into short-term production tests one after the other; modified isochronal tests and the following pressure buildup tests were conducted in these wells. Combined with results of modified isochronal tests, pressure buildup tests, and the short-term production test of these wells, the dynamic models of these wells were primarily established, a series of related study work was done, and basic understandings for gas wells in the SLG gas field were finally obtained.

1. He-8 Formation is low permeability sandstone

Transient test analysis of the aforementioned production test wells showed that the permeability of most wells is less than 2 mD. Well S-6 is an exception whose permeability is as high as 7 mD. By the end of 2007, more than 1000 gas wells penetrated He-8 Formation; the permeability values of them are all less than that of Well S-6, which further confirms understanding about the formation and determines the characteristics of low deliverability of the gas wells in this area.

2. AOFPs of wells in the field are slightly low to medium

Deliverability tests were conducted with the traditional test method during the production test of the field; the measured AOFP values obtained were smaller than $10 \times 10^4 \, \text{m}^3/\text{day}$ in general, much less than that obtained from well tests initially. The AOFP of Well S-6 was as high as $120 \times 10^4 \, \text{m}^3/\text{day}$ from its initial well test, but the value was only $25 \times 10^4 \, \text{m}^3/\text{day}$, one-fifth of the initial value, from the deliverability test conducted with the normative test method.

3. Stimulation of fracturing treatment is quite effective

The flow rates of most gas wells in sandstone formations with extremely low permeability completed by perforation cannot reach the standard of the industrial gas flow rate. However, those wells completed with fracturing treatment were confirmed by a transient test possessing a hydraulic fracture and the half-length of which can be about 60—100 meters and so much increasing their flow rates. The extremely large-scale fracturing treatment was expected at the very beginning to be able to create an extremely long fracture to penetrate the lithologic boundaries of the effective sandstone layer and so that to increase the flow rate and enhance the controllable reserves of the well significantly, but it is proved infeasible by field practices.

4. Existence of no-flow lithologic boundary near the well

It is forecasted by a sedimentary facies study that the effective reservoir of the SLG gas field in the Shihezi Formation in Permian locates in the local area of the high energy channel center in the river channel complex. The width of the effective sandstone layer of a single complex is no more than several hundred meters, and wells with a penetrated effective sand body are surely very close to the lithologic boundaries.

The transient test interpretation corroborated the aforementioned forecast results and quantitatively determined the distance to the boundaries from the well, and the widths of the river channels above five wells were calculated to be only 100—300 meters and lengths of about 1000—2000 meters, which implies that the area of the reservoir and the dynamic reserves controlled by every single wells are very limited and so nonconventional development methods have to be adopted for these gas reservoirs.

5. Study on reservoir pressure gradient of SGL gas field

The initial static pressure gradient analysis of exploration wells penetrating the Upper Paleozoic formation was performed in 2001 and 2002, and the adopted analysis methods were introduced in Chapter 4. First, all initial static pressure data of all exploration wells were collected, including ① initial static pressure data of 26 exploration wells drilled at the very beginning of exploration of the SLG gas field and ② initial static pressure data of 162 exploration wells penetrating the Upper Paleozoic Formation in the Ordos Basin by the end of 2002. Second, a static pressure gradient graph was plotted. Knowledge gained from analysis of the initial static pressure plot of the SLG gas field is that although the sand bodies of the major blocks of this field developed continuously, the measured static pressure data points are very scattered and do not locate on a "natural gas static pressure gradient line" at all, which confirmed that the SLG gas field is not an uncompartmentalized gas reservoir. The initial static pressure plot of 162 exploration wells in the Ordos Basin is approximately a static hydrostatic pressure gradient line, and the pressure coefficient is about 1 from regression analysis, which reflects the pressure characteristics formed during the deposition and migration process of natural gas.

4.3.2 Tracing Study of Long-term Production Test in 2003

1. Introduction

Just as illustrated at the beginning of this chapter, the production test is the most effective approach in solving important crucial problems in the development of special lithologic gas fields. After connecting to the surface pipeline, putting the representative gas wells in the gas field into a long-term production test, acquiring data of pressure and gas flow rate carefully, and then performing a study on dynamic description based on these data, thorough knowledge of the gas reservoir can certainly be gained, and so the effective approaches to dealing with gas field development problems can be found consequently.

This was what has been done in the SLG gas field. Since September 2002, 16 gas wells had been connected to the surface line and put into the long–term production test gradually. By the end of August 2003, the maximum cumulative output of a single well among them reached 1700×10^4 m^3; most of the test wells reached an unflowing limit and their bottom-hole pressures declined to 6−9 MPa; some wells appeared as depletion signs and had to be produced with a wellhead blowdown mode; the single well output decreased to about 1×10^4 m^3/day even lower than that.

2. Dynamic tracing study

   (1) Purpose and method of dynamic tracing study

   The purposes of dynamic tracing study are as follow.

   a. To realize and analyze on real time the transient performance of the gas wells, including their dynamic deliverability, dynamic AOFP, and dynamic formation pressure.

   b. To the dynamic models of the gas wells: modify and improve the preliminary model parameters by means of simulation, comparison, and match of theoretical calculation results and acquired pressure data and so modify and improve the models or make them more perfect gradually and then predict the future dynamic trend with the improved dynamic model.

   During the dynamic tracing study in 2003, model verification, modification, and improvement of model parameters were performed one by one for all key gas wells in the SLG gas field.

(2) Numerical well test analysis

Numerical well test analysis is a new well test analysis method developed in recent years; it can freely set boundary configuration, heterogeneity area distribution, and well completion status of the block in which the tested gas well locates with the numerical method, which makes the well test model vividly consistent with actual conditions of the formation. The study on gas wells in the SLG gas field with the numerical test method was the first application of this method in China.

The research group in which the author participated encountered unprecedented difficulties caused by the special heterogeneity distribution and specific boundary configuration near its bottom hole in the study of the key well, Well S-6. Fortunately, the problems were eventually solved successfully with the numerical well test method. The dynamic models of Well S-6, Well S-7, and other wells were established with the numerical well test method and then applied sufficiently in the following long-term dynamic description, verification, and prediction; the practice has proved their validities.

Numerical well test interpretation results showed that areas controlled by the tested gas wells were enclosed by rectangular no-flow boundaries—the width of which is 320 meters from east to west and the length 800 meters from south to north. It was proposed at the well test result report meeting based on the interpretation to drill inspection wells with well spacing of 500—1000 meters on both east and west sides of Well S-6 so as to verify the reliability of interpretation results.

4.3.3 Enhancement of Tracing Study in 2005

The research group in which the author participated traced and studied the SLG gas field continuously and made some new progress in 2005.

1. Agreement of conclusions obtained from facies analysis and dynamic description

The geological study fruits concerning the fluvial deposits in the Upper Paleozoic of the Ordos Basin were recognized in 2005. The fruits indicate that the sedimentary facies of the Shihezi Formation is dominated by braided river deltaic plain subfacies and that there are seven or eight types of microfacies, including channel floor lag, channel bar, and channel aggradational deposit. The effective sand bodies of the major gas producing layers are in the coarse-grained facies belts of complicated braided river channels with high energy, and because of strong lateral migration of the plain subfacies braided river channel, these coarse-grained facies belts disconnect each other laterally, and effective sand bodies formed in different sedimentary times scatter over different lateral positions, all of which cause the penetrated effective sand bodies of each single well to be less in quantity, thin in thickness, small in area, and unstable distribution between wells as well.

The aforementioned fruits of the geological study are very consistent with the results of several years of dynamic description and tracing study. Both of them reflect the structural characteristics of the reservoir from different aspects, but dynamic description results are more specific and quantitative even that from which how much gas can be produced by each gas well can be predicted.

2. The results of dynamic description were verified by drilling exploration along the east–west direction through Well S-6

As the best well in the SLG gas field, Well S-6 has drawn great attention of various different parties concerned. It was predicted on the basis of a numerical well test in 2003 that the well is located in a rectangular reservoir with lithologic boundaries and this rectangular reservoir is narrow in the east–west direction and long in the south–north direction; it was hoped that this prediction could be verified directly by drilling exploration. By 2005 there were 12 wells, including Wells S-6 and S-410, and 10 new drilled wells with an average well spacing of 800 meters along the east–west direction through Well S-6. Figure 8.45 is the subzone profile map plotted after all 12 wells were drilled.

As shown clearly in Figure 8.45, the gas-bearing sand layers penetrated by these wells were irrelevant to each other, and absolutely independent channel sand bodies formed in different geological times, despite the very short distances between them. It verified the conclusion on the Shihezi reservoir structural characteristics obtained and confirmed by several dynamic description studies from 2001 to 2003 and then to 2005.

3. Application of stable point LIT deliverability equation

The deliverability of each single well in this gas field is quite low and declines quite quickly so that the cumulative gas production of each single well is very limited; these conditions mean that normative deliverability test methods (e.g., modified

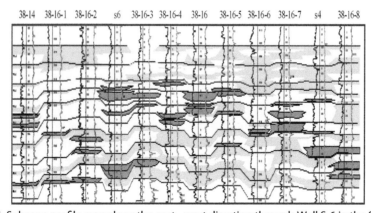

Figure 8.45 Subzone profile map along the east–west direction through Well S-6 in the SLG gas field.

isochronal test method) cannot be widely applied. In such a case, how to calculate the deliverability indices of each gas well on the principle of low cost and high efficiency? A "stable-point LIT deliverability equation" method is recommended based on careful theoretical derivation in the study at that time. The equation and its application have been discussed in detail in Chapter 3 and are not illustrated again here.

This method enables not only establishing the exact initial deliverability equation of each production well with only an acquired initial stable point of data, but also derives further the dynamic indexes such as dynamic deliverability equation, dynamic AOFP, dynamic IPR curve, and dynamic formation pressure of gas drainage boundary based on the initial deliverability equation.

4. Follow-up tracing study on dynamic model

A follow-up tracing study on key gas wells such as Well S-4 and Well S-6 was carried out with the gas flow rate and pressure histories newly acquired at a later time. Take Well S-6 for instance, its pressure history match verification showed that its actual flowing pressure declined more quickly than that calculated from its theoretical model, thereupon adjusting the distances to the outer boundaries, that is, reduced the length of the rectangular boundary from 800 to 700 meters under the condition of maintaining the basic formation parameters unchanged so that the bottom-hole pressure of the theoretical model was consistent with measured actual pressure data at the end of 2005, thus confirming that the dynamic reserves controlled by Well S-6 is slightly smaller than that estimated in 2003.

4.4 Dynamic Description Result of Typical Wells

As introduced in earlier parts of this chapter, the dynamic description research of the SLG gas field underwent several big stages. Perfect dynamic models were ultimately established for several dozens of wells based separately on study results at different stages. Take Well S-6 for instance again here to illustrate the specific operations at different study stages.

4.4.1 Primarily Establish the Dynamic Gas Well Model by Short-term Production Test

The short-term production test of Well S-6 is shown in Figure 8.46.

The transient deliverability equation of the well was established with data of the modified isochronal test performed at the beginning of the production test shown in Figure 8.46, and its stable point LIT deliverability equation was obtained from data of the extended test in the same figure. The equation establishment was introduced in detail in Chapter 3.

Following the production test was a pressure buildup test, and the dynamic model of the well was established based on the analysis results of the pressure buildup curve.

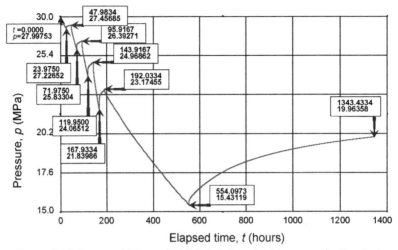

Figure 8.46 Pressure history of Well S-6 during short-term production test.

Model I of Well S-6 Established by Results of the Type-curve Match Analysis of Pressure Buildup Test Data

Figure 8.47 shows the analysis results of a log–log type curve match with well test interpretation software.

Figure 8.47 shows that the log–log curve match seems quite good, and the following conclusion and model parameters were obtained.

1. The well locates in a round composite reservoir, the flow conditions in the inner zone are better than that in the outer zone, and there is no boundary effect being detected
2. The permeability in the inner zone is $k_1 = 7$ mD, whereas that in the outer zone is $k_2 = 0.49$ mD

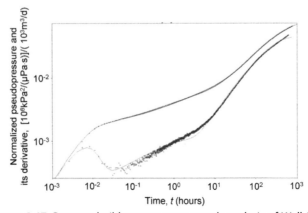

Figure 8.47 Pressure buildup type curve match analysis of Well S-6.

3. Radius of inner zone is $r_{1,2} = 101$ meters

4. Total skin of the well $S_t = -5.9$ (a hydraulic fracture exists)

5. Non-Darcy flow coefficient $D = 0.028$ $(10^4 \text{ m}^3/\text{day})^{-1}$

6. Wellbore storage coefficient $C = 2.2 \text{ m}^3/\text{MPa}$

The semilog plot of measured pressure buildup matches that of the model quite well. However, a great deviation is observed when comparing the pressure history of model I with measured pressure data in the short-term production test (see Figure 8.48), indicating that model I cannot represent the real formation conditions of Well S-6 correctly and completely, especially the boundary distribution.

Improved Model II of Well S-6 Obtained by Adjusting Model Parameters of Model I

Pressure history match is the final verification of the correctness of the model. But among those well test models, which can be obtained from the aforementioned pressure log–log plot, no one can make a better pressure history match than model I. According to the flowing pressure decline trend, a peripheral area in the reservoir exists with much worse flow conditions, and a more complicated composite model with three zones with different flow conditions and the permeability is worse and worse from inner to outer is finally constructed accordingly. The model and its parameters include the following

1. The gas well is in the center of the composite reservoir, which consists of three zones with different flow conditions and the permeability is worse and worse from the inner zone to the middle one and the outer one; there is no boundary effect

2. The permeability of the inner zone, middle zone, and outer zone is $k_1 = 7$ mD, $k_2 = 0.15$ mD, and $k_3 = 0.03$ mD, respectively

3. The radius of the inner zone is $r_{1,2} = 71$ meters, the outer zone is in the area of $r \geq r_{2,3} = 240$ meters, and the middle zone is in the area of $r_{1,2} < r < r_{2,3}$

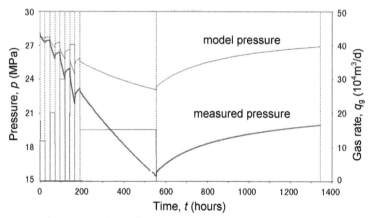

Figure 8.48 Pressure history match verification in the short-term production test of Well S-6 (model I).

4. Total skin factor is $S_t = -5.9$; non–Darcy flow coefficient $D = 0.028(10^4 \, m^3/day)^{-1}$, and wellbore storage coefficient $C = 2.2 \, m^3/MPa$

The pressure history match verification during the production test is shown in Figure 8.49.

It can be seen from Figure 8.49 that model II can match the measured pressure history during the short-term production test much better. Therefore, this model represents actual formation characteristics more accurately than model I, but because it still does not match the pressure variation very well in the period just before and after shutting in, it may not be able to match a long production history.

Model III of Well S-6 Modified with Production Test Data Acquired in 2003

The second production test of Well S-6 was initiated after connecting the well to the surface pipelines in September 2002 and ended in July 2003 when the wellhead pressure had seriously declined, and then the well was shut in and the pressure buildup test was performed.

Model II was verified in the extended pressure history, and the result is shown in Figure 8.50. Because bottom-hole pressure data in the pressure buildups were measured by electronic pressure gauges, pressure history match verification was done mainly with these sections of data.

As shown in Figure 8.50, a quite good match of model pressure and measured pressure in the initial pressure history period has been obtained, but the model pressure becomes far higher than the measured bottom-hole pressure acquired onsite when the test is nearly terminated, which indicates that model II is not corresponding perfectly to the actual reservoir conditions yet and its parameters need to be modified further.

However, the analytical test solution method can do nothing in this case: very complicated heterogeneity in the composite reservoir exists, which consists of three

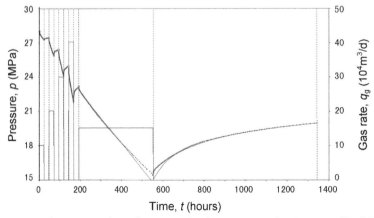

Figure 8.49 Pressure history match verification in the short-term production test of Well S-6 (model II).

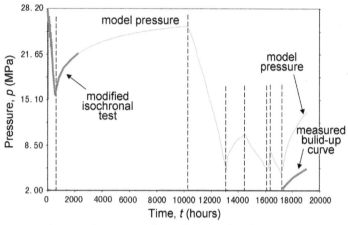

Figure 8.50 Pressure history match verification of Well S-6 in 2003 (model II).

zones with very different flow conditions, and the permeability of the zones is worse and worse from the inner to outer zones in the area Well S-6 locates. In addition, sedimentary facies analysis showed that the formation seems to be narrow and rectangular, surrounded by no-flow lithologic boundaries. These two conditions are impossible to represent simultaneously in an analytical well test model. To deal with this problem, a numerical well test model, which was named model III of Well S-6, was built.

Results of the numerical well test confirmed that Well S-6 is located in the central part of the composite formation surrounded by rectangular no-flow lithologic boundaries, as shown in Figure 8.51.

Figure 8.51a shows the geometry of heterogeneity distribution, and Figure 8.51b shows the grid division for numerical well test analysis.

1. A local high permeability zone exists like a channel bar near the bottom hole of the well, that is, Zone 1, in which permeability k is as high as 5.81 mD
2. In the area outside the high permeability zone, that is, Zone 2, permeability k decreases to 0.0387 mD, and in the area outside this zone, that is, in Zone 3, permeability k decreases to as low as 0.0116 mD
3. The area controlled by the well is enclosed by rectangular no-flow boundaries, and the rectangle area is 320 meters wide from east to west and 800 meters long from south to north

These interpretation results have been confirmed by the pressure history match verification of Well S-6 shown in Figure 8.52.

Figure 8.52 shows that the model pressure matches quite well with the measured pressure from the beginning to the end; although measured pressures are absent in the middle part, bottom-hole pressures converted from wellhead pressures are basically consistent with model pressures.

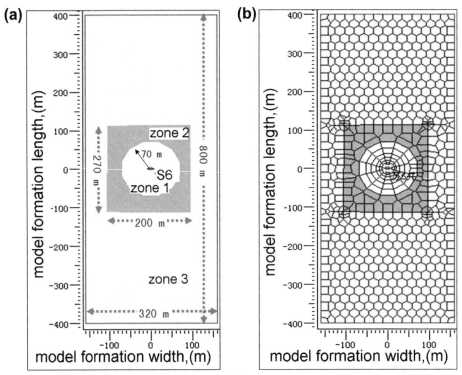

Figure 8.51 Results of numerical well test analysis of Well S-6.

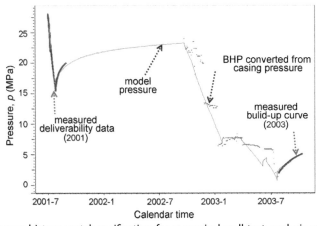

Figure 8.52 Pressure history match verification for numerical well test analysis results of Well S-6.

The other parameters of model III of Well S-6 include:

4. Parameters about the heterogeneity: radius of Zone 1 is 70 meters; the area of Zone 2 is a rectangle of 200×270 meters excluding Zone 1

5. The total skin factor of the well is $S_t = -5.9$; non-Darcy flow coefficient $D = 0.028(10^4 \text{ m}^3/\text{day})^{-1}$; and wellbore storage coefficient $C = 2.2 \text{ m}^3/\text{MPa}$.

It can be seen that, compared with the former two models, model III of Well S-6 reflects formation structure around the well much better and more faithfully and can be regarded as the more perfect dynamic model of Well S-6: it describes parameters such as formation permeability k and well completion parameters in the area near the bottom hole, parameters about the formation heterogeneity variations far away from the wellhole, and especially the type and distance to the well of the outer boundaries. This model can be used to predict future pressure variations and other dynamic performances of the well.

Continuing Improvement of Model III by Adjusting Parameters with Production Test Data in 2005

Recorded flow rate and pressure histories in the production test of Well S-6 up to 2005 are shown in Figure 8.53.

It can be seen from Figure 8.53 that the well produced intermittently at a gas flow rate of about 1×10^4 m^3/day, during which the wellhead tubing pressure decreased to lower than export pressure, and so blow-down production had to be

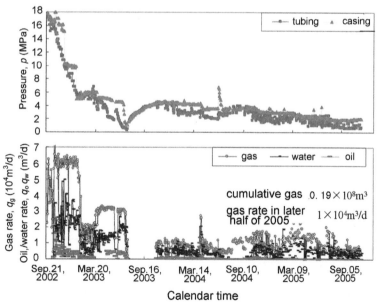

Figure 8.53 Production curve of Well S-6 from 2002 to 2005.

adopted, and the cumulative gas production amounted to about 0.19×10^8 m^3 by the end of 2005.

The tracing study on the dynamic model of Well S-6 was carried out according to the pressure history in the period, and the reservoir model was further improved consequently as shown in Figure 8.54.

Figure 8.54 shows that there is no change in near-hole formation parameters in the improved model III$_1$ of Well S-6, but slight modifications for outer no-flow boundaries have been made: The length of the rectangle reservoir decreased from 800 meters in original model III to 700 meters in improved model III$_1$. The pressure history match verification based on new model III$_1$ is shown in Figure 8.55.

It can be seen from Figure 8.55 that a very good match of theoretical model pressure history of model III$_1$ and measured pressure history throughout the whole duration has been obtained; it can be said that model III$_1$ is basically capable of reappearing exactly the dynamic performance of Well S-6 and the reservoir and is capable of representing correctly the formation structure around this well and describing the reservoir characteristics faithfully.

At that time, Well S-6 had been almost depleted and of no commercial value, despite being able to produce by blow-down intermittently.

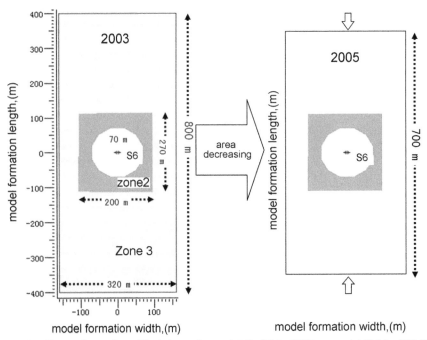

Figure 8.54 Comparison of modified dynamic model III$_1$ (b) in 2005 to model III (a) of Well S-6.

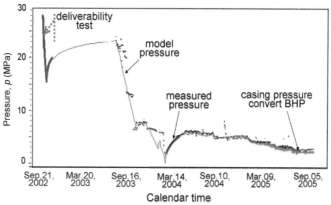

Figure 8.55 Pressure history match verification for dynamic model III₁ of Well S-6.

4.4.2 Calculation of Average Reservoir Pressure and Dynamic Reserves of Well S-6

Based on verification of the limited constant-volume dynamic model for Well S-6, and with data acquired during the process of natural gas producing continuously, well test interpretation software can calculate the average formation pressure of block Well S-6; block Well S-6 locates, at any time, and the calculation result can be plotted in a pressure history plot as shown in Figure 8.56.

After acquiring the average formation pressures of a constant-volume block, draw a plot of these average formation pressures vs the cumulative gas production, and dynamic reserves controlled by Well S-6 can be calculated as shown in Figure 8.57.

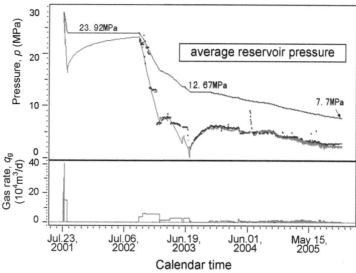

Figure 8.56 Calculation result of average reservoir pressure of Block S-6 from 2001 to 2005.

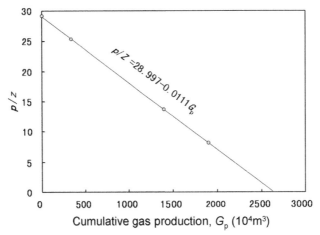

Figure 8.57 Reserves calculation of Block S-6 through the p/Z-G_p approach.

As obtained by regression analysis, the dynamic reserves controlled by Well S-6 is $2612 \times 10^4 \, \mathrm{m}^3$ (Figure 8.57), and the cumulative produced gas so far amounts to $1887 \times 10^4 \, \mathrm{m}^3$, 72% of the total dynamic reserves.

4.5 Knowledge Obtained From the Dynamic Description of SLG Gas Field

The study on dynamic description of the SLG gas field had been carried on until this book was reprinted. The main work was completed during the period of 2001−2005. In 2001, preliminary dynamic models were established separately for a batch of wells, and primary conclusions of the reservoir structure were made on the basis of short-term production test data acquired from the key area of the field; by 2003 the models of the wells had been further improved and verified by test data obtained from more wells in the duration for more than a year; it was recognized gradually from practice that the extracted conclusions from those dynamic descriptions were not only applicable for a few individual wells, but also a representative general rule applicable for the wells in this field. The typical gas well dynamic model parameters acquired from the SLG gas field by the author are listed in Table 8.16.

The following conclusions can be drawn from the listed data in Table 8.16.

1. The initial formation pressures are close to 30 MPa, indicating that the formation pressure, the one among three important factors of the deliverability of gas wells, is normal.
2. Another important factor of deliverability of gas wells is gas well completion quality indices. Table 8.16 shows that the half-lengths of fractures caused by fracturing treatment are as long as 50 meters or so, and the skin factors are about −5.2; both are favorable indices under the present technical conditions.

Table 8.16a Dynamic Model Parameters of Production Test Wells in SLG Gas Field (Part 1)

Well name	Initial pressure from match analysis, p_i MPa	Interpreted formation flow capacity, kh mD·m	Interpreted formation permeability, kmD	Interpreted half-length of fracture, x_f m	Interpreted total skin factor, S
S4	28.329	24.5	1.63	151	−6.56
S5	28.740	28.5	1.7	5.1	−3.59
S10	27.350	25.8	2.08	48.6	−5.85
T5	29.330	6.97	0.78	43	−5.73
S14	29.766	15.3	1.1	63	−6.11
S20	28.906	5.5	0.466	40	−5.26
S25	28.300	6.45	1.17	93	−6.36
S16–18	26.680	1.19	0.143	64	−6.13
S37–7	32.477	1.15	0.12	35	−5.52
S39–17	31.657	1.57	0.094	17	−4.8
S37–15	29.950	3.6	0.391	11	−4.36
S33–18	30.324	2.45	0.302	56	−5.99
S38–16	33.400	4.2	0.276	11	−4.36
Average	29.631	9.74	0.784	49.0	−5.17

3. Formation flow capacities (kh), the most important factor of the initial deliverability of gas wells, were as low as less than 10 mD·m; the average permeability of the effective sand layers was less than 0.8 mD. The production test wells whose data were used in the statistical analysis were all high-quality ones in the gas field, so the permeability of the gas reservoirs of most gas wells should be lower than that, even is as low as a few hundred mD, which decides that the initial deliverability of gas wells in this block is very low.

4. The key factor influencing stabilization of the deliverability of a gas well is the effective area and dynamic reserves controlled by this gas well. It can be seen from Table 8.16 that the average effective area controlled by a single well is about 0.2 km^2, and converted dynamic reserves controlled by a single well is about 0.2×10^8 m^3. Taking into account that the production test wells listed in Table 8.16 were optimized wells in the block, these indices are actually the upper limit of their value in the block.

Therefore it is easy to understand that the cumulative production of a single well in the SLG gas field averaged less than 0.1×10^8 m^3 when the gas well had been nearly depleted. For such a gas field incised seriously by lithologic boundaries and with a width of less than 110 meters and a length of 1.7 km, if a development program is formulated by adopting a line well pattern for homogeneous sandstone formations and common numerical simulation analysis suitable only for wide and connected reservoirs is done, the obtained development indices may be very optimistic but it is absolutely not applicable in guiding the development of the gas field.

Table 8.16b Dynamic Model Parameters of Production Test Wells in SLG Gas Field (Part 2)

Well No.	Reservoir geometry	Distances to boundaries of rectangular constant-volume block, m				Length—width ratio	Area of constant-volume block, km^2	Remarks
		L_{b1}	L_{b2}	L_{b3}	L_{b4}			
S4	Rectangular constant-volume block	20	505	100	995	12.5:1	0.18	Numerical well test
S5	Rectangular constant-volume block	33	338	91	561	7.25:1	0.111	
S10	Rectangular constant-volume block	68	281	59	1500	14.0:1	0.226	
T5	Rectangular constant-volume block	33	700	66	1890	26.2:1	0.256	
S14	Rectangular constant-volume block	29.5	480	25.5	92	10.4:1	0.031	
S20	Rectangular constant-volume block	14.1	210	49.5	1000	19.0:1	0.077	
S25	Rectangular constant-volume block	111	1000	52.5	1000	12.2:1	0.327	
S16-18	Rectangular constant-volume block	50	1500	50	1500	30.0:1	0.300	
S37-7	Rectangular constant-volume block	20	90	250	2500	12.6:1	0.699	
S39-17	Rectangular constant-volume block	60	1000	40	1000	20:1	0.200	
S37-15	Rectangular constant-volume block	42	1000	67	1000	18.3:1	0.218	
S33-18	Rectangular constant-volume block	17	400	37	400	14.8:1	0.043	
S38-16	Rectangular constant-volume block	50	1000	23	1000	27.4:1	0.146	
Average		42.1	654.1	70.0	1111	15.7:1	0.217	

During the period of 2001–2003 when the field was just discovered and the field became a public focus with great interest, to bring forth the aforementioned conclusions and opinions was undoubtedly risky and needed to bear great pressure. However, because the charm of science resides in her strict theoretical derivation, reasoning, and calculation, it will not be difficult to obtain correct conclusions as long as there is a bit more attitude of seeking truth from facts and sense of responsibility; the conclusions will be ultimately understood and accepted. Initial judgment on the SLG gas field was confirmed step by step by gradual development of the field.

In conclusion, a special development mode must be chosen for those special lithologic gas fields such as the SLG gas field.

5 DYNAMIC RESERVOIR DESCRIPTION OF YL GAS FIELD

5.1 Overview of YL gas field

5.1.1 Geographical Location and Geological Situation of YL Gas Field

The YL gas field is located at the eastern side of the Yishan Slope, Ordos Basin, east of the SLG gas field and south of the Changbei Contract Block. It can be divided into six production blocks in accordance with lateral reservoir development conditions.

The major pay zones of the YL gas field are Shanxi Formations in the Permian system of the Mesozoic group. The main producing interval, Shan-2, produced dry gas containing no formation water. Based on geological research results it is thought that the formation originated from frontal subfacies deposits of fluvial facies of braided river delta and that their microfacies are composed mainly of subaqueous distributary channels, channel mouth bars, and subaqueous distributary bays. The coarse sandstone layers are usually over 15 meters thick. As a whole and compared with upstream plain subfacies, the gas occurrence conditions of the reservoir here are better, the encounter ratio of the sand layer in drilling is higher, and the proportion of the effective thickness of sand is also higher. The channel sandstones were also commonly cut by lithologic boundaries such as the SLG gas field; only the geometric sizes of the effective permeable formation of it and Shihezi Formation in the SLG gas field were different.

Static geological data analysis results of Shan-2 Interval show that its average formation porosity is 6%, average permeability ranges about 3–5 mD, gas saturation is about 70%, and this is a lithologic sandstone gas field with low porosity and medium-low permeability.

These geological characteristics of the gas reservoir have been reflected sufficiently by the dynamic performance variation of the gas wells; however, the dynamic description of the gas reservoir enriched, quantized, and verified further the geological characteristics of the gas reservoir.

5.1.2 Commissioning and Production Test of YL Gas Field

The OGIP of the YL gas field is nearly 1000×10^8 m^3, and the field was constructed completely and put into production in 2004. The author participated in the research on the dynamic description of the field in 2005, collected initial gas production data of 30 production wells, and analyzed them as production test data. Among them, 21 wells had been produced over a year, and their pressure histories were used to ascertain and confirm the parameters of its improved dynamic model; the other wells had run deliverability tests and pressure buildup tests so that their preliminary dynamic models were established successfully.

Thanks to that, detailed wellhead tubing and casing pressure data were recorded during production, and no packer was installed in the production string, the casing pressure can be converted to bottom-hole pressure easily, and so that the pressure history over the whole production process was. In order to verify the feasibility of converting wellhead pressure to bottom-hole pressure, Well Y-17 was specially selected to monitor the bottom-hole pressure, wellhead casing pressure, and tubing pressure with electronic pressure gauges; the comparison and verification of these pressure data proved that the converted bottom-hole pressure from casing pressure is exact and reliable.

By the end of 2005, the average cumulative gas production of a single well for those 21 long-term production test wells was 0.34×10^8 m^3; the maximum cumulative gas production of a single well was 1.03×10^8 m^3, which meant that the conditions for the dynamic description study for this gas reservoir had been sufficient.

5.1.3 Process of Dynamic Description For the Gas Reservoir

The dynamic description of the YL gas field started after the dynamic description study of the SLG gas field. The study work was fully on the basis of lessons and experiences drawn from the study of the SLG field, as the two fields are of the same formation of the Upper Paleozoic in the Ordose Basin. From the very beginning, the study had been carried out thoroughly and comprehensively in accordance with "New thoughts in gas reservoir dynamic description" summarized in Chapter 1. The study mainly includes the following.

Gas Well Deliverability Study

The deliverability study is the core of dynamic reservoir description. It consists mainly of three parts.

1. Precise analysis of deliverability test data

    In the study, modified isochronal test data in nine wells were all collected, the deliverability equations of these wells were constructed, and AOFP values of these wells were calculated. However, during drilling, these wells penetrated different intervals, for example, Shan-2, Mawu, and Shiqianfeng, most of these wells were outlaying exploration wells, and the quality of data acquisition in some wells is not good enough; it is impossible to recognize the specific deliverability of the main production areas and major pay zones and to meet the requirements of the reservoir study neither from these data.

2. Establishing the LIT deliverability equation for production wells

    The initial stable-point LIT deliverability equations of 37 selected production wells in the main production area of the field were established respectively, and a "general deliverability equation" applicable to this area was induced consequently. The initial deliverability equation of any production well subsequently put into production can be generated instantly as long as data of a stable production point of the well, that is, the one stable flowing pressure and its corresponding stable gas flow rate, are acquired during its initial flowing periods.

3. Tracing study of dynamic deliverability

    With flowing pressure and flow rate data of the production wells monitored in 2006, dynamic deliverability equations of those wells were established and studied, dynamic AOFP and dynamic formation pressure were calculated, and dynamic IPR curves were plotted accordingly.

Establish Dynamic Analysis Models For Gas Wells

Perfect dynamic models of 21 important production test gas wells within the gas reservoir were established and then a tracing study was conducted continuously in 2006 and 2007.

Numerical Well Test Study

In order to ascertain the connectivity between Well Y46-9, Y45-10, S215, and Y47-10 and the uncompartmentalized conditions of the block above wells locate, numerical well test analysis was performed. The final conclusion from their dynamic performance is that the possibility of connecting between those wells is extremely small, although they are judged in the same widespread sand layer by geological research.

Pressure Gradient Analysis

Static pressure gradient analysis of the YL Block was carried out with initial static pressure data of the gas wells. Several static pressure gradient graphs were plotted separately for the whole block and the six subdivided main production areas. It is recognized that neither the whole field nor any of the subdivided areas could form an uncompartmentalized gas reservoir.

5.2 Deliverability Analysis For Production Wells in Main Gas Production Area

5.2.1 Initial Stable-Point LIT Deliverability Equation

General expression of the initial stable-point LIT deliverability equation of the major pay zone in theYL gas field, Shan-2 Formation, was established and written as

$$p_{\mathrm{R}}^2 - p_{\mathrm{wf}}^2 = \frac{198.943}{kh}q_{\mathrm{g}} + \frac{0.9377}{kh}q_{\mathrm{g}}^2. \tag{8.23}$$

In derivation of this equation, average physical property parameters in this gas block were applied, and completion and other parameters were determined: formation temperature $T_f = 378K$; in situ gas viscosity $\mu_g = 0.022$ mPa·s; deviation factor of natural gas $Z = 0.933$; skin factor $S_t = -5.5$; gas drainage radius of gas wells $r_e = 500$ meters; equivalent radius $r_w = 0.07$ meters; and non-Darcy flow coefficient $D = 0.01(10^4 \text{ m}^3/\text{day})^{-1}$.

The initial LIT equation of each well can be established by Equation (8.23) if only one stable data point of this well, that is, stable flow rate q_g and its corresponding stable flowing pressure p_{wf}, and the static formation pressure p_R at the early stage of the flowing were acquired.

For instance, for Well Y47-5: $p_R = 26.75$ MPa, $p_{wf} = 22.74$ MPa, and $q_g = 16 \times 10^4$ m³/day, the deliverability equation of this well is

$$p_R^2 - p_{wf}^2 = 11.547q_g + 0.0544q_g^2.$$

The calculated AOFP is $q_{AOF} = 50.13 \times 10^4$ m³/day, and the initial IPR curve is shown in Figure 8.58.

With the method, deliverability equations of all 37 production gas wells in the major pay zone in the YL gas field were established.

5.2.2 Dynamic Deliverability Equation and Dynamic Deliverability

Also take Well Y47-5 for example. A tracing study on the variation of dynamic deliverability was performed after the well had been put into production, and stable production data points in 2006 and 2007 were chosen, respectively: $p_{wf06} = 21.05$ MPa and $q_{g06} = 9.3141 \times 10^4$ m³/day in 2006; $p_{wf07} = 20.58$ MPa and $q_{g07} = 9.0233 \times 10^4$ m³/day in

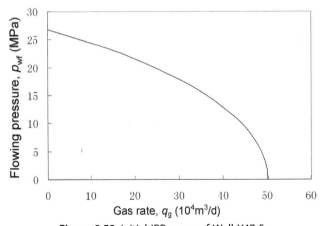

Figure 8.58 Initial IPR curve of Well Y47-5.

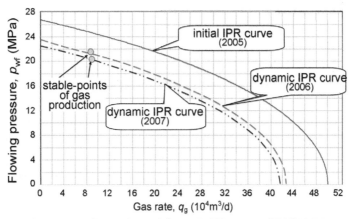

Figure 8.59 Comparison of dynamic IPR curves of Well Y47-5.

2007. The obtained dynamic deliverability equations and dynamic deliverability indices were as follow:

In 2006,

$$p_{R06}^2 - p_{wf}^2 = 10.61 q_g + 0.050 q_g^2,$$

$$p_{R06} = 23.37 \text{ MPa},$$

$$q_{AOF06} = 42.8275 \times 10^4 \text{ m}^3/\text{day}.$$

In 2007,

$$p_{R07}^2 - p_{wf}^2 = 10.43 q_g + 0.050 q_g^2,$$

$$p_{R07} = 22.84 \text{ MPa},$$

$$q_{AOF07} = 41.7816 \times 10^4 \text{ m}^3/\text{day}.$$

The corresponding dynamic IPR curves are shown in Figure 8.59.

Similar deliverability equations can also be constructed for other production wells, and then the corresponding tracing comparative graphs of IPR curves can be plotted in the same way and these results can be taken as the basis for production planning onsite.

5.3 Establishing the Dynamic Models of Gas Wells and Carrying on the Tracing Study

5.3.1 Establish Archives For Gas Well Production History

High-precision pressure history and exact flow rate history throughout the whole production process are the bases of establishing the dynamic model of a gas well.

Method of Establishing Gas Well Pressure History

In an offshore gas field of the China National Offshore Oil Corporation (CNOOC), the pressure history of each gas well began to be recorded continuously by permanent downhole high-precision electronic pressure gauges since it was put into production and recorded pressure data included the variation of flowing pressure and buildup pressure of every shut in during several years of the whole production process. This pressure history, combined with automatic recorded gas flow rate history in this field, is optimal basic data acquired onsite for the dynamic description study that has been seen so far.

In general, electronic pressure gauges are used to record all flowing pressure data during the deliverability well test process and all shut-in pressure data during the buildup test; but during the other long time production processes, the point measurement method is applied to acquire the flowing pressure discontinuously and wellhead tubing and casing pressures are acquired continuously to be converted to bottom-hole pressure so that a complete pressure history can be ultimately obtained by merging those converted and measured bottom-hole pressures. This is what has been done in the YL gas field.

Process and Verification of Converting Wellhead Casing Pressure to Bottom-hole Pressure

Many calculation formulae have been used for a long time to convert wellhead pressure to bottom-hole pressure in the field practice of gas reservoir engineering. The book entitled "Applied Calculation Methods of Gas Production" by Yang Jisheng and Liu Jianyi [79] introduced calculation methods and their application conditions. Usually, those formulae are compiled into computer software and then applied directly onsite. They can be classified into two categories for gas wells, especially for dry gas wells: one is used for conversion of shut-in pressures. Relatively exact bottom-hole pressure can be converted from wellhead pressure as long as the relative density of the gas can be determined section by section based on the well conditions and the state equation of gas so that the converted pressure gradient of the gas column in the wellbore can be obtained; the other is used for conversion of flowing pressures, and both the gas gravity gradient in the wellbore and the frictional resistance to the gas flow must be taken into account in this case. The fact that there are many influencing factors needs to be considered for the latter affects the accuracy of the converted results.

The conversion of wellhead pressure to bottom-hole pressure was verified in the YL gas field. Figure 8.60 shows three pressure histories of Well Y17 acquired during continuous monitoring with electronic pressure gauges: ① measured bottom-hole pressure history, ② measured wellhead tubing pressure history, and ③ measured casing pressure history. Converted bottom-hole pressure from the casing pressure is also plotted in the same graph for comparison.

It can be seen from Figure 8.60 that

1. The continuity of three monitored pressure histories is very good and of high precision.

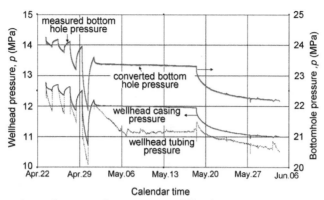

Figure 8.60 Comparison of pressure history acquired by electronic pressure gauge and converted pressure history of Well Y17.

2. The variations of bottom-hole pressure and casing pressure are very alike in curve shape but different in value, and their value differences at every point are approximately a constant.
3. Wellhead tubing and casing pressures are coincided when the vertical conduit flow ceased during shutting in; while during every flowing period, they are different: resulting from wellbore frictional resistance to flowing, wellhead tubing pressures are lower than casing pressures.
4. The bottom-hole pressures converted from the casing pressures with the formula for vertical conduit flow and the well condition when flow ceases are consistent with the measured bottom-hole pressures.

It was with this conversion method that the whole-process bottom-hole pressure histories of gas production wells of the YL gas field were established successfully.

5.3.2 Establish Dynamic Models of the Gas Wells

Take Well Y46-9 for instance, its dynamic model was established with its pressure history from 2002 to 2005 and pressure buildup test data were measured in early 2003. Model parameters obtained from aforementioned data are as follow:

1. Well storage coefficient $C = 3.48$ m^3/MPa
2. Formation flow capacity $kh = 18.9$ mD·m
3. Formation permeability $k = 1.28$ mD
4. Half-length of hydraulic fracture $x_f = 77.4$ meters
5. Skin factor of hydraulic fractures $S_f = 0.1$
6. Total skin factor $S_t = -6.0$
7. Matched initial pressure $p_i = 26.8916$ MPa
8. Boundary type: rectangular constant-volume block
9. Boundary distances: $L_{b1} = 500$ meters, $L_{b2} = 1800$ meters, $L_{b3} = 500$ meters, and $L_{b4} = 1800$ meters
10. Area of constant-volume block: 3.6 km^2

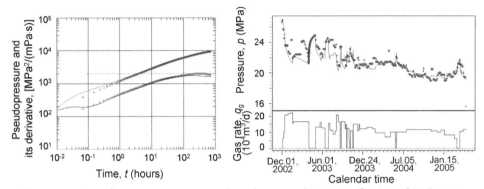

Figure 8.61 Log–log pressure curve match and pressure history verification of Well Y46-9.

The pressure log–log curve match and pressure history match verification in well test interpretation are shown in Figure 8.61.

The dynamic models of the other 20 gas production wells were established with the aforementioned method separately, which laid the foundation for the next dynamic analysis of the YL gas field.

5.3.3 Tracing Studys

The tracing study on the dynamic model of a gas well focusing on the following aspects can be carried on as soon as its improved dynamic model has been established.

1. Prediction of the performance of the well

Putting the planned flow rate into the established dynamic model of the well in the well test interpretation software, the variation of the flowing pressure of the well in a year or several years can be predicted, and with which the following strategic decision on the production plan can be made or modified.

2. Tracing, verifying, and improving the dynamic model

The dynamic model can be verified further by the following monitored pressure history. The verification method is inputting the following production history to the model to calculate the theoretical pressure variation of the model in the corresponding period and comparing the result with the measured pressure history; if both are still basically consistent, the original model is proved correct and reliable, but if any obvious difference exists between them, the model parameters should be modified to make the dynamic model more perfect, as shown in Figure 8.62.

Figure 8.62 shows that the pressures of the model are slightly lower than the actual measured pressures in 2007 in the pressure history, based on which the boundary parameters of the dynamic model of this well were modified accordingly to $L_{b1} = 550$ meters, $L_{b2} = 2100$ meters, $L_{b3} = 600$ meters, and $L_{b4} = 2100$ meters, and the other

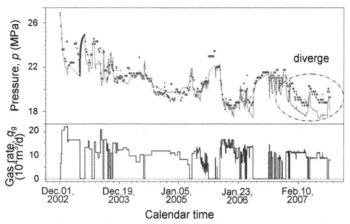

Figure 8.62 Tracing verification of dynamic model of Well Y46-9 in the period from 2005 to 2007.

model parameters remained unchanged. Consequently, an improved pressure history match verification shown in Figure 8.63 was obtained.

As shown in Figure 8.63, after improvement of the model, that is, boundary distances were lengthened slightly, on the basis of the result of the last tracing verification, the theoretical pressures of the new model were more consistent with the measured pressure than those of the original model. This indicates that the new model reflects the reservoir controlled by Well Y-46-9 much better or more truly. Also, this kind of tracing dynamic performance study should be the principal method of dynamic analysis and research in this area in the future: continuously verifying and improving the models of the wells, along with development of the field going on, and predicting the future dynamic performance of the wells in the block with the corresponding model.

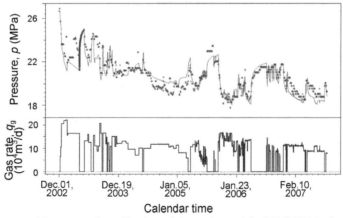

Figure 8.63 Pressure history match verification of dynamic model of Well Y46-9 after improvement.

5.4 Analysis of Reservoir Pressure Gradient of YL Gas Field

5.4.1 Analysis of the Initial Static Pressure Gradient

The initial static pressures of 78 gas production wells penetrating the San-2 Formation were collected for pressure analysis with the method described in Chapter 4; the initial static pressure gradient plot is shown in Figure 8.64.

It can be seen from Figure 8.64 that the pressure points somewhat scattered. Therefore, the linear characteristics of the static pressure gradient of an uncompartmentalized gas field should be found out from the physical properties of the gas in this area first. The representative physical properties of the gas in this area include

Average formation pressure, $p_R = 27$ MPa

Formation temperature, $T = 80°C$ or 353 K

Relative density, $\gamma_g = 0.59$

Z factor of the gas, $Z = 0.94$

The reservoir gas density was then calculated to be

$$\rho_g = \frac{27 \times 0.59 \times 28.97}{0.94 \times 8.3143 \times 10^{-3} \times 353} = 170.98 \ \text{kg/m}^3.$$

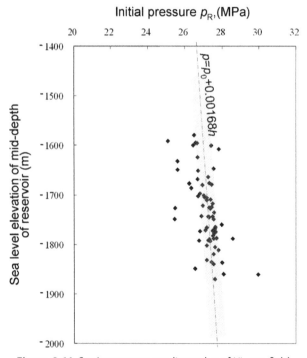

Figure 8.64 Static pressure gradient plot of YL gas field.

The static pressure gradient of the gas reservoir was

$$G_{DS} = 170.98 \text{ kg/m}^3 \times 9.80665 \text{ m/s}^2 = 0.00168 \text{ MPa/m}.$$

The equation of static pressure gradient line was set up based on the aforementioned pressure gradient value:

$$p = p_0 - 0.00168h.$$

The plot static pressure gradient line in Figure 8.64 is marked with a dashed line; simultaneously draw an area within ± 0.25 MPa from the static pressure gradient straight line there with shadow, then it can be seen from Figure 8.64 that

1. Even though a majority of measured pressure points are in the banded area with the shadow, a unique static pressure gradient line for gas cannot be created by the regression, which meant that YL Block was impossible to be a single uncompartmentalized gas reservoir.
2. Several parallel static gradient lines can be obtained if different values of p_0 are selected in the just given equation; on each gradient straight line there should be several measured pressure points of the test wells; this provides a necessary condition of being connected with each other for those wells on a same gradient straight line.
3. Static pressure gradient analyses like these were carried on further for six areal distribution zones of the gas wells marked in the geological map of the YL gas field, which confirmed that the gas zones were hardly possible to be connected to each other in the subdivided blocks of the field based on verifying the well location repeatedly and studying the relationship between the adjacent gas wells there.

5.4.2 Dynamic Formation Pressures of Gas Wells and Reservoir

Determination and analysis of dynamic formation pressure are important parts in dynamic performance research of a reservoir. In the YL gas field, dynamic formation pressures during the production process were obtained by actual measurement and by the calculation methods of the dynamic model and dynamic deliverability equation as well.

1. Estimating average formation pressure with dynamic model

    Quite perfect dynamic models were built separately for a number of gas wells in the Upper Paleozoic gas reservoir in the YL gas field by dynamic description of gas wells, and the distribution of no-flow boundaries near the gas wells was confirmed by long-time pressure history match verifications; among those wells, 21 were confirmed to be located in limited closed reservoirs. The established dynamic models in this case would generate average formation pressure variation curves automatically when plotting the pressure of the model, as shown in Figure 8.65.

    The average formation pressure value at a given specific moment read from Figure 8.65 can be used for the study of formation pressure distribution; if the

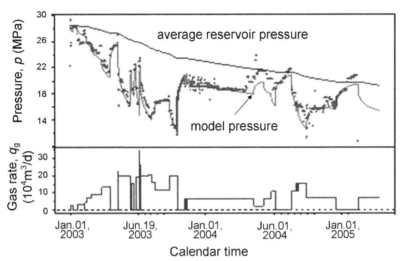

Figure 8.65 Calculating average formation pressure with dynamic model method in YL gas field.

correspondent gas cumulative production is obtained, the study of dynamic reserves controlled by the single well can be done.

2. Estimating formation pressure at gas drainage boundary with the method of dynamic deliverability equation

   The method of establishing the dynamic deliverability equation was introduced in Chapter 3. It is derived from the initial stable-point LIT deliverability equation and with a selected dynamic stable deliverability point in the follow-up production, and is also plotted in the comparative graph of the IPR curve decrement as shown in Figure 8.66.

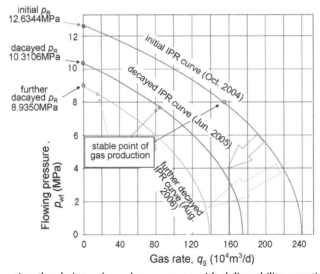

Figure 8.66 Estimating the drainage boundary pressure with deliverability equation in YL gas field.

As shown in Figure 8.66, dynamic drainage boundary pressure at any period can be estimated with the dynamic deliverability equation graph; the dynamic formation pressure of many gas wells of the YL gas field were estimated with this method. Analysis with this method can be performed for more gas wells because their perfect dynamic models are not required.

Once the dynamic formation pressures of gas wells and the reservoir during the production were available, dynamic reserves calculation with the material balance method can be done; the method and practical operation have been introduced in detail in the section of the study of the SLG gas field in Section 8.4 and are not necessary to be elaborated here.

5.5 Comparison of Reservoir Characteristics Between YL Gas Field and SLG Gas Field

YL and SLG are both important gas fields penetrating the major pay zones in Upper Paleozoic in the Ordos Basin and had been almost put into production at the same time; the OOIP of each of them is more than 1000×10^8 m^3. However, the difference between them appeared in their individual dynamic description. Table 8.17 lists quantitative descriptions of their characteristics.

1. Although the two gas fields are composed of the Upper Paleozoic Formation, they penetrated different horizons: Shanxi Formation in Lower Permian for the YL gas field and Shihezi Formation in Middle Permian for the SLG gas field. Their sedimentary conditions were different due to a different horizon and depositional age.

2. One of three important factors influencing initial gas well deliverability is formation pressure. The formation pressures of these two gas fields are close to the static water column pressure and nearly the same. Upper Paleozoic is a western-oriented monocline, and the SLG field is located in the west of the YL field, so Shihezi Formation penetrated in the SLG gas field is deeper than Shanxi Formation penetrated in the YL gas field and the initial formation pressure of the SLG gas field is higher than that of the YL gas field.

3. Average formation flow capacity (*kh*) of the YL gas field is twice as much as that of the SLG gas field. Table 8.17 shows that the *kh* value of the YL gas field is 21.9 mD·m, but that of the SGL gas field is only 9.74 mD·m; because the effective formation thicknesses of the two fields are nearly the same, the formation permeability of the YL gas field is calculated twice as much as that of the SLG gas field. This also makes the initial deliverability of the YL gas field higher than the other.

4. The well completion quality in two fields is nearly the same. Both gas fields adopted prefracturing completion, and the half-lengths of fractures range from about 50 to 60 meters, which makes the skin factor lower than −5, and the goal of reservoir

Table 8.17 Comparison of Well and Reservoir Parameters Between YL and SLG Gas Fields in Ordos Basin

Gas field Parameter	YL gas field	SLG gas field
Main pay zone	Interval Shan-2 of Shanxi formation in Permian	Interval He-8 of Shihezi formation in Permian
Test wells in statistics	21	13
Initial formation pressure (p_i), MPa	27.5587	29.631
Average formation flow capacity (kh), mD·m	21.9	9.74
Average effective formation thickness (h), m	11.7	12.4
Average permeability (k), mD	1.863 (averaged by well)	0.784
Average half-length of fracture (x_f), m	62.2	49.0
Average fracture skin factor (S_f)	0—0.1	0.15
Average total skin factor (S)	−6.0	−5.2
Average block width, km	1.372	0.112
Average block length, km	3.425	1.765
Average block area, km^2	4.70	0.22
Average width/length ratio	1:2.5	1:15.7
Estimated dynamic reserves controlled by a single well, 10^8 m^3	5.05	0.22

stimulation and deliverability enhancement in a low permeability sandstone gas field has been achieved.

5. Great differences exist in area and dynamic reserves controlled by a single well.

The area controlled by a single well in the YL gas field is about 5 km^2 and so the dynamic reserves controlled by a single well is as high as 5.0×10^8 m^3, whereas the area and the dynamic reserves controlled by a single well in the SLG field are only 0.2 km^2 and 0.2×10^8 m^3, respectively. The shapes of these two blocks are quite different: the length/width ratio of the block in the SLG gas field reaches 12.5:1 and so forms a reservoir like a strip and the percolation condition of which is worse than that of the YL gas field. It has great influence over the production stabilization and the cumulative production of each single well in these two gas fields: the YL field can meet the requirements of stabilized production and good economical benefits of each single well, and a considerable scale gas field has been built and put into normal production, whereas the single well in the SLG gas field can hardly maintain stabilized production during the production test. Even these wells selected in the dynamic description study are all with favorable production conditions in this area, and so the average index must be somewhat lower than these study results, which is why special development and operation modes are needed for the SLG gas field.

These studies on the two gas fields were implemented totally following the "New thoughts in gas reservoir dynamic description" illustrated in Chapter 1 and have become the representative paradigms.

6 STUDY ON DYNAMIC DESCRIPTION OF GAS RESERVOIR IN DF GAS FIELD

6.1 Overview of DF Gas Field

Dynamic gas reservoir description of the DF gas field of the CNOOC can be rated as the model; the reasons are (1) there are some outstanding characteristics in this study result, which are unavailable in gas field dynamic performance analysis in any other gas fields of China; and (2) there are several innovations in the comprehensive description of gas reservoirs of this field, which not only have successfully resolved various problems encountered in the field development, but also have enriched the methods of the dynamic description of the gas reservoir. This fact confirms again that, being one of three key techniques of a gas reservoir study, the modern well test and the dynamic description of the gas reservoir technique are closely integrated with field practice and are indispensable and have great vitality.

6.1.1 Characteristics and Difficulties in Development of DF Gas Field
All Production Wells are Horizontal Wells
The DF gas field is a medium-scale field in the Yinggehai Depression in the South China Sea. Field development had faced great challenges, for its structure was cut by faults, the planar distribution of its effective sandstone reservoirs was seriously heterogeneous, and the physical properties of natural gas seriously vary vertically and laterally. It can be called another special lithologic gas field.

Being an offshore gas field, gas is produced from the whole field in a few platforms and on each of which there are more than 10 horizontal wells around.

The acquisition and analysis of well test data in horizontal wells, especially for the dynamic performance analysis of so many horizontal wells, are unprecedented in China and are surely and severely challenging.

Monitoring of Bottom-hole Pressure History with High-precision Permanent Pressure Gauges
Facing such a complicated study object, the decision-making department of the CNOOC took important and effective measures for dynamic data monitoring from the very

beginning: high-precision permanent pressure gauges were installed in the horizontal section in nearly half of the wells in the field to monitor and acquire bottom-hole pressure data, including flowing pressure and pressure buildup test data, so that the filed pressure history of all of those wells can be displayed to management and research personnel, and thus a solid foundation for a further dynamic description of the gas reservoir is laid.

Complicated Follow-up Production Allocation

The previous production allocation for onshore gas fields was to roughly predict productions of the single well and the gas block on the basis of development program and then adjust the production after the field has put into development based on the actual performance of each gas well to maintain the total production of the field. The production adjustment is simple because the component of the natural gas produced from every gas well in the block and the requirement of downstream clients are nearly the same.

In the DF gas field, however, the component of the gas produced from different wells varies greatly, and the requirements of downstream clients are quite different, which make production allocation significantly difficult.

Therefore, deliverability tests were performed frequently since the commencement of being put into production of the gas field. The back-pressure test was run for each well not only before its putting into production of the well, but also during its production process in order to learn about the variation of its deliverability indices during the decrement process of each well.

Whereas it was proved by repeated practices that application of the conventional back-pressure test method was not only hard to measure and determine the deliverability indices during the decrement process, but also difficult to determine the initial deliverability indices, and so was incapable of meeting the requirement of the follow-up production allocation, not to mention meeting that of production planning of the next year or in the future.

Uncertainties in Transient Test Analysis of Horizontal Wells

Production wells in the DF gas field are all horizontal wells. There are many difficulties in the well test interpretation of horizontal wells according to the author's experiences drawn from well test analysis practices for years.

1. Data acquisition is difficult: in horizontal wells, a pressure gauge is quite difficult to install at the position of the reservoir, and load water in the bottom hole very often distorts the pressure buildup curve consequently. In addition, data of various different important even key flow regimes of transient flow in horizontal wells are very hard to acquire during the test due to not long enough test duration. In fact, successful acquisition of the whole pressure history from a horizontal well had never been seen before. However, all of these problems were resolved in the DF gas field; the whole bottom-hole pressure history was acquired in more than 10 wells there, including

data of several complete pressure buildup tests in each of these wells. It not only provides very good preconditions for data interpretation, but also puts forward more strict requirements for the reliability of interpretation results.

2. Ambiguity of well test analysis results: After good data have been acquired, the ambiguity problem in the well test interpretation of horizontal wells is waiting to be resolved. By practicing repeatedly, this problem is resolved primarily in the DF gas field study by the new thoughts of gas reservoir dynamic description illustrated repeatedly in Chapter 1 and other sections in this book.

6.1.2 Innovations in Dynamic Performance Study of DF Gas Field

With data acquired during 4 years of continuous development of the DF gas field from 2003 to 2007, a thorough dynamic performance analysis was performed and successfully resolved multiple puzzles in gas field development, among them three problems that had never been resolved before in China.

1. Establishment and application of "stable-point LIT equation" for horizontal wells

    The establishment of this equation resolved various puzzles encountered in the data acquisition and analysis with the conventional back-pressure test method, reestablished the initial deliverability equations for all horizontal gas wells in the DF gas field, drew their initial deliverability plots, and calculated their initial AOFP (q_{AOF}). All these results were verified to be correct by comparative analysis with the actual performance of the gas wells after putting into production.

2. Establishment and application of dynamic deliverability equation of horizontal gas wells

    Explicitly put forward the concept of the dynamic deliverability equation for horizontal gas wells, derived the dynamic deliverability equation at different stages for every well, plotted their dynamic IPR curves, estimated their dynamic AOFP, and put forward the method and procedure of obtaining the dynamic deliverability; all of these provided a reliable basis for production allocation of the gas wells at different development stages.

3. Resolve the problems of transient test interpretation and dynamic description of horizontal wells

    The established dynamic horizontal well models are all unparallel in the great amount of the well, comprehensiveness and reliability of description on the geological characteristics in the area controlled by the gas well, and practicability of the predicted dynamic trend in the future.

6.2 Evaluation of Initial Deliverability and Dynamic Deliverability

6.2.1 Back-Pressure Tests and Analyses

An initial back-pressure test was conducted for every horizontal well before putting into production in the DF gas field. Test data of all of the dozens of wells were complete, but it was regrettable and surprising that more than half of test data were unable to give the applicable deliverability equation in the regression analysis when being analyzed with various well test interpretation software. Some of them gave LIT equations with a negative value of coefficient B, whereas some gave an exponential equation with the exponent being greater than 1. Reasons causing these phenomena are very complex and those had been discovered include the following: ① there is load water in the bottom-hole and pressure gauges are installed above the load water level, resulting in different flowing pressure drifts measured under different gas flow rate conditions, ② the durations of test points in some wells are too short and not equal so that the measured flowing pressures have not yet been stable, and ③ formation pressure has declined during the test. These problems were discussed in Chapter 3.

The test for the initial deliverability equation is unrepeatable. If the opportunity is missed, it can never be obtained and will cause great difficulty in subsequent studies, and when dealing with the deliverability evaluation, other solutions have to be found out.

6.2.2 Establishment of "Initial Stable-Point LIT Equation" to Reevaluate the Initial Deliverability of Gas Wells

As introduced in Chapter 3, the stable-point LIT equation can be constructed for either vertical wells or horizontal wells if only the initial formation pressure and a pair of data of one stable test point in the initial flowing period (p_{wf0}, q_{g0}) are acquired.

The pressure history throughout the whole production in nearly half of the gas wells in the DF gas field has been acquired, and the initial deliverability expressed as Equation (3.59) in Chapter 3 of any well can be obtained as soon as the pair of data of one stable production point in the initial flowing period of the well is selected and read. The method of establishing the equation was elaborated in detail in Chapter 3. For those gas wells without a complete pressure history of the whole production process, the deliverability equation can also be constructed if only a measured deliverability point in the initial backpressure test, for example, the last measured point with the maximum gas flow rate, is selected.

Such initial stable-point LIT deliverability equations of dozens of production wells were set up one by one in the DF gas field, the initial IPR curves of those wells were plotted, and the initial AOFPs of those wells were calculated. All of these have been described with examples in Chapter 3 as shown in Equation (3.64) and Figure 3.55.

6.2.3 Derivation of Dynamic Deliverability Equation

It was necessary to know the deliverability, including the dynamic deliverability equation and the calculated indices of each well in DF gas field during its production history or its depletion process proposed in this book, as differences of flow rates and/or gas components among different wells caused by complex geological conditions are quite serious. Fortunately, the initial stable-point LIT equation established previously can just be the basis and starting point of deriving the dynamic deliverability equation.

It is for the study of the DF gas field and the application in this field that the method of establishing the dynamic deliverability equation for horizontal wells illustrated in Chapter 3 was created [refer to Equation (3.66) and Figure 3.58]. With the dynamic deliverability equation, the dynamic formation pressure and dynamic AOFP can be obtained, and it was these two indices that reflect factually the real deliverability of the gas well after periods of production and can also be the basis for production allocation of the wells.

6.2.4 Deliverability Test Design Under Dynamic Monitoring Conditions

Since the initial stable-point LIT equations of every well in the DF gas field have been established, and on the basis of which the dynamic deliverability equations of every well have also been established, the dynamic monitor of deliverability indices of every well in the field can be realized, and the procedure is as follows.

1. Select specific time intervals during which the dynamic deliverability of the selected wells needed to be estimated according to the gas field production program.
2. During the select specific time intervals, maintain a stabilized production rate in the selected well for a duration of 5—10 days— the fluctuation of the flow rate should be within 5%—and run a downhole pressure gauge into the well to measure the stable flowing pressure in this period.
3. Select a proper time interval for gas wells with permanent pressure gauges at their bottom hole to read the stabilized flow rate and the flowing pressure in this interval.
4. Establish the dynamic deliverability equation, calculate dynamic deliverability indices with the method described in Chapter 3, and send them to the departments concerned.

Deliverability monitoring has been doing so methodically in the DF gas field following this new thought.

6.3 Dynamic Description of Gas Wells and Gas Reservoirs

Dynamic description of the gas wells is the basis of a dynamic description of the gas reservoir. Dynamic descriptions of all horizontal wells were completed and analyzed first, and then a numerical well test study was performed for the well groups in which interference tests between wells had been run.

Field Example No. 1 is used for presenting the study process.

6.3.1 Establishment of Gas Well Dynamic Model with Pressure Buildup Test Data

The complete pressure history over the whole production of Field Example No. 1 has been acquired. During the initial production for more than 3 years, the well was shut in several times, and pressure buildup data of every shut in were recorded. Pressure buildup test data acquired in the earliest and lasting the longest duration for interpretation were selected, and obtained interpretation results are shown in Figure 8.67.

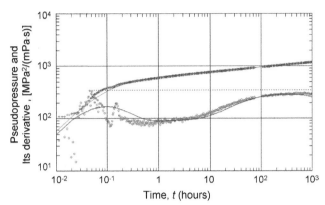

Figure 8.67 Log–log plot of pressure buildup in a horizontal well, Field Example No. 1.

The formation and completion parameters of the well were calculated primarily by type-curves match: ① wellbore storage coefficient $C = 5.0$ m³/MPa; ② skin factor $S = 0.03$; ③ length of effective horizontal well section $L_e = 318$ meters; ④ distance between horizontal well section and the bottom of the gas zone $Z_w = 20$ meters; ⑤ effective formation thickness $h = 36$ meters; ⑥ matched initial pressure $p_R = 13.96$ MPa; ⑦ formation flow capacity $kh = 402$ mD·m; ⑧ formation permeability $k = 11.2$ mD; ⑨ ratio of vertical permeability to lateral permeability $k_z/k_r = 0.389$; and ⑩ type of the block controlled by the gas well: constant-volume square block with a side length of 1600 meters. The dynamic reserve controlled by the well was calculated to be about 60×10^8 m³.

Figure 8.67 shows that the dynamic model curves of the well are very similar to those of typical flow characteristic curves of the horizontal well shown in Figure 5.161 of Chapter 5 and consist of all flow characteristic sections, that is, afterflow, vertical radial flow, linear flow, and pseudo radial flow sections. This kind of test data is rarely seen in test practice in horizontal wells. Many variations of the test curve shape in horizontal wells would be caused sometimes due to limited extension of drilling, penetrating the horizontal section with irregular trajectory, or inadequate combination of different parameters, which make the test curve look like a super penetrating well or fractured well sometimes. Readers can refer to the description in Chapter 5 or do some demonstrations with well test design software.

6.3.2 Model Reliability Verification Through Pressure History Match

Figure 8.68 shows pressure history match verification of the dynamic model of Field Example No. 1 established on the basis of transient test interpretation results.

As shown in Figure 8.68, the pressure variation of the dynamic model and measured pressure variation of the well seem perfectly consistent, which proves the reliability of the model parameters from the interpretation.

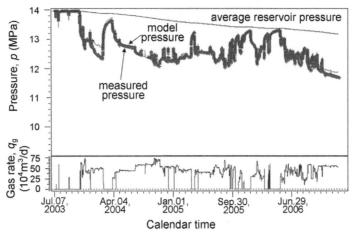

Figure 8.68 Dynamic model verification through a pressure history match for a horizontal well, Field Example No. 1.

6.3.3 Knowledge Obtained From Dynamic Description

Some explicit knowledge has been obtained for gas reservoirs from the dynamic descriptions of long-term production wells, especially from those with permanent downhole pressure gauges, in the DF gas field.

1. Formation flow capacity *kh* of the area controlled by every gas well is obtained

   The value of formation flow capacity *kh* of a gas well decides its initial AOFP. The *kh* values of different wells in the DF gas field are in a range of 100–700 mD·m, which indicates that the DF gas field possesses medium to slightly high gas deliverability.

2. Effect of constant-volume boundary exists in most of the wells

   It was acknowledged from dynamic descriptions of the wells that only limited areas are controlled by most of the wells, the areas are relatively independent from each another, and the average dynamic reserve controlled by a single well is about 20×10^8 m^3, which decides the decline rate of the deliverability of the wells in the future.

3. Interwell interference exists among a few wells in the field

   Gas wells with interwell interference were studied by the numerical well test method and the correlation of those wells was confirmed quantitatively, and so the migration direction of natural gas during production and the areas and dynamic reserves were controlled by the well groups as well.

4. Prediction of the dynamic performance

   It is the dynamic models of the wells verified by their pressure history matches that will become the basis for predicting the future dynamic performance of these wells. Models built with well test interpretation software not only provide a literal description of geological and well completion conditions for each gas well, but, more importantly, also act as a tool supported by mathematical equations and can reappear the subsurface percolating pattern during gas production of the well, which makes predictions of the bottom-hole flowing pressure decline and decreasing rules of gas flow rate possible, and so the countermeasures for dealing with these problems in production can be found out.

5. Modification and improvement of dynamic model

   The dynamic model built previously was impossible to be perfect. Some inevitable defects would be discovered in the further pressure history match verification. For instance, boundary distances, especially the distances of the boundaries, even the shape of them far away from the well, determined primarily are somewhat unreasonable, and the model needs will be improved further again and again by continuous parameters modification.

6.4 Long-term Dynamic Performance Analysis in DF Gas Field

6.4.1 Overview of Dynamic Performance Analysis Methods

Various semiempirical analysis methods, such as Arps method, Fetkovich type-curve matching method, Blasingame method, Agarwal–Gardner matching method, and normalized pressure integral method (NPI) method, had been developed in the field of gas reservoir dynamic analysis before the model well test was developed and commonly applied.

1. Arps method

   The Arps method is based on empirical observations of gas wells. No reservoir or well conditions are required for analysis. The method assumed that there are three pseudo-steady flow rate decline patterns or three mathematical modes: decline exponentially, decline harmonically, or decline hyperbolically. The mode of flow rate declining of a well is determined by matching and comparing its actual measured flow rate data with the theoretical modes, and then the gas flow rate variation is predicted with the best matched mode.

2. Fetkovich type-curve matching method

   This method connects the transient decline pattern in the early production period and the pseudo-steady decline pattern in the late production period together and the type curves are created; it then matches measured data with the type curves to confirm the model and make a prediction.

3. Blasingame method

   The Fetkovich production decline type-curve analysis method cannot be used in case the bottom-hole flowing pressure changes, and the Fetkovich method requires all PVT properties of the fluid being constant, so its application is limited. Blasingame, Palacio, and others introduced the material balance method for correction and consequently obtained two functions of pseudo pressure and pseudo time, integrated production analysis methods for gas wells and for oil wells in one, and brought forward curve match and analysis method of "production integral" and "derivative of production integral" and effectively decreased the ambiguity of interpretation results as a result.

4. Agarwal—Gardner decline type-curve match method

Based on the analytical solution of equivalent constant flow rate and equivalent constant pressure in the transient test analysis theory, Agarwal and Gardner developed an integrated type-curve analysis method and their type curves were similar to Fetkovitch's. There are three groups of type curves, that is, curves of flow rate vs time, cumulative production vs time, and flow rate vs cumulative production, being used in the application of this method. The prominent characteristic of this method is to introduce the ideas and concepts of transient pressure analysis into the production analysis of oil and gas wells so as to make the distinction of transient flow and flow affected and controlled by outer boundaries more obvious.

5. Normalized pressure integral method

The NPI method was first put forth by Blasingame and others for the purpose of eliminating the abnormal variation of the curve caused by production data fluctuation with the integral method so as to diagnose and select the model more easily and conveniently. The type curves used in this method are quite similar to the common log—log plots in transient test analysis, with normalized pressure (pressure normalized by production) and its derivative as the ordinate and material balance equivalent time as the abscissa. In the type curves, a unit-slope straight line will occur in the flow section controlled by boundaries, which is the same as that in the pseudo-steady flow section in the common log—log plot in transient test interpretation, and a horizontal straight line occurs on the normalized pressure derivative curve in the transient flow section, which is the same as that in the radial flow section in the common log—log plot in transient test interpretation.

The dynamic analysis methods mentioned earlier have been quoted, developed, and compiled in some software, for example, the module of "Topaze" is compiled in well test interpretation software "Saphir" to support the application of these analysis methods.

However, those methods were mainly generated and developed from statistical analysis and induction of practical production data and lacked a strict theoretical basis of percolation mechanics, which were the foundation of modern well test analysis and dynamic description, so the reservoir parameters cannot be calculated from these methods. Therefore, it would be much better that they are applied together with a modern well test analysis. In fact, the module Topaze in well test interpretation software Saphir has possessed such function; it can transmit the dynamic model containing detailed reservoir information obtained from the modern well test analysis into the module Topaze for reverification and reconfirmation of the model and its reliability.

6.4.2 Dynamic Performance Analysis of DF Gas Field

Software Topaze was also applied to analyze dynamic data of long-term production while studying the modern well test analysis method in the DF gas field. For instance, model parameters obtained from the dynamic description of well Example No. 1 were input to software Topaze for further match verification, and the results are shown in Figure 8.69.

As shown in Figure 8.69, with all parameters of the model determined by the modern well test method, verification done by Topaze is perfect: the flow rate calculated from the measured pressure and the measured flow rate are consistent and the pressure calculated from the measured flow rate and the measured pressure are also consistent, as are calculated cumulative production and measured cumulate production. It indicates that the model and its parameters are all correct.

Analysis and verification of the model with the model parameters obtained from modern well test analysis and with the Fetkovich type-curve matching method by application of software Topaze were performed and the flow rate curve of the model and the actual measured flow rate curve are matched very well, as are the cumulative production curve of the model and the actual cumulative production curve, which is shown in the result plot of the Fetkovich analysis of the model with Topaze software (Figure 8.70). It confirms again that the model and its parameters are reliable.

6.5 Comprehensive Knowledge of DF Gas Field

Comprehensive knowledge of such a complicated geologic object, the DF gas field, was obtained comprehensively by means of several different dynamic analysis methods.

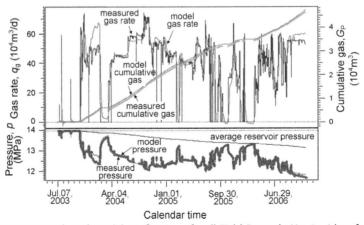

Figure 8.69 History match and model verification of well Field Example No. 1 with software Topaze.

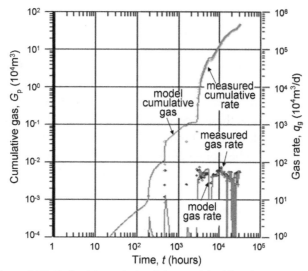

Figure 8.70 Fetkovich analysis of the model with software Topaze.

1. The initial deliverability equation and the dynamic deliverability equations in the following key production periods were established for every gas well, and on this basis the deliverability decline of each well was studied.

2. The dynamic model was built for each gas well based on transient test analysis results, and the geological features of the area around the well and the gas reservoir were identified; with the dynamic model, the dynamic performance trend of the gas well in the future could be predicted.

3. Correlations of the gas wells with interwell interference were confirmed quantitatively with numerical well test analysis, and the dynamic model of each area of the acting and interfering wells locate was established.

4. The reliability of the parameters of the dynamic model was reconfirmed from dynamic analysis with the software Topaze.

    Complete convincing and comprehensive knowledge of the gas field was obtained this way; and as mentioned at the beginning of this section, study on the dynamic description of the gas reservoir in the DF gas field is a model of this kind of study and can be used as a reference in the dynamic performance study of gas field development.

7 DYNAMIC PERFORMANCE STUDY OF EXTREMELY COMPLICATED LITHOLOGIC GAS FIELD

7.1 Extremely Complicated Special Lithologic Gas Field

Several example fields discussed in detail earlier in this chapter are of great significance in scale and influence and have played a key role in the natural gas industrial developments of China in the past years. The study on dynamic description of these fields was a concern of multiple parties in China, and some evaluations about them aroused controversy at the same time. Most of these fields have been put into production now, and the understandings of them have been deepened and become more and more perfect gradually.

However, there are some more special lithologic gas fields whose variation patterns are more difficult to conclude; these special gas fields are more difficult to develop and so are called extremely complex special lithologic gas fields. Therefore, many earlier stage studies on the fields have to be done before their development, and many human and financial resources have to be put into there. For example, favorable gas–bearing areas must be found out from reservoir prediction in geophysical prospecting, the internal structures and development patterns of the gas reservoir must be made clear from logging analysis and geological study, and the percolation parameters of the formation, completion parameters, initial deliverability, dynamic deliverability decline patterns of gas wells, interwell connectivity, and the effective area and dynamic reserves controlled by each single well must be made clear based on the dynamic reservoir description.

Such gas reservoirs can be classified into at least two categories according to the author's experiences accumulated onsite for years.

7.1.1 Buried Hill Carbonate Gas Fields

The QMQ gas field is a typical one of such kind of fields. The marine carbonate rock of Ordovician in Paleozoic was encountered during drilling in the field. Gas is stored in the weathering crust of the Upper Majiagou Formation and Fengfeng Formation within 200 meters under the top surface of the buried hill, where there are developed fissures and solution cavities and the reservoirs are seriously heterogeneous. The development rules of the reservoirs in the exploration area cannot be found out determinately even with great efforts, including three-dimensional (3D) seismic data processing, coring and core verification, and thorough and comprehensive studies with advanced geological research methods such as conventional and special logging analyses so that the adequate development well location cannot be determined in the preparatory stage of development. In addition, there are some other challenges in field development, for instance, the field is a deep condensate gas field with high formation pressure and extremely high formation temperature, which is as high as $180°C$, and bottom water exists and scale formation happens in some wells. Finally, field development had to be suspended, and a thorough and comprehensive preliminary study had to be done again.

7.1.2 Volcanic Rock Gas Fields

Volcanic rock gas reservoirs were discovered successfully in the deep formation of the Songliao Basin in Cretaceous Formation at the beginning of this century, and the test gas flow rate of the discovery wells was nearly one million cubic meters per day, which made a breakthrough in gas exploration there. After several years of reservoir geological research, 1000×10^8 m^3 of OGIP was submitted, and bright prospects had been presented. However, it was found from later thorough and comprehensive geological research that the geological structures of gas reservoirs in volcanic rock are more complicated, and the main gas pay zones are eruptive facies volcanic rock bodies and are formed by the accumulation of multiple volcanic products after many volcanic eruptions; there are various types of reservoirs, including volcanic conduit facies, volcanic explosion facies, lava magma effusion facies, and volcanic overflow facies, and these volcanic rock bodies were then cut by the faults. Both 3D seismic fine interpretation and logging lithology identification analysis were done but failed to conclude the reliable distribution patterns of effective reservoirs. Therefore, such gas fields in volcanic rock are regarded as extremely complicated special lithologic gas fields.

In addition to the aforementioned representative field examples in the two regions, some other deeper gas fields with more complicated geological structures have gradually come into the view of exploration and development technical personnel in recent years. More new challenges are continuously brought forward in the study of special lithologic gas fields.

7.2 Double Porosity Concepts in Geology and Dynamics

7.2.1 Double Porosity Medium Concept in Some Geology

Geologists have been trying to use the concept of double porosity medium to deal with the aforementioned complicated geologic objects. They discovered from core observation that fissures developed widely and commonly in the QMQ carbonate reservoir and so judged the reservoir as a double medium, estimated the reserves stored in the matrix system and fissure system separately with only a few core analysis and statistical data, calculated the ω value from them, obtained the reserves of the whole gas block by multiplying the ω value by the gas-bearing area, and finally made the numerical simulation analysis based on this double medium model and obtained quite good development indices. But the more than ten years of development practice in the QMQ gas field has clearly shown that this research result was totally wrong and has brought a serious aftereffect: the dynamic reserves of the main production wells confirmed by production test was less than 10% of the primarily estimated OGIP, and even suitable well locations were very difficult to determine when normal implementation of the development program started.

7.2.2 Definition of Double Porosity System From Well Test and Dynamic Performance

The definition and the percolation mechanics concept of a double medium were introduced by model graphs M-2 and M-3 in Chapter 5. This concept was initiatively put forward by Balanbulate, a former Soviet Union specialist, in the early 1960s; he also designed a numerical model of it, obtained an analytical solution of the partial differential equation, and gave a detailed and thorough description to the characteristics of the flow in double medium.

There are two important parameters deciding double medium flow characteristics: storativity ratio ω and interporosity-flow coefficient λ; their definition and expression were introduced in Section 5.5 of Chapter 5 [see Equations (5.57) and (5.58)]. There are some important notes about the basic concepts ω and λ:

1. Determination of double medium parameters ω and λ

During the description process of a gas reservoir, for example, the QMQ gas field or volcanic rock gas field, how to identify if it is truly a double porosity medium gas reservoir? Learned from the definition of double porosity medium, it is decided by the percolation process in the medium, concretely speaking, it is decided by the configuration of pressure variation during the transient test in gas wells, and is especially represented by configuration of the pressure derivative curve in the log–log plot. Using a metaphor, to evaluate a person's professional ability is to judge his work performance and achievements other than his appearance.

There are no measured pressure buildup curves showing the characteristics of double medium in either the QMQ gas field or the volcanic rock gas field so far; in this case, how can the reservoir be identified to be a double medium one?

2. Influence of ω value on gas field development

The following field example reflected that a few researchers misunderstand seriously the value of ω. A puzzle was once encountered in a well test interpretation of wells in an ultradeep buried hill carbonate gas field: the quality of the acquired pressure buildup curve was poor, there was no typical characteristic transition portion of the flow from the matrix system to the fissure system on the derivative curve at all, and, in fact, only data of the afterflow stage had been acquired. Some operators used the "advantage" of ambiguity in well test interpretation and set ω and λ values for each well subjectively and intentionally: set higher ω values for gas wells with better geological conditions and lower ω values for those with worse geological conditions. Doing so just made the result upside down. It should be at least known for researchers that the smaller the value of ω is, the better formations it indicates. It is just like the more deposits a man has, the richer he is. These gas wells, which cannot obtain a sufficient follow-up gas supply from the matrix, will affirmatively deplete quickly and with a low accumulative production. They cannot be good wells.

3. Influence of interporosity-flow coefficient λ on flow in the double medium formation

Compared with storativity coefficient ω, interporosity-flow coefficient λ draws less engineers' attention, but in fact is also an important parameter.

For carbonate formation in the QMQ gas field and other volcanic rock formations, geologists seemed not to pay attention to the λ value when calculating the ω

value with statistics of core observation. Equation (5.58) in Chapter 5 showed that the λ value is defined as the product of the ratio of matrix permeability k_m to fissure permeability k_f and shape factor α:

$$\lambda = \alpha \frac{k_m}{k_f}.$$

If the matrix permeability is very low, the λ value will become very small and the flow of natural gas from the matrix to the fissure will be very difficult—sometimes this process may last several years; in this case, even if there is natural gas stored in the matrix, it is very difficult to produce during the industrial production period. The λ value must be set in numerical simulation, but there is no real or proper value obtained from well test interpretation due to poor quality data so the only way is to select one from the values without any reliable basis; it is certain that there are inevitable uncertain factors existing in the predicted development indices.

4. Unreasonable assumption of double medium is unhelpful in the development of special lithologic gas reservoirs

In the past, when faced with special lithologic gas fields, geologists always looked for favorable areas for natural gas distribution through the analysis of their geologically genetic causes, evolution process, and some other static description methods; this is really the key means for resolving problems in exploration and development. Trying to resolve the problems in more complicated special lithologic gas fields by means of assuming the formations of which are double medium lacks a theoretical basis and will make the problems more confusing.

7.3 Production Test and Dynamic Description of Gas Reservoir are Effective Research Approaches For Gas Fields

The more complicated special lithologic gas fields did bring totally new challenges in static geological description, but on the other side there could be some advantages in the dynamic description.

7.3.1 Production Test and Reservoir Description of QMQ Gas Field

The QMQ gas field was discovered at the end of the 20th century. At that time, study of the production test and dynamic description of the gas field had not started in practice yet. Although the discovery well penetrated the reservoir in Ordovician, Well BS-7, and later drilled important gas wells BS-8, BS-6, Q12-18, and Q18-18 were not tested according to the design made in advance as introduced in Section 8.1, gas production of them was arranged completely in line with the idea of a production test and dynamic description, and a thorough dynamic description study associated with repeated tracing investigations was performed.

"Characteristic graph and field examples of fissured zone with boundaries," that is, Section 5.5.7.2 in Chapter 5, mainly discussed the important gas wells in the QMQ gas field. Some knowledge obtained based on those analyses is as follows.

1. Gas wells in the field locate in "fracture-developed zones cut by no-flow boundaries"

As shown in Figures 5.132, 5.133, 5.136, 5.137, 5.141, and 5.143 in Chapter 5, the log–log pressure buildup curves of those wells showing gas-bearing areas where those gas wells locate are characterized by a complicated fractured zone, pressure derivative curves go up and down and rise sharply in late time, suggesting that limited no-flow boundaries exist in the fracture-developed zone. In Chapter 5, such reservoirs were compared to beaded bands or more complicated fissure-developed area consisting of several connecting fissured zones. It is consistent with the ground feature of a carbonate karst landform observed on the surface, which is difficult to exactly describe and predict by such approaches as geophysical prospecting and logging analysis currently, and also difficult to mirror faithfully by the present percolation mechanics and well test methods.

However, the dynamic analysis method can exactly calculate the areas and dynamic reserves controlled by the tested gas wells with a dynamic pressure variation during the production test and can estimate future changes with the analogous model; this is really irreplaceable by any other methods available so far.

The dynamic reserves of areas controlled by two key gas production wells, Well BS-7 and BS-8, calculated with the material balance method were 1.27 and $6.02 \times 10^8 \, \text{m}^3$, respectively, and the sum of them accounts for about 2.5% of the approved OGIP of the whole area ($305 \times 10^8 \, \text{m}^3$). The cumulative production of all other gas production wells was less than $0.5 \times 10^8 \, \text{m}^3$ when the production test nearly finished, which shows a big difference between the actual development effect and the expected one in the development program of the QMQ gas field.

2. Initial deliverability and dynamic deliverability of gas wells

It was very difficult to determine the initial deliverability of some gas wells in the QMQ gas field. For instance, the measured pressure variation during the back-pressure test of Well BS-7 is as shown in Figure 8.71.

It can be seen from Figure 8.71 that the well was flowed with chokes of 7.94 mm, then changed to 10 mm and then back to 7.94 mm, the flowing pressure decreased from 40.3 MPa at the end of the first flow with a 7.94-mm choke to 38 MPa at the end of the second flow with the same choke; then smaller and smaller chokes were used in order to stabilize the flowing pressure but failed: the flowing pressure could not be stabilized but was continuously declining, resulting in that the deliverability equation was unable to be established from the back-pressure test run after a large amount of natural gas had been blown down; the initial AOFP for reference could not be calculated either. The main reason was concluded from analysis that the formation pressure in the gas drainage area declined rapidly during the not very long

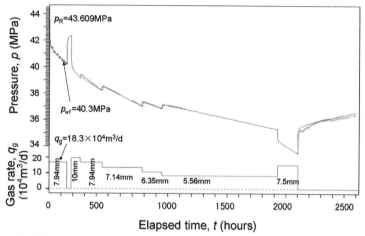

Figure 8.71 Flow rate and pressure history of Well BS-7 during back-pressure test.

flowing production period, which caused the flowing pressure to decline along with it; this was obviously observed from the followed pressure buildup test.

Now reanalyze data when this book is to be reprinted, a primary deliverability calculation can be done with the one stable-point LIT deliverability equation method. An initial deliverability equation is established with the initial test point in Figure 8.71, that is, $p_{wf} = 40.3$ MPa, $q_g = 18.3 \times 10^4$ m³/day, and initial formation pressure $p_R = 43.609$ MPa:

$$p_R^2 - p_{wf}^2 = 14.3323q_g + 0.0459q_g^2$$

and the calculated initial AOFP is $q_{AOF} = 100.4 \times 10^4$ m³/day.

Based on the initial deliverability equation, the dynamic deliverability variation and dynamic formation pressure variation can be obtained by selecting several test points after the back-pressure test.

3. Dynamic characteristics of condensate gas well

In addition to the complicated reservoir structures introduced previously, many problems were encountered in development of the QMQ gas field, among them distortion of the pressure buildup curve caused by natural gas retrograde condensation, and a series of consequent difficulties in data acquisition and analysis were prominent. These problems were discussed in detail with an example of Well BS-8 in Section 5.5 of Chapter 5 entitled "Characteristic graphs and field examples of condensate gas well."

7.3.2 Production Test and Dynamic Description of Reservoir in Volcanic Rock Gas Field

Since a volcanic rock gas field was first discovered in the Songliao Basin at the beginning of this century, the complication of this kind of field has been demonstratively reflected in its static geological characteristics; hence production testing was immediately decided to be performed by the oil field management administration, attempting to disclose key points of the field development with the reservoir description and find out the solutions to these problems as soon as possible.

1. Production test design of XS gas field

   The written production test designs, numerical simulation designs, and geological design reports were all made for Wells X1, X1-1, X6, and some other important gas production wells with the method introduced in Figure 8.1, and production tests consisting of a deliverability test, short-term production test, pressure buildup test, production test with constant production rate and declining flowing pressure, and production test with constant flowing pressure and declining production rate of these wells were arranged; the production tests were then run according to the designs.

2. Preliminary knowledge obtained from production tests

   (1) Well X1. The cumulative gas production of Well X1 was about 0.5×10^8 m^3 during the production test, and the AOFP $q_{AOF} = 87.8 \times 10^4$ m^3/day was verified with deliverability data acquired in the early stage of the production test. The dynamic reserves controlled by the well were calculated with the material balance method to be 4.9×10^8 m^3. The dynamic model of this well was built primarily based on the analysis results of the pressure buildup test during the production test: the formation flow capacity was $kh = 61$ mD·m, half-length of the fracture was $x_f = 35$ meters, and the well locates in a banded fissured zone, the width of which is only 120 meters, so the effective area controlled by the well in the fissured zone is limited. The pressure buildup log–log plot of this well is shown in Figure 8.72, which clearly shows linear flows formed by the fracture near the bottom hole and no-flow boundaries of banded fissured zone.

   (2) Well X1-1. This well is only 1 km away from Well X1, but known from geological data analysis, the main pay zones of them are in different volcanic rock bodies. The dynamic characteristics of both wells indicate that there is no pressure interference

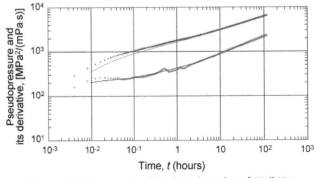

Figure 8.72 Pressure buildup log–log plot of Well X1.

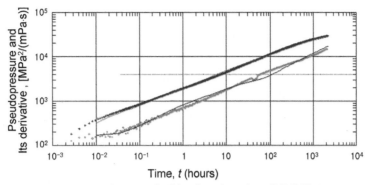

Figure 8.73 Pressure buildup log—log plot of Well X1-1.

from each other, and they are in different percolation areas. The cumulative gas production of the well during the production test was about $0.15 \times 10^8 \text{ m}^3$. The initial AOFP verified with deliverability test data in the early stage of the production test was $q_{AOF} = 37.4 \times 10^4 \text{ m}^3/\text{day}$. The dynamic reserve controlled by the well calculated with the material balance method was about $1.0 \times 10^8 \text{ m}^3$. The dynamic model of the well was established primarily on the basis of the analysis results of the pressure buildup test during the production test, and the formation flow capacity was obtained as $kh = 13.5 \text{ mD·m}$, the half-length of fractures was $x_f = 27$ meters, and the purpose of stimulation has been met. The result of model analysis showed that the well locates in a banded fissured zone, the width of which is only 110 meters, which means that the effective area controlled by the well was even more limited. The log—log plot of the pressure buildup test of the well is shown in Figure 8.73, which clearly shows linear flows formed by the near bottom-hole fracture and no-flow boundaries of the banded fissured zone.

(3) Well X6. The cumulative gas production of Well X6 in the production test was about $0.6 \times 10^8 \text{ m}^3$, and the initial AOFP verified with deliverability test data in the early stage of production test was $q_{AOF} = 68.3 \times 10^4 \text{ m}^3/\text{day}$. The dynamic reserve controlled by the well calculated with the material balance method was about $5.0 \times 10^8 \text{ m}^3$. The dynamic model of the well was established based on the analysis result of the pressure buildup test during the production test, and the formation flow capacity of the well was calculated as $kh = 62 \text{ mD·m}$, the half-length of the fracture caused by fracturing was $x_f = 59$ meters, and the purpose of stimulation has been met. The result of model analysis showed that the well locates in a banded fissured zone, the width of which is only 400 meters. Partial regions of the volcanic rock gas field have been put into production now, there will be more initial production data available for dynamic description study, and more thorough and comprehensive understandings on the reservoir characteristics of the region will be obtained.

8 SUMMARY

The title of this chapter corresponds to that of this book and is the overall embodiment of the purport of this book. Natural gas, as being one of clean energy resources, has been developed and utilized at a high speed, and along with which the technologies of natural gas development should certainly be enhanced in China. A dynamic description of a gas reservoir is key in understanding and developing gas reservoirs. And its importance has been proved by the previous successful practices.

In the first edition of this book, this chapter focused on successful field examples of two gas blocks, that is, the application of a dynamic reservoir description in JB and KL2 gas fields, from which the great practical significance of the dynamic description method promoted in this book had already been proved. When the second edition of this book is published, the dynamic description method will have been applied in more moderate to large gas fields such as the SLG gas field, YL gas field, DF gas field, and QMQ gas field and even in many different types of special lithologic gas fields (e.g., volcanic rock gas fields) discovered in China in the past 10 years.

In recent years, the new research idea of dynamic description, which was introduced in Chapter 1 and expressed in Figure 1.4, has been embodied in the operations of gas fields considerably. This new idea centers on deliverability evaluation and prediction of gas wells, not only to measure and evaluate the initial deliverability, including initial deliverability equation, initial AOFP, and initial IPR curve of different gas wells in the gas block, but also to evaluate dynamic deliverability of the wells, including dynamic deliverability equation, dynamic IPR curve, dynamic AOFP, and dynamic formation pressure at the drainage boundary during the follow-up dynamic description. It is these dynamic deliverability indices that form the reliable basis for studying the production performance of the gas wells and the gas fields. Another important job in the dynamic description of gas wells and reservoirs is to establish the dynamic models of the related wells. Establishment of the dynamic model reappears as the actual situation of gas wells and reservoirs with computer software. This situation is not only reflected by such real formation conditions as reservoir parameters, completion parameters, reservoir boundaries, reservoir size, and dynamic reserve controlled by each single well, but also is able to vividly and actively reappear the whole percolation process of gas wells in the production process from the very beginning to the end, which means that the model can not only reappear the previous performance but also predict the variation of the dynamic performance in the future. Moreover, if combined, these predictions of flow rate and pressure with dynamic deliverability evaluation the prediction of the rules of deliverability variation can be forecasted.

The JB gas field is a large-area and low-abundance fissured limestone gas field. There has been no experience obtained from previous practices for how to develop such a field. For the purpose of early intervention of development, management personnel and reservoir engineers attended to dynamic description and regarded it as a vital job at the very beginning. A study on dynamic testing and analysis lasting 6 years was performed from 1991; the study successfully resolved a series of key problems seriously influencing field development in the initial period, such as reservoir structural characteristics, formation connectivity, reserves calculations with static methods, and deliverability and deliverability stabilization of each well so that a gas field could be put into production successfully and very good development effects were achieved. The results of the study were praised and confirmed by a postevaluation by a specialist team in 2002. The dynamic description of the JB gas field is not only of important value to the development of a field but also has more far-reaching significance as a new study method, a new study stylization, and a new concept.

The KL-2 gas field is another typical field example given in this chapter to illustrate how to gain a basic knowledge of gas reservoirs by means of a short-term production test. This field is an extremely good one, for it possesses almost all characteristics to be a high-quality gas field: high abundance, high formation pressure, high permeability, very thick reservoir, and high deliverability. The test method for such a field never encountered before in China at the very beginning was not good. First, a layered well test procedure for each sublayer was conducted in Wells KL-2 and KL-201. The test provided qualitative understandings of high deliverability characteristics, but failed to determine the total commingled deliverability of the well and the reservoir parameters. The commingled well test was conducted in Well KL-203, but the reservoir parameters could not be determined because of insufficient consideration in the testing design. With lessons and experiences obtained from previous operations, a careful design of a well test for Well KL 205 was made and the test was very successful: it confirmed that the main pay zones were in a massive gas reservoir and communicate with each other vertically; it provided the commingled flow capacity of the whole well, Kh, which was as high as 1000−7000 mD·m, and the AOFP, which could be as high as $1000-6000 \times 10^4 \, m^3/day$; in addition, small faults in the zone could not hinder gas flow, well completion quality was good, and skin factor was decreased to zero after the wellbore had been cleaned up over a period of time. All of these laid a foundation for putting into production the gas field successfully. However, probably because the conditions of the KL-2 gas field were too favorable, the dynamic tracing study did not draw sufficient attention and was not performed timely after 10 production wells were put into production, and it was really a pity.

The SLG gas field is another typical field example. The geological basis of this field was quite special: the reservoir is composed of fluvial deposit of plain subfacies, effective sand layers are thin, formation permeability is low, and the reservoirs are cut into many elongated shivers by a lithologic boundary, which bring great difficulties in gas production. In the initial static geological evaluation, the sand layers were pieced together to be regarded as effective formation, which resulted in a very optimistic calculation result of reserves. But after putting into production for a very short time, the performance of the gas wells showed that the effective area of the reservoir drained by each single well was very limited. Afterward a certain amount of production tests and a dynamic description study lasting 7 years were performed, and the overall characteristics of the gas reservoir were gradually disclosed step by step. This is really a good lesson and experience for the readers.

Both the YL gas field and the SLG gas field are in the Ordose Basin, and the pay zones of both of them are in the same formation in the Upper Paleozoic, but their formations locate in different lateral locations and are in different horizons. The experience drawn from the SLG gas field was applied in the YL gas field; the new method of dynamic description has been adopted and the thorough related study was started from the very beginning. Although the both fields are quite similar in geological conditions, it is proved finally that the area of the reservoir controlled by each well of SLG gas field is much larger than that of YL gas field, and the width and the length of SLG gas field are more close each other; therefore, the overall development effect will be much better.

Dynamic description research on the DF gas field could be considered the model of dynamic research. The field is an offshore gas field penetrating marine sedimentary formation. All development wells in this field are horizontal. The study in such a field is inevitably very difficult. Fortunately, field management personnel had paid much attention to dynamic data acquisition and analysis from the very beginning of the field development, high-precision permanent pressure gauges were installed in the horizontal sections in nearly one-half of the wells in the field, and all flowing pressure data throughout the production periods and all pressure buildup data have been recorded successfully so that the pressure histories of all of those wells could be displayed to management and research personnel. Such an overall study on dynamic description based on so much dynamic data is unparalleled in the comprehensiveness of study content, innovation of the research method, and the number of wells participating in dynamic description.

Some other more complicated special lithologic gas fields, such as buried hill carbonate gas field and eruptive facies volcanic rock gas fields, are also important research targets of dynamic description. Even though the study is very difficult and takes considerable risks, no other effective methods other than the production test and dynamic description are feasible at present.

Well Test Design

Contents

It is self-evident that well test design is very important—just like when constructing a building, it is absolutely necessary to prepare a set of blueprints carefully. To construct a building without blueprints is unimaginable, and just the same, without a careful design, a test, especially a test for a special well, cannot achieve satisfactory results. Well test design is a special topic in itself. Understanding thoroughly the geological conditions of the tested reservoir, determining the purposes of the test definitely and what are expected to be obtained from the test, and mastering the methods and techniques of well test are necessary for making a successful well test design.

It is actually true that many successful examples use high-quality well test data to resolve the problems met during gas field development in China; the JB gas field and KL-2 gas field introduced in Chapter 8 are very good examples. The example of the KL-2 gas field is an especially excellent one. At the very beginning, due to lack of knowledge regarding this brand new large-scale gas field with extra thickness, methods involving layered testing of different gas-bearing formation were adopted; later, commingle tests were performed, but there were drawbacks in the well test techniques and tools. Eventually, a suitable well test method with optimal test techniques was applied

Dynamic Well Testing in Petroleum Exploration and Development   © 2013 Petroleum Industry Press. Published by Elsevier Inc.
ISBN 978-0-12-397161-6, http://dx.doi.org/10.1016/B978-0-12-397161-6.00009-7   All rights reserved.

successfully in Well KL-205, and high-quality data for establishment of the reservoir description were acquired successfully.

However, sometimes we are so accustomed to available procedures in the planning of well test well site operations that invalid data are acquired repeatedly. Although these data can meet the requirements quantitatively, a large portion of them cannot be used after they are delivered to the data bank. These abnormal situations are no longer acceptable.

Causes of these abnormal situations include mistakes in both management method and techniques; and the technique problem is mainly insufficient understanding of the methods of well test design. Accordingly, this chapter focuses on how to make a good well test design, combined with the author's own field experience.

1 PROCEDURE OF WELL TEST DESIGN AND DATA ACQUISITION

1.1 Procedure of Well Test Design

1.1.1 Collect Relational Geologic Data and Casing Program of the Tested Well

Following data must be obtained accurately in detail (take Well KL-205 as an example):

Basic Data of Tested Well
1. Well name: Well KL-205
2. Well type: Appraisal well
3. Geographic location: 50 km northeast to Baicheng County, Xinjiang, and 2.5 km southwest to Well KL-201
4. Structural location: Northwest wing of KL-2 Structure in Kelasu Structural Belt in northern part of the Kuche Depression
5. Coordinates: Ordinate (X) x xxx xxx.40 Abscissa (Y) x xxx xxx.70
6. Ground elevation: 1520.38 m
   Bushing elevation: 1529.52 m
   Bushing height: 9.14 m
7. Drilling data
   Date of spudding: Oct. 6, 2000
   Date of finishing drilling: June 16, 2001
   Date of completion: July 15, 2001
   Drilled well depth: 4050.0
   Drilled bottom horizon: Cretaceous system
8. Casing program
   Surface casing: $20'' \times 147.46$ m, 13 3/8$''$ $\times$ 2589.30 m
   Intermediate casing: 9 5/8$''$ $\times$ 3753.14 m
   Production casing: $7'' \times 4050$ m
   Plug back total depth: 4033.64 m
   (Attachment: Diagram of the casing program)

Basic Geologic Data of Tested Well

1. Logging data interpretation (attach the relational tables)
2. Tested intervals:
   4026.0–4029.5 m (for identifying gas–water contact)
   4011.0–4015.5 m (for identifying gas–water contact)
   3789.0–3952.5 m (for deliverability test and pressure buildup test)
3. Middepth of the tested intervals: 3870.75 m
4. Perforated horizons: (E+K) Lower Tertiary and Cretaceous system
5. Perforated interval: 3789.0–3952.5 m
6. Net pay thickness: 163.1 m
   Perforated net pay thickness: 147.1 m
7. Lithology of the producing layer: sandstone
8. Perforation density: 10 holes/m
9. Perforating gun and charge: 127-type gun and 127-type bullets
10. Average porosity of the tested zone: 13.64% (from core analysis)
11. Average permeability of the tested zone: 13 mD (from logging analysis)
12. Temperature at middepth of the tested zone: approximately 100 °C
13. Initial formation pressure: approximately 74.35 MPa
14. Initial gas saturation: 69.0% (core)
15. Initial water saturation: 31.0% (core)
16. Total compressibility: approximately 1.35×10^{-2} MPa^{-1}
17. Distance to the gas–water contact from the tested well: approximately 500 m

Well Test Data of Tested Well

If there is any available drill stem test or well test data of the tested well, they should be collected. However, they are not listed here as Well KL-205 has not been tested.

1.1.2 Objectives of the Test

The objectives are determined generally by the client. Test objectives for Well KL-205 include the following.

1. Make clear the commingled deliverability of the well and absolute open flow potential (AOFP) of the main producing pays to provide basic data for the design of the development program of the KL-2 gas field
2. Make sure the adaptability of the well structure to high deliverability of the well by the deliverability test and the suitability of the well structure to gas-producing techniques in wells producing with a rate over 2 million cubic meters per day
3. Identify gas–water contact of the gas reservoir
4. Measure the initial formation pressure of the gas reservoir
5. Calculate the effective permeability, k, of the gas reservoir

6. Calculate the skin factor, S, of the well to estimate the damages caused by drilling and completion and provide basic information for further modification of drilling and completion operations and stimulation of the reservoir
7. Make clear the hindering effect of nearby faults to the gas flow and the existence of gas—water contact and the distance from it to the well
8. Analyze how the reservoir pressure declines during the well test
9. Sample gas and water for analyzing their properties
10. Perform other tests, such as gas production profile and vertical interference tests, if possible.
   For oil or gas wells of other types, the following test objectives may also be included.
1. Interference test or pulse test run for identifying sealing of fault or interwell connectivity and calculating connectivity parameters or interlayer interference
2. Make clear the controlled dynamic reserves of the tested well by measuring the static formation pressure
3. Clarify the blocks and layer series of development by static pressure gradient test in the gas field

1.1.3 Simulation of Pressure Variation Trend and Log—Log Plot for Well Test Design with Well Test Interpretation Software

Take Well KL-205 as an example. Because the formation capacity of reservoir kh is quite high according to test data analysis of the nearby Well KL-203, it was proposed to run a back-pressure deliverability test. After running simulations in various conditions it was recommended to use the following conditions and parameters for simulation of pressure variation trend and log—log plot:

1. Back-pressure well test to be run with the rates of 50, 100, 150, 200, and 250×10^4 m^3/day
2. Stabilized flow with each choke should last 24 hours
3. Flow the well with the rate of 100×10^4 m^3/day for extended test, and take samples for laboratory analysis
4. Finally, after an extended test, shut in the well to conduct a pressure buildup test
5. Select a homogeneous formation model for the well test design
   Parameters for simulation include:
   - Permeability $k = 30$ mD
   - Formation capacity $kh = 4500$ mD·m
   - Net thickness $h = 147.1$ meters
   - Wellbore storage coefficient $C = 0.2$ m^3/MPa
   - Skin factor $S = 0$
   - Non-Darcy coefficient $D = 3 \times 10^{-3}$ $(10^4 \text{m}^3/\text{day})^{-1}$

   The pressure history obtained from simulation is shown in Figure 9.1. The pressure log—log plot is shown in Figure 9.2. Simulated deliverability plot and inflow

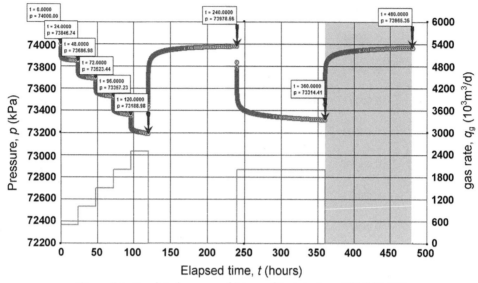

Figure 9.1 Simulated pressure history of test design of Well KL-205.

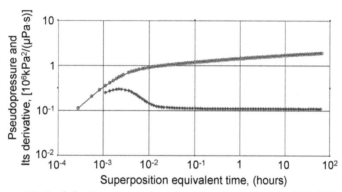

Figure 9.2 Final shut-in pressure log–log plot of test design of Well KL-205.

performance relationship curve are shown in Figures 9.3 and 9.4, respectively. From the simulation the predicted AOFP of Well KL-205 is $q_{AOF} = 6900 \times 10^4$ m^3/day.

1.1.4 Putting Forward the Geological Design Report of Well Test Project
Integrating aforementioned basic data of the tested well, test objectives, and simulation results, the geologic design report of the well test project is prepared, submitted to relevant departments for approval, and then executed.

1.1.5 Detail Operation Design of Well Test
Test tools, techniques, and procedures are keys for the successful implementation of test operations. As seen during testing of Well KL-203, only due to utilizing a drilling collar

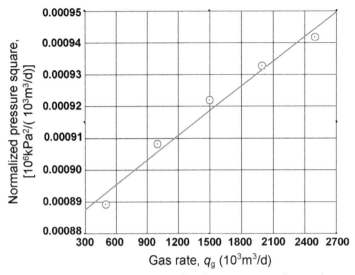

Figure 9.3 Simulated deliverability plot of Well KL-205 (LIT with pseudo pressure).

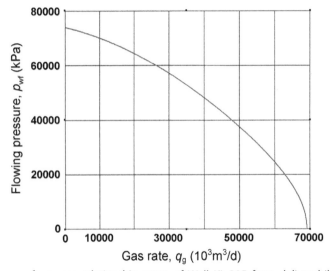

Figure 9.4 Inflow performance relationship curve of Well KL-205 from deliverability test (LIT with pseudo pressure).

with a smaller diameter within the tool assembly, through ensuring safety of the operation, the possibility of obtaining directly the AOFP of the well was missed, and many abnormalities were caused in acquired pressure data.

The design of test equipment, test techniques, and procedures requires special expertise and intensive technical considerations, especially for gas wells with high pressure and high productivity or those whose gas contains corrosive materials because they require some very special equipment, tools, techniques, and procedures of the test. It is not possible to discuss all these requirements in detail here, but care needs to be taken. A detailed operation design is outlined as follows.

Accuracy and Resolution of Pressure Gauges

Electric pressure gauges being used should meet the following requirements:
- Accuracy: 0.02% of FS
- Resolution: 0.01 psi (0.00007 MPa)
- Stability of pressure acquisition and hysteresis error <0.001% of FS
- Minimum data sample rate: one per second
- Maximum pressure or/and temperature data points acquired in one run: $>5 \times 10^5$ points
- Range of pressure gauges: determined in accordance with bottom-hole conditions of the tested well or the tested reservoir. Generally, it should be 50% higher than the maximum shut-in pressure of the tested well.
- Temperature range of the gauges: determined in accordance with requirements of the tested well or the tested reservoir. Generally, it should be slightly higher than the maximum static temperature of the tested reservoir.

High-quality electric pressure gauges should be used for key tested wells. All these pressure gauges should have been calibrated recently, possess proper calibration certificates, and functionally checked in advance.

Location where Pressure Gauges are Installed in Tested Well

To measure pressures in the well, these gauges shall be installed in the middepth of the major oil- or gas-producing zones. If there is load water or other liquid in the well, gauges should be placed below the liquid level, especially for condensate gas wells.

To Acquire Pressure Buildup Data with Downhole Shut-in Tools as Far as Possible

For gas wells with severe phase changes in wellbore or gas-bearing formations with extremely low permeability, special care shall be taken to acquire pressure data with the downhole shut-in tool.

1.2 Essential Requirements for Data Acquisition

For quite a long time and in some oil fields, people have been accustomed to acquiring "only pressure buildup test" data, that is, a transient well test without flowing pressure data during pressure drawdown before the buildup. Testing staff brought pressure gauges to the well site, run them into the well, and then shut in the well immediately to measure the pressure buildup. This way of testing is incorrect and may bring about many difficulties in data interpretation.

The correct way of testing is to monitor or acquire the flowing pressure variation as long as possible during production of the tested well first, and the duration of flowing pressure acquisition should be at least as long as that of pressure buildup, so that the pressure history of both drawdown and buildup are recorded. Only if the "whole pressure history" has been acquired can the pressure history match verification for confirming the correctness of model analyses be done. As this has been introduced in detail in Chapter 2 and Chapter 5 in this book, it is not discussed here. In short, the essential requirements for pressure data acquisition are as follow.

1. Pressure gauges should be placed in the bottom hole of the tested well to record continuously the pressure history of both drawdown and buildup
2. Pressure buildup data, especially those during the initial period of shutting in, should be acquired at a higher sample rate
3. Detailed flow rate data during the whole testing period and before the test must be recorded accurately
4. Detailed operation during the testing must be reported accurately and timely. The operation report should include following contents:
   * Programming of pressure gauges
   * Time when gauges are run in the hole and pulled out of the hole
   * Time when test tools are run in the hole and pulled out of the hole; the procedures of key operations such as setting the packer and opening or closing the test valve
   * Time when opening and closing the wellhead valves and wellhead pressure readings
   * Events during testing such as wellhead leakage and its elimination and other unexpected or important incidents

2 KEY POINTS OF SIMULATION DESIGN OF TRANSIENT WELL TEST FOR DIFFERENT GEOLOGIC OBJECTIVES

2.1 Well Test Design for Wells in Homogeneous Formation

Homogeneous formations are the most common and encountered most frequently. It seems quite simple, so the importance of well test design for the wells in them is often ignored.

The characteristics of well test plots of wells in homogeneous formations and the "position analysis method" have been discussed in Chapter 5. For wells in homogeneous formation, the most important characteristic of the valid test data is that the radial flow section is obtained. Well test curves with a radial flow section can be used for the interpretation of various formation parameters, whereas those without this section are basically useless.

Take a gas reservoir with the following parameters, for example:

- $p_R = 40$ MPa
- $k = 3$ mD
- $h = 5$ meters
- $C = 0.65$ m^3/MPa
- $S = 1$

A simulated log–log plot is shown in Figure 9.5.

Log–log plots like this are normal and can be interpreted and used to calculate parameters. However, in case the formation permeability is very low, say $k = 0.1$ mD, with other parameters remaining unchanged, the predicted plot will be like Figure 9.6.

In this type of well test in formations with extremely low permeability, it is impossible to acquire data of the radial flow period even if the wellbore storage coefficient is reduced to 0.03 m^3/MPa using downhole shut-in tools. The plot in this type of well test looks like Figure 9.7.

Geologic data for Figure 9.7 include

- $k = 0.1$ mD
- $h = 5$ meters
- $C = 0.03$ m^3/MPa

The situations shown in Figures 9.6 and 9.7 make calculation of the formation parameters very difficult and are often met in well testing in reservoirs with very low permeability. Certain improvements can be obtained only if the test time is extended greatly.

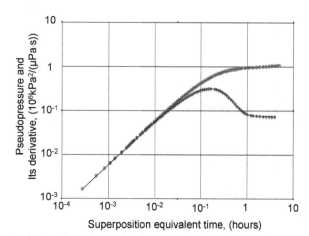

Figure 9.5 Design simulation chart No. 1 of homogeneous formation (obtained buildup data contain radial flow period).

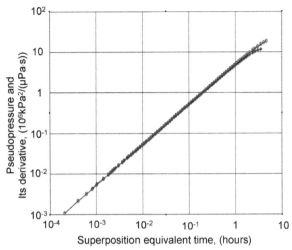

Figure 9.6 Design simulation chart No. 2 of homogeneous formation (obtained buildup data consist of afterflow period only).

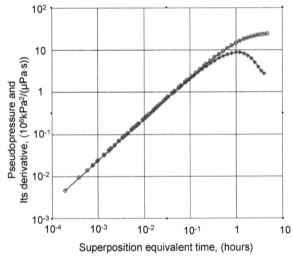

Figure 9.7 Design simulation chart No. 3 of homogeneous formation (obtained buildup data consist of afterflow period only).

2.2 Well Test Design for Double Porosity Formation

Well test design for double porosity formation is quite difficult due to significant uncertainty in the prediction of double porosity parameters ω and λ. Even whether the flow in double porosity happens or not in the fissured reservoir is uncertain.

Just as described in detail in Section 5.4 of Chapter 5, if a typical plot of the well in a double porosity reservoir looks like the model graph of M-2 or M-3, there is a transition period that is characterized by a downward concave curve in the derivative plot.

Major factors affecting simulation of the transition period in the test design include the following.

1. Before shutting in, the well should produce as long as the drawdown has gone through both the transition period and the period of radial flow in the total system of the reservoir.
2. The value of interporosity flow coefficient, λ, should be suitable. A very high λ may make the transition period appear very early and the transitional period be merged in the afterflow period consequently; a very low λ may delay the presence of the transition period so that to exceed the time limit of data acquisition or may make it be merged in the outer boundary reflections occurred at a later time.
3. The value of the storativity ratio, ω, should not be too high. In case ω is approximately 0.32–0.5, or even larger than 0.5, the downward concave on the derivative curve of the transition period may be too gentle to be distinguished.
4. The wellbore storage coefficient, C, of the tested well should not be too high, as otherwise the characteristics of the transition period may be concealed by the wellbore storage effect.

It can be seen that the test arrangement is very difficult, which is the reason why quite a few successful field examples of well tests in double porosity reservoirs are available in China since development of the modern well test method. Figure 9.8 shows an example of design for reference. Readers may make further quests in practice.

Parameters for the design are

- $K_f = 1$ mD
- $h = 10$ meters
- $\omega = 0.05$

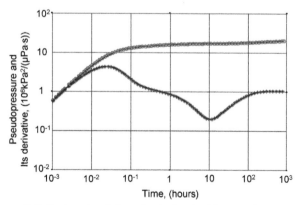

Figure 9.8 Design simulation chart of a double porosity reservoir.

- $\lambda = 1.0 \times 10^{-6}$
- $C_D = 50$
- $S = 1$

It can be seen from Figure 9.8 that the buildup test should last at least 200 hours.

2.3 Well Test Design for Fractured Well in Homogeneous Formation

For tests in fractured wells, fracture half-length x_f, formation permeability k, and skin factor of fracture S_f are needed to be calculated from the well test interpretation. Therefore, the designed test duration should be so long that the well test curve contains a pseudo radius flow period.

For example, the simulated test data plot of a fractured well in a homogeneous formation is as shown in Figure 9.9, and the parameters are as follow:

- $k = 3$ mD
- $h = 5$ meters
- $x_f = 50$ meters
- $k_f W = 3000$ mD·m
- $S_f = 0$
- $C_D = 1000$

It can be seen from Figure 9.9 that the test duration of this well should be at least 20−100 hours in order to record data of the pseudo-radial flow period, obtain the horizontal straight line on the pressure derivative curve, and do data analysis and calculate the parameters of the formation and the tested well.

2.4 Well Test Design for Formation with Flow Barrier

There are various different types of flow barrier or boundaries in various reservoirs. Before performing simulation for a test design for wells in formations with a flow barrier(s), it is necessary to study detailed geologic data to understand the nature and

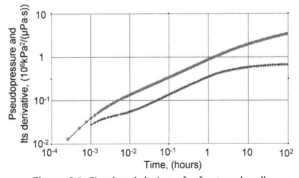

Figure 9.9 Simulated design of a fractured well.

configuration of the barrier(s) and the distance to the tested well from the barrier(s) as much as possible.

When simulation designing with well test design software, the duration of the pressure drawdown should be long enough. Only if the duration of the pressure drawdown prior buildup test is long enough and the influence range of pressure drawdown has extended to cover the boundaries can the following buildup test detect the reflections of those boundaries. In addition, only if the pressure history consisting of drawdown and buildup is long enough can pressure history match verification eliminate ambiguity as much as possible and verify the reliability of the interpretation results with the pressure trend.

2.5 Deliverability Test Design for Gas Wells

The deliverability test design for gas wells was introduced in detail in Chapter 3.

2.6 Multiple-Well Test Design

The multiple-well test design was illustrated in detail in Chapter 6.

2.7 Duties and Principles of Well Test Designers

As there are various different kinds of formation, it is impossible to cover every aspect with some simplified principles of well test design for wells in various kinds of formation. In practice, designers should make clear the geological conditions of the reservoir and study available data carefully to find the most practical scheme.

Well tests with too short a duration or too simplified a procedure may not be able to acquire enough data about the target reservoir. At the same time, field conditions may not allow well testing to last long enough.

Selecting the most suitable design is the most important work for a successful well test. A successful test design can only be fulfilled through in-depth and repeated research. Those reservoir engineers who can make a well test design successfully for wells in oil or gas fields with complicated conditions are called a "well test designer"; such well test designers are now in great shortage in oil or gas fields. Great efforts need to be made by both clients and operators when training well test designers.

APPENDIX 1

Nomenclature (with China Statutory Units CSU)

A area, m^2

A ordinate difference between the pressure and pressure derivative curves on log–log plot, mm

A_D dimensionless ordinate difference between the pressure and pressure derivative curves on log–log plot, $A_D = A/L_C$

A coefficient of fluid accumulation rate, m/hr

A coefficient of laminar flow term in LIT deliverability equation, $MPa/(10^4 \ m^3/day)$ or $MPa^2/(10^4 \ m^3/day)$ or $[MPa^2/(mPa \cdot s)]/(10^4 \ m^3/day)$

A_h coefficient of laminar flow term in stable-point LIT deliverability equation of horizontal well, $MPa/(10^4 \ m^3/day)$ or $MPa^2/(10^4 \ m^3/day)$ or $[MPa^2/(mPa \cdot s)]/(10^4 \ m^3/day)$

B formation volume factor, m^3/m^3

B coefficient of turbulent flow term in LIT deliverability equation, $MPa/(10^4 \ m^3/day)^2$ or $MPa^2/(10^4 \ m^3/day)^2$ or $[MPa^2/(mPa \cdot s)]/(10^4 \ m^3/day)^2$

B_h coefficient of turbulent flow term in stable-point LIT deliverability equation of horizontal wells, $MPa/(10^4 \ m^3/day)^2$ or $MPa^2/(10^4 \ m^3/day)^2$ or $[MPa^2/(mPa \cdot s)]/(10^4 \ m^3/day)^2$

B_g, B_o, B_w formation volume factors of gas, oil and water, respectively, m^3/m^3

B_{gi}, B_{oi} initial formation volume factors of gas and oil, respectively, m^3/m^3

C constant

C coefficient of exponential deliverability equation, $(10^4 \ m^3/day)/MPa^n$ or $(10^4 \ m^3/day)/ MPa^{2n}$ or $(10^4 \ m^3/day)/[MPa^2/(mPa \cdot s)]^n$

C wellbore storage constant, or wellbore storage coefficient, m^3/MPa

C compressibility, MPa^{-1}

C_g gas compressibility, MPa^{-1}

C_o oil compressibility, MPa^{-1}

C_r rock compressibility, MPa^{-1}

C_f effective rock compressibility, MPa^{-1}

C_w water compressibility, MPa^{-1}

C_t total compressibility, MPa^{-1}

C_d coalbed desorption compressibility, MPa^{-1}

C_D dimensionless wellbore storage constant, or dimensionless wellbore storage coefficient

C_{Dxf} dimensionless (based on x_f) wellbore storage constant (or dimensionless wellbore storage coefficient) of a fractured well

$C_D e^{2S}$ shape parameter group of log–log pressure and pressure derivative curves

D non-Darcy flow coefficient, $(10^4 \ m^3/day)^{-1}$ or $(m^3/day)^{-1}$

DR damage ratio, dimensionless

e 2.71828

Ei exponential integration function

E_{go} gas–oil ratio (GOR), m^3/m^3 or m^3/t

f_{CK} friction resistance factor of flow, dimensionless;

F_{CD} conductivity of fracture, $F_{CD} = k_{fD} \cdot W_{fD}$

FE flow efficiency, dimensionless

F' ratio of pulse length to the length of whole cycle, dimensionless

g gravitational acceleration, m/s^2

G original gas in place (OGIP), 10^8 m^3

G_p cumulative gas production, 10^8 m^3

G_{Dh} static pressure gradient in wellbore, MPa/m

G_{Dfl} flowing pressure gradient in wellbore, MPa/m

G_{DS} static reservoir pressure gradient, MPa/m

h formation thickness, m

h depth, m

h_b distance between the bottom of the perforation and that of the layer in partially perforated well, m

h_p perforation thickness of a partially perforated well, m

h_{top} distance between the tops of perforation and the layer in partially perforated well, m

h_{wD} dimensionless formation thickness of a partially perforated well

H distance between peak in wellbore storage period and horizontal straight line in radial flow period of pressure derivative curve, mm

H_D dimensionless distance between peak in wellbore storage period and horizontal straight line in radial flow period of pressure derivative curve

ΔH_N height difference between horizontal straight lines of the inner and outer areas on pressure derivative curve of a condensate gas well, mm

ICD retrograde diffusivity of a coalbed gas reservoir, mPa·s/(MPa·mD)

J_o, J_w, J_L oil, water, and liquid productivity indexes, respectively, m^3/(MPa·day)

k formation permeability, mD

k_a air permeability, mD

k_f fracture permeability, mD

k_{fD} dimensionless fracture permeability, $k_{fD} = k_f/k$

k_H horizontal formation permeability, mD

k_V or k_Z vertical formation permeability, mD

k_V/k_H ratio of vertical permeability to horizontal permeability of the formation

k_o effective permeability to oil, mD

k_r radial formation permeability, mD

kh formation flow capacity, mD·m

kh/μ formation flow coefficient, mD·m/(mPa·s)

$kh/(\mu C)$ flow-storage capacity ratio (location parameter), mD·MPa/(m^2·mPa·s)

lg common logarithm

ln natural logarithm

L_b distance between tested well and fault, m

L_C logarithm cycle length in ordinate direction of log–log paper, mm

L_e length of horizontal section of a horizontal well, m

L_M height difference between two horizontal straight lines reflecting, respectively, inside and outside areas of a composite system on pressure derivative curve, mm

L_{MD} dimensionless height difference between two horizontal straight lines reflecting, respectively, inside and outside areas of a composite system on pressure derivative curve, $L_{MD} = L_M/L_C$

L_p height difference between two horizontal straight lines on pressure derivative curve of a partially perforated well, mm

L_{pD} dimensionless height difference between two horizontal straight lines on a pressure derivative curve of a partially perforated well, $L_{pD} = L_p/L_C$

L_θ height difference between two horizontal straight lines on pressure derivative curve of a well located in an area in an included angle, mm

$L_{\theta D}$ dimensionless height difference between two horizontal straight lines on pressure derivative curve of a well located in an area in an included angle, $L_{\theta D} = L_\theta/L_C$

m mass, g or kg

m straight line slope in radial flow regime on semilog transient pressure plot, MPa/log cycle

m_g, m_o straight line slope in radial flow regime on semilog build-up curve of gas well and oil well, respectively, MPa/log cycle

m' maximum slope of after flow segment in semilog plot of a well in homogenous formation, MPa/cycle

$m^\star$ straight line slope of pseudo-steady section in Cartesian pressure drawdown plot, MPa/hr

M mobility ratio, dimensionless

M_C mobility ratio of composite reservoir, $M_C = \lambda_1/\lambda_2$

M_d mass of absorbed gas contained in unit volume of coal, kg

M_g molar mass of gas, kg/kmol, $M_g = \gamma_g \times 28.96$

M_N ratio of mobility of inner zone to that of outer zone in condensate gas reservoir

M_P ratio of flow coefficient of gross formation to that of perforated formation

n exponent of exponential deliverability equation, i.e., the exponent reflecting the degree of gas turbulent flow, dimensionless

n_g quantity of gas, kmol

N original oil in place (OOIP), 10^4 m^3 or 10^4 t

N_p cumulative oil production, 10^4 m^3 or 10^4 t

p pressure, MPa

p_C Langmuir pressure of coalbed gas layer, MPa

p_D dimensionless pressure

$p_D{}'$ derivative of dimensionless pressure p_D with respect to t_D/C_D

p_{DG} normalized pressure, $p_{DG} = p_D/p_D{}'$

p_{DG} dimensionless pressure square difference in deliverability equation of single-point test method

p_{Di} dimensionless integral pressure or mean pressure

$p_{Di}{}'$ derivative of dimensionless integral pressure

p_{Did} difference of dimensionless integral pressure

p_{Did1} the first type of normalized integral pressure

p_{Did2} the second type of normalized integral pressure

$(p_D)_M$ matched value of dimensionless pressure, i.e., the value of dimensionless pressure at the match point on type curve

p_{DMDH} dimensionless pressure in MDH plot

p_{DMBH} dimensionless pressure in MBH plot

p_e pressure at external boundary, MPa

p_H wellhead pressure, MPa

p_i initial pressure, MPa

p_c critical pressure of gas, MPa

p_{pc} pseudo-critical pressure of gas, MPa

p_R reservoir pressure, MPa

p_r reduced pressure of gas, dimensionless

p_{pr} pseudo reduced pressure of gas, dimensionless

p_{sc} standard pressure condition, $p_{sc} = 0.101325$ MPa

p_w bottom-hole pressure, MPa

p_w bottom-hole pressure of coalbed gas well during the process of injection and pressure falloff testing when the wellhead pressure drops to 0, MPa

p_{wf} bottom-hole flowing pressure, MPa

p_{ws} bottom-hole shut-in pressure, MPa

$\bar{p}$ average pressure, MPa

$p^\star$ extrapolated pressure, i.e., reservoir pressure extrapolated from Horner plot, MPa

Δp pressure difference, MPa

Δp_n pressure offset caused by liquid loading in wellbore, MPa

Δp_p ordinate offset between two semilog straight lines of a well in double porosity reservoir, MPa

Δp_s additional pressure drop caused by skin effect, MPa

Δp_{1h} pressure at 1 hour on straight line of semilog buildup curve, MPa

q daily flow rate, m^3/day or t/day

q_{AOF} absolute open flow potential (AOFP) of a gas well, 10^4 m^3/day

q_f bottom-hole flow rate, i.e., sandface flow rate, m^3/day

q_h surface flow rate, m^3/day

q_g daily gas flow rate, 10^4m^3/day

q_L daily liquid flow rate, m^3/day

q_o daily oil flow rate, m^3/day or t/day

q_w daily water flow rate, m^3/day

q_r quantity of desorbed gas from coalbed gas reservoir, m^3/day

q_{sc} gas flow rate of a gas well under standard conditions, 10^4 m^3/day

q_n, q_N flow rate in the nth and Nth period, respectively, during multi-rate testing, m^3/day

r radial distance from a well, m

r_d drainage radius, m

r_D dimensionless radius, $r_D = r/r_w$

r_e drainage radius, m

r_{eh} equivalent gas drainage radius of a horizontal well, m

r_i radius of influence, m

r_M radius of the inner area in radial composite reservoir, m

r_{MD} dimensionless radius of the inner area in radial composite reservoir, $r_{MD} = r_M/r_w$

r_N radius of condensate gas area, m

r_S radius of damaged area, i.e., radius of skin zone, m

r_w wellbore radius, m

r_{we} equivalent wellbore radius, m

r_{wh} equivalent wellbore radius of horizontal well, m

R universal gas constant, MPa·m^3/(kmol·K); $R = 8.3143 \times 10^{-3}$MPa·m^3/(kmol·K)

R_{eCK} Reynolds number, dimensionless

S skin factor, dimensionless

S_a apparent skin factor or pseudo skin factor, dimensionless

S_f fracture skin factor of fractured well, dimensionless

S_N apparent skin factor caused by two-phase flow in condensate gas wells, dimensionless

S_p apparent skin factor of partial penetrated well, dimensionless

S_g, S_o, S_w gas, oil, and water saturations, respectively, fraction

S_{gi}, S_{oi}, S_{wi} initial gas, oil, and water saturations, respectively, fraction

$SUPF$ time superposition function [see Equation (5.20)]

t time, hr

t_D dimensionless time, $t_D = \dfrac{3.6 \times 10^{-3} kt}{\mu \varphi C_t r_w^2}$

t_{DA} dimensionless time, $t_D = \dfrac{3.6 \times 10^{-3} kt}{\mu \varphi C_t A}$

t_{De} dimensionless time, $t_D = \dfrac{3.6 \times 10^{-3} kt}{\mu \varphi C_t r_e^2}$

t_{DL} dimensionless time, $t_D = \dfrac{3.6 \times 10^{-3} kt}{\mu \varphi C_t L^2}$

t_{Drf} dimensionless time, $t_D = \dfrac{3.6 \times 10^{-3} kt}{\mu \varphi C_t r_f^2}$

t_{Dxf} dimensionless time, $t_D = \dfrac{3.6 \times 10^{-3} kt}{\mu \varphi C_t x_f^2}$

t_{we} dimensionless time, $t_D = \dfrac{3.6 \times 10^{-3} kt}{\mu \varphi C_t r_{we}^2}$

t_f time corresponding to the peak value on pressure derivative curve in log–log plot, hr

$(t_D/C_D)_M$ matched value of dimensionless time, i.e., dimensionless time at the match point on type curve

$(t_D)_{pss}$ dimensionless time when pseudo-steady state flow begins

t_{DG} time in normalized pressure type curves

t_{DxfG} dimensionless time in normalized pressure type curves

t_M, $(\Delta t)_M$ matched value of time, i.e., time of the match point in data plot, hr

t_L lag time used in pulse testing analysis, hr

t_p production time before shut in, hr

t_{ps}, t_{pss} time when pseudo steady-state flow begins, hr

t_{ss} time when steady-state flow begins, hr

Δt elapse or shut-in time, hr

Δt_p pulse time in pulse testing, hr

Δt_C cycle length of pulse in pulse testing, hr

$\Delta t\star$ time corresponding to the intersecting point of two straight lines on MDH plot, hr

T temperature, K or °C

$\overline{T}$ Average temperature, K

T_f temperature of gas reservoir, K or °C

T_r reduced temperature

T_{pr} pseudo reduced temperature

T_c critical temperature, K

T_{pc} pseudo-critical temperature, K

T_{sc} standard temperature, K or °C, $T_{sc} = 20°C = 293.16K$ (China standard)

v velocity of fluid flow, m/hr

V volume, m^3

V_d volume of gas desorbed from coalbed at the standard conditions, m^3

V_f fissures volume per unit reservoir volume, dimensionless

V_m matrix volume per unit reservoir volum, dimensionless

V_L Langmuir volume of coalbed gas, m^3

V_p pore volume of rock, m^3

V_R grain volume of rock, m^3

V_u volume per unit length of wellbore, m^3/m

V_w wellbore volume, m^3

ΔV_0 critical volume of formation cell body, m^3

W width of fracture, m

W_{fD} dimensionless width of c fracture, $W_{fD} = W/x_f$

W_c humidity of coalbed rock, fraction

W_p cumulative water production, m^3

x_e distance from the well to the boundary in x direction in closed rectangular reservoir, m

x_f half-fracture length, m

y_e distance from the well to the boundary in y direction in closed rectangular reservoir, m

Z real gas deviation factor, i.e., Z-factor, dimensionless

$\overline{Z}$ real gas deviation factor at average test pressure, dimensionless

Z_w distance from a horizontal well section to the lower reservoir boundary, m

Z_{wD} dimensionless distance from a horizontal well section to the lower reservoir boundary, $Z_{wD} = Z_w/r_w$

α ash content in coalbed, fraction

α shape factor (used for calculating the interporosity flow coefficient in double medium system), dimensionless

α parameter in single-point test deliverability formula, $\alpha = A/(A + Bq_{AOF})$

β included angle

β turbulent flow velocity coefficient of gas, m^{-1}

β heterogeneity correction coefficient of gas reservoir, $\beta = \sqrt{k_H/k_V}$

β' parameter of transient flow between two media within double medium formation

γ relative density (liquid relative to water, and gas relative to air), dimensionless

γ_g relative density of gas, dimensionless

γ_o relative density of stock tank oil, dimensionless

δ coefficient of laminar−inertial−turbulent flow, dimensionless

δ correction factor of non-Darcy flow coefficient, dimensionless

∇ divergence operator

η diffusivity, $\eta = k/(\mu\varphi C_t)$

θ included angle of formation

λ mobility, $\lambda = k/\mu$, mD/(mPa·s)

λ interporosity flow coefficient of double porosity reservoir, dimensionless

μ viscosity, mPa·s

μ_g reservoir gas viscosity, mPa·s

μ_o reservoir oil viscosity, mPa·s

μ_w formation water viscosity, mPa·s

$\overline{\mu}$ viscosity at average test pressure, mPa.s

ρ_g, ρ_o, ρ_w densities of gas, oil, and water, respectively, g/cm^3 or kg/m^3

τ adjustment coefficient of desorption time of coalbed gas, dimensionless

ϕ porosity, fraction

ϕC_t storability coefficient, MPa^{-1}

$\psi(p)$ pseudo pressure of real gas corresponding to pressure p, MPa^2/(mPa·s)

ω storativity ratio of double porosity reservoir, dimensionless

ω_c storativity ratio of composite formation, dimensionless

$\kappa(\textbf{kappa})$ formation coefficient ratio of a double permeability formation

CSU China statutory units

SUBSCRIPTS

b pressure buildup

c coalbed

d pressure drawdown

D dimensionless

e external boundary

f formation

f fracture or front

F interpreted with the software FAST

g gas

H horizontal

i initial or the number of sequence

inj injection

L liquid

m matrix

max maximum

min minimum

M matched or type curve match point

M composite formation

o oil

R reservoir

S skin effected zone, i.e., damaged zone

t total

V vertical

w well bottom or water

ws wellbore and shut in

wf wellbore and flowing

z vertical

ψ calculated with pseudo-pressure method

2 calculated with pressure square method

ω double porosity formation

1h 1 hr

Commonly used units in different unit systems

Parameters	Basic units of SI	China statutory units	Darcy units	British field units
Pressure, p, p_e, p_i, p_o, p_R, p_{sc}, p_w, p_{ws}, p_{wf}, $p^\star$, Δp, Δp_s, Δp_M and Δp_{1h}, etc.	Pa or kPa	MPa	atm	psi
Pseudo gas pressure, ψ	$Pa^2/(Pa \cdot s)$	$MPa^2/(mPa \cdot s)$	atm^2/cP	psi^2/cP
Gas flow rate, q_g, q_{AOF} and q_{sc}	m^3/s or m^3/day	$10^4\ m^3/day$	cm^3/s	MMscfd
Oil or other liquid flow rate, q, q_o, q_w, q_L, q_n, q_N, q_r, q_h and q_f	m^3/s or m^3/day	m^3/day	cm^3/s	bbl/day
Cumulative oil production, N_p	m^3	$10^4\ m^3$	cm^3	bbl
Cumulative gas production, G_p	m^3	$10^4\ m^3$	cm^3	MMscf
OOIP controlled by an oil well, N	m^3	$10^8\ m^3$	cm^3	bbl
OGIP controlled by a gas well, G	m^3	$10^8\ m^3$	cm^3	MMscf
Time, t, t_M, t_p, t_{ps}, t_{ss}, Δt, Δt_M, $\Delta t^\star$, Δt_p and Δt_C	s or hr	hr	s	hr
Thickness, h, h_b, h_p, and h_{top}	m	m	cm	ft
Distance, L, L_b, L_e, r, r_d, r_e, r_i, r_f, r_M, r_N, r_s, r_w and r_{we} etc.	m	m	cm	ft
Area, A	m^2	m^2 or km^2	cm^2	ft^2
Permeability, k, k_H, k_V, k_r, k_f, k_m and k_o etc.	m^2	mD (or $10^{-3}\ \mu m^2$)	D (darcy)	mD
Porosity, ϕ	Fraction	Fraction	Fraction	Fraction
Saturation, S, S_o, S_w, S_g, S_{oi}, S_{gi} and S_{wi} etc.	Fraction	Fraction	Fraction	Fraction

(*Continued*)

Parameters	Basic units of SI	China statutory units	Darcy units	British field units
Compressibility, C, C_o, C_w, C_g, C_f, C_t, C_r and C_d etc.	Pa^{-1}	MPa^{-1}	atm^{-1}	psi^{-1}
Wellbore storage coefficient, C	m^3/Pa	m^3/MPa	cm^3/atm	bbl/psi
Fluid viscosity, μ, μ_o, μ_w and μ_g etc.	$Pa \cdot s$	$mPa \cdot s$	cP	cP
Temperature, T, T_f, T_c, T_{pc} and T_{sc} etc.	K	K or $°C$	$°C$	$°F$ or $°R$
Slope of straight line on semilog plot, m	$Pa/cycle$	$MPa/cycle$	$atm/cycle$	$psi/cycle$
Formation volume factor, B, B_o, and B_w etc.	m^3/m^3	m^3/m^3	cm^3/cm^3	RB/STB
Gas formation volume factor	m^3/m^3	m^3/m^3	cm^3/cm^3	RB/scf
Universal gas constant, R	$J/(mol \cdot K)$ or $Pa \cdot m^3/ (kmol \cdot K)$	$MPa \cdot m^3/ (kmol \cdot K)$	$atm \cdot cm^3/ (mol \cdot K)$	$psi \cdot ft^3/ (lbmol \cdot °R)$
Coefficient of non-Darcy gas flow, D	$(m^3/s)^{-1}$	$(10^4 m^3/day)^{-1}$	$(cm^3/s)^{-1}$	$(MMscfd)^{-1}$
Fluid density, ρ, ρ_g, ρ_o and ρ_w	kg/m^3	g/cm^3 or kg/m^3	g/cm^3	lb/ft^3

Unit conversion from China Statutory Unit System (CSU) to other unit systems

1. **Length**
   1 m = 100 cm = 10^3 mm = 3.281 ft = 39.37 in.
   1 ft = 0.3048 m = 30.48 cm = 304.8 mm = 12 in.
   1 km = 0.6214 mile
   1 mile = 1.609 km

2. **Area**
   $1 \text{ m}^2 = 10^4 \text{ cm}^2 = 10.76 \text{ ft}^2 = 1550 \text{ in.}^2$
   $1 \text{ km}^2 = 10^6 \text{ m}^2 = 100 \text{ ha} = 247.1 \text{ acre}$
   $1 \text{ mile}^2 = 2.590 \text{ km}^2 = 259 \text{ ha} = 640 \text{ acres}$
   $1 \text{ acre} = 4.356 \times 10^4 \text{ ft}^2 = 0.4046 \text{ ha} = 4046 \text{ m}^2$

3. **Volume**
   $1 \text{ m}^3 = 10^3 \text{ liter} = 10^6 \text{ ml} = 10^6 \text{ cm}^3 = 35.31 \text{ ft}^3 = 6.290 \text{ bbl} = 264.2 \text{ gal}$
   $1 \text{ liter} = 10^{-3} \text{ m}^3 = 3.531 \times 10^{-2} \text{ ft}^3 = 61.02 \text{ in.}^3 = 0.2642 \text{ gal}$
   $1 \text{ ft}^3 = 2.832 \times 10^{-2} \text{ m}^3 = 28.32 \text{ liter} = 2.832 \times 10^4 \text{ cm}^3 = 7.481 \text{ gal}$
   $1 \text{ bbl} = 5.615 \text{ ft}^3 = 42 \text{ gal} = 0.1590 \text{ m}^3 = 159 \text{ liter} = 158,988 \text{ cm}^3$

4. **Mass**
   $1 \text{ kg} = 10^3 \text{ g} = 2.205 \text{ lbm}$
   1lbm = 0.4536 kg = 453.6 g
   $1\text{t} = 10^3 \text{ kg} = 2205 \text{ lbm}$

5. **Density**
   $1\text{kg/m}^3 = 10^{-3} \text{ g/cm}^3 = 10^{-3} \text{ t/m}^3 = 6.243 \times 10^{-2} \text{ lbm/ft}^3$
   $1 \text{ lbm/ft}^3 = 16.02 \text{ kg/m}^3 = 1.602 \times 10^{-2} \text{ g/cm}^3$

6. **Force**
   $1\text{N} = 10^5 \text{ dyne} = 0.1020 \text{ kgf} = 0.2248 \text{ lbf}$
   $1\text{kgf} = 9.807 \text{ N} = 9.807 \times 10^5 \text{ dyne} = 2.205 \text{ lbf}$
   1lbf = 4.448 N = 0.4536 kgf

7. **Pressure**
   $1\text{MPa} = 10^6\text{Pa} = 9.8692 \text{ atm} = 10.197 \text{ at (kgf/cm}^2) = 145.04 \text{ psi}$
   $1\text{atm} = 0.10133 \text{ MPa} = 1.0332 \text{ at (kgf/cm}^2) = 14.696 \text{ psi} = 760 \text{ mm Hg}$
   $1\text{psi} = 6.8948 \times 10^{-3} \text{ MPa} = 6.8948 \text{ kPa} = 6.8046 \times 10^{-2} \text{ atm} = 7.0307 \text{ at (kgf/cm}^2)$

8. **Permeability**
   $1\mu\text{m}^2 = 10^{-12} \text{ m}^2 = 10^{-8} \text{ cm}^2 = 1.0133 \text{ D} = 1.0133 \times 10^3 \text{ mD}$
   $1\text{mD} = 10^{-3} \text{ D} = 0.98692 \times 10^{-3} \text{ } \mu\text{m}^2 = 9.8692 \times 10^{-16} \text{ m}^2$

9. Kinetic viscosity
$$1 \text{mPa·s} = 10^{-3}\text{Pa·s} = 10^3 \text{ μPa·s} = 1 \text{ cP}$$

10. Temperature
(T_c: °C, T_K: K, T_F: °F, T_R: °R)

(1) Conversion formulae of temperature in different units:

$$T_C = \frac{T_F - 32}{1.8}$$

$$T_K = \frac{T_F - 459.67}{1.8}$$

$$T_K = \frac{T_R}{1.8}$$

$$T_R = T_F + 459.67$$

$$T_K = T_C + 273.15$$

(2) Conversion relations of temperature in different units:

$$1\text{K} = 1°\text{C} = 1.8°\text{F} = 1.8°\text{R}$$
$$1°\text{F} = 1°\text{R} = 0.55556\text{K} = 0.55556°\text{C}$$

11. Wellbore storage coefficient
$$1 \text{m}^3/\text{MPa} = 4.3367 \times 10^{-2} \text{ bbl/psi} = 9.8068 \times 10^{-2} \text{ m}^3/\text{at}$$
$$1 \text{bbl/psi} = 23.059 \text{ m}^3/\text{MPa} = 2.2614 \text{ m}^3/\text{at} = 2.3365 \text{m}^3/\text{atm}$$

12. Compressibility
$$1 \text{MPa}^{-1} = 6.8948 \times 10^{-3} \text{ psi}^{-1} = 9.8068 \times 10^{-2} \text{ at}^{-1} = 0.10133 \text{ atm}^{-1}$$
$$1 \text{psi}^{-1} = 145.04 \text{ MPa}^{-1} = 14.223 \text{ at}^{-1} = 14.696 \text{ atm}^{-1}$$
$$1 \text{atm}^{-1} = 6.8046 \times 10^{-2} \text{ psi}^{-1} = 9.8692 \text{ MPa}^{-1}$$
$$1 \text{atm}^{-1} = 7.0307 \times 10^{-2} \text{ psi}^{-1} = 10.197 \text{ MPa}^{-1}$$

13. Production
$$1 \text{ m}^3/\text{day} = 6.2898 \text{ bbl/day} = 1.1574 \times 10^{-5} \text{ m}^3/\text{s} = 11.574 \text{ cm}^3/\text{s}$$
$$1 \text{bbl/day} = 0.15899 \text{ m}^3/\text{day} = 1.8401 \times 10^{-6} \text{ m}^3/\text{s} = 1.8401 \text{ cm}^3/\text{s}$$
$$1 \text{MMcfd} = 10^6 \text{ ft}^3/\text{day} = 2.831685 \times 10^4 \text{ m}^3/\text{day}$$

14. Relative density of stock tank oil, (γ_o) and °API

$$°\text{API} = \frac{141.5}{\gamma_0} - 131.5$$

$$\gamma_0 = 141.5/(131.5 + °\text{API})$$

15. Gas—oil ratio
$$1 \text{ m}^3/\text{m}^3 = 5.615 \text{ scf/STB}$$
$$1 \text{ scf/STB} = 0.1781 \text{ m}^3/\text{m}^3$$

Formulae commonly used in a well test under the China Statutory Unit System

I FORMULAS IN LOG–LOG PLOT ANALYSIS

1 Dimensionless pressure

$$p_D = \frac{2.714 \times 10^{-5} khT_{sc}\Delta\psi}{q_g T_f p_{sc}} \quad \text{or} \quad p_D = \frac{2.714 \times 10^{-5} khT_{sc}[\psi(p_i) - \psi(p_{wf})]}{q_g T_f p_{sc}}$$

(for pseudo pressure of gas wells)

$$p_D = \frac{2.714 \times 10^{-5} khT_{sc}\Delta(p^2)}{q_g \overline{\mu_g} \overline{Z} T_f p_{sc}} \quad \text{or} \quad p_D = \frac{2.714 \times 10^{-5} khT_{sc}(p_i^2 - p_{wf}^2)}{q_g \overline{\mu_g} \overline{Z} T_f p_{sc}}$$

(for pressure square of gas wells)

$$p_D = \frac{0.5428 kh\Delta p}{q\mu B} \quad \text{(for oil wells)}$$

2 Dimensionless pressure derivative

$$p_D' = \frac{2.714 \times 10^{-5} khT_{sc}\Delta\psi'}{q_g T_f p_{sc}} \quad \text{(for pseudo pressure of gas wells)}$$

$$p_D' = \frac{0.5428 kh\Delta p'}{q\mu B} \quad \text{(for oil wells)}$$

3 Dimensionless time

$$\frac{t_D}{C_D} = 2.262 \times 10^{-2}\frac{kh\Delta t}{\mu C}$$

(Applicable to Gringarten homogeneous formation type curves and Bourdet homogeneous formation type curves)

$$t_D = \frac{3.6 \times 10^{-3} k}{\phi\mu C_t r_w}\cdot\Delta t \quad \text{or} \quad t_D = \frac{3.6 \times 10^{-3} k}{\phi\mu_g C_t r_w^2}\cdot t$$

(Applicable to Agarwal and Ramey homogeneous formation type curves)

$$\frac{t_{\mathrm{D}}}{r_{\mathrm{D}}^2} = \frac{3.6 \times 10^{-3}k}{\phi \mu C_{\mathrm{t}} r^2} \cdot \Delta t$$

(Applicable to homogeneous formation interference type curves)

$$\frac{t_{\mathrm{D}}}{r_{\mathrm{D}}^2} = \frac{3.6 \times 10^{-3}k}{\mu (\phi C_{\mathrm{t}})_{\mathrm{f}} r^2} \cdot \Delta t$$

(Applicable to Gringarten double porosity formation interference type curves)

$$\frac{t_{\mathrm{D}}}{r_{\mathrm{D}}^2} = \frac{3.6 \times 10^{-3}k}{\mu \left[(\phi C_{\mathrm{t}})_{\mathrm{f}} + (\phi C_{\mathrm{t}})_{\mathrm{m}} \right] r^2} \cdot \Delta t$$

(Applicable to Zhuang–Zhu double porosity formation interference type curves)

$$t_{\mathrm{Dxf}} = \frac{3.6 \times 10^{-3}k}{\phi \mu C_{\mathrm{t}} x_{\mathrm{f}}^{\,2}} \cdot \Delta t$$

(Applicable to type curves for vertical fractured wells with infinite conductivity in homogeneous formation or with uniform flow)

$$t_{\mathrm{Dre}} = \frac{3.6 \times 10^{-3}k}{\phi \mu C_{\mathrm{t}} r_{\mathrm{we}}^{\,2}} \cdot \Delta t$$

(Applicable to type curves for vertical fractured wells with finite flow conductivity in homogeneous formation)

$$t_{\mathrm{Drf}} = \frac{3.6 \times 10^{-3}k_{\mathrm{r}}}{\phi \mu C_{\mathrm{t}} r_{\mathrm{f}}^{\,2}} \cdot \Delta t$$

(Applicable to type curves for horizontal fractured wells with uniform flow in homogeneous formation)

4 Dimensionless well storage coefficient

$$C_{\mathrm{D}} = \frac{0.1592C}{\phi h C_{\mathrm{t}} r_{\mathrm{w}}^2}$$

$$C_{\text{Dxf}} = \frac{0.1592C}{\phi h C_t x_{\text{f}}^2}$$

5 Calculation of gas well parameters by type curve matching

$$k = 12.741 \frac{q_{\text{g}} T_{\text{f}}}{h} \cdot \frac{(p_{\text{D}})_{\text{M}}}{(\Delta \psi)_{\text{M}}}$$

$$C = 2.262 \times 10^{-2} \frac{kh}{\mu} \cdot \frac{t_{\text{M}}}{(t_{\text{D}}/C_{\text{D}})_{\text{M}}}$$

$$S = \frac{1}{2} \ln \frac{(C_{\text{D}} e^{2S})_{\text{M}}}{C_{\text{D}}}$$

6 Calculation of oil well parameters by type curve matching

$$k = 1.842 \frac{q\mu B}{h} \cdot \frac{(p_{\text{D}})_{\text{M}}}{(\Delta p)_{\text{M}}}$$

$$k = 1.842 \frac{q\mu B}{h} \cdot \frac{\left(p_{\text{D}}' \cdot \dfrac{t_{\text{D}}}{C_{\text{D}}}\right)_{\text{M}}}{(\Delta p' \cdot t)_{\text{M}}}$$

$$C = 2.262 \times 10^{-2} \frac{kh}{\mu} \cdot \frac{t_{\text{M}}}{(t_{\text{D}}/C_{\text{D}})_{\text{M}}}$$

$$S = \frac{1}{2} \ln \frac{(C_{\text{D}} e^{2S})_{\text{M}}}{C_{\text{D}}}$$

II FORMULAE IN SEMILOG PRESSURE ANALYSIS

1 Transient pressure drawdown equation

$$\psi(p_{\text{i}}) - \psi(p_{\text{wf}}) = 4.242 \times 10^4 \frac{p_{\text{sc}}}{T_{\text{sc}}} \cdot \frac{q_{\text{g}} T_{\text{f}}}{kh} \left(\lg \frac{8.091 \times 10^{-3} kt}{\phi \mu C_t r_{\text{w}}^2} + 0.8686 S_{\text{a}} \right)$$

$$p_i^2 - p_{wf}^2 = 4.242 \times 10^4 \frac{p_{sc}}{T_{sc}} \cdot \frac{q_g \overline{\mu}_g \overline{Z} T_f}{kh} \left(\lg \frac{8.091 \times 10^{-3} kt}{\phi \mu C_t r_w^2} + 0.8686 S_a \right)$$

2 Transient buildup equation (MDH)

$$\psi(p_{ws}) - \psi\left(p_{wf}\right) = 4.242 \times 10^4 \frac{p_{sc}}{T_{sc}} \cdot \frac{q_g T_f}{kh} \left(\lg \frac{8.091 \times 10^{-3} kt}{\phi \mu C_t r_w^2} + 0.8686 S_a \right)$$

$$p_{ws}^2 - p_{wf}^2 = 4.242 \times 10^4 \frac{p_{sc}}{T_{sc}} \cdot \frac{q_g \overline{\mu}_g \overline{Z} T_f}{kh} \left(\lg \frac{8.091 \times 10^{-3} kt}{\phi \mu C_t r_w^2} + 0.8686 S_a \right)$$

3 Transient buildup equation (Horner)

$$\psi(p_{ws}) = \psi(p_i) - 4.242 \times 10^4 \frac{p_{sc}}{T_{sc}} \cdot \frac{q_g T_f}{kh} \left(\lg \frac{t_p + \Delta t}{\Delta t} \right)$$

$$p_{ws}^2 = p_i^2 - 4.242 \times 10^4 \frac{p_{sc}}{T_{sc}} \cdot \frac{q_g \overline{\mu}_g \overline{Z} T_f}{kh} \left(\lg \frac{t_p + \Delta t}{\Delta t} \right)$$

4 Calculating formation permeability and skin factor with slope of semilog straight line on pressure drawdown curve

$$k = 42.42 \times 10^3 \frac{p_{sc}}{T_{sc}} \cdot \frac{q_g T_f}{m_{\psi d} h} = 14.67 \frac{q_g T_f}{m_{\psi d} h}$$

$$k = 42.42 \times 10^3 \frac{p_{sc}}{T_{sc}} \cdot \frac{q_g \overline{\mu}_g \overline{Z} T_f}{m_{2d} h} = 14.67 \frac{q_g \overline{\mu}_g \overline{Z} T_f}{m_{2d} h}$$

$$S_a = 1.151 \left\{ \frac{\psi(p_i) - \psi(p_{wf}(1h))}{m_{\psi d}} - \lg \frac{k}{\phi \mu_g C_t r_w^2} - 0.9077 \right\}$$

$$S_a = 1.151 \left\{ \frac{p_i^2 - p_{wf}^2(1h)}{m_{2d}} - \lg \frac{k}{\phi \mu_g C_t r_w^2} - 0.9077 \right\}$$

5 Calculating formation permeability and skin factor with slope of semilog straight line on buildup curve

$$k = 42.42 \times 10^3 \frac{q_g T_f}{m_{\psi b} h} \cdot \frac{p_{sc}}{T_{sc}} = 14.67 \frac{q_g T_f}{m_{\psi b} h}$$

$$k = 42.42 \times 10^3 \frac{\bar{\mu}_g \bar{Z} q_g T_f}{m_{2b} h} \cdot \frac{p_{sc}}{T_{sc}} = 14.67 \frac{\bar{\mu}_g \bar{Z} q_g T_f}{m_{2b} h}$$

$$S_a = 1.151 \left\{ \frac{\psi(p_{ws}(1h)) - \psi(p_{wf})}{m_{\psi b}} - \lg \frac{k}{\phi \bar{\mu}_g C_t r_w^2} - 0.9077 \right\}$$

$$S_a = 1.151 \left\{ \frac{p_{ws}^2(1h) - p_{wf}^2}{m_{2b}} - \lg \frac{k}{\phi \bar{\mu}_g C_t r_w^2} - 0.9077 \right\}$$

III GAS FLOW RATE FORMULAE

1 Production in steady flow regime

$$q_g = \frac{2.714 \times 10^{-5} k h T_{sc} [\psi(p_e) - \psi(p_{wf})]}{p_{sc} T_f \left(\ln \frac{r_e}{r_w} + S_a \right)}$$

$$q_g = \frac{2.714 \times 10^{-5} k h T_{sc} [p_e^2 - p_{wf}^2]}{p_{sc} \bar{\mu}_g \bar{Z} T_f \left(\ln \frac{r_e}{r_w} + S_a \right)}$$

2 Production in pseudo-steady flow regime

$$q_g = \frac{2.714 \times 10^{-5} k h T_{sc} [\psi(p_R) - \psi(p_{wf})]}{p_{sc} T_f \left(\ln \frac{0.472 r_e}{r_w} + S_a \right)}$$

$$q_g = \frac{2.714 \times 10^{-5} k h T_{sc} [p_R^2 - p_{wf}^2]}{p_{sc} \bar{\mu}_g \bar{Z} T_f \left(\ln \frac{0.472 r_e}{r_w} + S_a \right)}$$

IV GAS WELL DELIVERABILITY EQUATIONS

1 Pseudo-pressure LIT equation (pseudo-steady flow)

$$\psi(p_R) - \psi(p_{wf}) = A_\psi q_g + B_\psi q_g^2$$

$$A_\psi = \frac{3.684 \times 10^4 p_{sc} T_f}{kh T_{sc}} \left(\ln \frac{0.472 r_e}{r_w} + S \right)$$

$$B_\psi = \frac{3.684 \times 10^4 p_{sc} T_f D}{kh T_{sc}};$$

or

$$A_\psi = \frac{29.22 T_f}{kh} \left(\lg \frac{0.472 r_e}{r_w} + \frac{S}{2.302} \right)$$

$$B_\psi = \frac{12.69 T_f}{kh} \cdot D$$

when $p_{sc} = 0.101$ MPa and $T_{sc} = 293.2$ K.

2 Pressure square LIT equation (pseudo-steady flow)

$$p_R^2 - p_{wf}^2 = A_2 q_g + B_2 q_g^2$$

$$A_2 = \frac{3.684 \times 10^4 \overline{\mu}_g \overline{Z} T_f p_{sc}}{kh T_{sc}} \left(\ln \frac{0.472 r_e}{r_w} + S \right)$$

$$B_2 = \frac{3.684 \times 10^4 \overline{\mu}_g \overline{Z} T_f p_{sc} D}{kh T_{sc}}$$

or

$$A_2 = \frac{29.22 \overline{\mu}_g \overline{Z} T_f}{kh} \left(\lg \frac{0.472 r_e}{r_w} + \frac{S}{2.302} \right)$$

$$B_2 = \frac{12.69 \overline{\mu}_g \overline{Z} T_f}{kh} \cdot D$$

when $p_{sc} = 0.101325$ MPa and $T_{sc} = 293.16$ K.

3 Pseudo-pressure exponential deliverability equation (pseudo-steady flow)

$$q_g = C_\psi \left[\psi(p_R) - \psi(p_{wf}) \right]^n$$

$$C_\psi = \frac{2.714 \times 10^{-5} kh T_{sc}}{p_{sc} T_f \left(\ln \dfrac{0.472 r_e}{r_w} + S_a \right)}$$

or

$$C_\psi = \frac{3.422 \times 10^{-2} kh}{T_f \left(\lg \dfrac{0.472 r_e}{r_w} + \dfrac{S_a}{2.302} \right)}$$

when $p_{sc} = 0.101325$ MPa and $T_{sc} = 293.16$K.

4 Pressure square exponential deliverability equation (pseudo-steady flow)

$$q_g = C_2 (p_R{}^2 - p_{wf}{}^2)^n$$

$$C_\psi = \frac{2.714 \times 10^{-5} kh T_{sc}}{p_{sc} \bar{\mu}_g \overline{Z} T_f \left(\ln \dfrac{0.472 r_e}{r_w} + S_a \right)}$$

or

$$C_\psi = \frac{3.422 \times 10^{-2} kh}{\bar{\mu}_g \overline{Z} T_f \left(\lg \dfrac{0.472 r_e}{r_w} + \dfrac{S_a}{2.302} \right)}$$

when $p_{sc} = 0.101325$ MPa, $T_{sc} = 293.16$ K.

5 Pseudo-pressure LIT equation (unsteady flow)

$$\psi(p_R) - \psi(p_{wf}) = A q_g + B q_g^2$$

$$A = \frac{4.242 \times 10^4 p_{sc} T_f}{kh T_{sc}} \left[\lg \frac{8.091 \times 10^{-3} kt}{\phi \bar{\mu}_g C_t r_w^2} + 0.8686 S \right]$$

$$B = \frac{3.684 \times 10^4 T_f p_{sc} D}{kh T_{sc}}$$

or

$$A = \frac{14.61\,T_{\mathrm{f}}}{kh}\left[\lg\frac{8.091\times10^{-3}kt}{\phi\overline{\mu}_{\mathrm{g}}C_{\mathrm{t}}r_{\mathrm{w}}^2}+0.8686S\right]$$

$$B = \frac{12.69\,T_{\mathrm{f}}D}{kh}$$

when $p_{\mathrm{sc}} = 0.101325$ MPa, $T_{\mathrm{sc}} = 293.16$ K.

6 Pressure square LIT equation (unsteady flow)

$$p_{\mathrm{Ri}}^2 - p_{\mathrm{wf}}^2 = Aq_{\mathrm{g}} + Bq_{\mathrm{g}}^2$$

$$A = \frac{4.242\times10^4\overline{\mu}_{\mathrm{g}}\overline{Z}Tp_{\mathrm{sc}}}{khT_{\mathrm{sc}}}\left(\log\frac{8.091\times10^{-3}kt}{\phi\overline{\mu}_{\mathrm{g}}C_{\mathrm{t}}r_{\mathrm{w}}^2}+0.8686S\right)$$

$$B = \frac{3.684\times10^4\overline{\mu}_{\mathrm{g}}\overline{Z}T_{\mathrm{f}}p_{\mathrm{sc}}D}{khT_{\mathrm{sc}}}$$

or

$$A = \frac{14.61\overline{\mu}_{\mathrm{g}}\overline{Z}T_{\mathrm{f}}}{kh}\left[\lg\frac{8.091\times10^{-3}kt}{\phi\overline{\mu}_{\mathrm{g}}C_{\mathrm{t}}r_{\mathrm{w}}^2}+0.8686S\right]$$

$$B = \frac{12.69\overline{\mu}_{\mathrm{g}}\overline{Z}T_{\mathrm{f}}D}{kh}$$

when $p_{\mathrm{sc}} = 0.101325$ MPa, $T_{\mathrm{sc}} = 293.16$ K.

V PULSE TEST FORMULAE (BY KAMAL)

1 Definition of dimensionless parameters

$$\Delta p_{\mathrm{D}} = \frac{0.5429kh}{\mu Bq}\cdot\Delta p \quad \text{(for oil wells)}$$

$$\Delta p_D = \frac{2.714 \times 10^{-5} kh T_{sc}}{q_g T p_{sc}} \cdot \Delta\psi \quad \text{(for gas wells)}$$

$$[t_L]_D = \frac{3.6 \times 10^{-3} kt_L}{\mu\phi C_t r_w^2}$$

$$F' = \frac{\Delta t_p}{\Delta t_C}$$

2 Parameter calculation formulae

$$\frac{kh}{\mu} = \frac{1.842q\left[\Delta p_D (t_L/\Delta t_C)^2\right]_{\text{reference graph}}}{\Delta p[t_L/\Delta t_C]^2} \quad \text{(for oil wells)}$$

$$\frac{kh}{\mu_g} = \frac{3.684 \times 10^4 q_g T p_{sc}}{T_{sc}} \cdot \frac{\left[\Delta p_D (t_L/\Delta t_C)^2\right]_{\text{reference graph}}}{\Delta\psi[t_L/\Delta t_C]^2} \quad \text{(for gas wells)}$$

$$\phi h C_t = 3.6 \times 10^{-3} \frac{kh}{\mu} \cdot \frac{t_L}{r^2\left[(t_L)_D/r_D^2\right]_{\text{reference graph}}}$$

VI OTHER COMMON FORMULAE OF GAS WELLS

1 Radius of influence

$$r_i = 0.12\sqrt{\frac{kt}{\phi\mu_g C_t}}$$

2 Distance from tested well to the fault

$$L_b = 0.045\sqrt{\frac{k\Delta t_b}{\phi\mu_g C_t}}$$

3 Additional pressure loss in skin zone

$$S = \frac{542.8kh}{qB\mu} \cdot \Delta p_s$$

$$\Delta p_s = 0.8686mS$$

4 Effective wellbore radius

$$r_{we} = r_w \times e^{-S}$$

5 Non-Darcy flow coefficient of gas flow

$$D = \frac{7.18 \times 10^{-16} \beta k M p_{sc}}{hr_w T_{sc} \mu_{g,wf}}$$

6 Turbulent flow coefficient of gas flow

$$\beta = 1.88 \times 10^{10} k^{-1.47} \phi^{-0.53}$$

7 Friction resistance coefficient of gas flow

$$f_{CK} = \frac{64\Delta p}{\beta \rho v^2 \Delta x}$$

8 Reynolds number of gas flow

$$R_{eCK} = \frac{\beta k \rho v}{1.0 \times 10^9 \mu_g}$$

9 Converted skin factor of fractured wells

$$S_a = \ln \frac{2r_w}{x_f}$$

10 Fracture skin of fractured wells

$$S_f = \frac{\pi b_S}{2x_f}\left(\frac{k}{k_S} - 1\right)$$

11 Additional skin factor of partially perforated formation

$$S_c = \left(\frac{1}{b} - 1\right)\left(\ln(h_D) - G(b)\right)$$

where

$$b = \frac{h_p}{h}$$

$$h_D = \frac{h}{r_w}\sqrt{\frac{k_h}{k_v}} \quad \text{or} \quad h_D = \frac{h}{2r_w}\sqrt{\frac{k_h}{k_v}}$$

and

$$G(b) = 2.948 - 7.363b + 11.45b^2 - 4.675b^3$$

12 Compressibility coefficient of desorption of coalbed methane layers

$$C_d = \frac{\rho_g^0 R T Z V_L p_L}{\phi M \bar{p}(\bar{p} + p_L)^2 \tau}$$

Conversion method of coefficients in a formula from one unit system to another one

The "division rule" is used in conversion of units under different systems. For the theoretical derivation of this conversion, please refer to "How to convert units in formulas" by Zhu Yadong published in the magazine *Buried Hill* [72].

Here are two examples of the conversion.

I CONVERSION OF GAS FLOW RATE FORMULA

Converting the gas flow rate formula under the SI Unit System into that under the China Statutory Unit System (CSU) is shown as following.

The formula of gas flow rate under SI Unit System is

$$q_g = \frac{\pi k h T_{sc}[p_e^2 - p_{wf}^2]}{p_{sc}\overline{\mu}_g \overline{Z} T_f \left(\ln \dfrac{r_e}{r_w} + S_a \right)}$$

Quantitative relations of the variables in this formula under SI and CSU are as follow:

q_g: 1 m^3/s (SI) = 8.64 × 10^4 m^3/day (CSU)
k: 1 m^2 (SI) = 10^{15}(10^{-3} μm^2) ≈ 10^{15} mD (CSU)
h, r: 1 m (SI) = 1 m (CSU)
p: 1Pa (SI) = 10^{-6} MPa (CSU)
μ: 1 Pa·s (SI) = 10^3 mPa·s (CSU)
T: 1K (SI) = 1K (CSU)

Convert the variables under different unit systems by the division rule:

$$\frac{q_g}{8.64} = \frac{\pi \cdot \dfrac{k}{10^{15}} \cdot \dfrac{h}{1} \cdot \dfrac{T_{sc}}{1} \cdot \left[\left(\dfrac{p_e}{10^{-6}} \right)^2 - \left(\dfrac{p_{wf}}{10^{-6}} \right)^2 \right]}{\dfrac{p_{sc}}{10^{-6}} \cdot \dfrac{\overline{\mu}_g}{10^3} \cdot \overline{Z} \cdot \dfrac{T_f}{1} \cdot \left(\ln \dfrac{r_e/1}{r_w/1} + S_a \right)}$$

The gas flow rate formula under the CSU system is now obtained:

$$q_g = \frac{2.714 \times 10^{-5} k h T_{sc}[p_R^2 - p_{wf}^2]}{p_{sc}\overline{\mu}_g \overline{Z} T_f \left(\ln \dfrac{0.472 r_e}{r_w} + S_a \right)}$$

II CONVERSION OF DIMENSIONLESS TIME FORMULA

Converting the dimensionless time formula under the British field unit system into that under CSU is as follows.

The formula of dimensionless time under British field unit is expressed as

$$t_D = 2.6368 \times 10^{-4} \frac{k \Delta t}{\phi \mu C_t r_w^2}$$

The quantitative relations of the variables in this formula under the British field unit system and CSU are as follow:

t: 1 hr (British) = 1 hr (CSU)
k: 1 mD (British) = 0.98692(10^{-3} μm^2)(CSU)
h, r_w: 1 ft (British) = 0.3048 m (CSU)
C_t: 1 psi^{-1} (British) = 145.04 MPa^{-1} (CSU)
μ: 1 cP (British) = 1 mPa·s (CSU)
T: 1K (British) = 1K (CSU)

Make unit conversion by the division rule:

$$t_D = 2.6368 \times 10^{-4} \frac{\dfrac{k}{0.98692} \cdot \dfrac{\Delta t}{1}}{\dfrac{\mu}{1} \cdot \phi \cdot \dfrac{C_t}{145.04} \cdot \dfrac{r_w^2}{0.3048^2}}$$

The dimensionless time formula under the CSU system is now obtained:

$$t_D = \frac{3.6 \times 10^{-3} k}{\phi \mu C_t r_w} \cdot \Delta t$$

REFERENCES

[1] Jiang LS, Chen ZX. Theoretical Basis of Well Test Analysis. Beijing: Petroleum Industry Press; 1985.

[2] Liu NQ. Applied Modern Well Test Interpretation Method. 5th ed. Beijing: Petroleum Industry Press; 2008.

[3] GRI: Identification of Linear Flow Geometries and Implications for Natural Gas Reservoir Development.

[4] Barenblatt GE, Zheltov IP, Kochina IN. Basic Concepts in the Theory of Homogeneous Liquids in Fissured Rocks,. J Appl Math (USSR) 1960;24(5):1286–303.

[5] Louis Mattar RV, Hawkes MS, Santo, and Karel Zaoral, Fekete Assocs. Inc: Prediction of Long-Term Deliverability in Tight Formation, Paper SPE 26178, SPE Gas Technology Symposium held in Calgary, Alberta, Canada, 28–30 June 1993.

[6] Miller CC, Dyes AB, and Hutchinson, CA.: The Estimation of Permeability and Reservoir Pressure from Bottom-Hole Pressure Build-Up Characteristics, JPT, 1950, 2(4): 91–104. Trans., AIME(1950) 189, 91–104. Also Reprint Series, No.9—Pressure Analysis Methods, Society of Petroleum Engineers of AIME, Dallas (1967), 11–24.

[7] Horner DR.: Pressure Build-Up in Wells, Proc., Third World Pet. Cong., The Hague, The Netherlands, 1951, 503–523. Also Reprint Series, No.9—Pressure Analysis Methods, Society of Petroleum Engineers of AIME, Dallas (1967), 25–43.

[8] Agarwal Ram G, Al-Hussainy Rafi, Ramey Jr HJ. An Investigation of Wellbore Storage and Skin Effect in Unsteady Liquid Flow. I. Analytical Treatment,. Soc Pet Eng J Sept. 1970;10(3):279–90. Trans., AIME, 249.

[9] Gringarten AC, Bourdet D, Landel PA, and Kniazeff V.: A Comparison Between Different Skin and Wellbore Type-Curve for Early-Time Transient Analysis, Paper SPE 8205, SPE Annual Technical Conference and Exhibition, 23–26 September 1979, Las Vegas, NV.

[10] Bourdet D, and Gringarten AC.: Determination of Fissure Volume and Block Size in Fractured Reservoirs by Type-Curve Analysis, Paper SPE 9293, SPE Annual Technical Conference and Exhibition, 21–24 September 1980, Dallas, TX.

[11] Earlougher Jr RC. Advances in Well Test Analysis. New York: Dallas; 1977.

[12] Wang NJ, et al. Atlas of Natural Gas Development in China. CNPC; 1992.

[13] Ramey Jr HJ. Non-Darcy Flow and Wellbore Storage Effects in Pressure Build-Up and Drawdown of Gas Wells. J Pet Tech Feb. 1965;17(2):223–33.

[14] Cornell D, Katz DL. Flow of Gases through Consolidated Porous Media. Ind Eng Chem 1953;4(10):2145–52.

[15] Jones SC.: Using the Inertial Coefficient, β, To Characterize Heterogeneity in Reservoir Rock, Paper SPE 16949, SPE Annual Technical Conference and Exhibition, 27–30 September 1987, Dallas, TX.

[16] Lee WJ. Well Testing. Richardson, TX: Society of Petroleum Engineers; 1982. 87.

[17] Wright DE. Nonlinear Flow through Granular Media. J Hydraul Div Am Soc Civ. Eng Proceedings 1968;94(HY4). 1968: 851–872.

[18] Wattenbager RA, Ramey Jr HJ. Gas Well Testing With Turbulence, Damage, and Wellbore Storage. J Pet Tech 1968;20(8):877–87.

[19] Standing MB, Katz DL. Density of Natural Gases. Trans, AIME 1942;146:249.

[20] ERCB of CANADA: Gas Well Testing—Theory and Practice. 4th ed. (Metric). Calgary: Alberta; 1979.

[21] Al-Hussainy R. Transient Flow of Ideal and Real Gases through Porous Media, Ph.D. Thesis. Texas A and M Univ 1967.

[22] Al-Hussainy R. The Flow of Real Gases through Porous Media, M. Thesis. Texas A and M Univ 1965.

[23] Pierce HR, Rawlins EL. The Study of a Fundamental Basis for Controlling and Gauging Natural-Gas Wells, U.S. Dept. of Commerce—Bureau of Mines,. Serial 1929;2929.

[24] Rawlins EL, Schellhardt MA. Backpressure Data on Natural Gas Wells and Their Application to Production Practices, U.S. Bureau of Mines. Monograph 1936;7:1936.

[25] Cullender MH. The Isochronal Performance Method of Determining the Flow Characteristics of Gas Wells. Trans, AIME 1955;204:137—42.

[26] Katz DL, Cornell D, Kobayashi R, Poettmann FH, Vary JA, Elenbaas JR, Weinaug CF. Handbook of Natural Gas Engineering. New York: McGraw-Hill Book Co., Inc.; 1959.

[27] Winestock AG, Colpitts GP. Advances in Estimating Gas Well Deliverability. J Can Pet Tech 1965;4(3):111—9.

[28] Chen YQ. Calculation Methods of Reservoir Engineering. Beijing: Petroleum Industry Press; 1990.

[29] Edging AN, Cleland NE. Gas Field Deliverability Predictions and Development Economics, Australian Pet. Exploration Assoc J 1967;7(2):115—9.

[30] Wentink JJ, Goemans JG, Hutchinson CW. Deliverability Forecasting and Compressor Optimization for Gas Fields. Oilweek 1971;22(31):50—4.

[31] William D, McCain Jr . Petroleum Fluid Property (Chinese Version). Beijing: Petroleum Industry Press; 1984.

[32] Bourdet D, et al. A New Set of Type-Curves Simplifies Well Test Analysis. World Oil May, 1983:95—106.

[33] Bourdet D, et al. Use of Pressure Derivative in Well Test Interpretation. SPE Formation Evaluation June 1989;4(2):293—302.

[34] Matthews CS, Brons F, Hazebroek P. A Method for Determination of Average Pressure in Bounded Reservoir. Trans, AIME 1954;201:182—91.

[35] Zhuang HN, Zhu YD. Study of Pressure Interference between Wells in Double Porosity Media with Type Curves. ACTA PETROLEI SINICA July 1986;7(3):63—72.

[36] Gringarten AC, Ramey Jr HJ, Raghavan R. Unsteady-State Pressure Distributions Created by a Well with a Single Infinite-Conductivity Vertical Fracture. SPEJ Aug. 1974:347—60.

[37] Rodriguez F, Horne RN, and Cinco-Ley H: Partially Penetrating Vertical Fractures Pressure Transient Behavior of a Finite-Conductivity Fracture. Paper SPE 13057, SPE Annual Technical Conference and Exhibition, 16—19 September 1984, Houston, TX.

[38] Ramey Jr HJ, Agarwal RG, Martin I. Analysis of 'Slug Test' or DST Flow Period Data. JCPT July—Sept.1975;14(3):37—42.

[39] Yeh M, Agarwal RG. Development and Application of New Type Curve used for Transient Well Test Analysis. Paper SPE 17567, International Meeting on Petroleum Engineering November 1988:1—4. Tianjin, China.

[40] Onur M, Reynolds AA. New Approach for Construction Derivative Type Curve for Well Test Analysis. SPE Formation Evaluation March 1988;3(1):197—206.

[41] Duong A. A New Set of Type Curves for Well-Test Interpretation with the Pressure/Pressure Derivative Ratio. SPE Formation Evaluation June 1989;4(2):264—72.

[42] Blasingane TA, Johnxston JL, and Lee WJ: Type-Curve Analysis Using the Pressure Integral Method. Paper SPE 18799, SPE California Regional Meeting, 5—7 April 1989, Bakersfield, CA.

[43] Zhuang HN. Application of Interference Testing and Pulse testing in Ordovician Reservoir, Kenli Oilfield. Beijing: Collections of Theses of International Meeting on Petroleum Engineering; March 1982.

[44] Zhuang HN. Interference Testing and Pulse Testing in the Kenli Carbonate Oil Poo—A Case History. JPT June 1984;36(6):1009—17.

[45] Tarim Oilfield: Assessment Report on Well Test in Well Tazhong-111, Assessment of Well Test: SinoPetroleum Technology Inc. (SinoPetroleum): Li Hua-an.

[46] Bremer RE, Winston H, Vela S. Analytical Model for Vertical Interference Test Across Low-Permeability Zones. SPE Journal June, 1985;25(3):407—18.

[47] Gillund GN, Kamal MM. Incorporation of Vertical Permeability Test Result in Vertical Miscible Flood Design and Operation. J Can Pet Tech March-April 1984;23(2):54—9.

[48] Zhuang HN, Liu NQ. Application of Pressure Transient Test Normalized Graph Analysis in China's Oilfields, Paper SPE 30003, International Meeting on Petroleum Engineering, 14—17. Beijing: China; November 1995.

[49] Theis CV. The Relation Between the Lowering of the Piezometric Surface and the Rate and Duration OD Discharge of a Well Using Groundwater Storage. Trans, Am Geophys Union 1935:519—24.

[50] Jacob CE. Coefficients of Storage and Transmissibility Obtained from Pumping Tests in the Houston District, Texas. Trans Am Geophys Union 1941:744—56.

[51] King Hubbert M. Discussion of Papers by Messers. Jacob and Guyton. Trans Am Geophys Union 1941:770—2.

[52] Muskat M. Physical Principles of Oil Production. New York: Mc Graw-Hill Book Co.; 1949.

[53] Matthies, E. Peter. Practical Application of Interference Test. J Pet Tech March 1964;16(3):249—52.

[54] McKinley RM, Vela S, Carlton LA. A Field Application of Pulse-Testing for Detailed Reservoir Description. J Pet Tech March 1968;20(3):313—21. Trans., AIME, 243.

[55] Earlougher Jr RC, Ramey Jr HJ. Interference Analysis in Bounded Systems. J Can Pet Tech Oct.—Dec. 1973;12(4):33—45.

[56] Ogbe DO, Brigham WE. A Model for Interference Testing with Wellbore Storage and Skin Effects at Both Wells. SPE Formation Evaluation September 1989;4(3):391—6.

[57] Tongpenyai J, Raghavan R. The Effect of Wellbore Storage and Skin on Interference Test Data. J Pet Tech Jan. 1981:151—60. Trans., AIME, 271.

[58] Mousli NA, Raghavan R, Cinco-Ley H, Samanieco VF. The Influence of Vertical Fractures Interference Active and Observation Wells on Interference Tests. Soc Pet Eng J Dec 1982:933—44. Trans., AIME, 273.

[59] Deruyck BG, Bourdet DP, Daprat G, and Ramey HJ Jr.: Interpretation of Interference Tests in Reservoirs with Double Porosity Behavior-Theory and Field Examples, SPE Annual Technical Conference and Exhibition, 26—29 September 1982, New Orleans, LA.

[60] Streltsova TD. Buildup Analysis for Interference Tests in Stratified Formation. J Pet Tech Feb. 1984:301—10. Trans., AIME, 277.

[61] Chen CC, Yeh N, Raghavan R, Reynolds AC. Pressure Response at Observation Wells in Fractured Reservoirs. Soc Pet Eng J Dec. 1984:628—38. Trans., AIME, 277.

[62] Johnson CR, Greenkorn RA, Woods EG. Pulse-Testing: A New Method for Describing Reservoir Flow Properties Between Wells. J Pet Tech Dec. 1966:1599—604. Trans., AIME, 237.

[63] Kamal MM, Brigham WE. Pulse-Testing Response for Unequal Pulse and Shut-in Periods. Soc Pet Eng J Oct. 1975:399—410. Trans., AIME, 259.

[64] Abobise EO, and Tiab D: Determining Fracture Orientation and Formation Permeability from Pulse Testing, Paper SPE 11027, SPE Annual Technical Conference and Exhibition, 26—29 September 1982, New Orleans, LA.

[65] Ekie S, Hadinoto N, and Raghavan R: Pulse—Testing of Vertically Fractured Wells, paper SPE 6751, SPE Annual Technical Conference and Exhibition, 9—12 October 1977, Denver, CO.

[66] Prats M. Interpretation of Pulse Tests in Reservoir with Crossflow between Contiguous Layers. SPE Formation Evaluation October 1986;1(5):511—20.

[67] Chropra AK. Reservoir Description via Pulse Testing: A Technology Evaluation, SPE 17568. Tianjin, China: International Meeting on Petroleum Engineering; November 1988. 1—4.

[68] Editing Group of WTM: Well Test Manual, Vol I. Petroleum Industry Press; July 1991.

[69] Zhuang HN, Han YX, Tan ZG, Zhang MY. Research of Interference Test between Wells in Central Shan-gan-ning Gas Field. Well Test Theory and Practice, Petroleum Industry Pressure June 1996.

[70] Saulsberry J, Schafer PS, Schraufnagel RA. Guide to Coalbed Methane Reservoir Engineering. Gas Research Institute, No.GRI-94/0397 1996.

[71] Li HP, Han YX, Zhuang HN. Well Testing Evaluation Study of Special High Production Gas Well: Well KELA-203. Natural Gas Industry 2001;(2):35—41.

[72] Zhu YD. How to Conduct Unit Conversion in Formula. Buried Hill 1991;(1):86—95.

[73] Wang XH, Yang JH, Ji YL. Single Phase Flow Method of Transient Testing Analysis for Condensate Wells. Well Testing 1997;6(2):1—3.

[74] Liu NQ. Deconvolution and Its Application. Well Testing 2007;16(5):1—3.

[75] Gringarten AC. From Straight Lines to Deconvolution: The Evolution of the Art in Well Test Analysis,. SPE Reservoir Evaluation & Engineering 2008;11(1):41—62.

[76] Joshi SD. Augmentation of Well Productivity with Slant and Horizontal Wells (includes associated papers 24547 and 25308). Journal of Petroleum Technology June 1988;40(6):729—39.

[77] Li D, et al. Analysis of Productivity Formulae of Horizontal Well. Petroleum Exploration and Development 1977;24(5):76—9.

[78] He K. Application of Gas Well Deliverability Evaluation Data in Optimizing Horizontal Well Design. Natural Gas Industry 2004:118—9. Sup.

[79] Yang JS, Liu JY. Applied Calculation Methods of Gas Production. Petroleum Industry Press; 1994.

[80] Bourdet D. Well Test Analysis: The Use of Advanced Interpretation Models. Paris: Elsevier; 2002.

AUTHOR INDEX

Note: Page numbers with "*f*" denote figures; "*t*" tables.

Note: Page numbers with "*f*" denote figures; "*t*" tables.

Color Plates

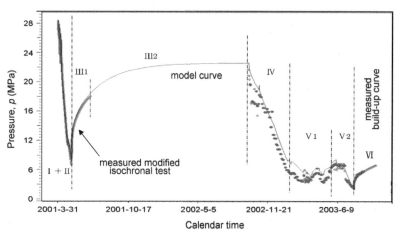

Figure 8.2 Bottom-hole pressure history of Well S-4.

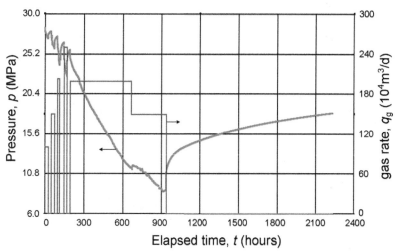

Figure 8.3 Gas flow rate and pressure history during deliverability test of Well S-4.

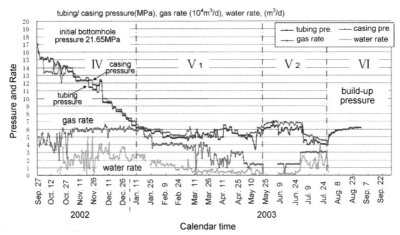

Figure 8.4 Monitored gas flow rate and pressure during the long-term production test of Well S-4.

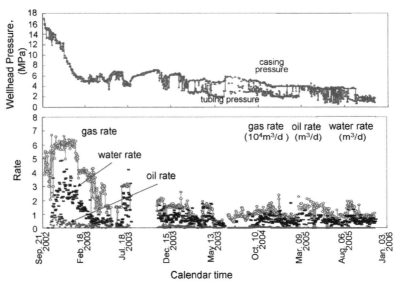

Figure 8.5 Whole process monitoring of flow rates and pressure during a long-term production test in Well S-4.

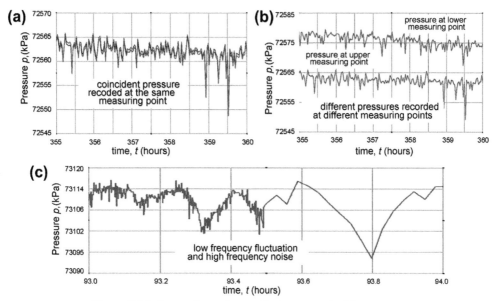

Figure 8.16 Comparison of magnified pressure buildup curves.

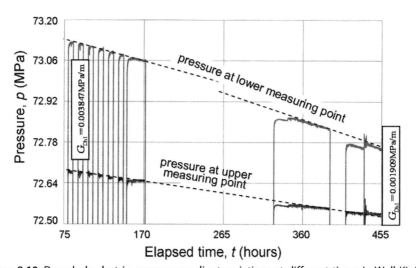

Figure 8.19 Downhole shut-in pressure gradient variations at different times in Well KL-203.

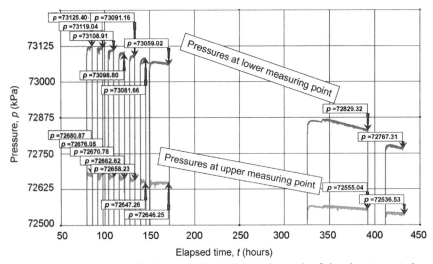

Figure 8.24 Pressures at both measuring points at the ends of the shut-in periods.

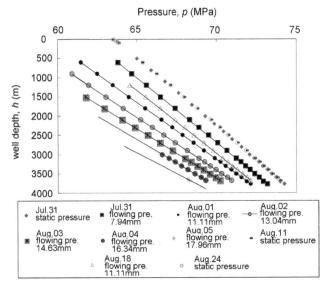

Figure 8.36 Flowing pressure and static pressure gradients of Well KL-205.

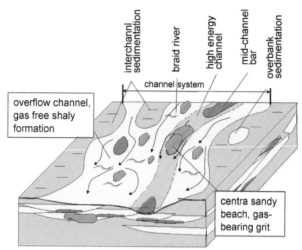

Figure 8.44 Schematic diagram of sedimentary model of Shihezi formation in the SLG gas field.

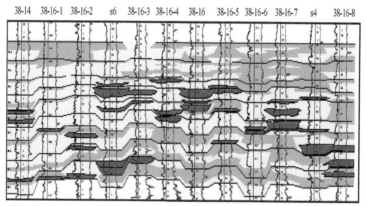

Figure 8.45 Subzone profile map along the east—west direction through Well S-6 in the SLG gas field.

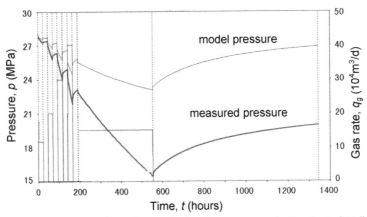

Figure 8.48 Pressure history match verification in the short-term production test of Well S-6 (model I).

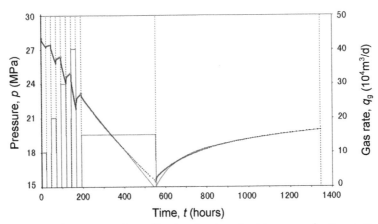

Figure 8.49 Pressure history match verification in the short-term production test of Well S-6 (model II).

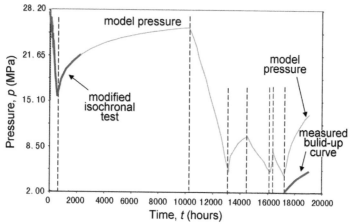

Figure 8.50 Pressure history match verification of Well S-6 in 2003 (model II).

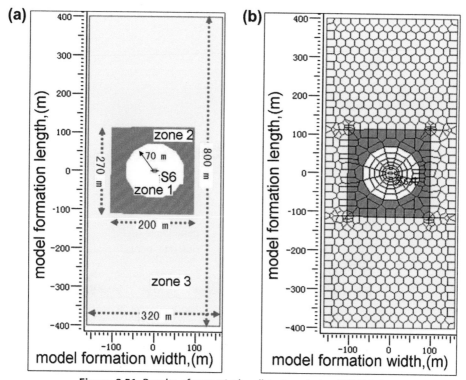

Figure 8.51 Results of numerical well test analysis of Well S-6.

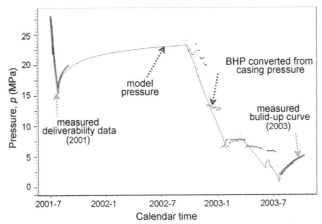

Figure 8.52 Pressure history match verification for numerical well test analysis results of Well S-6.

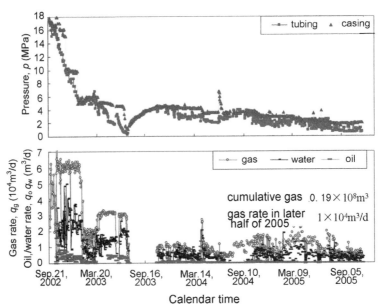

Figure 8.53 Production curve of Well S-6 from 2002 to 2005.

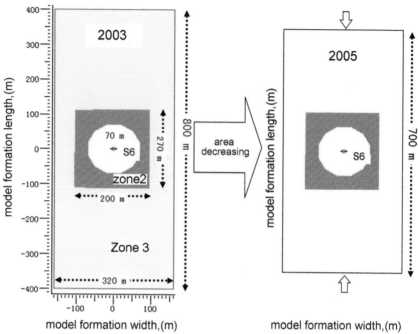

Figure 8.54 Comparison of modified dynamic model III₁ (b) in 2005 to model III (a) of Well S-6.

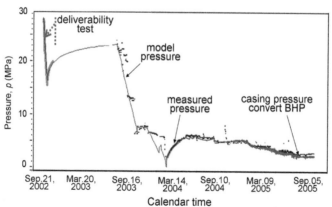

Figure 8.55 Pressure history match verification for dynamic model III₁ of Well S-6.

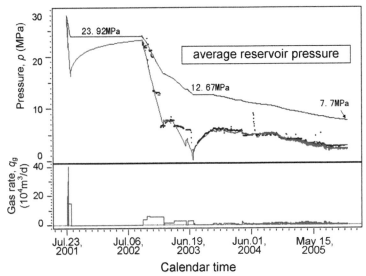

Figure 8.56 Calculation result of average reservoir pressure of Block S-6 from 2001 to 2005.

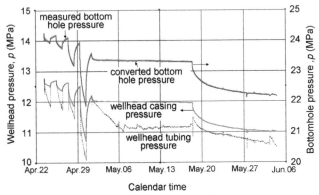

Figure 8.60 Comparison of pressure history acquired by electronic pressure gauge and converted pressure history of Well Y17.

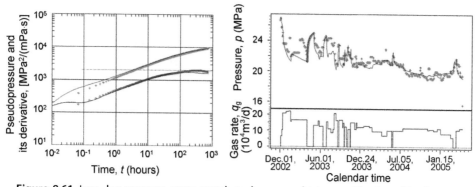

Figure 8.61 Log–log pressure curve match and pressure history verification of Well Y46-9.

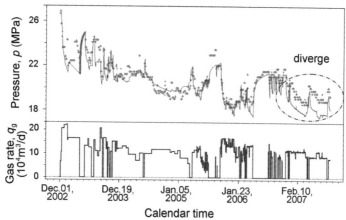

Figure 8.62 Tracing verification of dynamic model of Well Y46-9 in the period from 2005 to 2007.

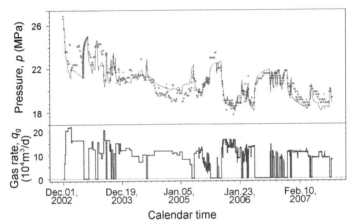

Figure 8.63 Pressure history match verification of dynamic model of Well Y46-9 after improvement.

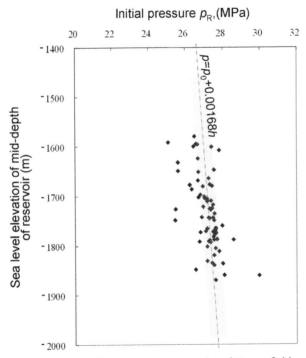

Figure 8.64 Static pressure gradient plot of YL gas field.

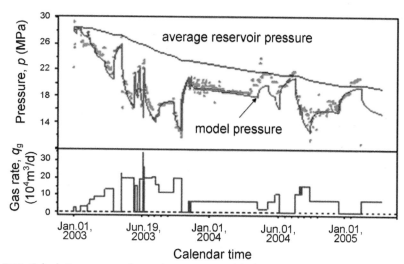

Figure 8.65 Calculating average formation pressure with dynamic model method in YL gas field.

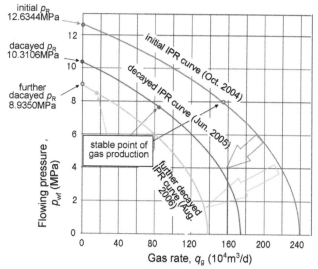

Figure 8.66 Estimating the drainage boundary pressure with deliverability equation in YL gas field.

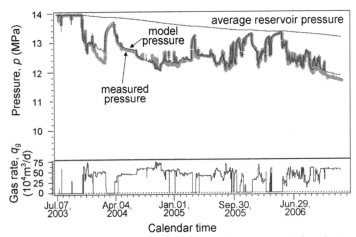

Figure 8.68 Dynamic model verification through a pressure history match for a horizontal well, Field Example No. 1.

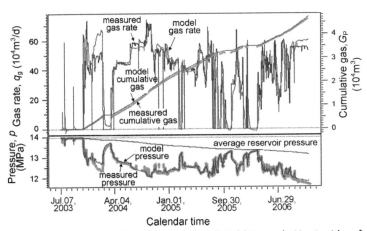

Figure 8.69 History match and model verification of well Field Example No. 1 with software Topaze.

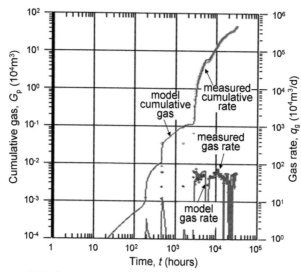

Figure 8.70 Fetkovich analysis of the model with software Topaze.

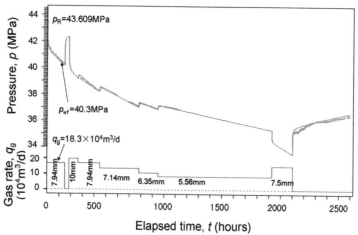

Figure 8.71 Flow rate and pressure history of Well BS-7 during back-pressure test.

Printed and bound by CPI Group (UK) Ltd, Croydon, CR0 4YY

08/05/2025

01864851-0001